Ciliary and
Flagellar Membranes

Ciliary and Flagellar Membranes

Edited by
Robert A. Bloodgood
University of Virginia School of Medicine
Charlottesville, Virginia

PLENUM PRESS • NEW YORK AND LONDON

Library of Congress Cataloging-in-Publication Data

Ciliary and flagellar membranes / edited by Robert A. Bloodgood.
       p.   cm.
   Includes bibliographical references.
   ISBN 0-306-43279-X
   1. Cell membranes. 2. Cilia and ciliary motion. 3. Flagella
(Microbiology)   I. Bloodgood, Robert A.
   [DNLM: 1. Cell Membrane--physiology. 2. Cell Movement. 3. Cilia-
-physiology. 4. Flagella--physiology. 5. Protozoa--physiology.
QS 532.5.E7 C572]
QH601.C55 1989
574.87'64--dc20
DNLM/DLC
for Library of Congress   89-23227
                                                               CIP

© 1990 Plenum Press, New York
A Division of Plenum Publishing Corporation
233 Spring Street, New York, N.Y. 10013

Printed in the United States of America

Contributors

Joseph C. Besharse   Department of Anatomy and Cell Biology, Emory University School of Medicine, Atlanta, Georgia 30322; *present address:* Department of Anatomy and Cell Biology, The University of Kansas Medical Center, Kansas City, Kansas 66103

Robert A. Bloodgood   Department of Anatomy and Cell Biology, University of Virginia School of Medicine, Charlottesville, Virginia 22908

Emmanuelle Boisvieux-Ulrich   Centre de Biologie Cellulaire C.N.R.S., 94205 Ivry sur Seine Cedex, France

G. B. Bouck   Department of Biological Sciences, University of Illinois, Chicago, Illinois 60680

Richard A. Cardullo   Worcester Foundation for Experimental Biology, Shrewsbury, Massachusetts 01545

Bernadette Chailley   Centre de Biologie Cellulaire C.N.R.S., 94205 Ivry sur Seine Cedex, France

William L. Dentler   Department of Physiology and Cell Biology, University of Kansas, Lawrence, Kansas 66045

Cynthia J. Horst   Department of Anatomy and Cell Biology, Emory University School of Medicine, Atlanta, Georgia 30322

Edna S. Kaneshiro   Department of Biological Sciences, University of Cincinnati, Cincinnati, Ohio 45221-0006

F. M. Klis   Department of Molecular Cell Biology, University of Amsterdam, 1098 SM Amsterdam, The Netherlands

P. J. Levasseur   Department of Biological Sciences, University of Illinois, Chicago, Illinois 60680

A. Musgrave   Department of Molecular Cell Biology, University of Amsterdam, 1098 SM Amsterdam, The Netherlands

Robin R. Preston   Laboratory of Molecular Biology, University of Wisconsin, Madison, Wisconsin 53706

T. K. Rosiere   Department of Biological Sciences, University of Illinois, Chicago, Illinois 60680

Yoshiro Saimi   Laboratory of Molecular Biology, University of Wisconsin, Madison, Wisconsin 53706

Daniel Sandoz   Centre de Biologie Cellulaire C.N.R.S., 94205 Ivry sur Seine Cedex, France

R. E. Stephens   Marine Biological Laboratory, Woods Hole, Massachusetts 02543

Laurence Tetley   Department of Zoology, University of Glasgow, Glasgow G12 8QQ, Scotland, United Kingdom

Elaine Tuomanen   The Rockefeller University, New York, New York 10021

H. van den Ende   Department of Molecular Cell Biology, University of Amsterdam, 1098 SM Amsterdam, The Netherlands

Keith Vickerman   Department of Zoology, University of Glasgow, Glasgow G12 8QQ, Scotland, United Kingdom

Tsuyoshi Watanabe   Department of Biological Science, Tohoku University, Kawauchi, Sendai 980, Japan

Norman E. Williams   Department of Biology, University of Iowa, Iowa City, Iowa 52242

George B. Witman   Cell Biology Group and Male Fertility Program, Worcester Foundation for Experimental Biology, Shrewsbury, Massachusetts 01545

David E. Wolf   Worcester Foundation for Experimental Biology, Shrewsbury, Massachusetts 01545

Preface

While there have been many recent books on the cell surface and a few on the topic of cilia and flagella, this is the first volume that attempts to bring together the available information on ciliary and flagellar membranes. This reflects a slow awakening by cell biologists and other scientists to the significance of ciliary and flagellar surfaces. When Michael Sleigh edited an excellent book entitled *Cilia and Flagella* in 1974, not one of the sixteen chapters was devoted to ciliary or flagellar surfaces. When W. B. Amos and J. G. Duckett edited the very fine 25th Symposium of the Society for Experimental Biology on *Prokaryotic and Eukaryotic Flagella* in 1982, only two of the twenty chapters on eukaryotic cilia and flagella were devoted to ciliary and flagellar surfaces. Only in 1989 has the timing become right to produce a volume entirely devoted to the nonaxonemal structures and functions of eukaryotic cilia and flagella. The fifteen chapters in this volume cover a wide spectrum of organisms (from protozoa and algae to birds and mammals) and an equally wide spectrum of topics (from sexual interactions in the algae to the binding of pathogens in the lung). It is hoped that almost any biological or medical scientist will find at least a few chapters of interest and I will bet that few who peruse this book will fail to discover at least a couple of roles for ciliary and flagellar surfaces that they had never imagined. For those readers who may be relatively new to the subject, Dr. George Witman has prepared an introductory chapter that introduces the reader to the general features of cilia and flagella and in particular to those aspects of their structure and function that influence or are influenced by the phenomena occurring at the ciliary or flagellar surface.

Eukaryotic cilia and flagella are essentially specialized cytoplasmic domains containing a highly structured cytoskeleton incompletely bounded by a specialized domain of the plasma membrane. For that reason, they are excellent systems in which to study membrane–cytoskeletal interactions and indeed many of the functions of ciliary and flagellar surfaces will only be understood as deriving from a functional interaction of the membrane with the underlying cytoskeleton. Eukaryotic cilia and flagella also constitute ideal systems for studying a defined plasma membrane domain because the act of deciliation or deflagellation immediately isolates a specialized domain of the plasma membrane; there is little contamination with other membrane material because these organelles lack any internal membranes.

While cilia and flagella have been recognized as motile organelles from the time of their discovery by Leeuwenhoek in approximately 1676, it is only recently that biologists have come to appreciate that cilia and flagella perform many other roles aside from the movement of cells through a liquid medium or the movement of fluid (or mucus) across a ciliated epithelium. Many of these additional functions derive from properties of ciliary and flagellar surfaces, here defined as the ciliary or flagellar membrane along with all of its associated appendages (coats, scales, mastigonemes). In fact, the first paper that dealt with the surface of eukaryotic cilia or flagella was probably that published in 1889 by Loeffler; he included a clear photograph of the mastigonemes (or Flimmergeissel) on the flagellar surface of a chrysophycean algal cell. The first person to seriously study the flagellar surface using the electron microscope was probably Manton (see Manton, 1952, for a review).

One of the biggest technical advances in the study of cilia and flagella came from the ability to isolate cilia and flagella, extract them with nonionic detergents, and obtain axonemes that could be reactivated to produce patterns of bending motility strikingly similar to those seen *in vivo* (Gibbons, 1981). Ironically, at the same time that this opened up much detailed study of the motile functions of ciliary and flagellar axonemes, it drew attention away from the study of ciliary and flagellar surfaces. In fact, many scientists studying cilia and flagella routinely discard all of the membrane and soluble components of these organelles down the laboratory sink as so much unnecessary junk!

Although the naked axoneme can exhibit motile behavior, the normal regulation of the motile activity of cilia and flagella is dependent upon the ciliary or flagellar membrane. An elegant combination of genetic, biochemical, and electrophysiological approaches using the ciliate protozoan *Paramecium* has demonstrated how the behavior of the ciliary axoneme (and the entire organism) is controlled by a complicated array of ion channels and pumps within the ciliary membrane (see chapter by Preston and Saimi in this volume).

In fact, the general field of sensory transduction and membrane signaling is one of the more exciting ways in which ciliary and flagellar membranes are being utilized to study basic biological problems. From the avoidance response in *Paramecium* to the initial sexual contact in *Chlamydomonas* gametes to the very human appreciation of an expensive perfume, signal transduction mechanisms in cilia and flagella are at work.

While the chapters in this book range widely in a phylogenetic and functional sense, there is no claim of completeness. For lack of space, a few interesting functions of ciliary and flagellar membranes have not been reviewed in this volume. For example, the ciliary membranes of olfactory receptor cells are responsible for binding the odorant molecules (Snyder *et al.*, 1989); this binding initiates a transmembrane signaling process involving G protein activation of a membrane-associated adenylate cyclase, an increase in intraciliary cyclic nucleotide level, and a direct cyclic nucleotide gating of ciliary membrane sodium channels (Nakamura and Gold, 1987; Lancet, 1988). The adhesion of certain free-living marine invertebrates is known to occur by means of specializations of the ciliary surface (Tyler, 1973). A third topic not represented in the present volume is a consideration of the evolution of ciliary and flagellar surface specializations, a topic that has been studied by Bardele (1981, 1983).

Despite these few omissions, the vast majority of research related to ciliary and flagellar membranes is reviewed within the present volume. My sincere thanks go to all of the chapter authors; their excellent and timely contributions have made this unique volume possible. Special thanks go to Mary Phillips Born and Susan Woolford at Plenum Publishing for their efforts on behalf of this volume.

Robert A. Bloodgood

Charlottesville, Virginia

References

Bardele, C. F., 1981, Functional and phylogenetic aspects of the ciliary membrane: A comparative freeze-fracture study, *BioSystems* **14**:403–421.

Bardele, C. F., 1983, Comparative freeze-fracture study of the ciliary membrane of protists and invertebrates in relation to phylogeny, *J. Submicrosc. Cytol.* **15**:263–267.

Gibbons, I. R., 1981, Cilia and flagella of eukaryotes, *J. Cell Biol.* **91**:107s–124s.

Lancet, D., 1988, Molecular components of olfactory reception and transduction, in: *Molecular Neurobiology of the Olfactory System* (F. L. Margolis and T. V. Getchell, eds.), Plenum Press, New York, pp. 25–50.

Loeffler, F., 1889, Eine neue Methode zum Färben der Mikroorganismen im besonderen ihre Wimperhaare und Geisseln, *Zentralbl. Bakteriol.* **6**:209–224.

Manton, I., 1952, The fine structure of plant cilia, *Symp. Soc. Exp. Biol.* **6**:306–319.

Nakamura, T., and Gold, G. H., 1987, A cyclic-nucleotide-gated conductance in olfactory receptor cilia, *Nature* **325**:442–444.

Snyder, S. H., Sklar, P. B., Hwang, P. M., and Pevsner, J., 1989, Molecular mechanisms of olfaction, *Trends Neurosci.* **12**:35–38.

Tyler, S., 1973, An adhesive function for modified cilia in an interstitial turbellarian, *Acta Zool.* **54**:139–151.

Contents

3. *Euglena gracilis:* **A Model for Flagellar Surface Assembly, with Reference to Other Cells That Bear Flagellar Mastigonemes and Scales**

G. B. Bouck, T. K. Rosiere, and P. J. Levasseur

4. **Gliding Motility and Flagellar Glycoprotein Dynamics in** *Chlamydomonas*

Robert A. Bloodgood

5. The Role of Flagella in the Sexual Reproduction of *Chlamydomonas* Gametes

H. van den Ende, A. Musgrave, and F. M. Klis

6. The Role of Ciliary Surfaces in Mating in *Paramecium*

Tsuyoshi Watanabe

7. **Calcium Ions and the Regulation of Motility in *Paramecium***

    Robin R. Preston and Yoshiro Saimi

8. **Structure, Turnover, and Assembly of Ciliary Membranes in *Tetrahymena***

    Norman E. Williams

9. **Ciliary Membrane Tubulin**

    R. E. Stephens

10. Lipids of Ciliary and Flagellar Membranes

Edna S. Kaneshiro

11. Flagellar Surfaces of Parasitic Protozoa and Their Role in Attachment

Keith Vickerman and Laurence Tetley

12. The Sperm Plasma Membrane: A Little More Than Mosaic, a Little Less Than Fluid

Richard A. Cardullo and David E. Wolf

13. Structure and Assembly of the Oviduct Ciliary Membrane

Bernadette Chailley, Emmanuelle Boisvieux-Ulrich, and Daniel Sandoz

14. The Surface of Mammalian Respiratory Cilia: Interactions between Cilia and Respiratory Pathogens

Elaine Tuomanen

15. The Photoreceptor Connecting Cilium: A Model for the Transition Zone

Joseph C. Besharse and Cynthia J. Horst

Ciliary and
Flagellar Membranes

Introduction to Cilia and Flagella

George B. Witman

1. Introduction

Cilia and flagella of eukaryotes are generally long, whiplike appendages extending from the cell body; historically, the term *flagellum* has been used when these structures are present singly or in small numbers, whereas the term *cilium* has been used when the structures occur in larger numbers. Typically, a flagellum propagates nearly symmetrical bends from the base to the tip of the organelle, causing the fluid in which it is beating to flow parallel to the flagellar axis (Fig. 1). Cilia generally move with an asymmetrical beat consisting of an effective and a recovery stroke. During the effective stroke a large bend is formed at the base of the cilium, causing the cilium to slice rapidly through the medium; this is followed by the recovery stroke, during which the bend is propagated along the ciliary shaft until the cilium returns to the position it held before the beginning of the effective stroke. The result of this beat pattern is that fluid is moved parallel to the cell surface in the direction of the effective stroke (Fig. 1). Cilia and flagella are very similar if not identical in terms of their internal structures and mechanisms of movement; indeed, the flagella of many organisms can beat with both flagellar and ciliary type waveforms. Consequently, most of my general comments about these organelles may be applied equally well to either cilia or flagella. The eukaryotic flagellum should not be confused with the prokaryotic flagellum, which is a completely different structure having a different protein composition and a different mechanism of movement (see Macnab, 1987a,b, for reviews). The prokaryotic flagellum is an extracellular appendage, whereas the eukaryotic flagellum is an intracellular organelle, surrounded by an extension of the cell's plasma membrane. It is, of course, the interesting and frequently unique properties of ciliary and flagellar membranes that are the raison d'être of this book.

In this introductory chapter, I attempt to give the reader an overview of the organization and functioning of the internal components of cilia and flagella so that he or she may appreciate the properties and roles of the ciliary and flagellar surface in an appropriate

George B. Witman • Cell Biology Group and Male Fertility Program, Worcester Foundation for Experimental Biology, Shrewsbury, Massachusetts 01545.

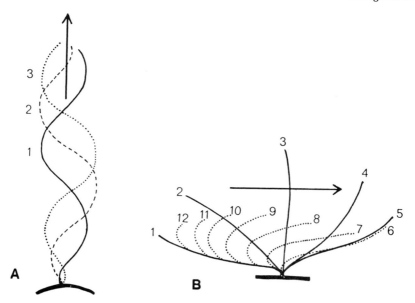

Figure 1. Diagrammatic illustration of the waveforms of a typical flagellum (A) and cilium (B). Successive positions of the flagellum or cilium during a beat cycle are numbered consecutively. In B, the positions of the cilium during the effective and recovery strokes are indicated by solid and dashed lines, respectively. The arrows indicate the resulting direction of movement of the medium. From Sleigh (1974).

structural and physiological context. Wherever possible, I have included references to books or review articles that may be consulted for a more complete discussion of a particular subject. Recent excellent reviews include Dentler (1987) and chapters in Warner *et al.*, (1989).

2. Occurrence and Function

Flagella are widely distributed among eukaryotes, being primitively present in every major group except the fungi (Ascomycetes, Basidiomycetes, and Zygomycetes); they are also absent in the red algae, from which they may have been lost secondarily (Cavalier-Smith, 1982). They are of course present on the surfaces of many unicellular organisms, where they are responsible for locomotion or generating water currents used during feeding. In some unicells, they also play roles in surface cleaning (Sleigh, 1962), mechanoreception (Machemer, 1985), and cell–cell recognition and adherence during mating; the latter function is dependent upon unique specializations of the ciliary or flagellar surface (see van den Ende *et al.,* this volume; Watanabe, this volume). Flagella also anchor certain parasitic protozoa to the surfaces of their hosts (see Vickerman and Tetley, this volume).

In invertebrate metazoans, cilia and flagella continue to have a prominent role in locomotion, particularly of larvae, and in the generation of water currents and mucus flow for respiration, feeding, and surface cleaning (Gray, 1928). They also serve a multitude of other functions, including excretion and water balance by virtue of their activity in flame cells and nephridia in many phyla (Sleigh, 1962), transportation of food through the alimentary canal in annelids, mollusks, tunicates, cephalochordates, and echinoderms (Eckert *et al.*, 1988), and circulation of coelomic fluid (Sleigh, 1962). Many invertebrate sensory receptors are derived from cilia (see below); in this regard, it is interesting to note that nematodes and crustaceans, which have dispensed with cilia and flagella for all other functions, have well-developed sensory cilia (Laverack and Ardill, 1965; Ward *et al.*, 1975; Perkins *et al.*, 1986). Of course, flagella also are important in the locomotion of sperm of many invertebrates, particularly in those species relying on external fertilization.

In vertebrates, muscle has taken over many of the functions carried out by cilia in lower organisms, but cilia and flagella still have important and essential roles. For example, in mammals, ciliated epithelial cells line the lower and upper respiratory tracts. The cilia of these cells move mucus, with entrapped foreign particles, out of the airways, thus preventing small particles from blocking the lower passages, and defending against respiratory tract infection by pathogenic organisms (Sleigh, 1977; Afzelius, 1979; Kelly *et al.*, 1984; Tuomanen, this volume). Fine fibrous projections extending from the tips of the cilia may increase the effectiveness of such mucociliary clearance (Jeffery and Reid, 1975). Similarly, cilia are present on the epithelium of the oviduct, where they are involved in fluid flow and may assist transport of the ovum (Verdugo *et al.*, 1980b; Villalon and Verdugo, 1982; Chailley *et al.*, this volume). It is generally believed that cilia on the fimbriae create currents which help draw the ovum into the ostium of the oviduct (Kelly *et al.*, 1984; Fawcett, 1986); however, women with immotile cilia syndrome, in which the cilia are paralyzed, are fertile, so the cilia must not be essential for this process (Afzelius *et al.*, 1978). The activity of cilia in both the respiratory tract and the female reproductive tract is under hormonal control (Verdugo *et al.*, 1980a,c; see Verdugo, 1982, for review). In at least some cases this control is mediated by Ca^{2+} (Verdugo, 1980). The ependymal cells lining the cavities of the spinal cord and brain have cilia that are believed to clear the ependymal surfaces of cellular debris (Afzelius, 1979); whether these cilia also help circulate the cerebrospinal fluid is not clear (Afzelius, 1976). Cilia are found on the epithelium of the eustachian tube and parts of the tympanic cavity of the middle ear (Fawcett, 1986), where they also function in clearance. In the male, cilia are present in the tiny efferent ducts which connect the testis to the epididymis. It has been stated that these cilia assist in the transportation of spermatozoa (Geneser, 1986), which are nonmotile in the testis. However, sperm transport appears to be normal in men with immotile cilia syndrome (Afzelius, 1979), so smooth muscle activity in the walls of the ducts, together with fluid flow from the testis to the epididymis, must suffice to transport the sperm through this region. Cilia are also present on embryonic epithelia, where they appear to influence morphogenesis; humans with congenitally immotile cilia frequently have *situs inversus* as a result of sinistral rather than dextral rotation of the archenteron (Afzelius, 1976). The mammalian sperm tail is of course a flagellum, and its motility is essential for fertilization.

Single nonmotile cilia, called *primary cilia,* are found in most cell types in both invertebrates and vertebrates (see Wheatley, 1982, for review). The widespread occurrence of primary cilia suggests that they have an important function; proposed roles include growth control (Tucker *et al.*, 1979) and sensory reception (Poole *et al.*, 1985; Roth *et al.*, 1988).

In both invertebrates and vertebrates, cilia or their derivatives are components in many different types of sensory receptors. The role of cilia in sensory reception will be discussed in more detail later in this chapter.

3. Structure

3.1. The Axoneme

The axoneme of cilia and flagella is a microtubular array that provides the structural support giving the cilium or flagellum its typical form. In motile cilia and flagella, the axoneme also constitutes the machinery responsible for movement. The typical "9 + 2" axoneme (Figure 2) consists of two central singlet microtubules surrounded by nine outer doublet microtubules. Each outer doublet microtubule itself consists of two conjoined tubules: an A tubule, which is circular in cross section, and a B tubule, which is C-shaped and shares part of the wall of the A tubule. Like cytoplasmic microtubules, the central and outer doublet microtubules are built up of α- and β-tubulin heterodimers that form 4-nm-diameter protofilaments running the length of the microtubules; the central microtubules and the A tubule of the outer doublets each contain 13 protofilaments, whereas the B tubule has 10 or 11 protofilaments (Ringo, 1967b; Warner and Satir, 1973; Tilney *et al.*,

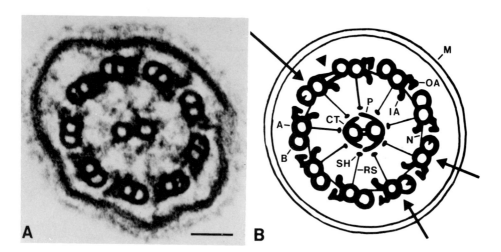

Figure 2. (A) Electron micrograph of a cross section of a *Chlamydomonas* flagellum viewed from base to tip. Bar = 0.05 μm. (B) Diagrammatic representation of A. M, flagellar membrane; A and B, A and B tubule, respectively, of outer doublet microtubule; OA and IA, outer and inner dynein arms, respectively; N, nexin link; RS, radial spoke; SH, spoke head; CT, central pair of microtubules; P, "central sheath" projections from central microtubules. Three of the outer doublets (arrows) have a beaklike projection in the lumen of their B tubule; the outer arm of one of these doublets is replaced by a two-part bridge (arrowhead). An amorphous link appears to connect each outer arm to the plasma membrane. The central tubules are joined by a central pair bridge.

1973). Within the protofilament, αβ-tubulin dimers repeat every 8 nm; this fundamental period serves as a basis for the positioning of the microtubule-associated structures along the microtubules (see below). Ciliary and flagellar tubulin is acetylated during assembly into the axoneme (L'Hernault and Rosenbaum, 1983) and deacetylated during axonemal disassembly (L'Hernault and Rosenbaum, 1985). The function of this posttranslational modification is not clear, but it may be involved in the stabilization of the flagellar microtubules. Nontubulin polypeptides also are intimately associated with ciliary and flagellar microtubules; of these, the best characterized are the tektins, a group of intermediate filamentlike proteins that form 2- to 3-nm-diameter filaments which run longitudinally in, or closely apposed to, the walls of the outer doublets (Linck *et al.*, 1985; Chang and Piperno, 1987).

The axoneme also contains a number of microtubule-associated structures:

3.1.1. Outer Dynein Arms. The *outer dynein arms* extend from the A tubule of an outer doublet toward the B tubule of the adjacent doublet microtubule at intervals of 24 nm along the doublet (Warner *et al.*, 1977; Witman and Minervini, 1982). The outer arm is compositionally complex, containing up to three heavy chains ($M_r > 400,000$) and several intermediate and light chains. Each heavy chain contains ATPase activity and forms the basis for the organization of the arm into structural and functional units (see Witman, 1989, for review). In many cilia and flagella, an amorphous link appears to connect each outer arm to the overlying membrane (Fig. 2; Escalier *et al.*, 1982). In at least some flagella, the outer arms are present on only eight of the nine doublets, being absent from one of the doublets that lies in the plane of beat (Hoops and Witman, 1983).

3.1.2. Inner Dynein Arms. Like the outer arms, the *inner arms* are attached to the A tubule of an outer doublet and extend toward the B tubule of the adjacent doublet. However, in contrast to the outer arms, there appear to be at least two different types of inner arms. One type, termed a *dyad,* contains two large globular domains or "heads"; the other type, termed a *triad,* contains three "heads" (Goodenough and Heuser, 1985). Diads and triads alternate along the doublet with one triad being followed by two diads, and each triad/diad/diad unit repeating every 96 nm along the doublet. The inner arms contain four or five heavy chains and ATPases and several intermediate and light chains; it is likely that each "head" corresponds to one heavy chain and one ATPase activity (reviewed in Witman, 1989). Interestingly, an actinlike protein is tightly associated with at least three different inner arm dynein subunit ATPases (Piperno and Luck, 1981; Piperno, 1988). The function of actin in the inner arms is unknown.

3.1.3. Nexin or Interdoublet Links. Thin fibers termed *nexin* or *interdoublet links* connect the A tubule of one outer doublet to the B tubule of the adjacent doublet at intervals of 96 nm (Stephens, 1970; Dallai *et al.*, 1973; Olson and Linck, 1977). Electron microscopic studies of partially disrupted axonemes revealed that the nexin links could be stretched up to ten times their normal length (Dallai *et al.*, 1973; Warner, 1976; Olson and Linck, 1977); this observation led to the proposal that the links were highly elastic and might limit the amount of sliding that could occur between doublets during axonemal bending. However, a more recent study concludes that nexin links do not stretch *in vivo,* but undergo detachment from one site and reattachment at another site farther down the tubule as the doublets slide past one another (Warner, 1983). The protein composition of the nexin links has not been determined.

3.1.4. Radial Spokes. *Radial links* or *spokes* extend from the A tubule of each outer doublet in toward the center of the axoneme, where each spoke terminates in a bulbous enlargement called the *spoke head*. Depending on the organism, the spokes repeat in groups of two or three at intervals of 96 nm (Hopkins, 1970; Chasey, 1972; Warner and Satir, 1974); the heads of the members of a group are attached to one another laterally (Witman *et al.*, 1978; Goodenough and Heuser, 1985). In *Chlamydomonas*, the radial spokes contain 17 polypeptides, of which 5 are components of the spoke head (see Luck, 1984, for review).

3.1.5. Central Tubule Projections. Two pairs of *projections* extend from each central microtubule; the projections differ in their lengths and may be used to distinguish the two microtubules (Hopkins, 1970; Olson and Linck, 1977; Witman *et al.*, 1978). The projections repeat longitudinally at intervals of 16 nm. One tubule also has *barbs* that repeat every 32 nm (Olson and Linck, 1977; Goodenough and Heuser, 1985); these structures are poorly understood and their relationship to the projections is not clear. Analyses of mutants of *Chlamydomonas* have shown that the central pair of microtubules contains as many as 23 nontubulin polypeptides (Witman *et al.*, 1978; Adams *et al.*, 1981; Dutcher *et al.*, 1984), of which 8 to 10 are uniquely associated with one tubule and 7 with the other (Dutcher *et al.*, 1984). Three of these are known to constitute one of the projections, and it is likely that many of the other polypeptides also are components of the projections and barbs. Historically, the projections have been referred to as the *central sheath* because in early electron micrographs they appeared to form a filament that wrapped around the central microtubules (Gibbons and Grimstone, 1960).

3.1.6. Other Axonemal Structures. Many other structures are associated with the axoneme. Some are general features of most cilia and flagella; others are present in a limited range of organisms. These include bridges between two of the doublets lying in the plane of beat (Afzelius, 1959; Gibbons, 1961; Satir, 1965; Hoops and Witman, 1983), and a *central pair bridge* that joins the two central tubules (Warner, 1976; Olson and Linck, 1977; Witman *et al.*, 1978). There are *beaklike projections* that extend into the lumen of the B tubule in three of the outer doublets of *Chlamydomonas* and many other green algae (Witman *et al.*, 1972); these structures are located in specific doublets in the plane of beat and appear to be involved in the control of flagellar waveform (Hoops and Witman, 1983; Segal *et al.*, 1984). Of particular interest in the context of this volume are the membrane–microtubule links that connect both the sides and tips of axonemal microtubules to the ciliary or flagellar membrane (see Dentler, this volume).

3.2. The Transition Zone

Between the base of the axoneme and the distal end of the basal body (see below) is a specialized region termed the *transition zone* or *transition region* (Fig. 3). This generally occurs at the level of the cell surface, where the cell membrane bends outwards to form the ciliary or flagellar membrane. The outer doublet microtubules are continuous through this region, but cease to have dynein arms, radial spokes, and nexin links. The central

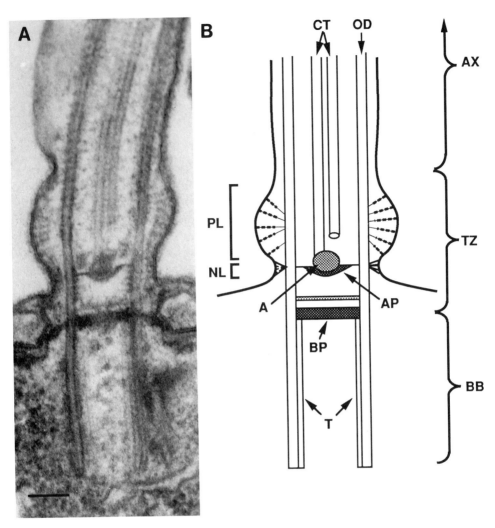

Figure 3. (A) Electron micrograph of longitudinal section through the base of a *Paramecium* cilium (Dute and Kung, 1978). Bar = 0.1 μm. (B) Diagrammatic representation of A. AX, axoneme; TZ, transition zone; BB, basal body. One of the central pair of microtubules (CT) originates in the axosome (A); the other central microtubule originates just above the axosome. The outer doublet microtubules (OD) continue through the transition zone to become the A and B tubules of the basal body triplet microtubules (T). Fine fibers or links extend from the outer doublets to the subunits of the ciliary necklace (NL) and plaque (PL). BP, basal plate; AP, axosomal plate.

tubules originate in the transition zone,* with one sometimes emerging from an electron-dense granule known as the *axosome* (Pitelka, 1974). Above the point of origin of the

---

*As first defined by Gibbons (1961; see also Ringo, 1967a), the distal boundary of the transition region occurs at that level where changes first become apparent in the links and arms associated with the outer doublets of the axoneme proper; this occurs distal to the point where the central microtubules originate. More recently, some authors (e.g., Pitelka, 1974; Huang *et al.*, 1982; Lewin and Lee, 1985) have considered the proximal end of the central pair to mark the distal end of the transition region. The original definition is used here.

central microtubules, the A tubule of each outer doublet is connected to the B tubule of the adjacent doublet by robust cross-bridges called *peripheral links* (Ringo, 1967a; Williams and Luft, 1968; Witman *et al.*, 1972), not to be confused with the thinner nexin or interdoublet links of the axoneme. The structure of the transition region below the point of origin of the central microtubules varies from organism to organism, but usually includes electron-dense *plates* that extend transversely across the transition zone (Fig. 3; Gibbons and Grimstone, 1960), or a *central cylinder* that appears closed at its proximal end (Ringo, 1967a). The transition regions of green algae and the sperm of lower plants contain a striking nine-pointed star or *stellate structure* formed of thin fibers that make a V-shaped connection between the A tubule of each outer doublet and the central cylinder (Ringo, 1967a). Just proximal to the level at which the central microtubules originate, the outer doublets of most cilia and flagella are connected to the surrounding membrane by prominent Y-shaped links (see Dentler, this volume; Besharse and Horst, this volume). Examination of replicas of freeze-fractured membranes reveals scalloped rows of intramembranous particles termed the *ciliary* or *flagellar necklace* (Gilula and Satir, 1972); these appear to correspond to the points of attachment or insertion of the Y-shaped links into the membrane. In protozoa, rectangular arrays of intramembranous particles are frequently seen just distal to the ciliary necklace (Wunderlich and Speth, 1972; Bardele, 1981; see Williams, this volume); these arrays are termed *plaques*. The particles of the plaques are connected to the underlying doublets by thin fibers (Dute and Kung, 1978).

It has been proposed that the intramembranous particles of the necklace and plaques are sites of Ca^{2+} entry into the cilium or flagellum (see Dentler, 1981, for review of cytochemical evidence). Physiological evidence that Ca^{2+} channels are located in the transition region recently has been obtained from studies of ctenophore (*Beroë*) macrocilia, which are composed of several hundred axonemes surrounded by a common ciliary membrane (Horridge, 1965). Individual axonemes retain their ciliary membranes only at the level of the transition zone, where the membranes of adjacent axonemes are fused together to form a tubular membranous reticulum termed the *ciliary rete* (Tamm, 1988b). The rete is continuous with the common ciliary membrane and therefore open to the surrounding seawater. Macrocilia are stimulated to beat by Ca^{2+} influx through voltage-sensitive Ca^{2+} channels (Tamm, 1988a), and intact macrocilia respond much more readily when Ca^{2+} is iontophoretically applied to their bases than to their tips or to the cell body, whereas demembranated macrocilia in an ATP-containing reactivation solution respond to Ca^{2+} applied to any region (Tamm, 1988b). These results strongly suggest that the ciliary rete serves to conduct Ca^{2+} to the bases of the axonemes deep within the macrocilium, and that the voltage-sensitive Ca^{2+} channels are located primarily in the membranous cisternae surrounding the transition zones. This is also the location of the ciliary necklaces, although this does not prove that the intramembranous particles of the ciliary necklace are the Ca^{2+} channels. The axonemal Ca^{2+} receptor is apparently distributed along the length of the axoneme. However, electrophysiological studies have shown that the voltage-sensitive Ca^{2+} channels of the giant ciliary comb plates of another ctenophore, *Pleurobrachia pileus,* are located over most of the length of the comb plate cilia (Moss and Tamm, 1987). Therefore, Ca^{2+} channels are not always confined to the membrane surrounding the transition zone.

Rupture of the ciliary or flagellar shaft during deciliation or deflagellation occurs within the transition region (Blum, 1971; see below); this process is Ca^{2+} dependent

(Kamiya and Witman, 1984; Huber *et al.*, 1986; Sanders and Salisbury, 1989) and may involve the Ca^{2+}-binding contractile protein "centrin," which has been localized to the stellate structure of the transition zone of *Chlamydomonas* (Sanders and Salisbury, 1989). The fibers of the stellate structure contract during flagellar abscission, and it has been proposed that this contraction causes severing of the outer doublet microtubules (Sanders and Salisbury, 1989). Centrin (also called "caltractin") is a 20,000-dalton protein closely related to calmodulin (Huang *et al.*, 1988).

During chemical deflagellation of green algae, the flagella abscise at a point just distal to the central cylinder (Lewin and Lee, 1985; Sanders and Salisbury, 1989); similarly, the cilia of *Paramecium* break at a point just distal to the axosome during deciliation (Kennedy and Brittingham, 1968). Therefore, the distal portion of the transition region, including the ciliary plaques if present, is isolated along with the cilia or flagella of these organisms, whereas the proximal portion, including the basal plate or central cylinder and the components of the ciliary necklace (Satir *et al.*, 1976), remain with the cell body. This is important to keep in mind when evaluating electrophysiological studies of deciliated cells and biochemical studies of isolated cilia or flagella.

3.3. The Basal Body and Associated Structures

At the base of the ciliary or flagellar shaft is the basal body, characterized by a ring of nine triplet microtubules. The A and B tubules of each triplet microtubule are continuous with the A and B tubules of the outer doublets; the third "C" tubule is present only in the basal body. The triplet microtubules are linked together by fibrous connectors that extend between the A tubule of one triplet and the C tubule of the adjacent triplet; the triplets of most basal bodies also are interconnected at their proximal ends by a distinctive cartwheel structure (Gibbons and Grimstone, 1960; Ringo, 1967a; Pitelka, 1974). A more detailed description of basal body morphology will be found in Anderson (1972) and in reviews by Stubblefield and Brinkley (1967), Wheatley (1982), and Dustin (1984). The basal body serves as a template for the growth of the outer doublet microtubules during ciliary and flagellar morphogenesis.

A number of different types of fibrous structures are associated with the outer surfaces of basal bodies, including basal feet, striated fibers and roots, and microtubule bands (Fig. 3; Wright *et al.*, 1985; reviews: Pitelka, 1974; Melkonian, 1980; Wheatley, 1982). These anchor the ciliary or flagellar apparatus to the cell body (Ringo, 1967a; Sleigh and Silvester, 1983), establish or maintain the correct rotational orientation of the basal bodies to ensure that the inherent functional polarity of the axoneme results in effective cell movement (Hoops and Witman, 1983; Hoops *et al.*, 1984), and may be important in establishing cell polarity and determining the positions of the nucleus and other organelles within the cell (Melkonian, 1978; Hoops and Witman, 1985; Salisbury *et al.*, 1987). Fibrous struts extending from the basal bodies to the plasma membrane immediately surrounding the bases of the mechanosensitive cilia of the statocyst of the nudibranch *Hermissenda* are believed to transmit mechanical pressure from the ciliary shaft to the plasma membrane, leading to membrane distortion and the opening of Na^+ channels that accompanies mechanotransduction (Kuzirian *et al.*, 1981; Alkon, 1983; see below).

Basal bodies serve as centrioles and vice versa; the relationship between basal bodies and centrioles is reviewed in detail by Wheatley (1982).

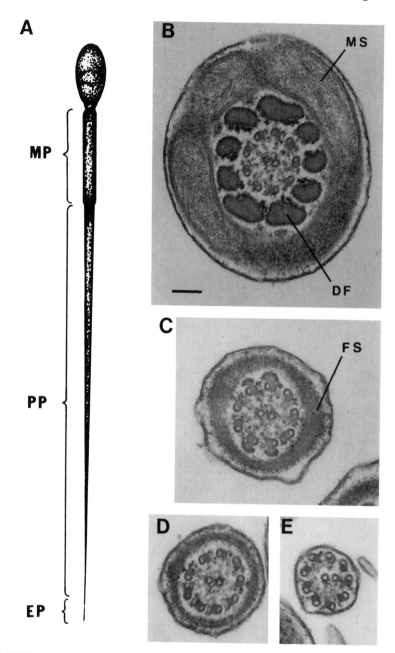

Figure 4. (A) Diagrammatic illustration of a typical mammalian (ram) sperm. The mid piece (MP), principal piece (PP), and end piece (EP) of the flagellum are indicated. (B–E) Electron micrographs of cross sections through the flagellum of a ram sperm at the level of (B) the mid piece, showing the dense fibers (DF) and the mitochondrial sheath (MS); (C) the principal piece, showing the fibrous sheath (FS) surrounding the axoneme and dense fibers, of which some have terminated and the remaining ones have become greatly reduced in size; (D) the distal portion of the principal piece, after termination of all the dense fibers; and (E) the end piece. Bar = 0.1 μm.

3.4. Accessory Structures of the Flagellar Shaft

In many organisms, structures of various types are present between the flagellar membrane and the axoneme. These structures can have important implications for flagellar surface architecture and function. For example, in *Euglena*, there is a paraflagellar rod, only slightly smaller in diameter than the axoneme, that runs longitudinally the length of the flagellum; there are also electron-dense sheets intercalated between the membrane and the axoneme or the paraflagellar rod. The rod and sheets are connected to the flagellar membrane and appear to be directly responsible for the positioning of the mastigonemes on the flagellar surface (Melkonian *et al.*, 1982; Bouck *et al.*, this volume.)

Prominent flagellar accessory structures are especially common in the sperm of organisms that have internal fertilization (Baccetti, 1982, 1985). In mammalian sperm, each outer doublet is associated with an *outer dense fiber* throughout most of the length of the sperm tail (Fig. 4). In the *mid piece* (the proximal portion of the tail), the axoneme and its associated dense fibers are completely surrounded by a helical *mitochondrial sheath*. Immediately distal to the mid piece is the *principal piece*, in which the mitochondrial sheath is replaced by a *fibrous sheath* which also surrounds the axoneme and dense fibers. The dense fibers and fibrous sheath taper distally, and are completely absent in the *end piece* or distal portion of the flagellum, the only part of the sperm tail where the flagellar membrane lies directly over the outer doublets of the axoneme. The functions of the dense fibers and fibrous sheath probably are to stiffen the flagellum so that its waveform is less likely to be deformed when swimming through the viscous environment of the female reproductive tract (Ishijima and Witman, 1987).

It is important to note that when such prominent accessory structures are present, they preclude any direct linkage between the outer doublet microtubules and the flagellar membrane. Thus, mechanisms that depend on direct attachment to the outer doublet microtubules for anchoring, organizing, or moving flagellar surface components in the simple flagella of organisms such as *Chlamydomonas* are unlikely to function, without profound modification, in the specialized flagella of sperm adapted for internal fertilization. Of course, this would not necessarily be the case for surface phenomena directly dependent upon a nonmicrotubular submembrane cytoskeleton (Bray *et al.*, 1986).

4. Production of Movement

4.1. The Sliding Microtubule Model

Early studies of isolated, permeabilized flagella demonstrated that flagellar movement is brought about by processes within the axoneme, and that the energy for motility is provided by ATP (Brokaw, 1961; see Bishop, 1962, and Sleigh, 1962, for reviews of the earlier literature). That this movement might result from an ATP-induced sliding of noncontractile filaments in the axoneme, in rough analogy with the sliding filament mechanism for muscle contraction (Hansen and Huxley, 1955), was immediately recognized as a possibility upon discovery of the outer doublet arms and spokes (Afzelius, 1959); both of these structures were in suitable locations to generate sliding forces between the axonemal microtubules. Support for such a mechanism was provided by the finding that the axonemal ATPase ("dynein") was contained in the arms (Gibbons, 1963,

1965). Shortly thereafter, detailed ultrastructural analyses of the termination levels of individual outer doublets in straight and bent cilia of the mussel *Elliptio* revealed that the outer doublets do indeed slide relative to one another during ciliary bending, and that they do not undergo appreciable contraction during bending (Satir, 1965, 1968); this work firmly established the "sliding microtubule" model for ciliary and flagellar movement. Sliding between axonemal microtubules has been demonstrated directly by limited tryptic digestion of fragments of sea urchin sperm axonemes; subsequent addition of ATP caused the microtubules to slide actively past one another, resulting in elongation and "disintegration" of the axonemal fragment (Summers and Gibbons, 1971). Recently, interdoublet sliding in actively bending axonemes was demonstrated by following the relative positions of 40-nm gold beads bound to the doublets of demembranated, reactivated sea urchin sperm (Brokaw, 1989a).

4.2. Active Sliding Is Produced by Dynein Arms Acting on Adjacent Outer Doublet Microtubules

Studies of the ATP-induced disintegration of trypsin-treated axonemal fragments of sea urchin sperm flagella (Summers and Gibbons, 1971) showed (1) that dynein ATPase activity and sliding disintegration had similar nucleotide and divalent cation specificities, i.e., both were highly specific for ATP and could utilize Mg^{2+} or Mn^{2+} equally well, with Ca^{2+} being used less effectively; (2) that the disintegrating axonemes elongated up to five times their original length, suggesting that active sliding occurred between adjacent outer doublets rather than between outer doublets and the central microtubules; and (3) that the trypsin digestion appeared to damage the radial spokes and nexin links while leaving the dynein arms relatively intact. These results strongly suggested that the active microtubule sliding responsible for ciliary and flagellar movement occurs between adjacent doublet microtubules and is powered by the dynein ATPases contained in the inner and outer arms of the outer doublets. More conclusive evidence for this subsequently came from the demonstration that axonemes of mutant *Chlamydomonas* lacking the radial spokes or central microtubules also undergo ATP-induced sliding disintegration following partial proteolysis (Witman *et al.*, 1978); in this case, the sliding could have come about only as a result of doublet–doublet interactions.

Electron microscopic examination of axonemes that had undergone ATP-induced sliding disintegration after partial digestion with trypsin revealed that the dynein arms that were attached to the A tubule of one outer doublet always pushed the adjacent doublet in a base-to-tip direction (Sale and Satir, 1977). This polarity of force generation appears to be an invariant feature of dynein arm activity, and has been observed to be the same in all organisms and under all conditions tested.

4.3. The Mechanism of Force Generation

Although it is certain that the dynein arms generate the forces that result in interdoublet sliding, the precise mechanism by which this occurs has not been determined. It is generally assumed that, concomitant with a cycle of ATP-binding and hydrolysis, the arm undergoes a cycle of attachment to and detachment from the B tubule of the adjacent doublet, causing the doublet to which the arm is permanently attached to walk over the surface of the opposing doublet (Satir *et al.*, 1981). The arrangement of the morphological domains in the outer arm *in situ* is quite different depending on whether the axoneme is prepared for electron microscopy in the presence or absence of ATP (Witman and Minervini, 1982; Goodenough and Heuser, 1982), and such images have been used to construct a model for how the arm might move during the force-generating step of its mechanochemical cycle (Avolio *et al.*, 1986). However, it is a matter of controversy whether these different morphologies represent different conformational states in the mechanochemical cycle of the arm, or different activity states of the arm (Johnson *et al.*, 1986). Further studies will be necessary to clarify this. In any case, there is as yet no conclusive evidence that an arm generates force by a "swinging cross-bridge" mechanism.

Recently, it has become possible to examine directly the mechanochemical transducing activity of purified axonemal dynein (Paschal *et al.*, 1987; Vale and Toyoshima, 1988). Isolated outer arm dynein is adsorbed to a glass coverslip, *in vitro*-assembled brain microtubules and ATP are added, and the microtubules are observed using video-enhanced differential interference contrast or darkfield microscopy. Under these conditions, the microtubules are seen to actively glide over the surface of the coverslip; this gliding is the direct result of the adsorbed dynein molecules acting on the microtubules. This *in vitro* system has already been of use in demonstrating that a purified subunit containing only one of the heavy chains of sea urchin outer arm dynein is capable of translocating microtubules (Sale and Fox, 1988). Further use of this compositionally simple and defined system should be very helpful in elucidating the mechanism of force production by the dynein ATPases.

4.4. Internal Resistances Convert Sliding into Bending

When the dynein-generated sliding between doublet microtubules is resisted by structures within the axoneme, axonemal bending results (Shingyoji *et al.*, 1977; see Fig. 5). The structures which provide this resistance to sliding have not been conclusively identified. Ultrastructural analysis of the time course of digestion of individual axonemal structures by trypsin showed that the protease disrupted the radial spokes and nexin links at the same rate as it sensitized axonemes to ATP-dependent sliding disintegration (Summers and Gibbons, 1973), suggesting that one or both of these structures were responsible for resisting interdoublet sliding in the intact axoneme. More recent studies of mutants of *Chlamydomonas* have shown that axonemal bending can occur in the absence of radial spokes (Huang *et al.*, 1982), leaving the nexin links as the most likely candidates for converting interdoublet sliding into axonemal bending. It should be noted, however, that protease treatment of axonemes also cleaves the dynein heavy chains (Bell and Gibbons, 1982; King and Witman, 1988), so the possibility exists that the dynein arms themselves are involved in resisting interdoublet sliding, perhaps during one stage of their mechanochemical cycle.

4.5. Coordination of Interdoublet Sliding

Because the dynein arms produce force in only one direction (Sale and Satir, 1977; see above), it is essential that the activity of the arms be controlled around the ring of outer doublets. For example, in a planar beating flagellum, active sliding must occur alternately

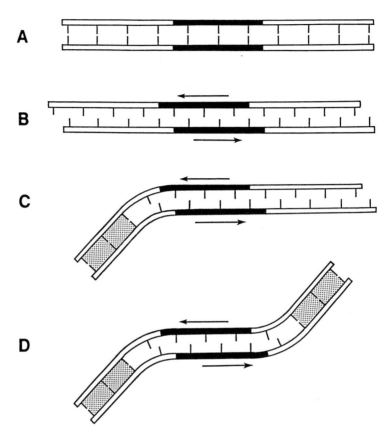

Figure 5. Diagrams showing how sliding between two flexible filaments such as outer doublet microtubules can be converted into bending. (A) Filaments before sliding. (B) Generation of shearing forces in the darkened regions leads to relative sliding of the filaments in the directions indicated by the arrows. If there is no resistance to the sliding, the filaments will remain straight and simply slide past one another. (C) If there is a region of resistance (stippling) to sliding on one side of the region of active force generation, then a bend will form between the two regions. (D) If there is resistance to sliding on both sides of the region of active force generation, then bends will be formed in opposite directions on each side of the region of force generation. From Shingyoji *et al.* (1977).

first in one half of the axoneme to produce a bend in one direction, and then in the other half of the axoneme to cause the flagellum to bend back the other way. The activity of the arms must also be controlled along the length of the axoneme to propagate a bend. There are several possibilities as to how this coordination might be achieved (reviewed by Brokaw, 1986, 1989b). Most likely, it is brought about by a feedback loop monitoring such physical parameters as curvature of the outer doublets, distance between outer doublets, shear displacement (the total amount of sliding that has accumulated between adjacent doublets), or changes in velocity of interdoublet sliding or in the rate of bending. It is also possible that the control involves a biochemical mechanism, such as phosphorylation of a dynein polypeptide. Although a great deal of experimental work and thought has been expended in an attempt to understand the basis for this coordination, the processes by which dynein arm activity and interdoublet sliding are controlled remain unclear.

Nevertheless, studies of mutants of *Chlamydomonas* have provided some clues to the structures involved in this control. Mutants lacking either the radial spokes (Witman *et al.*, 1976, 1978) or the central microtubules (Warr *et al.*, 1966) are paralyzed; however, when these mutations are combined with certain extragenic ''suppressor'' mutations, motility is restored *without* restoration of the missing structures (Huang *et al.*, 1982). Taken together, these results suggest that (1) in wild-type cells, radial spoke–central microtubule interactions are necessary for the activation of the dynein arms; (2) that in the absence of these interactions, dynein arm activity and interdoublet sliding are inhibited; and (3) that this inhibition can be released by mutationally altering another component of the regulatory system (see below).* It therefore seems likely that, in the wild-type flagellum, the location and timing of radial spoke–central microtubule interactions determine the location and timing of dynein arm activity.

The suppressors capable of restoring activity to the mutants fall into two classes. One class is capable of restoring motility to all radial spoke- and central pair-deficient strains, and is represented by at least one mutation that seems to alter one of the heavy chains of the outer arm (Huang *et al.*, 1982). The other class restores motility to only radial spoke-defective strains (Huang *et al.*, 1982), and is represented by mutations that affect several proteins, at least some of which are components of one of the inner arms (Luck and Piperno, 1989). Therefore, the dynein arms appear to be not only targets of this regulatory system, but also integral components of the system.

The motility of the suppressed mutants is not normal, but is more symmetrical than that of wild-type. Therefore, one function of the radial spoke–central microtubule regulatory system seems to be to control beat symmetry (Brokaw *et al.*, 1982; Brokaw and Luck, 1985). The fact that any type of beating at all is observed in the recombinants also indicates that a second, apparently more primitive, regulatory feedback mechanism remains functional even in the absence of the radial spokes or central microtubules. Indeed, cyclical bending has been observed in microtubule bundles, derived from frayed axonemes of *Chlamydomonas*, that apparently consisted of only a pair of outer doublet microtubules (Kamiya and Okagaki, 1986). This second regulatory system probably involves the dynein arms interacting with the outer doublets or the nexin links to monitor one of the physical parameters discussed above.

4.6. Central Pair Rotation

In any discussion of the regulation of dynein arm activity by radial spoke–central microtubule interactions, it should be noted that there is evidence suggesting that the central pair of microtubules rotates within the ring of outer doublet microtubules in the green algae *Synura* (Jarosch and Fuchs, 1975) and *Chlamydomonas* (Kamiya, 1982; Kamiya *et al.*, 1982) and in the ciliates *Opalina* (Tamm and Horridge, 1970) and *Paramecium* (Omoto and Kung, 1979, 1980). Perhaps the most dramatic example of central pair

---

*Proteolytic cleavage of a component of the regulatory system may also release the inhibition of dynein arm activity in axonemes lacking the radial spokes or central tubules, explaining why such axonemes undergo an ATP-dependent sliding disintegration after partial proteolysis (Witman *et al.*, 1978).

rotation is in the uniflagellate marine alga *Micromonas pusilla,* in which the central pair of microtubules extends several micrometers beyond the termination of the short ($\sim$ 1 μm long) outer doublets, takes on a helical configuration, and undergoes continuous rotation in one direction, pushing the cells through the medium like a propeller (Omoto and Witman, 1981). Although the function and generality of central pair rotation in normal "9 + 2" cilia and flagella are not certain (e.g., see Tamm and Tamm, 1981), it has been proposed that the central pair may act as a distributor to determine which outer doublets undergo active sliding at any particular phase in the beat cycle (Omoto and Kung, 1979).

5. Regulation of Movement

The controls discussed in the previous section must operate cyclically and over a short time scale to determine the timing, location, and extent of interdoublet sliding during each beat cycle. In addition, there are higher orders of control that operate sporadically and over longer time scales to regulate the overall activity of a cilium or flagellum and the frequency and form of its beat. These later regulatory mechanisms are important in the initiation of sperm motility and in behavioral responses, including hyperactivation of sperm in the female reproductive tract. These changes in motility are generally mediated via changes in Ca^{2+} or cyclic nucleotide concentration, and are usually triggered by or dependent upon events occurring at the cell surface.

5.1. Initiation of Motility

The sperm of most organisms are nonmotile or poorly motile until released from the male reproductive tract; this has the obvious advantage of conserving energy resources. There appear to be a variety of external factors that maintain sperm quiescence in the male. However, evidence is accumulating that the internal message that initiates motility in most if not all organisms is cAMP (reviews: Brokaw, 1987b; Morisawa, 1987; Stephens and Stommel, 1989).

The events at the level of the cell surface that lead to the initiation of motility have been particularly well studied in sea urchin sperm (reviews: Shapiro and Tombes, 1985; Trimmer and Vacquier, 1986). Sea urchin sperm in the testes are kept nonmotile by a low intracellular pH ($pH_i \leq 7.2$) that probably results from elevated CO_2 levels in the semen (Johnson *et al.,* 1983). Upon release into seawater, a rapid exchange of extracellular Na^+ for intracellular H^+ occurs, raising the pH_i and leading to activation of the dynein arms and motility (Christen *et al.,* 1982; Lee *et al.,* 1983). The Na^+–H^+ exchanger is contained in the sperm flagellar membrane and is voltage-sensitive (Lee, 1984); it probably is activated by hyperpolarization of the flagellar membrane as the sperm go from seminal plasma, containing $\sim$ 40 mM K^+, to seawater, containing only $\sim$ 10 mM K^+. The sperm plasma membrane also has Na^+/K^+-ATPase pumps that maintain a low intracellular Na^+ concentration and thus help regulate the internal pH (Gatti and Christen, 1985). Whether the dynein ATPase is turned on directly by elevated pH or through an indirect effect of pH on some other component is not known. However, when the sperm of some species of sea urchin are collected "dry" (without exposure to seawater) and then demembranated, they cannot be reactivated unless they are first exposed to cAMP (Brokaw, 1985, 1987a). It is therefore possible that the effect of pH is mediated by adenylate cyclase and cAMP-dependent protein phosphorylation.

cAMP is clearly important in the activation of trout sperm (Morisawa and Okuno, 1982) and tunicate (*Ciona*) sperm (Brokaw, 1985). It is also involved in the activation of mammalian sperm (reviews: Hoskins *et al.*, 1978; Tash and Means, 1983). Mammalian sperm are nonmotile in the testis, and they remain nonmotile when removed from the testis and placed in a buffered saline solution. However, when sperm are removed from the cauda epididymis or the vas deferens and placed in the same buffer, they do become motile. This acquisition of the capacity to be motile is part of a process known as *epididymal maturation*. The most direct evidence for a role of cAMP in this process has come from studies of reactivated demembranated sperm (Mohri and Yanagimachi, 1980; Yeung, 1984; Ishijima *et al.*, 1986). When ejaculated ram sperm were demembranated by treatment with nonionic detergents and then placed in a reactivation solution containing ATP, they began to beat with a waveform very similar to that of intact ejaculated sperm. When demembranated testicular sperm were placed in the same solution, a small percentage began beating after a lag of $\sim$ 3 min, but when cAMP in addition to ATP was present in the reactivation medium, the majority of sperm began beating within 30 sec (Ishijima *et al.*, 1986, and unpublished observations). The inclusion of phosphodiesterase in the reactivation solution lacking cAMP completely eliminated the low percentage of reactivation seen in the absence of added cAMP, indicating that the "baseline" reactivation was probably due to conversion of ATP to cAMP by sperm adenylate cyclase. These studies suggested (1) that the axoneme of the immature testicular sperm is fully assembled and awaiting a signal to become motile, and (2) that the signal involves cAMP. Elucidation of the pathway by which cAMP activates the sperm motile machinery will have to await identification and characterization of the proteins phosphorylated by cAMP-dependent protein kinase during maturation. It also is not known how the level of cAMP within sperm is regulated during epididymal transit. Maturation involves significant changes in the composition and properties of the sperm plasma membrane, including the flagellar membrane (Cardullo and Wolf, this volume; see Yanagimachi, 1988, for review), and some of these modifications may be important in controlling the activity of sperm adenylate cyclase.

5.2. Behavioral Responses

Most cilia and flagella are capable of changing their beat pattern in response to stimuli. For example, the cilia of *Paramecium* reverse the direction of their effective stroke in response to mechanical stimulation of the anterior end of the cell; this causes the cell to switch from forward swimming to transient backward swimming, enabling it to avoid the source of the stimulus. This response is triggered by cell depolarization and involves an inward flow of Ca^{2+} through voltage-sensitive channels in the ciliary membrane (Naitoh and Eckert, 1969; Machemer and Ogura, 1979; see Eckert and Brehm, 1979, for review). The electrophysiology of this response has been particularly well studied and is discussed in more detail in the chapter by Preston and Saimi. Increased intraciliary Ca^{2+} appears to act directly on the ciliary apparatus to bring about the change in beat direction (Naitoh and Kaneko, 1972). The cilia of *Paramecium* contain the Ca^{2+}-binding protein calmodulin (Maihle *et al.*, 1981; Walter and Schultz, 1981), and calmodulin antagonists block Ca^{2+}-induced backward swimming in demembranated cell models

(Otter *et al.*, 1984; Izumi and Nakaoka, 1987), suggesting that calmodulin may be the axonemal Ca^{2+} receptor (see Otter, 1989, for review).

Paramecium also increases the frequency of ciliary beating in the forward direction in response to mechanical stimulation of the posterior end of the cell; this response is elicited by hyperpolarization of the cell (Naitoh and Eckert, 1969) and involves cAMP (Nakaoka and Ooi, 1985; Bonini *et al.*, 1986). cAMP and cGMP antagonize the Ca^{2+}-induced backward swimming of demembranated models, suggesting that the Ca^{2+} and cyclic nucleotide regulatory pathways are integrated at the molecular level (Bonini and Nelson, 1988). Research is currently under way to identify the substrates phosphorylated by cAMP and cGMP-dependent protein kinases in *Paramecium* cilia.

cAMP and Ca^{2+} also appear to act antagonistically in regulating the motility of the gill cilia of the mussel *Mytilus* (Stommel and Stephens, 1985; see Stephens and Stommel, 1989, for review). When *Mytilus* is feeding, the lateral gill cilia are normally beating, producing a current which carries food particles over the gill, where the particles are trapped and then transported to the mouth. When the mussel is resting, the cilia are quiescent. The activity of the cilia is under neuronal control (Aiello, 1960; Aiello and Guideri, 1964); activation of the cilia is dependent upon cAMP, which is elevated by serotonin, whereas quiescence is presumably brought about by lowered cAMP levels due to the absence of neuronal stimulation (Stephens and Stommel, 1989). When beating lateral gill cilia are struck by a food particle, they briefly arrest due to an influx of Ca^{2+}; this arrest is mediated by calmodulin (Stommel, 1984). In demembranated cell models, Ca^{2+}-induced arrest can be overridden by the addition of cAMP (Stommel and Stephens, 1985). The activity of *Mytilus* gill cilia therefore appears to be regulated by a balance between cAMP and Ca^{2+}.

The green alga *Chlamydomonas* exhibits several behavioral responses, including positive and negative phototaxis, and the photophobic response, a transient, light-induced reversal of the swimming direction. During the photophobic response, the two flagella coordinately switch from a breast stroke-like, ciliary pattern of beating to an undulating, flagellar pattern (Ringo, 1967a). This response is dependent upon extracellular Ca^{2+} (Schmidt and Eckert, 1976). It has been proposed that, following photostimulation of a specialized region of the plasma membrane overlying the eyespot in the cell body, transmembrane ion fluxes are initiated that lead to increased intraflagellar Ca^{2+} (Schmidt and Eckert, 1976), which in turn causes the switch from forward to backward swimming. The change from a ciliary to a flagellar type of beat is induced in isolated demembranated axonemes when free Ca^{2+} is increased from 10^{-6} to 10^{-4} M (Bessen *et al.*, 1980), indicating that Ca^{2+} acts directly on an axonemal component to bring about the change in waveform. The axonemes of *Chlamydomonas* contain tightly bound calmodulin (Gitelman and Witman, 1980; Otter and Witman, manuscript in preparation), so calmodulin may mediate this process. Studies of the *in vitro* phosphorylation of isolated *Chlamydomonas* axonemes have shown that phosphorylation of an 85-kDa protein is repressed by $\geq 10^{-6}$ M Ca^{2+}, while that of a 95-kDa protein is stimulated by $> 10^{-6}$ M Ca^{2+} (Segal and Luck, 1985). The latter protein is lacking in a mutant that does not undergo the photophobic response (Segal *et al.*, 1984), suggesting that this protein is a link between the increase in intracellular Ca^{2+} and the photophobic response. It is possible that the phosphorylation of this protein is brought about by a Ca^{2+}–calmodulin-dependent protein kinase.

Phototaxis in *Chlamydomonas* is also dependent upon extracellular Ca^{2+} (Stavis and Hirschberg, 1973). However, in contrast to the photophobic response, phototactic steering must involve a differential response of the two flagella. Studies of demembranated models of *Chlamydomonas* have shown that the activity of the *trans*-axoneme (the axoneme farthest from the eyespot) is differentially inactivated by Ca^{2+} concentrations below 10^{-8} M, whereas the activity of the *cis*-axoneme is selectively inactivated at 10^{-6} or 10^{-7} M Ca^{2+} (Kamiya and Witman, 1984). Both axonemes remain relatively active at 10^{-8} M Ca^{2+}. The effect of Ca^{2+} is reversible and probably depends on biochemical modification of an axonemal component. This Ca^{2+}-dependent modulation of *cis*- or *trans*-axonemal activity may form the basis for phototactic steering. In this case phototaxis would not depend on differences between the flagellar membranes of the *cis*- and *trans*-flagella, but on inherent differences between the axonemes of the two flagella.

In contrast to the situation in most organisms, where cAMP stimulates ciliary or flagellar movement (see above), cAMP inhibits the motility of intact vegetative *Chlamydomonas* (Rubin and Filner, 1973; Hartfiel and Amrhein, 1976). The percent motility of isolated, demembranated axonemes of *Chlamydomonas* is significantly enhanced by the protein inhibitor of cAMP-dependent protein kinase (PKI), and by phosphodiesterase, which presumably hydrolyzes cAMP produced by endogenous adenylate cyclase (Hasegawa *et al.*, 1987). These results strongly suggest that the level of motility is determined by competing phosphorylation/dephosphorylation reactions. Inasmuch as these effects were observed in axonemes that had been washed to remove detergent-soluble components, the adenylate cyclase, cAMP-dependent protein kinase, protein phosphatase, and their phosphoprotein substrates all may be components of or intimately associated with the axoneme. Further studies will be necessary to determine if this effect is related to any of the behavioral responses of *Chlamydomonas*. It should be noted that cAMP is the primary signal for most if not all of the events occurring during the mating reaction in *Chlamydomonas* (Pasquale and Goodenough, 1987; see Bloodgood, this volume; van den Ende *et al.*, this volume). However, IBMX, which increases intracellular cAMP and inhibits flagellar motility in vegetative cells, is reported not to inhibit the motility of intact gametic cells, so the flagella of vegetative and gametic cells appear to respond to cAMP in different ways (Pasquale and Goodenough, 1987).

5.3. Hyperactivation of Sperm

Dramatic changes in mammalian sperm motility occur during capacitation in the female reproductive tract (see Cardullo and Wolf, this volume, and Yanagimachi, 1981, for reviews). At this time, the sperm switch from a normal, nearly symmetrical, pattern of beat to a "hyperactivated" pattern in which highly asymmetrical bends of large amplitude and large curvature are generated. Hyperactivation occurs in the ampulla of the oviduct, and is believed to be important in aiding sperm penetration of the egg investments (Katz and Yanagimachi, 1981; Ishijima and Mohri, 1985; see Katz *et al.*, 1989 for review). The molecular mechanism by which hyperactivation is brought about has not been determined, but undoubtedly involves dynamic events at the flagellar surface. The process can be induced *in vitro* by incubation in various capacitation media, usually containing serum albumin, and is preceded or accompanied by changes in the sperm flagellar membrane (Friend *et al.*, 1977; Wolf *et al.*, 1986). Hyperactivation *in vitro* requires extracellular

Ca^{2+} (Yanagimachi and Usui, 1974; Cooper, 1984), and is promoted by the ionophore A23187 (Cooper, 1984). These observations suggest that during hyperactivation *in vivo*, changes in the sperm plasma membrane lead to an influx of Ca^{2+}, causing an increase in the intraflagellar concentration of the ion, which then affects the flagellar apparatus to alter the form of beat. Whether Ca^{2+} acts on the motile machinery of the axoneme directly or indirectly is not known. It is perhaps relevant that increased intracellular Ca^{2+} appears to stimulate sperm adenylate cyclase, which in turn induces the acrosome reaction (Garbers, 1981), but it is not known if elevated cAMP plays any role in hyperactivation. Bicarbonate, which is present in high concentrations in oviductal fluid (Maas *et al.*, 1977), is required for the Ca^{2+}-induced elevation of cAMP in guinea pig sperm, and apparently acts via an effect on Ca^{2+} transport (Garbers *et al.*, 1982).

6. Sensory Reception

Cilia play an important role in sensory reception, including mechanoreception, olfactory reception, and photoreception. In most of these cases the sensory cilia lack the dynein arms and the central pair of microtubules and are nonmotile, and even when they are motile, motility per se is not required for transducing sensory stimuli into electrical signals for conduction in the nervous system. Sensory transduction appears always to occur at the level of the cell surface, involving either the ciliary membrane itself, which may be greatly modified, or the plasma membrane immediately surrounding the base of the cilium. Whether the axoneme has any role in sensory reception depends on the receptor. For example, in the statocyst of the nudibranch mollusk *Hermissenda crassicornis*, the axonemal shaft transmits pressure from the statoconia, in contact with the tip of the cilium, to its base, where the resulting deformation of the plasma membrane is believed to cause the opening of Na^+ channels, leading to a generator potential (Stommel *et al.*, 1980). The actual mechanotransducer therefore is likely to be located at the base of the cilium, and may correspond to the ciliary necklace (see above and Dentler, this volume; Chailley *et al.*, this volume). Similarly, the tail cilia of *Paramecium* passively transmit mechanical stimuli to the "soma" or cell body surface, which contains the mechanotransducers (Machemer and Ogura, 1979; Machemer and Machemer-Rohnisch, 1983). In contrast, mechanotransduction probably occurs at the distal tip of the modified cilium in the campaniform sensillum of the cockroach (Moran *et al.*, 1976).

In chemoreceptors, a major function of cilia probably is to increase the area of chemosensitive membrane available for binding odoriferous molecules. In the main olfactory organ of mammals, each olfactory cell gives rise to several cilia which lie flattened against the mucosa surface; in rats, the epithelial surface area available for chemoreception is increased 30-fold by the presence of cilia (Menco, 1980). Deciliation of olfactory neurons eliminates the electrophysiological response to odorants (Adamek *et al.*, 1984), and isolated olfactory cilia are capable of binding odorants (Rhein and Cagan, 1980), indicating that the odorant receptor molecules are contained in the ciliary membrane. The conventional view of sensory transduction in olfactory neurons is that interaction of the odorant with the receptor causes a voltage change across the membrane which then spreads down the ciliary shaft to the cell body (Kauer, 1983). More recently, evidence has accumulated that cAMP mediates olfactory transduction via a GTP-binding protein (Pace *et al.*, 1985; Nakamura and Gold, 1987; see Lancet, 1988, for review). The mechanism of sensory transduction in olfactory cilia is therefore similar to that in the rod outer segment, which is also a modified cilium (see below). It should be noted that odor is also detected in the vomeronasal (Jacobson's) organ, which lacks cilia, indicating that the cilium per se is not essential for chemoreceptive function (Kauer, 1983).

All metazoan photoreceptor organelles have evolved from either cilia or microvilli. Photoreceptors based on cilia are found in coelenterates, possibly ctenophores, some mollusks and annelids, invertebrate deuterostomes (echinoderms and chordates), and vertebrates, whereas microvilli-based photoreceptors are found in the protostomes (including flatworms, mollusks, annelids, and arthropods) (see Eakin, 1972, for review). In the cilia-based photoreceptors, the ciliary membrane frequently has undergone extensive branching (in some annelids), outward folding (in coelenterates, ctenophores, echinoderms, and cephalochordates), inward folding (ascidians, chaetognaths, and vertebrates), or other elaboration to increase the area of membrane containing the photosensitive molecule (Eakin, 1972). The best-studied examples of this are the rods and cones of vertebrates (see Besharse and Horst, this volume). The outer segment of the rod is a modified cilium in which the ciliary membrane has been greatly amplified and has undergone infolding to form internal flattened disks that cease to have continuity with the enveloping ciliary membrane. The disk membrane is packed with the photoreceptor molecule rhodopsin. The absorption of light by rhodopsin initiates a cGMP cascade that results in blockage of Na^+ channels in the ciliary membrane and consequent hyperpolarization of the outer segment (Stryer, 1986); the hyperpolarization is then passively transmitted to the more proximal parts of the cell through a narrow constriction occupied by a "9 + 0" ring of outer doublets and termed the *connecting cilium*. The membrane of the connecting cilium contains particles resembling those of the ciliary necklace over its entire length and thus appears to be analogous to the transition zone of motile cilia (Rohlich, 1975). The organization of the cone is similar, except that the membranous sacs remain continuous with the ciliary membrane from which they arise.

Many mechanoreceptors, including those in the lateral line of fish and the vestibular hair cells of the vertebrate inner ear, have a single nonmotile cilium, termed a *kinocilium*, at the edge of a bundle of specialized microvilli, unfortunately misnamed *stereocilia*. Studies have shown that a receptor potential is produced in response to displacement of the stereociliary bundle (Hudspeth, 1985) and does not require the presence of the kinocilium (Hudspeth and Jacobs, 1979), so the role of the latter is not clear. It is possible that the kinocilium plays a role in organizing the stereociliary bundle during development (Tilney *et al.*, 1988).

It is likely that cilia have been utilized in so many sensory receptors in such a wide variety of organisms because they have special properties that preadapted them for this purpose. First, they are surrounded by an extension of the cell membrane, so that they provide the cell with a ready means for expanding the amount of surface area available for signal reception, thus increasing the sensitivity of the receptor. This is a feature shared with microvilli, which, as noted above, also are utilized for sensory reception. Second, their membranes are equipped, and must have been equipped from very early times, with the ion gates and pumps and the metabolic enzymes, such as adenylate and guanylate cyclase, necessary for maintaining their internal environment and rapidly changing that environment in response to external or internal signals (see above and Preston and Saimi, this volume). Perhaps future studies will show that the membranes of microvilli have similar proteins. Third, ciliary membranes have very high transmembrane resistances, perhaps the highest of any cell membrane (Machemer and Ogura, 1979); this makes them exceptionally well suited for the conduction of electrical signals (Machemer, 1986). Finally, cilia have evolved mechanisms for membrane turnover and renewal (see Bloodgood, this volume; Williams, this volume; Bouck *et al.*, this volume); such turnover is important in the maintenance of many sensory receptors (see Besharse and Horst, this volume). Together, these properties made the cilium an ideal structure from which to fashion a sensory receptor.

7. Origin

The origin of the eukaryotic flagellum is still shrouded in mystery. One hypothesis, the symbiotic theory, proposes that it was derived from symbiotic bacteria containing microtubulelike structures (Margulis, 1970). In this case, all other microtubule-containing organelles would have evolved secondarily from the flagellum. A contrasting hypothesis, the autogenous theory, proposes that the flagellum originated as a result of the aggregation of preexisting cytoplasmic microtubules (Pickett-Heaps, 1974; Cavalier-Smith, 1975); this theory supposes that the evolution of the spindle apparatus preceded that of the flagellum. The latter theory is consistent with the fact that spindle microtubules are ubiquitous in eukaryotes whereas cilia and flagella are phylogenetically more restricted. However, compelling data in support of either hypothesis remain to be obtained. The recent discovery in the green alga *Chlamydomonas reinhardtii* of a unique circular linkage group containing a number of genes involved in flagellar development and function (Ramanis and Luck, 1986; Dutcher, 1986), and at least one gene required for cell division (Dutcher, personal communication), has renewed speculation that the basal bodies from which flagella grow may be associated with an extranuclear genome (see Wheatley, 1982, for a critical discussion of earlier work on this subject). Further characterization of this genetic material may clarify our understanding of the origin of these structures.

ACKNOWLEDGMENTS. I am grateful to Dr. Anthony Moss for discussion and helpful suggestions during preparation of the manuscript, and for calling my attention to published papers that I had not seen previously. I thank Dr. Robert Bloodgood for his patience, constructive criticisms, and careful editing of the manuscript. Work from this laboratory is supported by NIH Grants GM 30626, HD 23858, and CA 12708, and by a grant from the Mellon Foundation.

References

Adamek, G. D., Gesteland, R. C., Mair, R. G., and Oakley, B., 1984, Transduction physiology of olfactory receptor cilia, *Brain Res.* **310:**87–97.

Adams, G. M. W., Huang, B., Piperno, G., and Luck, D. J. L., 1981, Central-pair microtubular complex of *Chlamydomonas* flagella: Polypeptide composition as revealed by analysis of mutants, *J. Cell Biol.* **91:**69–76.

Afzelius, B., 1959, Electron microscopy of the sperm tail. Results obtained with a new fixative, *J. Biophys. Biochem. Cytol.* **5:**269–278.

Afzelius, B. A., 1976, A human syndrome caused by immotile cilia, *Science* **193:**317–319.

Afzelius, B. A., 1979, The immotile cilia syndrome and other ciliary diseases, *Int. Rev. Exp. Pathol.* **19:**1–43.

Afzelius, B. A., Camner, P., and Mossberg, B., 1978, On the function of cilia in the female reproductive tract, *Fertil. Steril.* **29:**72–74.

Aiello, E. L., 1960, Factors affecting ciliary activity on the gill of the mussel *Mytilus edulis, Physiol. Zool.* **33:** 120–135.

Aiello, E., and Guideri, G., 1964, Nervous control of ciliary activity, *Science* **146:**1692–1693.

Alkon, D. L., 1983, The role of statocyst sensory cilia in mechanotransduction, *J. Submicrosc. Cytol.* **15:**145–150.

Anderson, R. G. W., 1972, The three-dimensional structure of the basal body from the rhesus monkey oviduct, *J. Cell Biol.* **54:**246–265.

Avolio, J., Glazzard, A. N., Holwill, M. E. J., and Satir, P., 1986, Structures attached to doublet microtubules of cilia: Computer modeling of thin-section and negative-stain stereo images, *Proc. Natl. Acad. Sci. USA* **83:**4804–4808.

Baccetti, B., 1982, The evolution of the sperm tail, *Symp. Soc. Exp. Biol.* **35:**521–532.

Baccetti, B., 1985, Evolution of the sperm tail, in: *Biology of Fertilization,* Volume 2 (C. B. Metz and A. Monroy, eds.), Academic Press, Orlando, pp. 3–58.

Bardele, C. F., 1981, Functional and phylogenetic aspects of the ciliary membrane: A comparative freeze-fracture study, *BioSystems* **14:**403–421.

Bell, C. W., and Gibbons, I. R., 1982, Structure of the dynein-1 outer arm in sea urchin sperm flagella. II. Analysis by proteolytic cleavage, *J. Biol. Chem.* **257:**516–522.

Bessen, M., Fay, R. B., and Witman, G. B., 1980, Calcium control of waveform in isolated flagellar axonemes of *Chlamydomonas, J. Cell Biol.* **86:**446–455.

Bishop, D. W., 1962, Sperm motility, *Physiol. Rev.* **42:**1–59.

Blum, J. J., 1971, Existence of a breaking point in cilia and flagella, *J. Theor. Biol.* **33:**257–263.

Bonini, N. M., and Nelson, D. L., 1988, Differential regulation of *Paramecium* ciliary motility by cAMP and cGMP, *J. Cell Biol.* **106:**1615–1623.

Bonini, N. M., Gustin, M. C., and Nelson, D. L., 1986, Regulation of ciliary motility by membrane potential in *Paramecium:* A role for cyclic AMP, *Cell Motil. Cytoskel.* **6:**256–272.

Bray, D., Heath, J., and Moss, D., 1986, The membrane-associated "cortex" of animal cells: Its structure and mechanical properties, *J. Cell Sci. Suppl.* **4:**71–88.

Brokaw, C. J., 1961, Movement and nucleoside polyphosphatase activity of isolated flagella from *Polytoma uvella, Exp. Cell Res.* **22:**151–162.

Brokaw, C. J., 1985, Cyclic AMP-dependent activation of sea urchin and tunicate sperm motility, *Ann. N.Y. Acad. Sci.* **438:**132–141.

Brokaw, C. J., 1986, Future directions for studies of mechanisms for generating flagellar bending waves, *J. Cell Sci. Suppl.* **4:**103–113.

Brokaw, C. J., 1987a, A lithium-sensitive regulator of sperm flagellar oscillation is activated by cAMP-dependent phosphorylation, *J. Cell Biol.* **105:**1789–1798.

Brokaw, C. J., 1987b, Regulation of sperm flagellar motility by calcium and cAMP-dependent phosphorylation, *J. Cell Biochem.* **35:**175–184.

Brokaw, C. J., 1989a, Direct measurements of sliding between outer doublet microtubules in swimming sperm flagella, *Science* **243:**1593–1596.

Brokaw, C. J., 1989b, Operation and regulation of the flagellar oscillator, in: *Cell Movement,* Volume 1 (F. D. Warner, P. Satir, and I. R. Gibbons, eds.), Liss, New York, pp. 25–35.

Brokaw, C. J., and Luck, D. J. L., 1985, Bending patterns of *Chlamydomonas* flagella. III. A radial spoke head deficient mutant and a central pair deficient mutant, *Cell Motil.* **5:**195–208.

Brokaw, C. J., Luck, D. J. L., and Huang, B., 1982, Analysis of the movement of *Chlamydomonas* flagella. The function of the radial spoke system is revealed by comparison of wild-type and mutant flagella, *J. Cell Biol.* **92:**722–732.

Cavalier-Smith, T., 1975, The origin of nuclei and of eukaryotic cells, *Nature* **256:**463–468.

Cavalier-Smith, T., 1982, The evolutionary origin and phylogeny of eukaryote flagella, *Symp. Soc. Exp. Biol.* **35:**465–493.

Chang, X.-j., and Piperno, G., 1987, Cross-reactivity of antibodies specific for flagellar tektin and intermediate filament subunits, *J. Cell Biol.* **104:**1563–1568.

Chasey, D., 1972, Further observations on the ultrastructure of cilia from *Tetrahymena pyriformis, Exp. Cell Res.* **74:**471–479.

Christen, R., Schackmann, R. W., and Shapiro, B. M., 1982, Elevation of the intracellular pH activates respiration and motility of sperm of the sea urchin, *Strongylocentrotus purpuratus, J. Biol. Chem.* **257:** 14881–14890.

Cooper, T. G., 1984, The onset and maintenance of hyperactivated motility of spermatozoa from the mouse, *Gamete Res.* **9:**55–74.

Dallai, R., Bernini, F., and Giusti, F., 1973, Interdoublet connections in the sperm flagellar complex of *Sciara, J. Submicrosc. Cytol.* **5:**137–145.

Dentler, W. L., 1981, Microtubule–membrane interactions in cilia and flagella, *Int. Rev. Cytol.* **72:**1–47.

Dentler, W. L., 1987, Cilia and flagella, *Int. Rev. Cytol. Suppl.* **17:**391–456.

Dustin, P., 1984, *Microtubules,* 2nd ed., Springer-Verlag, Berlin.

Dutcher, S. K., 1986, Genetic properties of linkage group XIX in *Chlamydomonas reinhardtii,* in: *Extrachromosomal Elements in Lower Eukaryotes* (R. B. Wickner, A. Hinnelbusch, A. M. Lambowitz, I. C. Gunsalu, and A. Hollaender, eds.), Plenum Press, New York, pp. 303–325.

Dutcher, S. K., Huang, B., and Luck, D. J. L., 1984, Genetic dissection of the central pair microtubules of the flagella of *Chlamydomonas reinhardtti, J. Cell Biol.* **98:**229–236.

Dute, R., and Kung, C., 1978, Ultrastructure of the proximal region of somatic cilia in *Paramecium tetraurelia, J. Cell Biol.* **78:**451–464.

Eakin, R. M., 1972, Structure of invertebrate photoreceptors, in: *Handbook of Sensory Physiology,* Volume VII/1 (H. J. A. Dartnall, ed.), Springer-Verlag, Berlin, pp. 625–684.

Eckert, R., and Brehm, P., 1979, Ionic mechanisms of excitation in *Paramecium, Annu. Rev. Biophys. Bioeng.* **8:**353–383.

Eckert, R., Randall, D., and Augustine, G., 1988, *Animal Physiology,* 3rd ed., Freeman, San Francisco.

Escalier, D., Jouannet, P., and David, G., 1982, Abnormalities of the ciliary axonemal complex in children: An ultrastructural and cinetic study in a series of 34 cases, *Biol. Cell* **44:**271–282.

Fawcett, D. W., 1986, *Bloom and Fawcett: A Textbook of Histology,* 11th ed., Saunders, Philadelphia.

Friend, D. S., Orci, L., Perrelet, A., and Yanagimachi, R., 1977, Membrane particle changes attending the acrosome reaction in guinea pig spermatozoa, *J. Cell Biol.* **74:**561–577.

Garbers, D. L., 1981, The elevation of cyclic AMP concentrations in flagellaless sea urchin sperm heads, *J. Biol. Chem.* **256:**620–624.

Garbers, D. L., Tubb, D. J., and Hyne, R. V., 1982, A requirement of bicarbonate for Ca^{2+}-induced elevations of cyclic AMP in guinea pig spermatozoa, *J. Biol. Chem.* **257:**8980–8984.

Gatti, J.-L., and Christen, R., 1985, Regulation of internal pH of sea urchin sperm. A role for the Na/K pump, *J. Biol. Chem.* **260:**7599–7602.

Geneser, F., 1986, *Textbook of Histology,* Munksgaard, Copenhagen.

Gibbons, I. R., 1961, The relationship between the fine structure and direction of beat in gill cilia of a lamellibranch mollusc, *J. Biophys. Biochem. Cytol.* **11:**179–205.

Gibbons, I. R., 1963, Studies on the protein components of cilia from *Tetrahymena pyriformis, Proc. Natl. Acad. Sci. USA* **50:**1002–1010.

Gibbons, I. R., 1965, Chemical dissection of cilia, *Arch. Biol.* **76:**317–352.

Gibbons, I. R., and Grimstone, A. V., 1960, On the flagellar structure in certain flagellates, *J. Biophys. Biochem. Cytol.* **7:**697–716.

Gilula, N. B., and Satir, P., 1972, The ciliary necklace. A ciliary membrane specialization, *J. Cell Biol.* **53:** 494–509.

Gitelman, S. E., and Witman, G. B., 1980, Purification of calmodulin from *Chlamydomonas:* Calmodulin occurs in cell bodies and flagella, *J. Cell Biol.* **87:**764–770.

Goodenough, U. W., and Heuser, J. E., 1982, The substructure of the outer dynein arm, *J. Cell Biol.* **95:**798–815.

Goodenough, U. W., and Heuser, J. E., 1985, Substructure of inner dynein arms, radial spokes, and the central pair projection complex of cilia and flagella, *J. Cell Biol.* **100:**2008–2018.

Gray, J., 1928, *Ciliary Movement,* Cambridge University Press, London.

Hansen, J., and Huxley, H. E., 1955, The structural basis of contraction in striated muscle, *Symp. Soc. Exp. Biol.* **9**:228–264.

Hartfiel, G., and Amrhein, N., 1976, The action of methylxanthines on motility and growth of *Chlamydomonas reinhardtii* and other flagellated algae. Is cyclic AMP involved? *Biochem. Physiol. Pflanz.* **169**:S531–S556.

Hasegawa, E., Hayashi, H., Asakura, S., and Kamiya, R., 1987, Stimulation of in vitro motility of *Chlamydomonas* axonemes by inhibition of cAMP-dependent phosphorylation, *Cell Motil. Cytoskel.* **8**:302–311.

Hoops, H. J., and Witman, G. B., 1983, Outer doublet heterogeneity reveals structural polarity related to beat direction in *Chlamydomonas* flagella, *J. Cell Biol.* **97**:902–908.

Hoops, H. J., and Witman, G. B., 1985, Basal bodies and associated structures are not required for normal flagellar motion or phototaxis in the green alga *Chlorogonium elongatum*, *J. Cell Biol.* **100**:297–309.

Hoops, H. J., Wright, R. L., Jarvik, J. W., and Witman, G. B., 1984, Flagellar waveforms and rotational orientation in a *Chlamydomonas* mutant lacking normal striated fibers, *J. Cell Biol.* **98**:818–824.

Hopkins, J. M., 1970, Subsidiary components of the flagella of *Chlamydomonas reinhardii*, *J. Cell Sci.* **7**:823–839.

Horridge, G. A., 1965, Macrocilia with numerous shafts from the lips of the ctenophore *Beroë*, *Proc. R. Soc. London B* Ser. **162**:351–364.

Hoskins, D. D., Brandt, H., and Acott, T. S., 1978, Initiation of sperm motility in the mammalian epididymis, *Fed. Proc.* **37**:2534–2542.

Huang, B., Ramanis, Z., and Luck, D. J. L., 1982, Suppressor mutations in *Chlamydomonas* reveal a regulatory mechanism for flagellar function, *Cell* **28**:115–124.

Huang, B., Mengersen, A., and Lee, V. D., 1988, Molecular cloning of cDNA for caltractin, a basal body-associated Ca^{2+}-binding protein: Homology in its protein sequence with calmodulin and the yeast CDC31 gene product, *J. Cell Biol.* **107**:133–140.

Huber, M. E., Wright, W. G., and Lewin, R. A., 1986, Divalent cations and flagellar autotomy in *Chlamydomonas reinhardtii* (Volvocales, Chlorophyta), *Phycologia* **25**:408–411.

Hudspeth, A. J., 1985, The cellular basis of hearing: The biophysics of hair cells, *Science* **230**:745–752.

Hudspeth, A. J., and Jacobs, R., 1979, Stereocilia mediate transduction in vertebrate hair cells, *Proc. Natl. Acad. Sci. USA* **76**:1506–1509.

Ishijima, S., and Mohri, H., 1985, A quantitative description of flagellar movement in golden hamster spermatozoa, *J. Exp. Biol.* **114**:463–475.

Ishijima, S., and Witman, G. B., 1987, Flagellar movement of intact and demembranated reactivated ram spermatozoa, *Cell Motil. Cytoskel.* **8**:375–391.

Ishijima, S., McCracken, J. A., and Witman, G. B., 1986, Activation of intact and demembranated ram testicular spermatozoa and spermatids, *Dev. Growth Differ.* **28**:42.

Izumi, A., and Nakaoka, Y., 1987, cAMP-mediated inhibitory effect of calmodulin antagonists on ciliary reversal of *Paramecium*, *Cell Motil. Cytoskel.* **7**:154–159.

Jarosch, R., and Fuchs, B., 1975, Zur fibrillenrotation in der *Synura* - Geissel, *Protoplasma* **85**:285–290.

Jeffery, P. K., and Reid, L., 1975, New features of rat airway epithelium; a quantitative and electron microscopic study, *J. Anat.* **120**:295–320.

Johnson, C. H., Clapper, D. L., Winkler, M. W., Lee, H. C., and Epel, D., 1983, A volatile inhibitor immobilizes sea urchin sperm in semen by depressing the intracellular pH, *Dev. Biol.* **98**:493–501.

Johnson, K. A., Marchese-Ragona, S. P., Clutter, D. B., Holzbaur, E. L. F., and Chilcote, T. J., 1986, Dynein structure and function, *J. Cell Sci. Suppl.* **5**:189–196.

Kamiya, R., 1982, Extension and rotation of the central-pair microtubules in detergent-treated *Chlamydomonas* flagella, *Cell Motil. Suppl.* **1**:169–173.

Kamiya, R., and Okagaki, T., 1986, Cyclical bending of the two outer doublet microtubules in frayed axonemes of *Chlamydomonas*, *Cell Motil. Cytoskel.* **6**:580–585.

Kamiya, R., and Witman, G. B., 1984, Submicromolar levels of calcium control the balance of beating between the two flagella in demembranated models of *Chlamydomonas*, *J. Cell Biol.* **98**:97–107.

Kamiya, R., Nagai, R., and Nakamura, S., 1982, Rotation of the central pair microtubules in *Chlamydomonas* flagella, in: *Biological Functions of Microtubules and Related Structures* (H. Sakai, H. Mohri, and G. G. Borisy, eds.), Academic Press, New York, pp. 189–198.

Katz, D. F., and Yanagimachi, R., 1981, Movement characteristics of hamster and guinea pig spermatozoa upon attachment to the zona pellucida, *Biol. Reprod.* **25**:785–791.

Katz, D. F., Drobnis, E. Z., and Overstreet, J. W., 1989, Factors regulating mammalian sperm migration through the female reproductive tract and oocyte vestments, *Gamete Research* **22**:443–469.

Kauer, J. S., 1983, Surface morphology of olfactory receptors, *J. Submicrosc. Cytol.* **15**:167–171.

Kelly, D. E., Wood, R. L., and Enders, A. C., 1984, *Bailey's Textbook of Microscopic Anatomy,* 18th ed., Williams & Wilkins, Baltimore.

Kennedy, J. R., Jr., and Brittingham, E., 1968, Fine structure changes during chloral hydrate deciliation of *Paramecium caudatum, J. Ultrastruct. Res.* **22**:530–545.

King, S. M., and Witman, G. B., 1988, Structure of the α and β heavy chains of the outer arm dynein from *Chlamydomonas* flagella. Location of epitopes and protease-sensitive sites, *J. Biol. Chem.* **263**:9244–9255.

Kuzirian, A. M., Alkon, D. L., and Harris, L. G., 1981, An infraciliary network in statocyst hair cells, *J. Neurocytol.* **10**:497–514.

Lancet, D., 1988, Molecular components of olfactory reception and transduction, in: *Molecular Neurobiology of the Olfactory System* (F. L. Margolis and T. V. Getchell, eds.), Plenum Press, New York, pp. 25–50.

Laverack, M. S., and Ardill, D. J., 1965, The innervation of the aesthetasc hairs of *Panulirus argus, Q. J. Microsc. Sci.* **106**:45–60.

Lee, H. C., 1984, A membrane potential-sensitive $Na^+–H^+$ exchange system in flagella isolated from sea urchin spermatozoa, *J. Biol. Chem.* **259**:15315–15319.

Lee, H. C., Johnson, C. H., and Epel, D., 1983, Changes of internal pH associated with initiation of motility and acrosome reaction of sea urchin sperm, *Dev. Biol.* **95**:31–45.

Lewin, R. A., and Lee, K. W., 1985, Autotomy of algal flagella: Electron microscope studies of *Chlamydomonas (Chlorophyceae)* and *Tetraselmis (Prasinophyceae), Phycologia* **24**:311–316.

L'Hernault, S. W., and Rosenbaum, J. L., 1983, *Chlamydomonas* α-tubulin is posttranslationally modified in the flagella during flagellar assembly, *J. Cell Biol.* **97**:258–263.

L'Hernault, S. W., and Rosenbaum, J. L., 1985, Reversal of the posttranslational modification on *Chlamydomonas* flagellar α-tubulin occurs during flagellar resorption, *J. Cell Biol.* **100**:457–462.

Linck, R. W., Amos, L. A., and Amos, W. B., 1985, Localization of tektin filaments in microtubules of sea urchin sperm flagella by immunoelectron microscopy, *J. Cell Biol.* **100**:126–135.

Luck, D. J. L., 1984, Genetic and biochemical dissection of the eucaryotic flagellum, *J. Cell Biol.* **98**:789–794.

Luck, D. J. L., and Piperno, G., 1989, Dynein arm mutants of *Chlamydomonas reinhardtii,* in *Cell Movement,* Volume 1 (F. D. Warner, P. Satir, and I. R. Gibbons, eds.), Liss, New York, pp. 49–60.

Maas, D. H. A., Storey, B. T., and Mastroianni, L., Jr., 1977, Hydrogen ion and carbon dioxide content of the oviductal fluid of the rhesus monkey (*Macaca mulatta*), *Fertil. Steril.* **28**:981–985.

Machemer, H., 1985, Mechanoresponses in protozoa, in: *Sensory Perception and Transduction in Aneural Organisms* (G. Colombetti, F. Lenci, and P. S. Song, eds.), Plenum Press, New York, pp. 179–209.

Machemer, H., 1986, Electromotor coupling in cilia, in: *Membrane Control of Cellular Activity,* Volume 33 (H. C. Luttgau, ed.), Fortschritte der Zoologie, Fischer Verlag, Stuttgart, pp. 205–250.

Machemer, H., and Machemer-Rohnisch, S., 1983, Tail cilia of *Paramecium* passively transmit mechanical stimuli to the cell soma, *J. Submicrosc. Cytol.* **15**:281–284.

Machemer, H., and Ogura, A., 1979, Ionic conductances of membranes in ciliated and deciliated *Paramecium, J. Physiol. (London)* **296**:49–60.

Macnab, R. M., 1987a, Flagella, in: *Escherichia coli and Salmonella typhimurium. Cellular and Molecular Biology, Vol. 1.* (J. L. Ingraham, D. B. Low, B. Magasanik, M. Schaechter, and H. E. Umbarger, eds.) American Society for Microbiology, Washington, D. C., pp. 70–83.

Macnab, R. M., 1987b, Motility and Chemotaxis, in: *Escherichia coli and Salmonella typhimurium. Cellular and Molecular Biology,* Volume 1 (J. L. Ingraham, D. B. Low, B. Magasanik, M. Schaechter, and H. E. Umbarger, eds.), American Society for Microbiology, Washington, D.C., pp. 732–759.

Maihle, N. J., Dedman, J. R., Means, A. R., Chafouleas, J. G., and Satir, B. H., 1981, Presence and indirect immunofluorescent localization of calmodulin in *Paramecium tetraurelia, J. Cell Biol.* **89**:695–699.

Margulis, L., 1970, *Origin of Eukaryotic Cells,* Yale University Press, New Haven.

Melkonian, M., 1978, Structure and significance of cruciate flagellar root systems in green algae: Comparative investigations in species of *Chlorosarcinopsis (Chlorosarcinales), Plant Syst. Evol.* **130**:265–292.

Melkonian, M., 1980, Ultrastructural aspects of basal body associated fibrous structures in green algae: A critical review, *BioSystems* **12**:85–104.

Melkonian, M., Robenek, H., and Rassat, J., 1982, Flagellar membrane specializations and their relationship to mastigonemes and microtubules in *Euglena gracilis*, *J. Cell Sci.* **55**:115–135.

Menco, B., 1980, Qualitative and quantitative freeze-fracture studies on olfactory and nasal respiratory structures of frog, ox, rat, and dog. I. A general survey, *Cell Tissue Res.* **207**:183–209.

Mohri, H., and Yanagimachi, R., 1980, Characteristics of motor apparatus in testicular, epididymal and ejaculated spermatozoa, *Exp. Cell Res.* **127**:191–196.

Moran, D. T., Rowley, J. C., Zill, S. N., and Varela, F. G., 1976, The mechanism of sensory transduction in a mechano-receptor: Functional stages in campaniform sensilla during the molting cycle, *J. Cell Biol.* **71**: 832–847.

Morisawa, M., 1987, The process of initiation of sperm motility at spawning and ejaculation, in: *New Horizons in Sperm Cell Research* (H. Mohri, ed.), Japan Sci. Soc. Press, Tokyo, pp. 137–157.

Morisawa, M., and Okuno, M., 1982, Cyclic AMP induces maturation of trout sperm axoneme to initiate motility, *Nature* **295**:703–704.

Moss, A. B., and Tamm, S. L., 1987, A calcium regenerative potential controlling ciliary reversal is propagated along the length of ctenophore comb plates, *Proc. Natl. Acad. Sci. USA* **84**:6476–6480.

Naitoh, Y., and Eckert, R., 1969, Ionic mechanisms controlling behavioral responses of *Paramecium* to mechanical stimulation, *Science* **164**:963–965.

Naitoh, Y., and Kaneko, H., 1973, Control of ciliary activities by adenosine-triphosphate and divalent cations in Triton-extracted models of *Paramecium caudatum*, *J. Exp. Biol.* **58**:657–676.

Nakamura, T., and Gold, G. H., 1987, A cyclic nucleotide-gated conductance in olfactory receptor cilia, *Nature* **325**:442–444.

Nakaoka, Y., and Ooi, H., 1985, Regulation of ciliary reversal in Triton-extracted *Paramecium* by calcium and cyclic adenosine monophosphate, *J. Cell Sci.* **77**:185–195.

Olson, G. E., and Linck, R. W., 1977, Observations on the structural components of flagellar axonemes and central pair microtubules from rat sperm, *J. Ultrastruct. Res.* **61**:21–43.

Omoto, C. K., and Kung, C., 1979, The pair of central tubules rotates during ciliary beat in *Paramecium*, *Nature* **279**:532–534.

Omoto, C. K., and Kung, C., 1980, Rotation and twist of the central-pair microtubules in the cilia of *Paramecium*, *J. Cell Biol.* **87**:33–46.

Omoto, C. K., and Witman, G. B., 1981, Functionally significant central–pair rotation in a primitive eukaryotic flagellum, *Nature* **290**:708–710.

Otter, T., 1989, Calmodulin and the control of flagellar movement, in: *Cell Movement,* Volume 1 (F. D. Warner, P. Satir, and I. R. Gibbons, eds.), Liss, New York, pp. 281–298.

Otter, T., Satir, B. H., and Satir, P., 1984, Trifluoperazine-induced changes in swimming behavior of *Paramecium:* Evidence for two sites of drug action, *Cell Motil.* **4**:249–267.

Pace, U., Harski, E., Solomon, Y., and Lancet, D., 1985, Odorant-sensitive adenylate cyclase may mediate olfactory reception, *Nature* **316**:255–258.

Paschal, B. M., King, S. M., Moss, A. G., Collins, C. A., Vallee, R. B., and Witman, G. B., 1987, Isolated flagellar outer arm dynein translocates brain microtubules *in vitro, Nature* **330**:672–674.

Pasquale, S. M., and Goodenough, U. W., 1987, Cyclic AMP functions as a primary sexual signal in gametes of *Chlamydomonas reinhardtii, J. Cell Biol.* **105**:2279–2292.

Perkins, L., Hedgecock, J. N., Thomson, J. N., and Culotti, J., 1986, Mutant sensory cilia in *Caenorhabditis elegans, Dev. Biol.* **117**:456–487.

Pickett-Heaps, J., 1974, Evolution of mitosis and the eukaryote condition, *BioSystems* **6**:37–45.

Piperno, G., 1988, Isolation of a sixth dynein subunit adenosine triphosphatase of *Chlamydomonas* axonemes, *J. Cell Biol.* **106**:133–140.

Piperno, G., and Luck, D. J. L., 1981, Inner arm dyneins from flagella of *Chlamydomonas reinhardtii, Cell* **27**: 331–340.

Pitelka, D. R., 1974, Basal bodies and root structures, in: *Cilia and Flagella* (M. A. Sleigh, ed.), Academic Press, New York, pp. 437–469.

Poole, A. C., Flint, M. H., and Beaumont, B. W., 1985, Analysis of the morphology and function of primary cilia in connective tissues: A cellular cybernetic probe? *Cell Motil.* **5**:175–193.

Ramanis, Z., and Luck, D. J. L., 1986, Loci affecting flagellar assembly and function map to an unusual linkage group in *Chlamydomonas reinhardtii, Proc. Natl. Acad. Sci. USA* **83**:423–426.

Rhein, L. D., and Cagan, R. H., 1980, Biochemical studies of olfaction: Isolation, characterization, and odorant binding activity of cilia from rainbow trout olfactory rosettes, *Proc. Natl. Acad. Sci. USA* **77:**4412–4416.

Ringo, D. L., 1967a, Flagellar motion and fine structure of the flagellar apparatus in *Chlamydomonas, J. Cell Biol.* **33:**543–571.

Ringo, D. L., 1967b, The arrangement of subunits in flagellar fibers, *J. Ultrastruct. Res.* **17:**266–277.

Rohlich, P., 1975, The sensory cilium of retinal rods is analogous to the transitional zone of motile cilia, *Cell Tissue Res.* **161:**421–430.

Roth, K. E., Rieder, C. L., and Bowser, S. S., 1988, Flexible substratum technique for viewing cells from the side: Some *in vivo* properties of primary (9+0) cilia in cultured kidney epithelia, *J. Cell Sci.* **89:** 457–466.

Rubin, R. W., and Filner, P., 1973, Adenosine 3',5'-cyclic monophosphate in *Chlamydomonas reinhardtii.* Influence on flagellar function and regeneration, *J. Cell Biol.* **56:**628–635.

Sale, W. S., and Fox, L. A., 1988, Isolated β-heavy chain subunit of dynein translocates microtubules in vitro, *J. Cell Biol.* **107:**1793–1797.

Sale, W. S., and Satir, P., 1977, Direction of active sliding of microtubules in *Tetrahymena* cilia, *Proc. Natl. Acad. Sci. USA* **74:**2045–2049.

Salisbury, J. L., Sanders, M. A., and Harpst, L., 1987, Flagellar root contraction and nuclear movement during flagellar regeneration in *Chlamydomonas reinhardtii, J. Cell Biol.* **105:**1799–1805.

Sanders, M. A., and Salisbury, J. L., 1989, Centrin-mediated microtubule severing during flagellar excision in *Chlamydomonas reinhardii, J. Cell Biol.* **108:**1751–1760.

Satir, B., Sale, W. S., and Satir, P., 1976, Membrane renewal after dibucaine deciliation of *Tetrahymena, Exp. Cell Res.* **97:**83–91.

Satir, P., 1965, Studies on cilia. II. Examination of the distal region of the ciliary shaft and the role of filaments in motility, *J. Cell Biol.* **26:**805–834.

Satir, P., 1968, Studies on cilia. III. Further studies on the cilium tip and a sliding filament model of ciliary motility, *J. Cell Biol.* **39:**77–94.

Satir, P., Wais-Steider, J., Lebduska, S., Nasr, A., and Avolio, J., 1981, The mechanochemical cycle of the dynein arm, *Cell Motil.* **1:**303–327.

Schmidt, J. A., and Eckert, R., 1976, Calcium couples flagellar reversal to photostimulation in *Chlamydomonas reinhardtii, Nature* **262:**713–715.

Segal, R. A., and Luck, D. J., 1985, Phosphorylation in isolated *Chlamydomonas* axonemes: A phosphoprotein may mediate the Ca^{2+}-dependent photophobic response, *J. Cell Biol.* **101:**1702–1712.

Segal, R. A., Huang, B., Ramanis, Z., and Luck, D. J. L., 1984, Mutant strains of *Chlamydomonas reinhardtii* that move backwards only, *J. Cell Biol.* **98:**2026–2034.

Shapiro, B. M., and Tombes, R. M., 1985, A biochemical pathway for a cellular behaviour: pH_i, phosphorylcreatine shuttles, and sperm motility, *Bio Essays* **3:**100–103.

Shingyoji, C., Murakami, A., and Takahashi, K., 1977, Local reactivation of Triton-extracted flagella by iontophoretic application of ATP, *Nature* **265:**269–270.

Sleigh, M. A., 1962, *The Biology of Cilia and Flagella,* Macmillan Co., New York.

Sleigh, M. A., 1974, Introduction, in: *Cilia and Flagella* (M. A. Sleigh, ed.), Academic Press, New York, pp. 1–7.

Sleigh, M. A., 1977, The nature and action of respiratory tract cilia, in: *Respiratory Defense Mechanisms,* Part 1, (J. D. Brain, D. F. Procter, and L. M. Reid, eds.), Dekker, New York, pp. 247–288.

Sleigh, M. A., and Silvester, N. R., 1983, Anchorage functions of the basal apparatus of cilia, *J. Submicrosc. Cytol.* **15:**101–104.

Stavis, R. L., and Hirschberg, R., 1973, Phototaxis in *Chlamydomonas reinhardtii, J. Cell Biol.* **59:**367–377.

Stephens, R. E., 1970, Isolation of nexin—the linkage protein responsible for maintenance of the nine-fold configuration of flagellar axonemes, *Biol. Bull. (Woods Hole, Mass.)* **139:**438. (Abstr.).

Stephens, R. E., and Stommel, E. W., 1989, The role of cAMP in ciliary and flagellar motility, in: *Cell Movement,* Volume 1 (F. D. Warner, P. Satir, and I. R. Gibbons, eds.), Liss, New York, pp. 299–316.

Stommel, E. W., 1984, Calcium activation of mussel gill abfrontal cilia, *J. Comp. Physiol. A* **155:**457–469.

Stommel, E. W., and Stephens, R. E., 1985, Cyclic AMP and calcium in the differential control of *Mytilus* gill cilia, *J. Comp. Physiol. A* **157:**451–459.

Stommel, E. W., Stephens, R. E., and Alkon, D. L., 1980, Motile statocyst cilia transmit rather than directly transduce mechanical stimuli, *J. Cell Biol.* **87:**652–662.

Stryer, L., 1986, Cyclic GMP cascade of vision, *Annu. Rev. Neurosci.* **9**:87–119.

Stubblefield, E., and Brinkley, B. R., 1967, Architecture and function of the mammalian centriole, in: *Formation and Fate of Cell Organelles* (K. B. Warren, ed.) Academic Press, New York, pp. 175–218.

Summers, K. E., and Gibbons, I. R., 1971, Adenosine triphosphate-induced sliding of tubules in trypsin-treated flagella of sea urchin sperm, *Proc. Natl. Acad. Sci. USA* **68**:3092–3096.

Summers, K. E., and Gibbons, I. R., 1973, Effects of trypsin digestion on flagellar structures and their relationship to motility, *J. Cell Biol.* **58**:618–629.

Tamm, S. L., 1988a, Calcium activation of macrocilia in the ctenophore *Beroë*, *J. Comp. Physiol. A* **163**:23–31.

Tamm, S. L., 1988b, Iontophoretic localization of Ca-sensitive sites controlling activation of ciliary beating in macrocilia of *Beroë*. The ciliary rete, *Cell Motil. Cytoskel.* **11**:126–138.

Tamm, S. L., and Horridge, G. A., 1970, The relationship between the orientation of the central fibrils and the direction of beat in cilia of *Opalina*, *Proc. R. Soc. London Ser. B* **175**:219–233.

Tamm, S. L., and Tamm, S., 1981, Ciliary reversal without rotation of axonemal structures in ctenophore comb plates, *J. Cell Biol.* **89**:495–509.

Tash, J. S., and Means, A. R., 1983, Cyclic adenosine 3',5' monophosphate, calcium, and protein phosphorylation in flagellar motility, *Biol. Reprod.* **28**:75–104.

Tilney, L. G., Bryan, J., Bush, D. J., Fujiwara, K., Mooseker, M. S., Murphy, D. B., and Snyder, D. H., 1973, Microtubules: Evidence for 13 protofilaments, *J. Cell Biol.* **59**:267–275.

Tilney, L. G., Tilney, M. S., and Cotanche, D. A., 1988, Actin filaments, stereocilia, and hair cells of the bird cochlea. V. How the staircase pattern of stereociliary lengths is generated, *J. Cell Biol.* **106**:355–365.

Trimmer, J. S., and Vacquier, V. D., 1986, Activation of sea urchin gametes, *Annu. Rev. Cell Biol.* **2**:1–26.

Tucker, R. W., Pardee, A. B., and Fujiwara, K., 1979, Centriole ciliation is related to quiescence and DNA synthesis in 3T3 cells, *Cell* **17**:527–535.

Vale, R. D., and Toyoshima, Y. Y., 1988, Rotation and translocation of microtubules *in vitro* induced by dyneins from *Tetrahymena* cilia, *Cell* **52**:459–469.

Verdugo, P., 1980, Ca^{2+}-dependent hormonal stimulation of ciliary activity, *Nature* **283**:764–765.

Verdugo, P., 1982, Introduction: Mucociliary function in mammalian epithelia, *Cell Motil. Suppl.* **1**:1–5.

Verdugo, P., Johnson, N. T., and Tam, P. Y., 1980a, Beta-adrenergic stimulation of respiratory ciliary activity, *J. Appl. Physiol.* **48**:868–871.

Verdugo, P., Lee, W. I., Halbert, S. A., Blandau, R. J., and Tam, P. Y., 1980b, A stochastic model for oviductal egg transport, *Biophys. J.* **29**:257–270.

Verdugo, P., Rumery, R. E., and Tam, P. Y., 1980c, Hormonal control of oviductal ciliary activity: Effect of prostaglandins, *Fertil. Steril.* **33**:193–196.

Villalon, M., and Verdugo, P., 1982, Hormonal regulation of ciliary function in the oviduct: The effect of beta-adrenergic agonists, *Cell Motil. Suppl.* **1**:59–65.

Walter, M. F., and Schultz, J. E., 1981, Calcium receptor protein calmodulin isolated from cilia and cells of *Paramecium tetraurelia*, *Eur. J. Cell Biol.* **24**:97–100.

Ward, S., Thomson, N., White, J. G., and Brenner, S., 1975, Electron microscopical reconstruction of the anterior sensory anatomy of the nematode *Caenorhabditis elegans*, *J. Comp. Neurol.* **160**:313–338.

Warner, F. D., 1976, Ciliary inter-microtubule bridges, *J. Cell Sci.* **20**:101–114.

Warner, F. D., 1983, Organization of interdoublet links in *Tetrahymena* cilia, *Cell Motil.* **3**:321–332.

Warner, F. D., and Satir, P., 1973, The substructure of ciliary microtubules, *J. Cell Sci.* **12**:313–326.

Warner, F. D., and Satir, P., 1974, The structural basis of ciliary bend formation. Radial spoke positional changes accompanying microtubule sliding, *J. Cell Biol.* **63**:35–63.

Warner, F. D., Mitchell, D. R., and Perkins, C. R., 1977, Structural conformation of the ciliary ATPase dynein, *J. Mol. Biol.* **114**:367–384.

Warner, F. D., Satir, P., and Gibbons, I. R. (eds.), 1989, *Cell Movement*, Volume 1, Liss, New York.

Warr, J. R., McVittie, A., Randall, J., and Hopkins, J. M., 1966, Genetic control of flagellar structure in *Chlamydomonas reinhardtii*, *Genet. Res.* **7**:335–351.

Wheatley, D. N., 1982, *The Centriole: A Central Enigma of Cell Biology*, Elsevier, Amsterdam.

Williams, N. E., and Luft, J. H., 1968, Nitrogen mustard derivative in fixation for electron microscopy and observations on the ultrastructure of *Tetrahymena*, *J. Ultrastruct. Res.* **25**:271–292.

Witman, G. B., 1989, Perspective: Composition and molecular organization of the dyneins, in: *Cell Movement*, Volume 1 (F. D. Warner, P. Satir, and I. R. Gibbons, eds.), Liss, New York, pp. 25–35.

Witman, G. B., and Minervini, N., 1982, Dynein arm conformation and mechanochemical transduction in the eukaryotic flagellum, *Symp. Soc. Exp. Biol.* **35**:203–224.

Witman, G. B., Carlson, K., Berliner, J., and Rosenbaum, J. L., 1972, *Chlamydomonas* flagella. I. Isolation and electrophoretic analysis of microtubules, matrix, membranes, and mastigonemes, *J. Cell Biol.* **54**:507–539.

Witman, G. B., Fay, R., and Plummer, J., 1976, *Chlamydomonas* mutants: Evidence for the roles of specific axonemal components in flagellar movement, in: *Cell Motility,* Book C (R. D. Goldman, T. D. Pollard, and J. L. Rosenbaum, eds.), Cold Spring Harbor Laboratory, Cold Spring Harber, N.Y., pp. 969–986.

Witman, G. B., Plummer, J., and Sander, G., 1978, *Chlamydomonas* flagellar mutants lacking radial spokes and central tubules. Structure, composition, and function of specific axonemal components, *J. Cell Biol.* **76**:729–747.

Wolf, D. E., Hagopian, S. S., and Ishijima, S., 1986, Changes in sperm plasma membrane lipid diffusibility following hyperactivation during *in vitro* capacitation in the mouse, *J. Cell Biol.* **102**:1372–1377.

Wright, R. L., Salisbury, J., and Jarvik, J. W., 1985, A nucleus–basal body connector in *Chlamydomonas reinhardtii* that may function in basal body localization or segregation, *J. Cell Biol.* **101**:1903–1912.

Wunderlich, F., and Speth, V., 1972, Membranes in *Tetrahymena.* I. The cortical pattern, *J. Ultrastruct. Res.* **41**:258–269.

Yanagimachi, R., 1981, Mechanism of fertilization in mammals, in: *Fertilization and Embryonic Development in Vitro* (L. Mastroianni, Jr., and J. D. Biggers, eds.), Plenum Press, New York, pp. 81–182.

Yanagimachi, R., 1988, Mammalian fertilization, in: *The Physiology of Reproduction* (E. Knobil, J. Neill, L. L. Ewing, G. S. Greenwald, C. L. Markert, and D. W. Pfaff, eds.), Raven Press, New York, pp. 135–185.

Yanagimachi, R., and Usui, N., 1974, Calcium dependence of the acrosome reaction and activation of guinea pig spermatozoa, *Exp. Cell Res.* **89**:161–174.

Linkages between Microtubules and Membranes in Cilia and Flagella

William L. Dentler

1. Introduction

In comparison with ciliary and flagellar axonemes, relatively little is known about the structure, composition, or function of ciliary membranes or about the bridge structures that link the membranes to the microtubules. Bridges link the membrane to the microtubules at the ciliary bases, along the length of the axoneme, and to the distal tips of the doublet and central pair microtubules (Fig. 1). The role of each type of bridge structure in ciliary function is not understood but they appear to anchor the membrane to the microtubules, bind proteins or other structures to the external surface of the membrane, and may be involved with the transport of proteins up and down the flagellum. The capping structures linking the distal tips of the microtubules to the membrane may be responsible for the initiation and regulation of microtubule growth. This chapter will review some aspects of microtubule–membrane interactions in cilia and flagella with an expectation that the mechanisms regulating these interactions will provide insight into ciliary and flagellar function as well as aspects of microtubule function and assembly in the cytoplasm.

2. Linkage of Basal Bodies and Transition Regions to the Membrane

2.1. Structural Studies

The basal bodies are attached to the membrane by numerous structures, including transitional fibers, strut arrays, alar sheets, and a flagellar "bracelet," most of which attach to the sides of the basal body microtubules (see Figs. 1 and 2; Dentler, 1981a,

William L. Dentler • Department of Physiology and Cell Biology, University of Kansas, Lawrence, Kansas 66045.

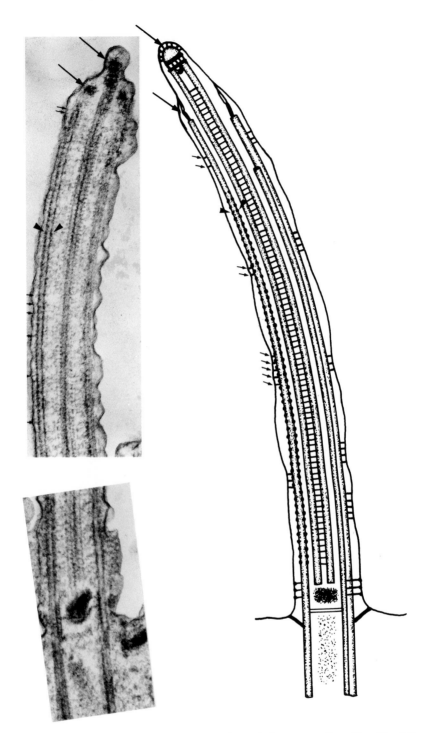

Figure 1. Bridges linking the membrane to microtubules in a typical protozoan cilium. Transitional fibers link the basal bodies to the membrane, microtubule–membrane bridges (small arrows) link the doublet microtubules to the membrane, and the microtubule capping structures (large arrows) link the distal ends of the A and central microtubules to the membrane.

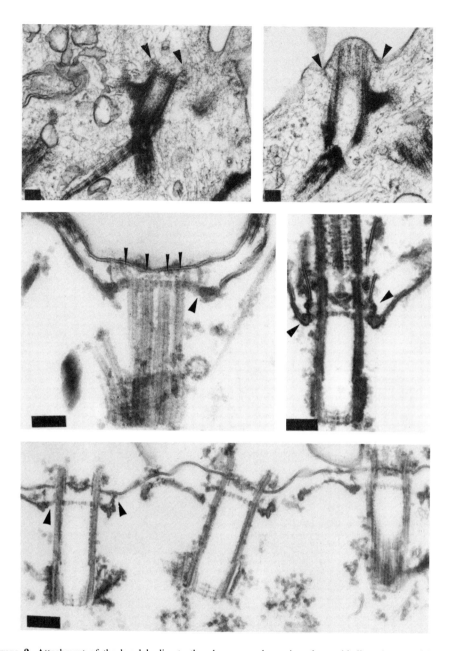

Figure 2. Attachment of the basal bodies to the plasma membrane in palate epithelium (top panels) and *Tetrahymena thermophila* (middle and lower panels). Basal bodies in ciliated epithelia appear to be completely assembled in the cytoplasm (top left) and contain the transition fibers necessary to link the basal body to the membrane (arrowheads). Basal body microtubules do not nucleate microtubule assembly until they are attached to the plasma membrane (top right), which means that inhibitors of assembly must be removed from the distal ends of basal body microtubules when they dock with the membrane or that additional structures or proteins necessary for assembly are recruited from the membrane upon docking. The sides (large arrowheads) and distal microtubule tips (small arrowheads) of *Tetrahymena* basal bodies are linked to the membrane by bridge structures seen in the middle and bottom panels. These basal bodies are attached to the membrane in detergent-extracted *Tetrahymena* cells. Bars = 0.1 μm.

1987; Arima *et al.*, 1984, 1985). Once the basal bodies are attached to the membrane and initiate microtubule assembly, the sides of the doublet microtubules become linked to the membrane by bridges in the newly formed transition region between the basal body and the "9 + 2" cilium or flagellum.

The bridges in the transition region are well defined and link the microtubules to membrane proteins (see Bardele, 1981; Dentler, 1981a; Besharse and Horst, this volume). The "champagne glass" or "Y"-shaped bridge structures first described by Gilula and Satir (1972) are the most commonly seen, but other structures, like the filamentous row of particles seen in ascidian branchial gill cilia, have been described (Dallai *et al.*, 1985). In some, but not all cilia, many of the linkages near the ciliary base are connected to intramembrane particles, including the ciliary necklace and ciliary granule plaques. The structure and, in a few cases, the composition, of the bridges in the transition region have been identified and are reviewed elsewhere (Dentler, 1981a, 1987; Menco, 1980; Bardele, 1981; Boisvieux-Ulrich *et al.*, 1977; Besharse and Horst, this volume) and will not be reviewed in detail here.

2.2. Functions of the Bridges

The linkages between the membrane and the basal body or transition region microtubules provide structural support for the cilium, are essential for the initiation of microtubule assembly, may aid in the selection of proteins to be transported into the growing and fully formed cilium or flagellum, and may be sites for ion transport across the ciliary membrane (see Dentler, 1981a). Evidence for structural support comes largely from electron micrographs. In contrast to the membrane surrounding the cilium proper, which usually appears flexible and is not always tightly adherent to the doublet microtubules, the membrane surrounding the transition region appears rigid, like a tight collar, and each doublet microtubule is linked to the membrane by prominent bridge structures. These bridges bind the membrane tightly to the microtubules during and after deflagellation (Lewin and Lee, 1985). The tight collar of membrane in the transition region is resistant to solubilization by nonionic detergents and is attached to isolated cilia and basal bodies after detergent extraction, which suggests that the bridges are bound to protein-rich regions of the membrane (Fig. 2; Dentler, 1981a, 1987; Snell *et al.*, 1974; Stephens, 1977; Horst *et al.*, 1987).

The bridges at the ciliary bases may help orient or maintain orientation of the cilia during beating (Dallai *et al.*, 1985; Martinucci *et al.*, 1987). However, although the bridges may anchor the microtubules to the membrane, they do not prevent the cilia from moving relative to the membrane. During ciliogenesis in metazoan epithelial cells, basal bodies attach to the membrane with random orientation [Sorokin, 1968; Frisch and Farbman, 1968; Boisvieux-Ulrich *et al.*, 1985; Holley and Afzelius, 1986; Tamm and Tamm, 1988 (type II ciliogenesis); Portman and Dentler, 1989] and orient *after* the ciliary microtubules have grown and started to beat (Boisvieux-Ulrich *et al.*, 1985; Portman and Dentler, 1989). This means that the axonemes must rotate while attached to the membrane and that the microtubule–membrane bridges must either make and break their associations (with the microtubule or membrane) or must be able to freely float in the lipid bilayer. There is no particular reason to believe that the basal body and transition region bridges maintain the orientation of each cilium, since the maintenance of ciliary and flagellar orientation is largely due to the associations of basal bodies and their rootlets with cytoskeletal microtubules, actin, and intermediate filaments (Pitelka, 1974; Dentler and LeCluyse, 1982b; Sleigh and Silvester, 1983; Hard and Rieder, 1983; Reed *et al.*, 1984; Dentler, 1987; Portman and Dentler, 1989). However, axonemal rotation may not occur in protozoans, since their basal bodies appear to be placed in the proper orientation prior to ciliary growth (Beisson and Sonneborn, 1965). They may attach to a "submembrane skeleton" that defines basal body binding sites, and, possibly, the orientation of the basal bodies (see Williams *et al.*, 1987; Williams, this volume).

The basal bodies are not essential for either structure or motility once axonemal assembly is completed. Basal bodies in the alga *Chlorogonium* detach from their flagella during cell division without altering flagellar position or motility, although the transition region at the base of the flagellum attaches to an apical mitochondrion when the basal bodies are lost (Hopps and Witman, 1985). Additionally, some nonmotile sensory cilia lack or have rudimentary basal bodies but carry out their normal functions (see Perkins *et al.*, 1986). Similarly, the distal centriole is attached to the flagellar microtubules during early spermiogenesis in rodent sperm but the centriole begins to disintegrate during later spermiogenesis and appears to be completely absent in mature sperm (Fawcett and Phillips, 1969; Woolley and Fawcett, 1973). The doublet microtubules in rodent sperm end blindly but are associated with a striated connecting piece that appears to link the flagellum with the base of the sperm head (Fawcett and Phillips, 1969; Woolley and Fawcett, 1973).

Probably the most important function of the interaction between the ends of the basal body microtubules and the membrane is the initiation of microtubule assembly. This is particularly evident in developing ciliated epithelial cells, in which free basal bodies within the cytoplasm cannot nucleate microtubules but basal bodies attached to the plasma membrane do nucleate ciliary microtubule assembly (Fig. 2; Dirksen and Crocker, 1965; Steinman, 1968; Sorokin, 1968; Dirksen, 1971; Anderson and Brenner, 1971; Portman and Dentler, 1987). The initiation of microtubule assembly by the basal bodies may require the addition of proteins associated with the microtubule capping structures that link the ends of ciliary and flagellar microtubules to the membrane or the removal of inhibitors (possibly basal body capping structures?) of microtubule assembly from the ends of the basal body microtubules. It may be relevant that *Chlamydomonas* stumpy flagellar mutants are characterized by the presence of short, poorly assembled microtubules, the presence of amorphous electron-dense material in the flagellar tips, and a membrane that has altered sensitivity to detergent solubilization (Jarvik and Chojnacki, 1985). Growing epithelial cilia have amorphous material between the membrane and (often poorly fixed) short microtubules until the formation of capping structures is completed (Portman and Dentler, 1987). A possible defect in the stumpy mutants may include abnormal assembly of capping structures and the linkage of the microtubules with the membrane. The role of the membrane and, possibly, the interactions between the membrane and microtubules may be partially understood by studying these or similar flagellar mutants.

The bridges linking the sides of the doublet microtubules to the membrane in the transition region may also be involved in ciliary and flagellar assembly (Dentler, 1981a). The microtubules assemble at the distal tips, so all proteins must be transported through the transition region to the tips prior to assembly. Since the basal body, transition region microtubules, and the microtubule–membrane bridge complexes are the only structures that separate the cytoplasm from the axoneme, all proteins destined for the axoneme or membrane must pass through this region. The specializations at the ciliary base may filter or sort out proteins destined for transport up the ciliary shaft and eventual assembly into the axonemes and membrane. In *Chlamydomonas eugametos*, a barrier has been discovered that limits the diffusion of cell-surface proteins between the cell body and the flagellum (Musgrave *et al.*, 1986). The nature of the barrier is not known but it does maintain different sets of glycoproteins on the cell body and flagellar surfaces in this species of *Chlamydomonas*. It would be interesting to determine the site of flagellar membrane growth relative to the barrier in order to determine if the barrier can sort out proteins destined for the flagellum from those that remain on the cell body.

3. Microtubule Capping Structures Attach the Ends of Microtubules to the Membrane

The distal tips of the central and A tubules are bound to the membrane by microtubule capping structures in most, if not all, cilia and flagella (Fig. 3, Table I). The structure of the caps has been reviewed elsewhere (Dentler, 1981a, 1987; Suprenant and Dentler, 1988) and only an overview of their possible functions will be considered here. Ringo (1967) was the first to point out the presence of structures at the microtubule tips in thin-sectioned *Chlamydomonas* flagella but higher-resolution images of the capping structures were obtained in demembranated and negatively stained *Tetrahymena* cilia (Sale and Satir, 1977a) and *Chlamydomonas* flagella (Dentler and Rosenbaum, 1977). These papers provided the first identification of the prominent cap linking the two central microtubules to a common plate structure and a bead that links the microtubules and plate to the ciliary membrane. This is appropriately named the "central microtubule cap." When the central microtubule caps are released from the microtubules, short plug structures that attach the cap into the lumen of each central microtubule are revealed (Dentler, 1984).

The distal filament plugs are attached to each A tubule (Dentler and Rosenbaum, 1977; Dentler, 1980a). The "plug" portion inserts into the microtubule lumen (Dentler, 1980a; Suprenant and Dentler, 1988), similar to the attachments of the plugs on the central microtubule caps to the central microtubules. The other end of the structure, the "distal filament," is composed of thin filaments tied into two bundles by a bead-shaped structure (Dentler, 1980a; Suprenant and Dentler, 1988) and bound to the ciliary membrane by short bridge structures (Fig. 3; Dentler, 1980a). Both the central microtubule cap and distal filaments appear to be common to all protozoan cilia and flagella (Table I).

Complex capping structures are typically found at the tips of cilia on the epithelium of trachea, oviduct, and palate (see Fig. 3, Table I). In these cilia, the central pair and A tubules are bound to a common cap that is bound to the membrane. Some epithelial cilia contain one set of A tubules and the central microtubules bound to a large cap that is linked to the ciliary membrane while the rest of the A tubules are bound to a smaller cap that can be either immediately adjacent to the large cap, as in frog palate cilia (Fig. 3; LeCluyse and Dentler, 1984), or ~ 1 μm proximal to the central pair cap, as occurs in acoel flatworm cilia (Tyler, 1973, 1979). The positioning of the capping structures on the microtubules is precise and the same doublet microtubules are attached to the large or

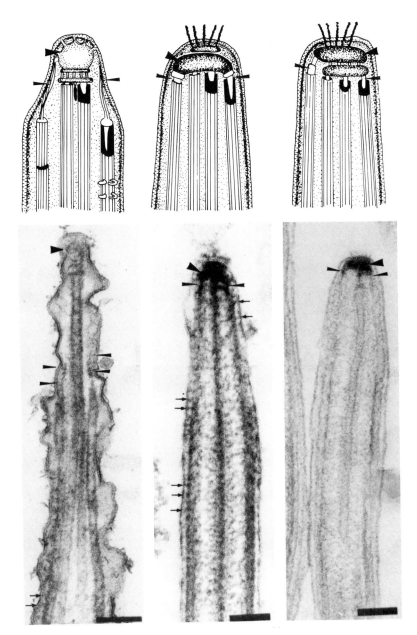

Figure 3. Microtubule capping structures typically found in *Tetrahymena* (top), a ciliated protozoan (left), rabbit trachea (middle), and frog palate (right). The membrane is attached to the tips of protozoan A tubules by distal filaments (small arrowheads) and to central microtubules by the central microtubule cap (large arrowhead). In contrast to the tips of protozoan cilia, the A and central tubules are bound to a common, complex, capping structure in trachea and palate and this structure is often attached to the ciliary crown filaments that project from the membrane opposite the site of the cap. In all cilia, the capping structures are linked to the microtubules by plug structures (small arrowheads), often better resolved in negatively stained cilia (see Dentler, 1980b, 1987). The frog palate cilia have asymmetrical capping structures in which the central microtubules and doublet microtubules Nos. 1, 2, 3, 8, 9 are bound to a small proximal shelf and doublets Nos. 4, 5, 6, 7 are bound to the upper distal shelf (see LeCluyse and Dentler, 1984). Microtubule–membrane bridges are also seen to link the A tubules to the membrane in the tracheal cilium. Bars = 0.1 μm.

Table I. Occurrence of Microtubule Capping Structures

	Cap + DF/P[a]	Complex cap[b]	Crown[c]	Reference
Mammals				
Human trachea, nose		X	X	Dalen (1986)
Rabbit trachea		X	X	Dentler and LeCluyse (1982a)
Rabbit oviduct		X	X	Anderson and Hein (1977)
Guinea pig trachea		X	X	Dalen (1983)
Mouse trachea		X	X	Kuhn and Engleman (1978)
Mouse oviduct		X	X	Dirksen and Satir (1972)
Mouse thymic cysts		X	X	Cordier (1975)
Rat trachea		X	X	Jeffery and Reid (1975)
Rat oviduct		X	X	W. L. Dentler (unpublished observations)
Whale trachea		X	X	W. L. Dentler (unpublished observations)
Sheep trachea		X		W. L. Dentler (unpublished observations)
Cow trachea		X		Dentler and LeCluyse (1982a)
Primary cilia, 3T3; rat testis		X		E. L. LeCluyse (unpublished observations), P. R. Burton (personal communication)
Birds				
Chick trachea		X	X	Dentler and LeCluyse (1982a)
Quail oviduct		X	X	Chailley et al. (1981)
Amphibians				
Xenopus laevis (toad) palate, embryo surface		X[a]	X	E. L. LeCluyse (unpublished observations)
Bombina orientalis (frog) palate		X[a]	X	LeCluyse and Dentler (1984)
Pseudacris triseriata (frog) palate		X[a]	X	E. L. LeCluyse (unpublished observations)
Rana pipiens (frog) palate		X[a]	X	P. R. Burton (personal communication)
Rana berlandieri (leopard frog) palate		X[a]	X	E. L. LeCluyse (unpublished observations)
Hyla cinerea (barking tree frog) palate		X[a]	X	E. L. LeCluyse (unpublished observations)
Hyla gratiosa (green tree frog) palate		X[a]	X	E. L. LeCluyse (unpublished observations)
Hyla chrysoscelis palate		X[a]	X	E. L. LeCluyse (unpublished observations)
Pachymedusa dacnicolor palate		X[a]	X	E. L. LeCluyse (unpublished observations)
Scaphiopus couchi palate		X[a]	X	E. L. LeCluyse (unpublished observations)
Caecilia		X[a]	X	E. L. LeCluyse (unpublished observations)
Ambystoma mexicanum embryo surface		X[a]	X	W. L. Dentler (unpublished observations)

			Reference
Pleurodeles walt (newt) palate	X[a]	X	W. L. Dentler (unpublished observations)
Ambystoma texanum (salamander) palate	X[a]	X	W. L. Dentler (unpublished observations)
Invertebrates			
Mollusks			
Gastropods			
Land snail	X[a]	X	W. L. Dentler (unpublished observations)
Slug "foot"	X[a]	X	W. L. Dentler (unpublished observations)
Pelecypods			
Scallop gills		X	W. L. Dentler (unpublished observations)
Mussel gills		X	W. L. Dentler (unpublished observations)
Worms			
Turbellarians			
Acoels			
Neochildia fusca	X[a]		Tyler (1979)
Endocincta punctata	X[a]		Tyler (1979)
Paratomella rubra	X[a]		Tyler (1979)
Archaphanastoma sp.	X[a]		Tyler (1979)
Hesiolicium inops	X[a]		Tyler (1979)
Nemerteans			
Procephalothrix spiralis		X	Tyler (1979)
Tetrastemma sp.		X	Tyler (1979)
Protozoans			
Tetrahymena		X	Dentler (1980)
Chlamydomonas		X	Dentler and Rosenbaum (1977)
Stentor		X	D. Diener (personal communication)
Diplodinium		X	Roth and Shigenaka (1964)
Leishmania	X(?)		E. Krug (personal communication)
Sperm			
Hamster		X	Woolley and Nickels (1985)
Rat		X[e]	
Acanthocephales		X[e]	Marchand and Mattei (1976)
Arbacia punctulata		X	W. L. Dentler (unpublished observations)
Saltwater mussel		X	W. L. Dentler (unpublished observations)

[a] A central microtubule cap binds the central microtubules to the membrane and distal filament/plugs attach individual A-tubules to the membrane.

[b] Plug structures attached to individual A-tubules and central microtubules are attached to a single large cap that is linked to the cytoplasmic face of the membrane.

[c] Filamentous structures attached to a complex cap and extend through the membrane to project from the tip of the cilium.

[d] Complex bipartite caps with "shelf."

[e] Central microtubules are fused to a structure at the distal tips.

small caps throughout the ciliated epithelium. The attachment of the caps to the microtubules is strong, and is not dissociated by reactivation of ciliary beating (Dentler and LeCluyse, 1982a) or by chemical conditions shown to dissociate protozoan capping structures from the microtubules (Dentler, unpublished results). Similar to the protozoan capping structures, the caps attached to tracheal and frog palate cilia bind to the microtubules by plug structures (Dentler and LeCluyse, 1982a,b; LeCluyse and Dentler, 1984).

The distal tips of microtubules in sea urchin, mussel, and hamster sperm also are capped with electron-dense structures, comparable to the plug structures observed in cilia (Table I; Dentler, unpublished data; Woolley and Nickels, 1985). A modified capping structure is present in annelid sperm, in which the distal tips of the nine doublet microtubules end in a transverse plate that is attached to a hollow structure called a "sting" in sperm flagella (Afzelius, 1983). The structures attaching the microtubules to the transverse plate were not identified, but the plate is at least superficially similar to the capping structures found at the ends of cilia. Finally, the distal tips of *Acanthocephales* sperm tails appear to be joined by a structure similar to the capping structure found in tracheal and oviduct cilia (Marchand and Mattei, 1976).

The microtubule capping structures can be considered as two structures. The caps are bound to microtubules by plug structures, which are common to all cilia and flagella. The distal portion of the capping structures can take a variety of forms, from the thin filaments in *Tetrahymena* and *Chlamydomonas* to the dense caps in tracheal and palate cilia. Although it is tempting to assign a different name for each of the different shapes of caps, I prefer to simplify the nomenclature for the time being and simply refer to these structures as "microtubule capping structures." The "central microtubule caps" and the "distal filament-plugs" will refer to the capping structures attached to the central microtubules and the A tubules, respectively.

To date, the capping structures have not been purified so that biochemical studies remain to be carried out but methods have been developed to selectively release the central microtubule cap and distal filaments from *Tetrahymena* axonemes (Suprenant and Dentler, 1988). The distal filaments and central microtubule cap are released under different salt conditions, which suggests that their association with the A or central microtubules is precisely controlled *in vivo*. The capping structures are present in only 11 copies per axoneme; they thus comprise a very small proportion of the total axonemal protein and, judging from their morphology, are complex structures composed of many different proteins. No specific proteins associated with the capping structures have been identified but it is reasonable to expect significant progress toward their purification in the near future.

Each of the microtubule capping structures discussed above is attached to microtubules by plug structures. In contrast to these structures, caps of amorphous material that *surrounds* the distal ends of the doublet and central microtubules have been described in *Beroë* macrocilia (Tamm and Tamm, 1985), cilia in cockroach chordotonal sensilla (Toh and Yokohari, 1985), and pallial cilia of *Lima hians* (Owen and McCrae, 1979). Due to the density of the material surrounding the microtubules, it is almost impossible to determine if the microtubules are capped by structures similar to those in other cilia and flagella. However, the distal ends of the microtubules in *Lima hians* (Owen and McCrae, 1979) appear to have dense material in the microtubule lumen, which may be similar to the plug structures discussed above. The ends of the central microtubules in growing

Beroë macrocilia terminate in a dense cap that appears to fill the ends of the central microtubules and bind them to the membrane, and dense material is seen at the tips of the growing doublet microtubules in these same cilia (Tamm and Tamm, 1985). However, the identity of these dense structures and their similarity to the microtubule capping structures remain to be determined. The dense caps surrounding the microtubules in sensory cilia may be involved in detecting pressure or other movements of the cilium by external forces (Owen and McCrae, 1979; Toh and Yokohari, 1985) and those in cteno-phore macrocilia are thought to help capture and ingest prey (Tamm and Tamm, 1985). The dense caps in oral cilia of *Glaucoma* may be sensory and/or may simply be a junction to bind adjacent cilia together (Montesano *et al.*, 1981).

3.1. Capping Structures and Motility

What effects could capping structures have on motility? One of the simplest explana-tions is that they are ''ball bearings'' that protect the ends of the microtubules. Since ciliary beating depends on the sliding of microtubules, the shearing of microtubules along the membrane might damage the end of each microtubule as it slides along or bumps into the ciliary membrane. This might explain why the ends of the A tubules and the central microtubules are capped, since their tips contact or attach to the membrane. In contrast, the B tubules of each outer doublet usually terminate proximal to the A tubule, do not contact the membrane, and lack recognizable capping structures. The ends of the A tubules also may be protected by bridges that link their sides to the membrane and must minimize the contact of the microtubule tips and the membrane (Figs. 3, 5).

A more active role in ciliary function may occur in the complex capping structures in tracheal, palate, and oviduct epithelial tissue. The function of these cilia is to transport mucus across the epithelial surface, toward the pharynx or uterus. The mucus is only penetrated by the distal portion of the cilium, so most of the force exerted to push the mucus must be concentrated at the ciliary tip. It may be significant that cilia whose function is to move mucus contain capping structures that link each of the plugs inserted into the A and central microtubules to a common cap (Table I). The complex cap may help the cilium to form a ''claw'' that can efficiently reach into the mucus and push it along (Jeffery and Reid, 1975).

In some, but not all, mucus-moving cilia, the cap is linked, through the membrane, to extraciliary hairs, called the ciliary crown (Fig. 3, Table I; Dirksen and Satir, 1972). The crown in oviduct cilia is negatively charged (Anderson and Hein, 1977; Kuhn and Engleman, 1978; Sandoz *et al.*, 1979; Chailley *et al.*, this volume) and, since negatively charged regions appear to be necessary for cilium-mediated transport of ova (Norwood and Anderson, 1980), the charged ciliary crowns may be important in mucus transport, possibly to help ciliary tips adhere to the mucus. However, although all respiratory cilia have the complex capping structures linking A and central microtubules to the membrane, not all respiratory cilia have ciliary crowns (Table I), so the crown may be useful, but not essential, for normal mucus transport.

The complex capping structures found in respiratory and oviduct cilia present some problems for the normal interpretation of ciliary microtubule sliding. Kuhn and Engleman (1978) first reported that the complex caps in respiratory cilia were tightly bound to the microtubules and were not released when the membranes were dissolved with nonionic

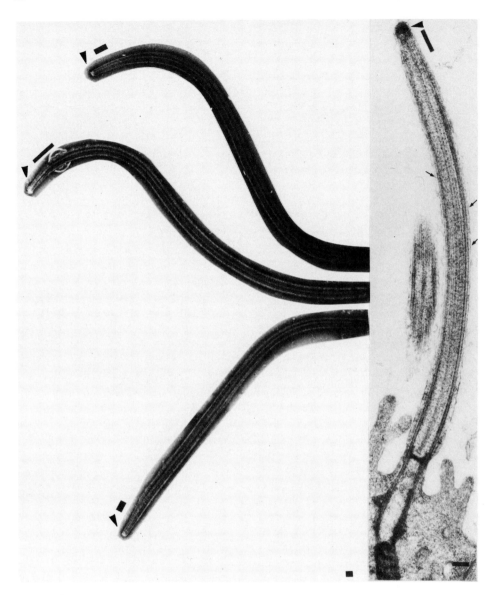

Figure 4. *Bombina* palate cilia showing the difference in the measured and predicted sliding expected for microtubules according to the sliding filament model. Each cilium is attached to a basal body, is capped by a typical capping structure (see Fig. 3), and is fixed in a bend. The sectioned cilium contains a basal body and a capping structure. The cilia at the left were isolated, demembranated, applied to a Formvar grid, and negatively stained. The cilia are flattened in one plane and contain the caps (large arrowhead) and the basal bodies. The total bend angle for each cilium is 133° (section), 114° (top left), 48° (middle left), and 128° (bottom left), and the amount of displacement of doublet microtubules on opposite sides of the axoneme is predicted to be 0.16, 0.23, 0.46, and 0.18 μm, respectively. The predicted displacement is indicated by the black bars at the ciliary tips but this displacement is not observed. Presumably, the difference between predicted and observed displacement (i.e., 0) is accommodated by stretching or compression of the microtubules. Small arrows also indicate microtubule–membrane bridges linking the doublets and membrane. Bars = 0.1 μm.

detergent. They also measured the diameter of the caps and reported that the caps were too small to accommodate the displacement of the microtubule tips predicted by the sliding filament model. Dentler and LeCluyse (1982a) showed that the caps remain attached to demembranated and reactivated rabbit tracheal cilia. Furthermore, they measured the bend angles of 36 rabbit and chick tracheal cilia, compared the expected displacement of microtubule tips with that observed in thin sections, and discovered that the amount of microtubule tip displacement predicted was far less than observed. The respiratory cilia do beat according to the sliding filament model, because sliding was observed in isolated and sheared cilia (which lacked capping structures) (Dentler and LeCluyse, 1982a). Three explanations for these results include (1) the caps stretch or shrink during bending, (2) the pattern of ciliary bending in cilia with the complex caps is different from that in cilia in which the A and central microtubules are not connected to a common cap, and (3) the microtubules stretch or shrink a certain amount during bending. The first possibility is unlikely because measurements of the complex caps have not revealed any stretching (Kuhn and Engleman, 1978; Dentler and LeCluyse, 1982a). The second possibility has not been observed, although mechanical models show that normal ciliary bends *can* be made without any net sliding of microtubules at the tips *if* the cilium is able to twist, and twisting of epithelial cilia is occasionally observed (Dentler and LeCluyse, 1982b). However, the predicted twisting is not seen in isolated and negatively stained frog axonemes whose microtubules are capped at the proximal and distal ends by the basal body and capping structures (Fig. 4). The third possibility may be the best explanation, since the maximum difference in predicted and measured displacements of the microtubules is ~ 0.5 µm, less than 10% of the ciliary length (Fig. 4; Dentler and LeCluyse, 1982b). It is reasonable to consider that the ciliary microtubules stretch or compress since axostyle microtubules can shrink as much as 25% of their length (McIntosh, 1973) and both the pitch and diameter of isolated outer doublet microtubules can change significantly *in vitro* (Miki-Noumura and Kamiya, 1979). Clearly, capped axonemes *can* bend without displacement of the caps or basal bodies. Isolated frog axonemes fixed at various angles and negatively strained assume a planar bend, show no twisting, and the microtubules are bound to the basal bodies and capping structures (Fig. 4). The sliding filament model predicts that microtubules on opposite sides of the axoneme should be displaced up to 460 nm but no such displacements are observed in these axonemes. At present, the best explanation is that the microtubules are compressed or stretched to accommodate sliding, but the apparent differences in measured and predicted tip displacements in cilia with the complex caps remain to be fully explained.

3.2. Capping Structures and Microtubule Assembly

Although it is clear that ciliary and flagellar assembly is well regulated, the regulation mechanisms are not understood (see Lefebvre and Rosenbaum, 1986). Microtubule assembly is not simply regulated by the exhaustion of a pool of the major ciliary proteins, since many cells contain a pool of protein sufficient to regenerate part or all of its complement of cilia or flagella in the absence of protein synthesis (Auclair and Siegel, 1966; Burns, 1973). The regulation of ciliary length must be localized within individual cilia, often at the level of individual microtubules. Cilia on epithelial cells (e.g., on the palate, trachea, and oviduct) and on protozoans grow asynchronously but all grow to equal lengths and protozoans frequently have two or more flagella or cilia that are of different, but precise, lengths. If there is a common pool of ciliary protein, then each axoneme must be able to regulate the quantity of protein it draws from the pool. An example of individual microtubule growth control is seen by comparing the growth of the central pair and outer doublet microtubules. The central microtubules can be several times longer than the doublets, as in *Micromonas* (Omoto and Witman, 1981), slightly longer than the doublets, as in *Chlamydomonas* (Dentler, 1980b), or of equal length to the doublets, as in most respiratory and oviduct cilia. There are several *Chlamydomonas* "9 + 0" flagellar mutants that lack central microtubules but have a normal complement of doublet microtubules (Randall, 1969). Only one central microtubule is present in the spermatozoid of the alga *Golenkinia minutissima* (Moestrup, 1972) and the male gametes of centric diatoms have normal doublet microtubules but lack both of the central microtubules (Manton and Von Stosch, 1966; Heath and Darley, 1972). The length of individual outer doublet microtubules is also precisely controlled. Each outer doublet in *Tetrahymena* appears to have a specific length (Sale and Satir, 1977b), as do individual doublets in epidermal cilia of flatworm cilia (Tyler, 1979) and frog palate (LeCluyse and Dentler, 1984).

A dramatic example of microtubule growth regulation occurs in the macrocilia of the ctenophore *Beroë*, in which 10–20 individual cilia fuse within a common membrane (except at the base) during ciliogenesis (Tamm and Tamm, 1988). During "Pattern II ciliogenesis," all axonemes have the same length at the time of membrane fusion but as the macrocilium grows, the axonemes on the oral side elongate, while those on the opposite, or aboral, side are arrested. Eventually, the axonemes on the oral side reach their full lengths and those on the aboral side start growing. The mechanisms that regulate elongation are not known but the pattern of growth could be due to unequal rates of elongation or different durations of elongation at a constant rate (Tamm and Tamm, 1988). Macrocilium growth in ctenophore cilia provides a dramatic example of the regulation of individual microtubule assembly within an organelle.

The role of the capping structures in microtubule assembly is speculative but is suggested by three principal observations: (1) the caps are present at the sites of microtubule assembly, (2) they are present during ciliary growth, and (3) they block tubulin addition to ciliary microtubules *in vitro*. Since the growth of each ciliary microtubule must be regulated, the presence of capping structures at the ciliary tips provides an excellent site at which tubulin addition, and microtubule growth, might be regulated. Implication of the capping structures in the regulation of ciliary microtubule growth is, therefore, based partly on "guilt by association" but it is certain that the capping structures are present at the right time and in the right place to have an important role in microtubule assembly.

3.3. Caps Are Bound to Assembling and Disassembling Microtubules

Based on pulse-labeling and autoradiographic studies, the site of tubulin addition to ciliary and flagellar microtubules is at the distal tip (Rosenbaum *et al.,* 1969; Witman, 1975). Microtubule assembly at the ends of the A tubules also occurs during flagellar tip activation in *Chlamydomonas* gametes (Mesland *et al.,* 1980). Although these studies provide the most direct evidence that microtubule assembly occurs at the distal tips, morphological studies reveal the presence of amorphous material near the tips of assembling ciliary microtubules, and the ends of growing microtubules of *Chlamydomonas* frequently splay apart when negatively stained and appear similar to the ends of rapidly assembling brain microtubules *in vitro* (Dentler, 1980a; Dentler and Rosenbaum, 1977). Finally, *in vitro* assembly studies have confirmed that structures are attached to the ''+,'' or favored, end of microtubule assembly (Allen and Borisy, 1974; Binder *et al.*, 1975; Sell *et al.*, 1974). Therefore, the microtubule capping structures are at the sites of tubulin addition to the microtubules, i.e., they are in the *right place* to play a role in microtubule assembly.

The capping structures are also present at the right time to play a role in microtubule assembly and disassembly. Studies of flagellar regeneration in *Chlamydomonas* revealed that the capping structures are bound to the microtubule ends by the time the flagella are 4 μm long and remain attached to the microtubule ends throughout growth (Dentler and Rosenbaum, 1977; Dentler, 1980a). The complex capping structures are present in developing frog palate cilia as short as 1.5 μm and are fully formed by the time the cilia have grown to 2.0 μm (Portman and Dentler, 1987). The capping structures remain attached to the microtubules throughout the rest of ciliary growth and are found in all fully grown cilia. The capping structures also remain attached to microtubules during flagellar disassembly *in vivo* (Dentler and Rosenbaum, 1977). Therefore, in addition to being in the *right place* (the sites of tubulin addition), the capping structures are also present at the *right time* for their having a role in ciliary microtubule assembly and disassembly. The proteins that compose the capping structures may be present during the initial 1–2 μm of microtubule growth but, lacking probes to identify these proteins, we can only say that they are present after 1–2 μm of growth.

3.4. Can Capping Structures Regulate Tubulin Addition to Microtubules?

If they are present at the site and time of microtubule assembly, the capping structures should not inhibit the addition of tubulin to the microtubules. Pieces of axonemes, basal bodies, or brain microtubules nucleate the assembly of brain microtubules *in vitro* (Binder *et al.*, 1975; Snell *et al.*, 1974; Dentler *et al.*, 1974) but the capping structures block *in vitro* microtubule assembly onto the ends of *Chlamydomonas* flagella (Dentler and Rosenbaum, 1977), rabbit tracheal cilia (Dentler and LeCluyse, 1982b), and isolated bovine tracheal cilia (Dentler, unpublished results), regardless of the tubulin concentration used. This means that the caps are tightly bound to the microtubules and that tubulin alone cannot compete for the cap–microtubule binding sites. However, microtubules can assemble onto free capping structures *in vitro,* so tubulin must be able to interact with the caps (Dentler and LeCluyse, 1982b), and preliminary evidence indicating that tubulin can be inserted between the caps and A or central microtubules in the presence of ATP and Ca^{2+} suggests that a specific event, possibly phosphorylation, of the capping proteins may be involved in the assembly process (Dentler, 1986).

At the very least, there must be a mechanism *in vivo* that permits tubulin addition to capped microtubules. The caps may passively float up the growing microtubules in association with the membrane or they could actively walk up the growing microtubules. Each of these mechanisms could be sensitive to specific cytoplasmic factors that regulate the association of the caps with the microtubule ends. Since ciliary tubulin is posttranslationally modified by acetylation (L'Hernault and Rosenbaum, 1983; Piperno and Fuller,

1985), the caps may be machines containing tubulin acetylase or they might select for acetylated tubulin to be assembled into the microtubules. In the absence of data, the mechanisms that release the caps and permit tubulin addition to microtubules are a matter of speculation but the evidence that they have an important function in the regulation of tubulin assembly is increasingly convincing.

3.5. Are Microtubule Caps Found in the Cytoplasm?

If ciliary and flagellar capping structures play a role in the regulation of tubulin assembly, it is not unreasonable to search for similar capping structures in other cytoskeletal structures. Capping proteins are bound to the ends of actin filaments and, in some cases, link the actin filament to membranes (see Pollard and Cooper, 1986; Stossel *et al.*, 1985) and a capping structure is bound to the tip of a bacterial flagellum (Ikeda *et al.*, 1985). The bacterial flagellar cap is found at the site of flagellar assembly and it blocks addition of flagellin to the flagellum *in vitro*. The ends of many cytoplasmic microtubules terminate at membranes, centrosomes, or kinetochores. The capping structures are similar to kinetochores in that they are tightly bound to the ``+'' ends of microtubules, at the site of microtubule assembly ad disassembly (Michison *et al.*, 1986). Both structures must ``walk'' along the microtubule as the tubulin is added or removed, and precise regulation of the time and amount of assembly at the site of microtubule caps (see above) and kinetochores (during prophase and anaphase) occurs *in vivo* (Gorbsky *et al.*, 1988). The structure of the kinetochore and the mechanism by which it binds to microtubules are unknown, although Huitorel and Kirschner (1988) suggested that it captures a microtubule by surrounding it with a collarlike structure. Since the ciliary capping structures are functionally similar to the kinetochore, it is not unreasonable to suggest that kinetochores and other structures that link the ends of microtubules to membranes or organelles may do so by a plug-shaped structure similar to the ciliary and flagellar microtubule capping structures.

4. Bridges Linking the Sides of Outer Doublet Microtubules to the Membrane

Doublet microtubules are commonly linked to the ciliary membrane or to a matrix of material immediately under the membrane but the structure and occurrence of these linkages vary somewhat depending on the type of cilium or flagellum and the location within a single cilium. For the most part, little is known about the structure, composition, or function of the bridges but they tether the membrane to the axoneme and occasionally bind extracellular structures to the microtubules. The most prominent bridge structures between the doublet microtubules and the membrane are found in the transition region, near the base of the cilium. The structure and occurrence of these bridges are briefly discussed above and in other reviews (Dentler, 1981a; Besharse and Horst, this volume).

The most common form of microtubule–membrane bridge is a thin filamentous structure that links the membrane to the doublet microtubules (Ringo, 1967; Allen, 1968; Sattler and Staehelin, 1974; Bloodgood, 1987). Evidence that the bridges actually bind the membrane to the microtubules comes from micrographs in which the membrane balloons outward from the axoneme, revealing bridges that retain patches of the membrane bound to the microtubules (Fig. 5; Dentler *et al.*, 1980; Toh and Yokohari, 1985). Detergent extraction of cilia partially dissolves ciliary membranes but frequently leaves patches of membrane linked to the cilia by bridge structures (Fig. 6). Although commonly seen, it is difficult to routinely observe the bridges. Toh and Yokohari (1985) reported that rapid freezing and freeze-substitution improved their ability to routinely visualize microtubule–membrane bridges in insect campaniform cilia and such a technique may help to preserve or visualize bridges in other cilia and flagella. Omission of OsO_4 from the fixation protocol also tends to preserve bridges in a variety of cilia (Dentler, unpublished results). Rapid freezing and deep etching as pioneered by Heuser (1981) should constitute excellent methods for visualizing microtubule–membrane bridges, but such methods have not as yet revealed the lateral bridges seen in thin sections (J. Heuser, personal communication). Conventional fixation methods often fail to preserve and stain the bridges, even when it is clear that the bridges are cross-linking membranes to the ciliary microtubules (Dentler *et al.*, 1980). The lack of methods to routinely preserve and stain the bridges has slowed progress in this area.

Campaniform sensilla in *Drosophila* halteres contain modified sensory cilia in which single microtubules are separated from the membrane by 13- to 25-nm bridges spaced with a longitudinal periodicity of 15–20 nm (Toh, 1985). The functions of the bridges are unknown, but they may have a sensory role since some, but not all, sensory cilia lack the organization and many of the structures necessary for ciliary beating. The role of the microtubules or the microtubule–membrane bridges in sensory functions has never been identified or proven, but the structural rigidity of the sensory cilia provided by the microtubules bound to the membrane and the fact that the sensory cilium is subjected to vibration, for example, during flight, suggest that the cytoskeletal structures may transmit vibrations to the ciliary membrane by the microtubule–membrane bridges (Toh, 1981, 1985; Toh and Yokohari, 1985). Active sliding of some insect sensory cilia may also be involved in the sensory function (Moran *et al.*, 1977), and the connections between membrane-associated bridges and the outer dynein arms (Crouau, 1980) suggest a direct interaction between dynein-based microtubule sliding and the microtubule–membrane bridges.

A variety of structures are interposed between the microtubules and the membranes in sperm tails. For example, single microtubules line the cytoplasmic face of frog lung, fluke sperm (Justine and Mattei, 1982), and the undulating membrane in a parasitic flatworm (Justine and Mattei, 1985). The cytoplasmic face of mammalian sperm membranes is lined with a dense matrix or sheath that is linked to doublet microtubules and to the flagellar membrane by large bridge structures that project from the external face of the membrane and bind to concanavalin A (Con A) (Enders *et al.*, 1983). A particularly impressive example showing the binding of the sheath and axoneme to the membrane by discrete sites is found in Enders *et al.* (1983), in which cross sections of six axonemes attached to their sheath are wrapped within one plasma membrane. Each sheath is attached to the membrane by a Concanavalin A-stainable patch that may be associated with a line of intramembrane particles called a "zipper." The function of the zipper line is not understood since it appears to be involved with the linkage of the membrane to the fibrous sheath in guinea pig sperm but not in mouse sperm (Enders *et al.*, 1983).

In addition to sperm flagella, other microtubule–membrane bridges are associated

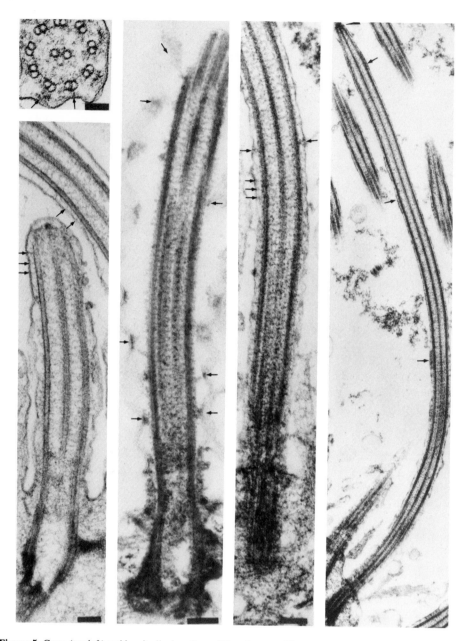

Figure 5. Cross (top left) and longitudinal sections of *Tetrahymena* cilia (top left panel) and rabbit tracheal cilia (remaining panels) showing microtubule–membrane bridges linking the membrane to the doublet microtubules. Bridges (arrows) are present throughout the length of the cilia, including along the A tubules near the ciliary tip (left). Dense material on the axonemal side of the membrane is frequently seen in association with the bridges. This is particularly prominent in mammalian tracheal and frog palate cilia and may be part of a submembrane skeleton associated with the microtubule–membrane bridges. The cilium in the right panel shows that bridges are present from the base to the cap at the distal tip of the cilium. Bars = 0.5 μm for the cross section and 0.1 μm for all others.

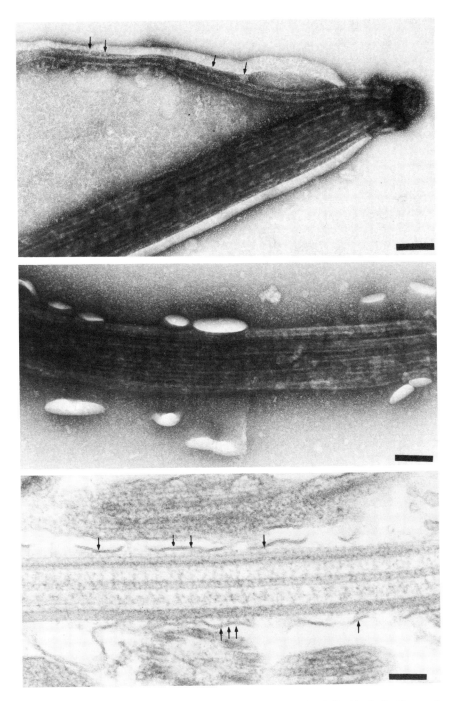

Figure 6. Isolated bovine tracheal cilia partially demembranated with 0.1% Nonidet P-40. The membranes frequently are seen to wrap around the axonemes and are clearly linked to the membrane by microtubule–membrane bridges (arrows) after detergent extraction and washing. The membranes often appear to be structured and peel off the axoneme without forming vesicles (middle panel). This suggests that there is a (skeletal) structure associated with the membrane but no structure has as yet been identified. Bars = 0.1 μm.

with glycosylated proteins exposed at the external surface of the ciliary membrane. Con A-binding proteins are present at the distal tip and at the base of cilia in anuran seminal vesicles (Chailley *et al.*, 1981). The receptors near the ciliary base are preferentially located opposite prominent bridges that link the doublet microtubules with the ciliary membrane. Not all lectin-binding proteins are associated with bridges, however, since wheat germ agglutinin receptors cover the entire ciliary shaft and show no preferential association with bridges or doublet microtubules. Lectins also bind to proteins associated with ciliary bridges in the transition region of retinal rods and cones (Horst *et al.*, 1987) and a major 350-kDa Con A-binding protein in *Chlamydomonas reinhardtii* flagella binds particles and moves them along the flagellar surface (Bloodgood, this volume). Additionally, a 240-kDa glycoprotein has been found to be tightly associated with the axoneme of *C. moewusii* (Reinhart and Bloodgood, 1988) and may be directly or indirectly associated with a microtubule–membrane bridge; the nature of its association with the microtubules has not been determined (see Reinhart and Bloodgood, 1988). Taken together, these studies suggest that cell-surface proteins associated with microtubule–membrane bridges may occur in many cilia and flagella. To date, several proteins, including the protozoan immobilization antigens (Williams *et al.*, 1985), have been labeled on the surface of cilia and flagella but the associations of these proteins with microtubule–membrane bridges, if any, have not been determined.

4.1. Bridges Linking Microtubules to Extraciliary Structures

The most prominent bridges observed between the doublet microtubules and the ciliary membranes are associated with mastigonemes, which are hairlike appendages attached to the external surface of a variety of protozoan flagellar membranes (Bouck, 1972; Bouck *et al.*, this volume; Dentler, 1981a). The structure of the mastigonemes varies from the ~ 10-nm-diameter filamentous hairs of variable lengths to the ~ 20-nm-diameter and 300- to 400-nm-diameter tubular mastigonemes (Bouck, 1972). The mastigonemes attach to the microtubules indirectly either through a network of filamentous material on the membrane surface that is linked, through the membrane, to a filamentous microtubule–membrane bridge, as in *Euglena* (Bouck *et al.*, 1978), or to a granular microtubule–membrane bridge structure on the cytoplasmic side of the membrane, as seen in *Ochromonas* (Markey and Bouck, 1977). The mastigonemes are tightly bound to the bridges and remain attached to the axonemes after solubilization of the flagellar membrane (see Dentler, 1981a). In *Ochromonas*, the bridges are periodically spaced at 200-nm intervals along the doublet microtubules (Markey and Bouck, 1977).

The mastigonemes are added to the membrane during flagellar growth (Bouck, 1971, 1972; Hill and Outka, 1974). They are assembled within cytoplasmic membrane vesicles, are added to the cell surface near the base of the flagellum, and appear to move along the flagellum until they reach the appropriate attachment sites. The attachments between mastigonemes and the microtubules are specific, since they are only associated with doublet microtubules on opposite sides of the axoneme, in a plane parallel to that formed by the central microtubules and perpendicular to the bending plane of the flagellum (Bouck, 1971; Hill and Outka, 1974). The assembly of the mastigonemes onto the flagellar membrane sites requires the preassembly of the prominent microtubule–membrane bridges on specific doublet microtubules. If the mastigonemes are attached to the cell membrane and transported up the ciliary membrane, as suggested by Hill and Outka (1974), they must be transported up the membrane in association with specific doublet microtubules and locked into place when they reach their place on a doublet. The movement of the mastigonemes within the membrane could occur by membrane flow that occurs as the growing flagellum stretches the plasma membrane, or could be an active process, involving a "motor" similar to that described by Bloodgood (1977) for cell surface motility in *Chlamydomonas,* which also contains mastigonemes.

Microtubule–membrane bridges also form junctions between adjacent cilia and between the cilia and cell bodies. Ciliary junctions link the flagellum to the cell body of trypanosomes (Vickerman, 1969; Hogan and Patton, 1976; Anderson and Ellis, 1965) and link neighboring cilia in ctenophore comb plates (Afzelius, 1961; Dentler, 1981b), neighboring axonemes in macrocilia (Tamm and Tamm, 1985), and laterofrontal mussel gill cilia (Warner, 1974). The linkages between adjacent cilia or axonemes appear to maintain the parallel orientation of the beating cilia, since these axonemes all beat in a plane perpendicular to that of the lamellar junctions. Interestingly, the microtubule caps can also be modified to form interciliary junctions at the ciliary tips of *Glaucoma* (Montesano *et al.,* 1981).

4.2. Sites of Attachment of the Bridges

Dentler *et al.* (1980) presented three models of microtubule–membrane bridges. Each bridge could be a microtubule-associated protein that (reversibly) contacts proteins or lipids in the membrane, it could be a membrane-associated protein that (transiently) contacts the doublet microtubules, or it could be a bifunctional molecule that can reversibly bind to either a membrane or a microtubule. Since there are a variety of structures seen between the microtubules and the ciliary membranes, it is likely that the microtubule–membrane bridges will not be identical in all cilia and flagella.

4.3. Attachment to the Doublet Microtubles

Cross sections of most cilia and flagella reveal that the bridges are attached along the midpoint of the doublet microtubules that face the membrane, i.e., to the junction of the A and B tubules of each doublet (Figs. 1, 3, 5; Sattler and Staehelin, 1974; Allen, 1968; Dentler, 1981a, 1987). However, there are examples in which the membrane appears to be linked to the outer dynein arms. This may be an artifact, since the outer dynein arms are kinked, with their "elbows" pointed toward the wavy membrane, and ciliary cross sections contain several arms and membrane profiles superimposed to give the appearance that an arm is associated with the membrane. However, cross sections of a *Chlamydomonas* flagellum in which the membrane was pulled away from the axoneme revealed highly stretched filamentous bridges linking the outer arm with the membrane (Bloodgood, 1987). Furthermore, Crouau (1980) showed insect mechanoreceptor cilia in which bridges extend from the ciliary membrane and to sites on the outer dynein arms and to the A–B junction of each doublet microtubule. He suggested that the linkages to the outer arms may be involved with mechanoreception by these cilia. Numerous bridge structures are attached to a dense matrix of material under the ciliary membrane, only a few of which are attached to the outer arms in any single image. Not all bridges attach to the junction of the A and B tubules, since *single* microtubules are bridged to the membrane in campaniform sensilla (Toh, 1985) and along the A tubule in protozoan and tracheal cilia (Figs. 3, 5).

The bridges and their attachment to the microtubules are best resolved in cross sections of cilia but attachments between the membrane and microtubules are seen in longitudinal sections. Reported periodicities are 200 nm for the attachment of mastigonemes to *Ochromonas* flagella (Markey and Bouck, 1977) and the membrane to sensory cilia (Toh and Yokohari, 1985), 15–20 nm for the attachment of bridges to sensory cilia (Toh, 1985), and 20–24 nm in scallop gill (Stephens *et al.,* 1987) and *Tetrahymena* cilia (Dentler *et al.,* 1980). However, the unambiguous identification of bridges in longitudinal sections is complicated by the superposition of the 20- to 24-nm periodicity of the outer dynein arms as well as the curving membrane in thin sections. Outer arms do not account for all periodic bridges seen, however, since the A-tubule extensions of the doublets to the tips of cilia lack arms but are bridged to the membrane (Figs. 3, 5). Less ambiguous identification of the longitudinal periodicity may be possible by studying axonemes that lack the outer arms. Outer arm-less flagellar mutants have been isolated from *Chlamydomonas* (Mitchell and Rosenbaum, 1985; Kamiya and Okamoto, 1985) but morphological analysis of the microtubule–membrane bridges has not been carried out.

4.4. Sites of Membrane Attachment

Since membranes are fluid structures, the attachment sites of bridges may not be as periodic as the microtubule-binding site and we do not know if the bridge structure is bound to the cytoplasmic face of the membrane or if it is a transmembrane protein. If the bridges do not span the lipid bilayer, they could be associated with proteins that do span the bilayer, since extraflagellar structures such as mastigonemes and ciliary junctions are attached through the membrane to the microtubules (see above) and lectin-binding sites on the membrane surface have been associated with microtubule–membrane linkages in anuran cilia (Chailley *et al.,* 1981), photoreceptor cilia (Horst *et al.,* 1987; Besharse and Horst, this volume), and sperm membranes (Enders *et al.,* 1983). Finally, movement of particles along the flagellar surfaces (Bloodgood, 1977, this volume) occurs in association with the microtubules. It may be unlikely that a single protein spans the ciliary membrane and the space between the doublets and the membrane. It is more likely that the bridge is formed from a complex of proteins part of which could include a proteinaceous submembrane skeleton that could include membrane-associated tubulin complexed with other detergent-soluble ciliary membrane proteins (Stephens *et al.,* 1987; Stephens, this volume). Such a skeleton could serve as a binding site for transmembrane proteins as well as microtubule-associated bridge proteins. The structure of a membrane skeleton in protozoan or molluscan gill cilia, if it exists, has not been described but there is evidence of amorphous protein on the cytoplasmic face of ciliary membranes that may attach to the microtubules via bridge structures. Dense plaques of material are associated with bridge structures in *Tetrahymena* cilia (Sattler and Staehelin, 1974), and in bovine tracheal cilia (Figs. 4, 5) the ciliary membrane often peels away as if it was associated with a submembranous structure (Fig. 6). One of the clearest examples of a submembrane skeleton in cilia is found in an insect mechanoreceptor cilium, which contains a regular array of paired "protuberances," or bridges, that project toward the microtubules (Crouau, 1980). The bridges are uniformly arranged around the cilium and linkages to the microtubules or outer dynein arms are formed when a doublet microtubule is closely apposed to a pair of bridges. Similar submembrane structures were found in *Drosophila* campaniform sensilla by Toh (1985), and Toh and Yokohari (1985) showed that the doublet microtubules are bridged to granular material attached to the cytoplasmic face of the ciliary membrane in swollen regions of the antennal chordotonal sensillum of the cockroach. A convincing case can be made for the presence of a submembrane skeleton capable of attaching to ciliary microtubules in insect sensory cilia but its structure and composition remain to be identified.

Fuzzy material on the cytoplasmic face of the membrane is quite evident in squid sperm (De Montrion, 1986), and a fibrous sheath, similar to a submembrane skeleton, is prominent in some sperm tails. The doublet microtubules either attach directly to the sheath (as seen in guinea pig sperm; Enders *et al.*, 1983) or attach to large accessory fibers that associate with the membrane (Dallai and Afzelius, 1982). The attachment of the microtubules or sheath to the membrane is often associated with short filaments, or spikes (Dallai and Afzelius, 1982), and these spikes may be associated with the rows of intramembrane particles, or "zippers," seen in many sperm (Dallai and Afzelius, 1982; Enders *et al.*, 1983).

The identity and functions of bridge proteins are not known. Reinhart and Bloodgood (1988) reported that the major 240-kDa surface-exposed glycoprotein in *Chlamydomonas moewusii* is tightly associated with the outer doublet microtubules, so this protein either is a transmembrane protein and a microtubule–membrane bridge or is a membrane-associated protein that is tightly associated with another protein(s) that together comprise the microtubule–membrane bridge. The use of a photoactivatable cross-linking reagent revealed the presence of a dyneinlike protein that linked membranes to the microtubules in isolated *Tetrahymena* and scallop gill cilia (Dentler *et al.*, 1980). The cross-linking experiments indicated that the bridge is linked to proteins that migrate at 55–60 kDa, some of which may be membrane-associated tubulin that composes a submembrane skeleton (Dentler *et al.*, 1980; Dentler, 1988; Stephens *et al.*, 1987). Fractionation of ciliary membranes with the nonionic detergent Triton X-114 reveals that tubulin is probably not an integral membrane protein but it may be associated with the membrane surface (Stephens, 1985, this volume; Dentler, 1988).

The presence of a submembrane skeleton composed of cytoskeletal proteins is not unprecedented. The best characterized "submembrane skeleton" contains spectrin, which is a major component of a filamentous matrix that lies beneath the red cell membrane. Spectrinlike proteins are associated with protozoan membranes (N. E. Williams, personal communication). They also compose part of the paraflagellar rod of *Trypanosoma brucei* (Schneider *et al.*, 1988). The paraflagellar rod is a prominent structure that lies along the axoneme of kinetoplastid flagellates and euglenoids and is attached to the membrane and to the A–B junctions of doublet microtubules by bridge structures (Dentler, 1981a; Hyams, 1982). Although the identity of the proteins associated with the bridge has not been resolved, good starts toward their characterization have been made and as we begin to obtain a better understanding of the cytoplasmic face of the ciliary membrane, the nature of the bridges and the proteins to which they bind will undoubtedly become more evident.

4.5. Nonciliary Microtubule–Membrane Bridges

To gain a better understanding of ciliary microtubule–membrane bridges, it may be useful to consider the way cortical microtubules bind to the plasma membrane in other cells. One of the most elegant structural studies of microtubule–membrane interactions was carried out on the cortex of the unicellular alga *Distigma proteus* (Murray, 1984a). A regular array of particles, possibly a "submembrane skeleton," is held to the membrane by strong hydrophobic interactions and stabilized by disulfide bonds. The microtubules lie ~ 12 nm from the lipid bilayer and the membrane particles lie less than 5 nm from the bilayer, so a bridging protein(s) must bridge the ~ 7-nm gap to link microtubules to the membrane. The particles on the cytoplasmic face of the plasma membrane are arranged in double rows and the particles are spaced such that microtubules could bind to the particles by "MAPs" if the microtubules are lined up at a 36° angle to the membrane particles and if MAPs are spaced along the microtubules with the 96-nm spacing measured by Amos (1977). The protein composition of the bridges in *Distigma* is unknown, but Murray (1984b) has shown that Ca^{2+} extraction of *Distigma* cortexes removes microtubules but leaves short projections (microtubule–membrane bridge proteins?) attached to the membrane skeleton. When exogenous microtubule protein (containing MAPs) is added to the extracted membranes, microtubules assemble into the sites formerly occupied by microtubules prior to Ca^{2+} extraction and appear to bind to the short projections connected to the membrane skeleton.

It is not known if the microtubule–membrane bridges in the trypanosome or in *Distigma* are similar to the ciliary microtubule–membrane bridges. However, the highly organized membrane-associated material that binds to microtubules in the *Distigma* pellicle may be similar to the membrane skeleton proposed by Stephens *et al.* (1987). If they are similar, future studies of the ciliary and flagellar membranes should examine the structure of the skeleton or material underlying the membrane and its ability to attach to axonemal structures. Finally, recent studies of the associations of other cytoskeletal structures with cell membranes have revealed a new class of amphitropic proteins, or proteins whose reversible association with membranes may involve the binding and dissociation of lipids which may be regulated by the phosphatidylinositol cycle (see Burn, 1988). This group of proteins currently includes the actin-binding proteins α-actinin, vinculin, profilin, gelsolin, p36, spectrin, erythrocyte protein 4.1, and a 110-kDa microvillar protein (Burn, 1988). Although no microtubule-associated proteins have been identified as amphitropic proteins, it is important to consider the possibility that the microtubule–membrane bridges are partially composed of amphitropic proteins.

4.6. Functions of the Bridges

The most obvious function of the microtubule–membrane bridge is to anchor the membrane to the axoneme, which helps to stabilize the fluid membrane and maintain the organization of membrane proteins relative to axonemal microtubules. The principal evidence for this comes from electron micrographs that clearly show that the bridges bind the membrane to the axoneme but, in the absence of visible linkages, the membrane frequently balloons outward from the axoneme (see above). The wavy membrane seen in thin sections may be partly a fixation artifact but the membranes are tightly bound to microtubules when bridges are seen. This is particularly evident in the transition region near the bases of cilia and flagella where pieces of membrane remain linked to axonemes even after extensive washing with nonionic detergents (Dentler *et al.*, 1980; Stephens, 1977; Snell *et al.*, 1974). Microtubule–membrane bridges also bind extracellular structures, including mastigonemes (Markey and Bouck, 1977), to the microtubules and link adjacent cilia via specialized junctions (Afzelius, 1971; Dentler, 1981b). Bridge-mediated physical contact between the membrane and axonemes has been proposed to serve sensory functions (Toh, 1981, 1985; Crouau, 1980; Toh and Yokohari, 1985) although the mechanisms by which they transmit sensory information have not been identified.

Child (1978) proposed that the interactions of growing doublet microtubules with the surrounding membrane, possibly via the microtubule–membrane bridges (or switch-linkages), may help to regulate ciliary and flagellar growth. In his model, the microtubules, which grow at their distal tips, interact with the membrane, which grows at its base. The interactions between these structures that assemble at opposite ends of the cilium affect the rate and final extent of ciliary assembly. As the cilium elongates, the number of microtubule–membrane linkages increases and the rate of growth slows down until the final length is reached. Growth is a function of the number of linkages that are either assembled or "closed" after assembly. One limitation of this model is that the membrane cannot be completely fluid, since a relatively rigid component must be present to interact with the microtubule–membrane linkages (Child, 1978). However, if there is a ciliary membrane skeleton (see above), linkages between the doublet tubules and the membrane skeleton may provide a sufficiently rigid skeleton to regulate growth. To date, however, the nature of the bridges and their role (if any) in growth regulation have not been experimentally determined.

The bridges may have a yet uncharacterized role in ciliary motility. Dentler *et al.* (1980) demonstrated that the linkages between the axoneme and ciliary membrane could be stabilized by a photoactivated cross-linking reagent and that the stabilization arrested ciliary beating in the scallop gill and slowed beating in *Tetrahymena* cilia. Although this suggests that the bridges have an important role in normal ciliary beating, their role is not understood. If the membrane is fluid, then one might not expect that the stabilization of a connection between a bridge protein and a protein or lipid in the membrane would arrest beating. On the other hand, if the bridge attaches to a submembrane skeleton (see above), then the stabilization of the connection may increase the rigidity of the ciliary shaft to bending or at least partly resist the normal sliding of individual microtubules by adding drag to the sliding microtubule.

The bridges may transport proteins within the plane of the flagellar membrane. The movements of mastigonemes (Hill and Outka, 1974), particles attached to flagellar surfaces (Bloodgood, 1977, 1987, this volume), gliding of flagella along solid substrates (Lewin, 1952, 1982; Bloodgood, 1981), and the redistribution of surface proteins in protozoan and sperm flagella have been well documented and reviewed elsewhere (Bloodgood, 1987, this volume) and will not be covered here. Some, but not all, of the movements of membrane proteins require energy, which suggests that ATPases may be involved in the membrane-associated motility and may be the motors directly responsible for their movements. Bloodgood (1987) suggested that the bridges could contain a calmodulin–Ca^{2+}-ATPase that could move *Chlamydomonas* glycoproteins, since Ca^{2+} appears to have a role in the membrane bead movement.

Ciliary and flagellar membranes contain several ATPases but their functions have not been thoroughly characterized. Ca^{2+}-stimulated ATPases have been identified (Baugh *et al.*, 1976; Satir, 1976; Doughty, 1978; Fay and Witman, 1977; Bessen *et al.*, 1980; Brugerolle *et al.*, 1980; Andrivon *et al.*, 1983; Doughty and Kaneshiro, 1985; Travis and Nelson, 1986) and a possible role of Ca^{2+} in the linkage of microtubules to membranes was proposed by Bloodgood (1987). *Tetrahymena* ciliary membranes contain a dyneinlike ATPase as well as an integral membrane Mg^{2+}- and Ca^{2+} stimulated ATPase (Dentler *et al.*, 1980; Dentler, 1988). The presence of ATPase activity in association with the microtubule–membrane bridges has been suggested by cytochemical methods, although these methods cannot identify specific ATPases or their function (Dentler, 1977; Lansing and Lamy, 1961; Nagano, 1965; Baccetti *et al.*, 1979). The most direct evidence for an ATPase associated with a microtubule–membrane bridge was obtained from cross-linking studies in which a ciliary dynein was found to be associated with membranes cross-linked to the ciliary microtubules (Dentler *et al.*, 1980). The dynein was released from "non-cross-linked" cilia with nonionic detergent and was found to sediment in sucrose gradients at 14 S. Additional support for the 14 S dynein being a cross-linking protein includes the staining of the doublet microtubules at or near the site of attachment of the membrane bridge by an antibody against 14 S ciliary dynein (Marchese-Ragona and Johnson, 1985). Although these studies suggest that a 14 S dynein is part of the microtubule–membrane bridge, we still do not know the role of the dynein or how it is associated with the ciliary membrane. If the structure of the membrane dynein is similar to ciliary outer arm dynein, we expect the ATPase site of the dynein to bind to tubulin (Porter and Johnson, 1983). The ATPase "head" of the dynein could either bind to the doublet microtubule, in which case the tail must bind to a membrane-associated protein, or the tail must bind to the doublet microtubule and the head to a membrane-associated protein, possibly to tubulin associated with the membrane or a submembrane skeleton. Lacking data to prove one or the other location for the membrane-associated dynein, endless speculations can be made for each model.

While the function of the bridge dynein, if it *is* 14 S dynein, is unknown, it is interesting that Vale and Toyoshima (1988) reported that *Tetrahymena* 14 S dynein adsorbed onto a glass surface induces a torque so that the axonemes rotate as they are propelled along the surface. They propose that this may provide a clue to the mechanism by which central pair microtubule rotation might modify doublet microtubule sliding to create three-dimensional beating patterns, but it is unclear how this could occur, since the doublet and central microtubules are too far apart to be linked by a dynein arm and dynein has not been identified as either a radial spoke or central microtubule component. It is interesting to speculate that a 14 S dynein is part of the microtubule–membrane bridge (Dentler *et al.*, 1980). The bridge could contact the doublet microtubule at a site similar to that accessible to the dynein bound to glass (Vale and Toyoshima, 1988), and the arrest of ciliary beating when the bridge is cross-linked to the axoneme (Dentler *et al.*, 1980) is consistent with the bridge having some important role in ciliary beating. If the bridge is bound to the lipid bilayer, it may not be able to produce the necessary force to alter microtubule sliding, since the bilayer is fluid, as shown by particle and protein transport along membrane surfaces (see above) and by basal body rotation during ciliogenesis (Portman and Dentler, 1989). However, if there is a submembrane skeleton, then this skeleton may provide structural rigidity for the bridge to produce force on the axoneme. The skeleton could be analogous to the glass slide used by Vale and Toyoshima (1988).

5. Summary

The interactions between ciliary and flagellar membranes and microtubules have been described but relatively little is understood about them. For the most part, the studies have been morphological since the bridges linking microtubules to membranes are present in relatively low quantities. For example, microtubule capping structures at the tips of flagellar microtubules are present in only 11 copies per axoneme. The bridges linking the sides of the doublets to the membrane are not seen in all sections of any given cilium, which indicates either that they are present in relatively low quantities or that we have failed to develop methods to routinely visualize them in cilia prepared for electron microscopy. However, biochemical studies of the bridges and capping structures are being carried out and it is reasonable to expect that characterization of the bridges linking the sides and distal tips of the microtubules will be forthcoming. This is important, because studies of microtubule–membrane interactions may provide valuable insight into the assembly and function of ciliary and flagellar microtubules as well as the function and assembly of cytoplasmic microtubules. The lateral interactions between the doublets and membrane are likely to be similar to those that form between microtubules and a variety of organelles or the plasma membrane in many types of cells. The microtubule–membrane bridges may provide structural support to the membrane and may be involved in active movement of cell-surface components or of the cytoskeleton. Definition of the role of the capping structures may reveal important information about the regulation of cytoplasmic or mitotic microtubule assembly. The major advantage of using cilia and flagella as model systems to study microtubule assembly and the interactions of microtubules with membranes is the relative ease of genetic analysis (Huang, 1986). As we begin to identify structures and specific proteins that link membranes with microtubules or the capping structures at the microtubule tips, genetic analysis will become increasingly useful and will ultimately provide the tools with which to identify the functions of these structures.

The function of the capping structures is still a matter of speculation and evidence for their role in assembly is discussed above. The presence of capping proteins that affect actin assembly and a capping structure at the tip of a bacterial flagellum that affects flagellin assembly is consistent with a general role of capping proteins in the regulation of cytoskeletal filament assembly. However, the role of the ciliary and flagellar capping structures and the identity of capping proteins remain to be discovered but the identification of capping proteins and the assembly of tubulin onto capped microtubules *in vitro* are under way. It is important to realize that the microtubule capping structures are only attached to the central and A tubules and no capping structure has been identified at the end of any B tubule so a complete description of the role of the capping structures may only shed light on the assembly of the A and central microtubules. Clearly, there is still much to be learned about the regulation of ciliary and flagellar microtubule assembly.

In order to understand the function of the bridges connecting the sides of microtubules to the ciliary and flagellar membrane, we must develop better methods to preserve the bridges for electron microscopy. A better understanding of the structure of the membrane is also essential. A reasonable argument can be made for the presence of a "membrane skeleton" composed of proteins associated with the cytoplasmic face of the membrane can be made. However, the structure of the membrane skeleton and the identification of proteins associated with the skeleton remain to be made. Relatively little is known about the cytoplasmic face of ciliary membranes but it will be important to better characterize its structure if we are to understand the way microtubules associate with the membrane and with flagellar surface structures or particles.

These are perhaps the most accessible areas of study and the ones in which the most progress is being made at this time. Another major problem that remains to be explored is the nature of the interactions of basal bodies with the membrane, with particular emphasis on the events that trigger the assembly of ciliary microtubules by basal body microtubules. It is clear that basal body microtubules must attach to the membrane before they can nucleate microtubule assembly but the mechanisms that regulate this are completely unknown although it may require the presence of membrane-associated capping structures or the unmasking of hydrophobic proteins that cap free basal bodies. Understanding the events that occur upon docking of the basal body with the membrane may reveal an important mechanism used to regulate specific microtubule assembly.

Note Added in Proof

Sandoz and colleagues recently reviewed linkages between basal bodies of growing cilia and the ciliary membrane (Sandoz et al., 1988). They also emphasized the importance of the association of membrane vesicles with basal bodies by showing that taxol treatment of quail oviduct could result in the nucleation of "9 + 2" axonemes onto cytoplasmic basal bodies but only if the basal bodies were associated with a membrane vesicle (Boisvieux-Ulrich et al., 1988). Thus, the association of membrane with the microtubule tips or the sides of the basal bodies is essential for the nucleation of ciliary microtubules.

We recently discovered that the microtubule capping structures at the tips of ciliary and flagellar microtubules share epitopes with antibodies that specifically recognize kinetochores in mammalian cells (Miller et al., 1988, 1989). A 97-kDa polypeptide uniquely fractionates with the ciliary capping structures and antibodies that recognize this polypeptide stain the ciliary capping structures and mammalian kinetochores. These results support our proposal that the capping structures share properties with the kinetochore and that it may be involved with the regulation of ciliary and flagellar microtubule growth (see Section 3).

ACKNOWLEDGMENT. This work was supported by USPHS Grant GM 32556 from the National Institute of General Medical Sciences.

References

Afzelius, B. A., 1961, The fine structure of cilia from ctenophore swimming plates, *J. Biophys. Biochem. Cytol.* **9**:383–394.

Afzelius, B. A., 1971, Fine structure of the spermatozoon of *Tubularia larynx*, *J. Ultrastruct. Res.* **37**:679–689.

Afzelius, B. A., 1983, The spermatozoon of *Myzostomum cirriferum* (Annelida, Myzostomida), *J. Ultrastruct. Res.* **83**:58–68.

Allen, C. A., and Borisy, G. G., 1974, Structural polarity and directional growth of microtubules of *Chlamydomonas* flagella, *J. Mol. Biol.* **90**:391–402.

Allen, R. D., 1968, A reinvestigation of cross-sections of cilia, *J. Cell Biol.* **37**:825–831.

Amos, L. A., 1977, Arrangement of high molecular weight associated proteins on purified mammalian brain microtubules, *J. Cell Biol.* **72**:642–654.

Anderson, R. G. W., and Brenner, R. M., 1971, The formation of basal bodies (centrioles) in the rhesus monkey oviduct, *J. Cell Biol.* **50**:10–34.

Anderson, R. G. W., and Hein, C. E., 1977, Distribution of anionic sites on the oviduct ciliary membrane, *J. Cell Biol.* **72**:482–492.

Anderson, W. A., and Ellis, R. A., 1965, Ultrastructure of *Trypanosoma lewisi*. Flagellum, microtubules, and the kinetochore, *J. Protozool.* **12**:483–499.

Andrivon, C., Brugerolle, G., and Delachambre, D., 1983, A specific Ca^{2+}-ATPase in the ciliary membrane of *Paramecium tetraurelia*, *Biol. Cell.* **47**:351–364.

Arima, T., Shibata, Y., and Yamamoto, T., 1984, A deep etching study of the guinea pig tracheal cilium with special reference to the ciliary transitional region, *J. Ultrastruct. Res.* **89**:34–41.

Arima, T., Shibata, Y., and Yamamoto, T., 1985, Three-dimensional visualization of basal body structures and some cytoskeletal components in the apical zone of tracheal ciliated cells, *J. Ultrastruct. Res.* **93**:61–70.

Auclair, W., and Siegel, B. W., 1966, Cilia regeneration in the sea urchin embryo: Evidence for a pool of ciliary proteins, *Science* **154**:913–915.

Baccetti, B., Burrini, A. G., Dallai, R., and Pallini, V., 1979, The dynein electrophoretic bands in axonemes naturally lacking the inner or the outer arm, *J. Cell Biol.* **80**:334–340.

Bardele, C. F., 1981, Functional and phylogenetic aspects of the ciliary membrane: A comparative freeze-fracture study, *BioSystems* **14**:403–421.

Baugh, L., Satir, P., and Satir, B., 1976, A ciliary membrane Ca^{++} ATPase: A correlation of structure and function, *J. Cell Biol.* **70**:66a.

Beisson, J., and Sonneborn, T. M., 1965, Cytoplasmic inheritance of the organization of the cell cortex in *Paramecium aurelia*, *Proc. Natl. Acad. Sci. USA* **53**:275–282.

Bessen, M., Fay, R. B., and Witman, G. B., 1980, Calcium control of waveform in isolated flagellar axonemes of *Chlamydomonas*, *J. Cell Biol.* **86**:446–455.

Binder, L. I., Dentler, W. L., and Rosenbaum, J. L., 1975, Assembly of chick brain tubulin onto flagellar microtubules from *Chlamydomonas* and sea urchin sperm, *Proc. Natl. Acad. Sci. USA* **72**:1122–1126.

Bloodgood, R. A., 1977, Motility occurring in association with the surface of the *Chlamydomonas* flagellum, *J. Cell Biol.* **75**:983–989.

Bloodgood, R. A., 1981, Flagella-dependent gliding motility in *Chlamydomonas, Protoplasma* **106**:183–192.

Bloodgood, R. A., 1987, Glycoprotein dynamics in the *Chlamydomonas* flagellar membrane, *Adv. Cell Biol.* **1**:97–130.

Boisvieux-Ulrich, E., Sandoz, D., and Chailley, B., 1977, A freeze-fracture and thin section study of the ciliary necklace in quail oviduct, *Biol. Cell.* **30**:245–252.

Boisvieux-Ulrich, E., Laine, M. C., and Sandoz, D., 1985, The orientation of ciliary basal bodies in quail oviduct is related to the ciliary beating cycle commencement, *Biol. Cell.* **55**:147–150.

Boisvieux-Ulrich, E., Laine, M. C., and Sandoz, D., 1988, *In vitro* effects of taxol on ciliogenesis in quail oviduct, *J. Cell Sci.* **92**:9–20.

Bouck, G. B., 1971, The structure, origin, isolation, and composition of the tubular mastigonemes of the *Ochromonas* flagellum, *J. Cell Biol.* **50**:362–384.

Bouck, G. B., 1972, Architecture and assembly of mastigonemes, in: *Advances in Cell and Molecular Biology* (E. J. DuPraw, ed.), Academic Press, New York, pp. 237–271.

Bouck, G. B., Rogalski, A., and Valaitis, A., 1978, Surface organization and composition of *Euglena*. II. Flagellar mastigonemes, *J. Cell Biol.* **77**:805–826.

Brugerolle, G., Andrivon, C., and Bohatier, J., 1980, Isolation, protein pattern, and enzymatic characterization of the ciliary membrane of *Paramecium tetraurelia*, *Biol. Cell.* **37**:251–260.

Burn, P., 1988, Amphitropic proteins: A new class of membrane proteins, *Trends Biochem. Sci.* **13**:79–83.

Chailley, B., N'Diaye, A., Boisvieux-Ulrich, E., and Sandoz, D., 1981, Comparative study of the distribution of fuzzy coat, lectin receptors, and intramembrane particles of the ciliary membrane, Eur. J. Cell Biol. **25:** 300–307.

Child, F. M., 1978, The elongation of cilia and flagella: A model involving antagonistic growth-zones, in: *Cell Reproduction: in Honor of Daniel Mazia* (E. R. Dirksen, D. M. Prescott, and C. F. Fox, eds.), Academic Press, New York, pp. 351–358.

Cordier, A. C., 1975, Ultrastructure of the cilia of thymic cysts in ''nude'' mice, *Anat. Rec.* **122:**227–250.

Crouau, Y., 1980, Comparison of a new structure associated with the membrane of 9 + 0 cilia of chordotonal sensilla with the central structure of motile cilia and flagella, *Biol. Cell.* **39:**349–352.

Dalen, H., 1986, An ultrastructural study of the tracheal epithelium of the guinea pig with special reference to the ciliary structure, *J. Anat.* **136:**47–67.

Dallai, R., and Afzelius, B. A., 1982, On zipper lines or particle arrays within the plasma membrane of hemipteran spermatozoa, *J. Ultrastruct. Res.* **80:**197–205.

Dallai, R., Burighel, P., and Martinucci, G. B., 1985, Ciliary differentiations in the branchial stigmata of the ascidian *Diplosoma listerianum, J. Submicrosc. Cytol.* **17:**381–390.

De Montrion, C. M., 1986, Zipper lines in the flagellar plasma membrane of the squid (Loligo) spermatozoon, *Tissue Cell* **18:**251–265.

Dentler, W. W., 1977, Fine structural localization of phosphatases in cilia and basal bodies of *Tetrahymena pyriformis, Tissue Cell* **9:**209–222.

Dentler, W. L., 1980a, Microtubule–membrane interactions in cilia. I. Isolation and characterization of *Tetrahymena* ciliary membranes, *J. Cell Biol.* **84:**364–380.

Dentler, W. L., 1980b, Structures linking the tips of ciliary and flagellar microtubules to the membrane, *J. Cell Sci.* **42:**207–220.

Dentler, W. L., 1981a, Microtubule–membrane interactions in cilia and flagella, *Int. Rev. Cytol.* **72:**1–47.

Dentler, W. L., 1981b, Microtubule–membrane interactions in ctenophore swimming plate cilia, *Tissue Cell* **13:**197–208.

Dentler, W. L., 1984, Attachment of the cap to the central microtubules of *Tetrahymena* cilia, *J. Cell Sci.* **66:** 167–173.

Dentler, W. L., 1986, Isolation of capped cilia from bovine trachea and the effect of caps on microtubule assembly *in vitro, J. Cell Biol.* **103:**279a.

Dentler, W. L., 1987, Cilia and flagella, *Int. Rev. Cytol. Suppl.* **17:**391–456.

Dentler, W. L., 1988, Fractionation of *Tetrahymena* ciliary membranes with Triton X-114 and the identification of a ciliary membrane ATPase, *J. Cell Biol.* **107:**2679–2688.

Dentler, W. L., and LeCluyse, E. L., 1982a, Microtubule capping structures at the tips of tracheal cilia: Evidence for their firm attachment during ciliary bend formation and the restriction of microtubule sliding, *Cell Motil.* **2:**549–572.

Dentler, W. L., and LeCluyse, E. L., 1982b, The effects of structures attached to the tips of tracheal ciliary microtubules on the nucleation of microtubule assembly *in vitro, Cell Motil. Suppl.* **1:**13–18.

Dentler, W. L., and Rosenbaum, J. L., 1977, Flagellar elongation and shortening in *Chlamydomonas.* III. Structures attached to the tips of flagellar microtubules and their relationship to the directionality of microtubule assembly, *J. Cell Biol.* **74:**747–759.

Dentler, W. L., Granett, S., Witman, G. B., and Rosenbaum, J. L., 1974, Directionality of brain microtubule assembly *in vitro, Proc. Natl. Acad. Sci. USA* **71:**1710–1714.

Dentler, W. L., Pratt, M. M., and Stephens, R. E., 1980, Microtubule–membrane interactions in cilia. II. Photochemical cross-linking of bridge structures and the identification of a membrane-associated ATPase, *J. Cell Biol.* **84:**381–403.

Dirksen, E. R., 1971, Centriole morphogenesis in developing ciliated epithelium of mouse oviduct, *J. Cell Biol.* **51:**286–302.

Dirksen, E. R., and Crocker, T. T., 1965, Centriole replication in differentiating ciliated cells of mammalian respiratory epithelium. An electron microscopic study, *J. Microsc. (Oxford)* **5:**629–644.

Dirksen, E. R., and Satir, P., 1972, Ciliary activity in the mouse oviduct as studied by transmission and scanning electron microscopy, *Tissue Cell* **4:**389–404.

Doughty, M. J., 1978, Ciliary Ca^{2+}-ATPase from the excitable membrane of *Paramecium.* Some properties and purification by affinity chromatography, *Comp. Biochem. Physiol.* **60B:**339–345.

Doughty, M. J., and Kaneshiro, E. S., 1985, Divalent cation-dependent ATPase activities in ciliary membranes and other surface structures in *Paramecium tetraurelia:* Comparative in vitro studies, *Arch. Biochem. Biophys.* **238:**118–128.

Enders, G. C., Werb, Z., and Friend, D. S., 1983, Lectin binding to guinea pig sperm zipper particles, *J. Cell Sci.* **60:**303–329.

Fawcett, D. W., and Phillips, D. M., 1969, The fine structure and development of the neck region of the mammalian spermatozoon, *Anat. Rec.* **165:**153–184.

Fay, R. B., and Witman, G. B., 1977, The localization of flagellar ATPases in *Chlamydomonas reinhardtii, J. Cell Biol.* **75:**286a.

Frisch, D., and Farbman, A. I., 1968, Development of order during ciliogenesis, *Anat. Rec.* **162:**221–232.

Gilula, N. B., and Satir, P., 1972, The ciliary necklace: A ciliary membrane specialization, *J. Cell Biol.* **53:** 494–509.

Gorbsky, G. J., Sammak, P. J., and Borisy, G. G., 1988, Microtubule dynamics and chromosome motion visualized in living anaphase cells, *J. Cell Biol.* **106:**1185–1192.

Hard, R., and Rieder, C. J., 1983, Mucociliary transport in newt lungs: The ultrastructure of the ciliary apparatus in isolated epithelial sheets and in functional Triton-extracted models, *Tissue Cell* **15:**227–243.

Heath, I. B., and Darley, W. M., 1972, Observations on the ultrastructure of the male gametes of *Biddulphia levis* Ehr., *J. Phycol.* **8:**51–59.

Heuser, J. E., 1981, Quick freeze, deep-etch preparation of samples for 3-D electron microscopy, *Trends Biochem. Sci.* **6:**64–68.

Hill, F. G., and Outka, D. E., 1974, The structure and origin of mastigonemes in *Ochromonas minute* and *Monas* sp., *J. Protozool.* **21:**299–312.

Hogan, J. C., and Patton, C. L., 1976, Variation in intramembrane components of *Trypanosoma brucei* from intact and X-irradiated rats: A freeze-cleave study, *J. Protozool.* **23:**205–215.

Holley, M. C., and Afzelius, B. A., 1986, Alignment of cilia in immotile-cilia syndrome, *Tissue Cell* **18:**521– 529.

Hoops, H. J., and Witman, G. B., 1985, Basal bodies and associated structures are not required for normal flagellar motion or phototaxis in the green alga *Chlorogonium elongatum, J. Cell Biol.* **100:**297–309.

Horst, C. J., Forestner, D. M., and Besharse, J. C., 1987, Cytoskeletal–membrane interactions: A stable interaction between cell surface glycoconjugates and doublet microtubules of the photoreceptor connecting cilium, *J. Cell Biol.* **105:**2973–2988.

Huang, B. P.-H., 1986, *Chlamydomonas reinhardtii:* A model system for the genetic analysis of flagellar structure and motility, *Int. Rev. Cytol.* **99:**181–215.

Huitorel, P., and Kirschner, M. W., 1988, The polarity and stability of microtubule capture by the kinetochore, *J. Cell Biol.* **106:**151–159.

Hyams, J. S., 1982, The *Euglena* paraflagellar rod: Structure, relationship to other flagellar components and preliminary biochemical characterization, *J. Cell Sci.* **55:**199–210.

Ikeda, T., Asakura, S., and Kamiya, R., 1985, "Cap" on the tip of *Salmonella* flagella, *J. Mol. Biol.* **184:**735– 737.

Jarvik, J. W., and Chojnacki, B., 1985, Flagellar morphology in stumpy-flagella mutants of *Chlamydomonas reinhardtii, J. Protozool.* **32:**649–656.

Jeffery, P. K., and Reid, L., 1975, New observations of rat airway epithelium. A quantitative and electron microscopic study, *J. Anat.* **120:**295–320.

Justine, J.-L., and Mattei, X., 1982, Reinvestigation de l'ultrastructure du spermatozoide d'*Haematoloechus* (Trematoda: Haematoloechidae), *J. Ultrastruct. Res.* **81:**322–332.

Justine, J.-L., and Mattei, X., 1985, A spermatozoon with undulating membrane in a parasitic flatworm, *Gotocotyla* (Monogena, Polyopisthocotylea, Gotocotylidae), *J. Ultrastruct. Res.* **90:**163–171.

Kamiya, R., and Okamoto, M., 1985, A mutant of *Chlamydomonas reinhardtii* that lacks the flagellar outer dynein arm but can swim, *J. Cell Sci.* **74:**181–191.

Kuhn, C., and Engleman, W., 1978, The structure of the tips of mammalian respiratory cilia, *Cell Tissue Res.* **186:**491–498.

Lansing, A., and Lamy, F., 1961, Localization of ATPase in rotifer cilia, *J. Cell Biol.* **11:**498–501.

LeCluyse, E. L., and Dentler, W. L., 1984, Asymmetric microtubule capping structures at the tips of frog palate cilia, *J. Ultrastruct. Res.* **86:**75–85.

Lefebvre, P. A., and Rosenbaum, J. L., 1986, Regulation of the synthesis and assembly of ciliary and flagellar proteins during regeneration, *Annu. Rev. Cell Biol.* **2:**517–546.

Lewin, R. A., 1952, Studies on the flagella of algae. I. General observations on *Chlamydomonas moewusii* Gerloff, *Biol. Bull. (Woods Hole, Mass.)* **103:**74–79.

Lewin, R. A., 1982, A new kind of motility mutant (non-gliding) in *Chlamydomonas*, *Experientia* **38:**348–349.

Lewin, R. A., and Lee, K. W., 1985, Autotomy of algal flagella: Electron microscope studies of *Chlamydomonas* (Chlorophyceae) and *Tetraselmis* (Prasinophyceae), *Phycologia* **24:**311–316.

L'Hernault, S. W., and Rosenbaum, J. L., 1983, *Chlamydomonas* α-tubulin is posttranslationally modified in the flagella during flagella assembly, *J. Cell Biol.* **97:**258–263.

McIntosh, J. R., 1973, The axostyle of *Saccinobaculus*. II. Motion of the microtubule bundle and a structural comparison of straight and bent axostyles, *J. Cell Biol.* **56:**324–339.

Manton, I., and Von Stosch, H. A., 1966, Observations on the fine structure of the male gamete of the marine centric diatom *Lithodesmium undulatum*, *J. R. Microsc. Soc.* **85:**119–134.

Marchand, B., and Mattei, X., 1976, La spermatogenese des *Acanthocephales*, *J. Ultrastruct. Res.* **54:**347–358.

Marchese-Ragona, S. P., and Johnson, K. A., 1985, Localization of 14S and 22S dyneins in *Tetrahymena* cilia, *J. Cell Biol.* **101:**278a.

Markey, D. R., and Bouck, G. B., 1977, Mastigoneme attachment in *Ochromonas*, *J. Ultrastruct. Res.* **59:**173–177.

Martinucci, G. B., Dallai, R., and Burighel, P., 1987, A comparative study of ciliary differentiations in the branchial stigmata of ascidians, *Tissue Cell* **19:**251–263.

Menco, B. P., 1980, Qualitative and quantitative freeze-fracture studies on olfactory and respiratory epithelial surfaces of frog, ox, rat, and dog. IV. Ciliogenesis and ciliary necklaces (including high-voltage observations), *Cell Tissue Res.* **212:**1–16.

Mesland, D. A. M., Hoffman, J. L., Caligor, E., and Goodenough, U. W., 1980, Flagellar tip activation stimulated by membrane adhesions in *Chlamydomonas* gametes, *J. Cell Biol.* **84:**599–617.

Michison, T., Evans, L., Schultze, E., and Kirschner, M. W., 1986, Sites of microtubule assembly and disassembly in the mitotic spindle, *Cell* **45:**515–527.

Miki-Noumura, T., and Kamiya, R., 1979, Conformational change in the outer doublet microtubules from sea urchin sperm flagella, *J. Cell Biol.* **81:**355–360.

Miller, J. M., Wang, W., and Dentler, W. L., 1988, Microtubule capping structures from *Tetrahymena* cilia contain antigens found in kinetochores, *J. Cell Biol.* **107:**29a.

Miller, J. M., Wang, W., Balczon, R., and Dentler, W. L., 1989, CREST antisera bind to microtubule capping structures in *Tetrahymena* cilia, *J. Cell Biol.* (in press).

Mitchell, D. R., and Rosenbaum, J. L., 1985, A motile *Chlamydomonas* flagellar mutant that lacks outer dynein arms, *J. Cell Biol.* **100:**1228–1234.

Moestrup, O., 1972, Observations on the fine structure of spermatozoids and vegetative cells of the green alga *Golenkinia*, *Br. Phycol. J.* **7:**169–183.

Montesano, R., Didier, P., and Orci, L., 1981, The ciliary junction: A unique membrane specialization in the ciliate, *Glaucoma ferox*, *J. Ultrastruct. Res.* **77:**360–365.

Moran, D. T., Carela, F. J., and Rowley, J. C., III, 1977, Evidence for active role of cilia in sensory transduction, *Proc. Natl. Acad. Sci. USA* **74:**793–797.

Murray, J. M., 1984a, Three-dimensional structure of a membrane–microtubule complex, *J. Cell Biol.* **98:**283–295.

Murray, J. M., 1984b, Disassembly and reconstitution of a membrane–microtubule complex. *J. Cell Biol.* **98:**1481–1487.

Musgrave, A., DeWildt, P., van Etten, I., Pijst, H., Scholma, C., Kooyman, R., Homan, W., and van den Ende, H., 1986, Evidence for a functional membrane barrier in the transition zone between the flagellum and cell body of *Chlamydomonas eugametos* gametes, *Planta* **167:**544–553.

Nagano, T., 1965, Localization of adenosine triphosphatase activity in the rat sperm tail as revealed by electron microscopy, *J. Cell Biol.* **25:**101–112.

Norwood, J. T., and Anderson, R. G. W., 1980, Evidence that adhesive sites on the tips of oviduct cilia membranes are required for ovum pickup in situ, *Biol. Reprod.* **23:**788–791.

Omoto, C. K., and Witman, G. B., 1981, Functionally significant central pair rotation in a primitive eukaryotic flagellum, *Nature* **290:**708–710.

Owen, G., and McCrae, J. M., 1979, Sensory cell/gland cell complexes associated with the pallial tentacles of the bivalve *Lima hians* (gmelin), with a note on specialized cilia on the pallial curtains, *Philos. Trans. R. Soc. London Ser. B* **287**:45–62.

Perkins, L. A., Hedgecock, E. M., Thompson, J. N., and Culotti, J. G., 1986, Mutant sensory cilia in the nematode *Caenorhabditis elegans, Dev. Biol.* **117**:456–487.

Piperno, G., and Fuller, M. T., 1985, Monoclonal antibodies specific for an acetylated form of a tubulin recognize the antigen in cilia and flagella from a variety of organisms, *J. Cell Biol.* **101**:2085–2094.

Pitelka, D. R., 1974, Basal bodies and root structures, in: *Cilia and Flagella* (M. A. Sleigh, ed.), Academic Press, New York, pp. 437–469.

Pollard, T. D., and Cooper, J. A., 1986, Actin and actin-binding proteins. A critical evaluation of mechanisms and functions, *Annu. Rev. Biochem.* **55**:987–1035.

Porter, M. E., and Johnson, K. E., 1983, Characterization of the ATP-sensitive binding of *Tetrahymena* 30 S dynein to bovine brain microtubules, *J. Biol. Chem.* **258**:6575–6581.

Portman, R., and Dentler, W. L., 1987, The development of microtubule capping structures in ciliated epithelia, *J. Cell Sci.* **87**:85–94.

Portman, R. P., and Dentler, W. L., 1989, Development of orientation of basal bodies during ciliogenesis in palate epithelium, *Cell Motil. Cytoskeleton* (in press).

Randall, J., 1969, The flagellar apparatus as a model organelle for the study of growth and morphopoiesis, *Proc. R. Soc. London Ser. B* **173**:31–62.

Reed, W., Avolio, J., and Satir, P., 1984, The cytoskeleton of the apical border of the lateral cells of freshwater mussel gill: Structural integration of microtubule and actin filament-based organelles, *J. Cell Sci.* **68**: 1–33.

Reinhart, F. D., and Bloodgood, R. A., 1988, Membrane–cytoskeleton interactions in the flagellum: A 240,000Mr surface-exposed glycoprotein is tightly associated with the axoneme in *Chlamydomonas moewusii, J. Cell Sci.* **89**:521–531.

Ringo, D. L., 1967, Flagellar motility and fine structure of the flagellar apparatus in *Chlamydomonas, J. Cell Biol.* **33**:543–571.

Rosenbaum, J. L., Moulder, J. E., and Ringo, D. L., 1969, Flagellar elongation and shortening in *Chlamydomonas, J. Cell Biol.* **41**:600–619.

Roth, L. E., and Shigenaka, Y., 1964, The structure and formation of cilia and filaments in rumen protozoa, *J. Cell Biol.* **20**:249–270.

Sale, W., and Satir, P., 1977a, The termination of the central microtubules from the cilia of *Tetrahymena pyriformis, Cell Biol. Int. Rep.* **1**:56–63.

Sale, W. S., and Satir, P., 1977b, Direction of active sliding of microtubules in *Tetrahymena* cilia, *Proc. Natl. Acad. Sci. USA* **74**:2045–2049.

Sandoz, D., Boisvieux-Ulrich, E., and Chailley, B., 1979, Relationships between intramembrane particles and glycoconjugates in the ciliary membrane of the quail oviduct, *Biol. Cell.* **39**:267–279.

Sandoz, D., Chailley, B., Boisvieux-Ulrich, E., Lemullois, M., Laine, M. C., and Bautista-Harris, G., 1988, Organization and functions of cytoskeleton in metazoan ciliated cells, *Biol. Cell.* **63**:183–193.

Satir, P., 1976, Local design of membranes in relation to cell function, *6th Eur. Congr. Electron Microsc.* Jerusalem, pp. 41–44.

Sattler, C. A., and Staehelin, L. A., 1974, Ciliary membrane differentiation in *Tetrahymena pyriformis. Tetrahymena* has four types of cilia, *J. Cell Biol.* **62**:473–490.

Schneider, A., Lutz, H. U., Marugg, R., Gehr, P., and Seebeck, T., 1988, Spectrin-like proteins in the paraflagellar rod structure of *Trypanosoma brucei, J. Cell Sci.* **90**:307–316.

Sleigh, M. A., and Silvester, N. R., 1983, Anchorage functions of the basal apparatus of cilia, *J. Submicrosc. Cytol.* **15**:101–104.

Snell, W. J., Dentler, W. L., Haimo, L., Binder, L. I., and Rosenbaum, J. L., 1974, Assembly of chick brain tubulin onto isolated basal bodies of *Chlamydomonas reinhardtii, Science* **185**:357–360.

Sorokin, S. P., 1968, Reconstructions of centriole formation and ciliogenesis in mammalian lungs, *J. Cell Sci.* **3**:207–230.

Steinman, R. M., 1968, An electron microscopic study of ciliogenesis in developing epidermis and trachea in the embryo of *Xenopus laevis, Am. J. Anat.* **122**:19–56.

Stephens, R. E., 1977, Major membrane protein differences in cilia and flagella: Evidence for a membrane-associated tubulin, *Biochemistry* **16**:2047–2058.

Stephens, R. E., 1985, Ciliary membrane tubulin and associated proteins: A complex stable to Triton X-114 dissociation, *Biochim. Biophys. Acta* **821:**413–419.

Stephens, R. E., Oleszko-Szuts, S., and Good, M. J., 1987, Evidence that tubulin forms an integral membrane skeleton in molluscan gill cilia, *J. Cell Sci.* **88:**527–535.

Stossel, T. P., Chaponnier, C., Ezzel, R. M., Hartwig, J. H., Janmey, P. A., Kwiatkowski, D. J., Lind, S. E., Smith, D. B., Southwick, F. S., Yin, H. L., and Zaner, K. S., 1985, Nonmuscle actin-binding proteins, *Annu. Rev. Cell Biol.* **1:**353–402.

Suprenant, K. A., and Dentler, W. L., 1988, Release of intact microtubule capping structures from *Tetrahymena* cilia, *J. Cell Biol.* **107:**2259–2269.

Tamm, S. L., and Tamm, S., 1985, Visualization of changes in ciliary tip configuration caused by sliding displacement of microtubules in macrocilia of the ctenophore *Beroë, J. Cell Sci.* **79:**161–179.

Tamm, S. L., and Tamm, S., 1988, Development of macrociliary cells in *Beroë*. II. Formation of macrocilia, *J. Cell Sci.* **89:**81–95.

Toh, Y., 1981, Fine structure of sense organs on the antennal pedicel and scape of the male cockroach *Periplaneta americana, J. Ultrastruct. Res.* **77:**119–132.

Toh, Y., 1985, Structure of campaniform sensilla on the haltere of *Drosophila* prepared by cryofixation, *J. Ultrastruct. Res.* **93:**92–100.

Toh, Y., and Yokohari, F., 1985, Structure of the antennal chordotonal sensilla of the American cockroach, *J. Ultrastruct. Res.* **90:**124–134.

Travis, S. M., and Nelson, D. L., 1986, Characterization of Ca^{2+}- or Mg^{2+}-ATPase of the excitable ciliary membrane from *Paramecium tetraurelia:* A comparison with a soluble Ca^{2+}-dependent ATPase, *Biochim. Biophys. Acta* **862:**39–48.

Trimmer, J. S., and Vacquier, V. D., 1988, Monoclonal antibodies induce the translocation, patching, and shedding of surface antigens of sea urchin spermatozoa, *Exp. Cell Res.* **175:**37–51.

Tyler, S., 1973, An adhesive function for modified cilia in interstitial turbellarian, *Acta Zool.* **54:**139–151.

Tyler, S., 1979, Distinctive features of cilia in metazoans and their significance for systematics, *Tissue Cell* **11:**385–400.

Vale, R. D., and Toyoshima, Y. Y., 1988, Rotation and translocation of microtubules in vitro induced by dyneins from *Tetrahymena* cilia, *Cell* **52:**459–469.

Vickerman, K., 1969, On the surface coat and flagellar adhesion in trypanosomes, *J. Cell Sci.* **5:**163–193.

Warner, F. D., 1974, The fine structure of the ciliary and flagellar axoneme, in: *Cilia and Flagella* (M. A. Sleigh, ed.), Academic Press, New York, pp. 11–38.

Williams, N. E., Doerder, F. P., and Ron, A., 1985, Expression of a cell surface immobilization antigen during serotype transformation in *Tetrahymena thermophila, Mol. Cell Biol.* **5:**1925–1932.

Williams, N. E., Honts, J. E., and Jaeckel-Williams, R. F., 1987, Regional differentiation of the membrane skeleton in *Tetrahymena, J. Cell Sci.* **87:**457–463.

Witman, G. B., 1975, The site of in vivo assembly of flagellar microtubules, *Ann. N. Y. Acad. Sci.* **253:**178–191.

Woolley, D. M., and Fawcett, D. W., 1973, The degeneration and disappearance of the centrioles during the development of the rat spermatozoon, *Anat. Rec.* **177:**289–302.

Woolley, D. M., and Nickels, S. N., 1985, Microtubule termination patterns in mammalian sperm flagella, *J. Ultrastruct. Res.* **90:**221–234.

Euglena gracilis
A Model for Flagellar Surface Assembly, with Reference to Other Cells That Bear Flagellar Mastigonemes and Scales

G. B. Bouck, T. K. Rosiere, and P. J. Levasseur

1. Introduction

The flagella of euglenoids consist of a rich assemblage of surface appendages precisely positioned with respect to each other and to axonemal components. The presence of these appendages (mastigonemes or flagellar hairs) and the occurrence of flagellar scales in certain "green" flagellates provide visible evidence that the flagellar surface in these forms is constructed and maintained to a large extent independently from the adjacent cell surface. Thus, flagellar assembly may require both the specific targeting of some proteins to and the exclusion of other proteins from the flagellar surface. In the case of some green flagellates, for example, there may be two layers of geometrically arranged scales on the flagella, each layer may consist of a different kind of scale, the layers do not intermix, and the flagellar scales differ markedly from the scales (up to three different kinds) found on the cell body. Organisms with scales and mastigonemes have served as useful models in helping to understand how flagellar surfaces are assembled and how surface molecules are stabilized or held in position along a presumably fluid flagellar membrane.

    This chapter will focus on flagellar surface organization, synthesis, and assembly in the euglenoids—particularly *Euglena gracilis*—with reference where appropriate to other flagellates. *Euglena* shares with *Chlamydomonas* many of the virtues of unicellular organisms for studying flagellar behavior; i.e., it is readily cultured axenically and has a single emergent flagellum which after amputation will readily regenerate within a reasonable period of time. A significant asset of working with euglenoids is that in addition to the

G. B. Bouck, T. K. Rosiere, and P. J. Levasseur • Department of Biological Sciences, University of Illinois, Chicago, Illinois 60680.

flagellar membrane, the adjoining plasma membrane can be isolated in pure form. This permits a direct comparison of the properties of two separate domains of the same continuous membrane system—a possibility not easily realized in most other flagellates. The apparent stability of the euglenoid flagellar surface and the absence of any demonstrable surface motility which might reorient flagellar proteins also favor these cells as models for exploring surface assembly.

Three basic questions will be addressed in this chapter: (1) What is the organization of the *Euglena* flagellar surface? (2) How is this surface stabilized against shedding, random rearrangement, and the centrifugal forces of the beating flagellum? (3) How is the flagellar surface assembled during flagellar development? Much of the earlier work on flagella of "lower" flagellates has been carefully and systematically reviewed by Moestrup (1982).

2. The Relationship of Euglenoids to Other Organisms

First it is relevant to ask where euglenoids fit in the phylogenetic spectrum, i.e., how do they relate to other organisms? The phylogenetic origin and systematic position of *Euglena* and its allies (along with many other lower eukaryotes) have been extensively discussed and reviewed in recent years. Euglenoids have been categorized by various authors as protozoa, algae, protists, or lower plants. On two points, however, there seems to be general agreement: (1) As a group the euglenoids have "a formidable array of clear cut diagnostic features" (Leedale, 1978) and, therefore, despite their "varied cytology, metabolism and ecology," they can probably be considered as a natural grouping or phylum (see below). (2) The group with closest affinity to euglenoids may not be the green photosynthetic forms (e.g., Moestrup, 1982) but rather the trypanosomes and bodonids, collectively known as the Kinetoplastida (reviewed in Willey *et al.*, 1988). This long-suspected relationship has been supported by evidence of cytological similarities (Kivic and Walne, 1984), buoyed by arguments that chloroplasts of euglenoids, when present, may be secondary acquisitions (Gibbs, 1978), and placed on still firmer grounds by recent RNA sequence data (Gunderson *et al.*, 1987; Sogin *et al.*, 1986; Wolters and Erdmann, 1986). Phylogenies based solely on ribosomal small subunit (16 S) RNA sequence similarities and differences (Fig. 1) indicate that *Euglena* and a trypanosome share more genetic similarities than a legitimate green form, *Chlamydomonas*, which appears to be only distantly related. A cladistic analysis incorporating 16 S and 5 S rRNA sequence data as well as flagellar properties affirms that euglenoids and trypanosomes stand apart from other groups (Wolters and Erdmann, 1986). These results add additional support to the earlier proposal that euglenoids and kinetoplastids together be designated as a supraphylum assemblage (Corliss, 1984), the Euglenozoa (Cavalier-Smith, 1982).

Are there comparable similarities between the flagella of trypanosomes and euglenoids? At the level of the flagellar surface the trypanosomes exhibit little that is directly comparable to the elaborate mastigonemes found in the euglenoids, although "simple" mastigonemes may be present on more primitive members of the Kinetoplastida (Mylnikov, 1986; see Moestrup, 1982, for review of earlier studies). A more convincing analogue that is common to flagella of both euglenoids and trypanosome-like organisms is the

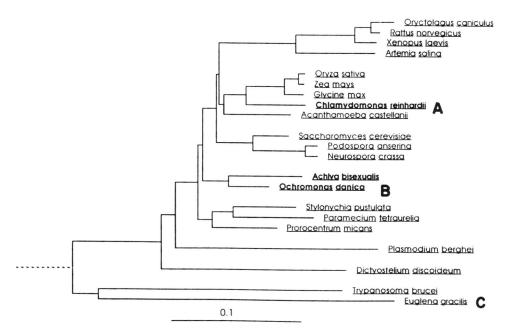

Figure 1. Phylogenetic tree inferred from 16 S ribosomal RNA similarities. Note that in this scheme, *Euglena* (C), *Ochromonas* (B) = a chromophyte, and *Chlamydomonas* (A) = a green flagellate are each positioned on separate branches and do not share extensive sequence similarities. Reproduced with permission from Gunderson *et al.* (1987).

paraxial rod. Paraxial rods of the two groups will be considered in more detail below, particularly with respect to their important role in anchoring euglenoid mastigonemes.

3. Flagellar Anatomy

Summarized here are details of flagellar surface organization and evidence that this organization is closely tied to specific axonemal components.

3.1. Mastigonemes and the Flagellar Sheath

Surface appendages of *Euglena gracilis* are of two types: long (≥ 3 μm) relatively simple mastigonemes unilaterally aligned along the long axis of the flagellum, and a series of complex, filamentous mastigoneme units (Bouck *et al.*, 1978) which collectively form a sheath, partially enveloping the flagellum (Bouck and Rogalski, 1982; Melkonian *et al.*, 1982). The sheath is not continuous laterally around the flagellum but appears as two half-sheaths separated by an unequal distance on either side of the flagellum (seen particularly clearly in the euglenoid *Peranema;* Hilenski and Walne, 1985). Despite the lack of sheath continuity and the probability that each half-sheath is anchored to a different internal component, there appears to be alignment of the tiers of mastigoneme units of one half-sheath with tiers of units of the other half-sheath (Melkonian *et al.*, 1982). This suggests that the two half-sheaths must be coordinately assembled or, alternatively, there may be continuity between the two half-sheaths during early flagellar development followed by later separation. The sheaths lie completely outside the flagellar membrane, but removal of the membrane with detergents (Bouck *et al.*, 1978) or organic solvents (Rogalski and Bouck, unpublished observations) does not detach the mastigonemes. Thus, unlike the scales of some green flagellates which appear to be electrostatically bonded to the flagellar surface (Melkonian, 1982), the mastigonemes of *Euglena* must either pass through the flagellar membrane and bind to the axoneme (here defined as the detergent-resistant components internal to the flagellar membrane) or, more likely, join the axoneme through intermediary links.

That the two half-sheaths of mastigonemes are attached to different axonemal components is particularly evident after flagella are extracted with detergent (Bouck *et al.*, 1978; Bouck and Rogalski, 1982). Some of these attachments can also be detected in sectioned untreated flagella (Melkonian *et al.*, 1982). The mastigoneme units of one half-sheath are bound to what has been termed the paraxial ribbon (Bouck and Rogalski,

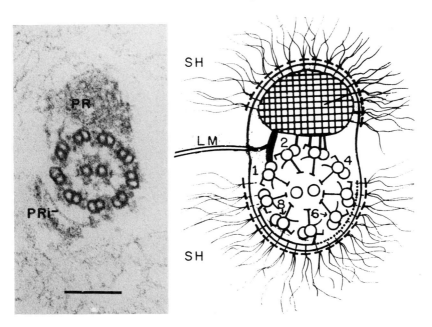

Figure 2. (Left) Section through the flagellum of *E. gracilis*. In this preparation the flagellar membrane was removed with the neutral detergent NP-40. Both the paraxial rod (PR) and the paraxial ribbon (PRi) remain attached to axonemal doublet microtubules. Bar = 100 nm. From Bouck and Rogalski (1982).

Figure 3. (Right) Diagrammatic representation of the flagellum of *E. gracilis*. One half-sheath is associated with the paraxial rod; the other half-sheath is attached in part to the paraxial ribbon. Dotted lines indicate regions where the paraxial ribbon and mastigoneme attachment sites have not been clearly documented. The heavy linker between the paraxial rod and doublet No. 1 is believed to be the "gobletlike" extensions from the paraxial rod seen in longitudinal section in Fig. 5. LM, unilateral row of long mastigonemes. Redrawn and modified from Melkonian *et al.* (1982).

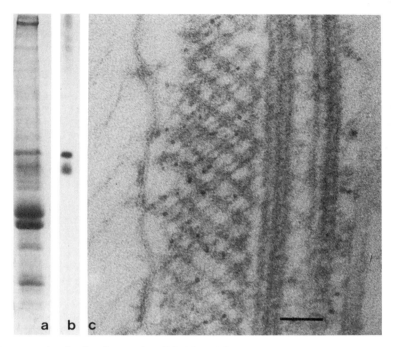

Figure 4. Demonstration that flagellar proteins of 77, 72, and 69 kDa are associated with the paraxial rod in *E. gracilis*. (a) Coomassie blue-stained acrylamide gel of flagellar polypeptides. (b) Similar gel transferred to nitrocellulose and incubated with the monoclonal antibody 1D4. An radioiodinated secondary antibody was used for detection. The monoclonal antibody recognized three bands of 77, 72, and 69 kDa. (c) The same monoclonal was incubated with whole flagella and the flagella were then treated with ferritin-conjugated secondary antibody. Ferritin particles are associated mostly with elements of the paraxial rod. Bar = 50 nm.

1982), or electron-dense "sheath" (Melkonian *et al.*, 1982). The paraxial ribbon is itself linked to axonemal doublet microtubule No. 8 (see below for numbering), probably No. 7, and possibly No. 9 in detergent-treated flagella (Figs. 2, 3). Additional links to No. 6 as diagrammed (Melkonian *et al.*, 1982) are problematic. The width of a paraxial ribbon in flagella sectioned either intact or after detergent treatment is consistently less than the width of the half-sheath of mastigonemes which it subtends. Thus, a portion of the ribbon is either labile to the preparative methods used or the ribbon does not extend the full width of the sheath and a portion of the sheath must therefore be anchored directly to an axonemal doublet microtubule (No. 6).

The other half-sheath of mastigonemes is attached directly or indirectly to the paraxial rod (Bouck *et al.*, 1978; Bouck and Rogalski, 1982) which lies offset about 30–45° counterclockwise (when viewed from the flagellar tip) from a position directly opposite the visible portion of the paraxial ribbon. The paraxial rod in *Euglena* extends most of the length of the flagellum; it is constructed of 22-nm filaments arranged in a seven-start helix with a pitch of 45° and has a periodicity of 54 nm (Hyams, 1982). "Gobletlike" extensions from the paraxial rod bind to a doublet microtubule (Hyams, 1982) which would appear to be microtubule No. 1. Microtubule doublet No. 1 has been so designated because of this prominent attachment to the paraxial rod (Melkonian *et al.*, 1982) and

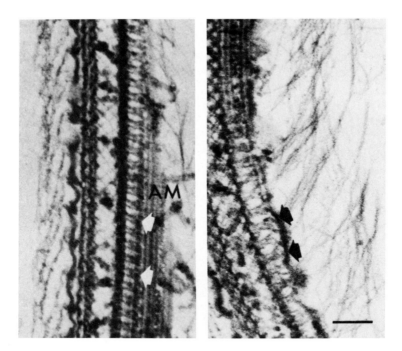

Figure 5. Gobletlike extensions of the paraxial rod are associated both with axonemal doublet microtubules (white arrows) and with the flagellar membrane (black arrows). Note that in the right panel the long mastigonemes are included in the same section, suggesting that their attachment sites are close to or at the gobletlike extensions from the paraxial rod (see also Fig. 3). Micrographs courtesy of Dr. A. A. Rogalski. Bar = 100 nm.

because of its proximity and linkage to the set of long, unilaterally arranged mastigonemes. The paraxial rod/microtubule doublet linkage is about 80 nm long as are the gobletlike extensions in longitudinal section (Fig. 5). Moreover, the latter sections often show close association of the 80-nm linkers with the flagellar membrane (Fig. 5). This membrane association would preclude an 80-nm linker attachment at doublets Nos. 2 and 3 which are some distance from the membrane, and implicates doublet No. 1 which is adjacent to the membrane. There are also shorter, fine linkages between the paraxial rod and axonemal doublets Nos. 2 and 3, and these seem to be stable to detergent extraction (Hyams, 1982).

It would be useful to know how the paraxial rod is oriented relative to the axis of flagellar bend, but the twist of the central pair of microtubules and the absence of doublet markers or missing dynein arms, which are useful markers in other cells (Hoops and Witman, 1983), preclude at present a determination of the plane of effective thrust. In the scaly green flagellate *Pyramimonas octopus*, Hori and Moestrup (1987) have recently been able to relate the bilateral rows of mastigonemes with the absolute plane of flagellar beat. They found that the longitudinal rows of mastigoneme (hair scale) attachment sites are slightly displaced from a line passing through the axis of the flagellum. It will be of considerable interest to see if rows of mastigonemes in other cells have the same relationship to the effective stroke, and can therefore be used to determine absolute doublet orientation.

3.2. The Paraxial Rod

The similarity between the paraxial rod of trypanosomids and euglenoids noted in a number of earlier studies (e.g., Russell *et al.*, 1983; reviewed in Moestrup, 1982; Willey *et al.*, 1988) has been verified to some extent by biochemical and immunological evidence (Gallo and Schrével, 1985). In the latter studies, mice were immunized with a Nonidet P-40-insoluble fraction from *Trypanosoma brucei,* and hybridomas were constructed from the immune spleen lymphocytes. One cloned hybridoma cell line produced monoclonal antibodies which were shown by immunofluorescence to bind to the trypanosome paraxial rod, and also recognized two polypeptide bands at 75 and 72 kDa in immunoblots of trypanosome flagellar proteins. Interestingly, the same monoclonal antibody also identified two polypeptide bands with slightly different sizes (76 and 67 kDa) in immunoblots of flagellar proteins of *E. gracilis*. On the basis of their prominence and their sensitivity to trypsin, Hyams (1982) had earlier suggested that a pair of polypeptide bands which migrated at 80 and 68 kDa were paraxial proteins of *E. gracilis*. The two bands were designated PFR1 and PFR2 (paraflagellar rod 1 and 2), which are probably more appropriately termed PR1 and PR2 (paraxial rod 1 and 2) in accord with Moestrup's (1982) reasoning. Table I summarizes the molecular mass of PR proteins as determined in several studies of trypanosomes and *E. gracilis*. In the trypanosomid *Crithidia fasciculata*, PR1 and PR2 have identical isoelectric points and display a single isoform in equilibrium IEF gels (Russell *et al.*, 1983; see, however, Russell *et al.*, 1984, for small differences in isoelectric points). PR1 and PR2 seem to be related to each other as indicated by the cross-reactivity of both bands with a single monoclonal antibody (Fig. 4; see also Gallo and Schrével, 1985, who used a different monoclonal antibody). Peptide mapping would help resolve the extent of the similarities and differences between the two PR bands but this has yet to be carried out. Russell *et al.* (1983) have pointed out some of the major differences between the PRs of euglenoids and trypanosomids. Particularly notable is the hollow cylindrical organization of the *E. gracilis* PR as opposed to the paracrystalline fibrous organization of the PR of trypanosomes. In some euglenoids, however, one of the two flagella houses a PR which appears to be paracrystalline and not hollow (Mignot, 1966; Walne *et al.*, 1986), but the composition and immunological properties of these para-

Table I. Reported Relative Mobilities of Paraxial Rod (PR) Polypeptides in *Euglena* and Trypanosomes

Organism	PR[a]	PR1	PR2	Reference
Euglena gracilis	+	80 kDa	69 kDa	Hyams (1982)
	+	76 kDa	67 kDa	Gallo and Schrével (1985)
	+	77 kDa	72, 69 kDa[b]	Rosiere, Levasseur, and Bouck
	+	75 kDa	66, 64 kDa[c]	(unpublished)
Trypanosoma brucei	+	75 kDa	72 kDa	Gallo and Schrével (1985)
Herpetomonas megaseliae	+	78 kDa	73 kDa	Cunha *et al.* (1984)
Crithidia fasciculata	+	76 kDa	68 kDa	Russell *et al.* (1983)
Crithidia oncopelta	−	[d]		Gallo and Schrével (1985)
Crithidia deanei	−	—	—	Cunha *et al.* (1984)

[a]Paraxial rod present (+) or absent (−).
[b]Determined from immunoblots using a monoclonal antibody.
[c]Determined from immunoblots and gels using polyclonal antibodies.
[d]PR antibody weakly recognizes an 87-kDa polypeptide in immunoblots.

crystalline PRs are not known. It would be premature to rely too heavily on the immuno-logical data or discount the small differences in molecular mass or structure of PRs from varied sources. Nonetheless, these results taken together encourage further assessment of possible homologies of this unique flagellar component found only in euglenoids, try-panosomids, and a few other flagellates (see Moestrup, 1982).

3.3. The Flagellar Membrane

Four more or less distinct regions of the continuous surface membrane can be identified in *Euglena:* the cell body membrane (referred to here as the plasma membrane for ease of discussion), the canal membrane, the reservoir membrane, and the flagellar membrane (reviewed in Bouck, 1982). Do these regionally separated membranes have different compositions?

3.3.1. Membrane Proteins. Both the plasma membrane and the flagellar mem-brane can be isolated as parts of larger complexes in which they are the only membrane present. For example, after gentle cavitation of deflagellated cells a surface complex is released which consists of the plasma membrane, the membrane skeleton, microtubules, and assorted interconnecting bridgework (Hofmann and Bouck, 1976; Murray, 1984; Dubreuil and Bouck, 1985). The surface complex can be purified on sucrose or Percoll gradients and the plasma membrane then separated from peripheral proteins. When the membrane is compared to the flagellar membrane, it is evident that it differs markedly in composition and organization: (1) The major integral membrane protein present in the plasma membrane is a 39-kDa monomer, its 68-kDa-dimer, or higher molecular weight oligomers (Dubreuil *et al.,* 1988). The flagellar membrane seems to lack these proteins as judged by the absence of anti-39-kDa antibody binding in immunoblots of whole flagella, and by the absence of a diagnostic shift of 68-kDa polypeptides to 39-kDa polypeptides when the SDS gels are run in the presence of small amounts of Nonidet P-40 (Dubreuil *et al.,* 1988). (2) The flagellar membrane is coated with a layer of fine 10-nm filaments (Fig. 6) which has been equated with a flagellar glycoprotein termed *xyloglycorein*. This glycoprotein is not detectable in SDS gels of cell bodies (Rogalski and Bouck, 1980) or surface isolates, nor is a surface coat visible on the plasma membrane of surface isolates of *E. gracilis* (Dubreuil and Bouck, 1985). (3) The membrane skeleton which underlies the plasma membrane in euglenoids (Dubreuil and Bouck, 1985) is absent in flagella. Moreover, reconstitution of the membrane skeleton on plasma membranes is readily accomplished (Dubreuil and Bouck, 1988), whereas membrane skeletal proteins do not bind to flagellar membranes (Rosiere and Bouck, 1987), indicating a lack of appropriate membrane skeletal binding sites in flagella. (4) After freeze-fracturing, the plasma mem-brane displays characteristic striations on the EF (ectoplasmic face) and noncomplemen-tary particles on the PF (protoplasmic face; Miller and Miller, 1978; Lefort-Tran *et al.,* 1980). Replicas of fractured flagellar membranes revealed no striations on the EF and rows of particles on the PF (Melkonian *et al.,* 1982).

The sum of all these observations clearly indicates that the flagellar membrane and plasma membrane of *E. gracilis* are structurally distinct and have significantly different integral and peripheral membrane proteins.

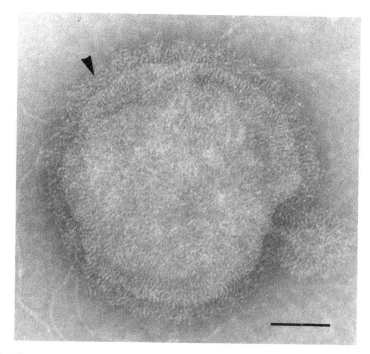

Figure 6. Flagellar membrane vesicle from *E. gracilis* showing surface filaments (arrowhead). The filaments have been equated with a flagellum-specific glycoprotein, "xyloglycorein." From Rogalski and Bouck (1980) with permission of Rockefeller University Press. Bar = 500 nm.

3.3.2. Membrane Lipids. Flagellar lipids of *E. gracilis* have been extracted with chloroform/methanol and separated by thin-layer chromatography (TLC) on silica gel plates (Rogalski, 1981). Four spots are clearly visible after exposure of the chromatogram to iodine vapors (Fig. 7). The two most polar spots can be equated with phosphatidyl-serine (PS) and phosphatidylethanolamine (PE) on the basis of their comigration with standards and on the basis of their reaction with ninhydrin. Two additional spots migrating closer to the solvent front (less polar) are tentatively identified as steryl glucosides based on the following evidence: (1) both spots stain with orcinol which detects carbohydrates, and (2) the most nonpolar spot comigrates with a cholesterol standard, and cholesterol as well as both spots react with ferric chloride (Fig. 7). The presence of sterols in the flagellar membrane has also been suggested from filipin binding experiments (Melkonian *et al.*, 1982), and from evidence for sterol glycosyltransferase activity in isolated flagella of *Euglena* (see below).

In consideration of the clear differences in protein composition between the flagellar membrane and the plasma membrane summarized above, it might be expected that lipid composition would also differ. As determined by Nakano *et al.* (1987), the plasma (cell body) membrane of *E. gracilis* consists of 12% neutral (nonpolar) lipids of which 7.8% are sterols. Melkonian *et al.* (1982) found no sterol–filipin complexes in the plasma

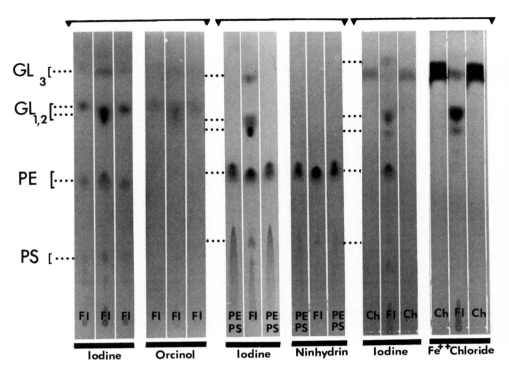

Figure 7. TLC separation of flagellar lipids and glycolipids. Portions of three chromatograms (brackets at top) are shown. Iodine was used for visualizing all classes of lipids, orcinol for glycolipids, ninhydrin for aminolipids, and ferric chloride for sterol-like lipids. Lane labels: F1, whole flagella; PE, phosphatidylethanolamine; PS, phosphatidylserine; CH, cholesterol. GL_3 and $GL_{1,2}$ are presumed to be glycolipids, probably steryl glucosides based on their staining reactions. Dotted lines link similar regions in the three separate chromatograms. Courtesy of Dr. A. A. Rogalski.

membrane. Exogenously added sugar nucleotides, however, are incorporated into a product of cell surfaces which migrates on TLC plates with properties similar to those of a steryl glucoside (Chen and Bouck, 1984). When polar lipids from plasma membranes were separated on two-dimensional TLC plates and the fatty acid composition of these lipids determined by gas–liquid chromatography (Nakano *et al.*, 1987), it was found that about 42% of the polar lipids consisted of phosphatidylglycerol and the more predominant PE. A number of less abundant phosphatides were reported, but no PS was found. This apparent difference in PS content of flagella (abundant) and that of plasma membranes of cell surfaces (absent) must be confirmed by simultaneous analysis of both membrane systems using identical methods before the significance of these findings can be evaluated. In *Chlamydomonas*, Gealt *et al.* (1981) found identical sterols in flagella and whole cells, but there were significant differences in fatty acid composition. Unfortunately, the plasma membrane cannot be readily separated from other membranes in these cells (Bloodgood, 1987), and the estimates of whole cell lipids in *Chlamydomonas* may not be indicative of plasma membrane composition.

3.3.3. Flagellar Membrane Enzymes. Relatively few enzymes have been found to be specifically associated with the flagellum, reinforcing the general view that most flagellar components arrive preassembled or in an assembly-competent form from the cell body. In *Chlamydomonas*, five enzymes involved with nucleotide transformations as well as a number of dyneinlike ATPases have been identified (reviewed in Huang, 1986). Maruta *et al.* (1986) have reported that a 67-kDa polypeptide is the monomer of a flagellum-associated α-tubulin acetyltransferase. Enzymes specifically associated with the flagellar membrane are listed in Table II. Evidence for guanylate cyclase activity has come mostly from sea urchin sperm flagella in which the enzyme was thought to be associated with egg/sperm reactions. Reports of a similar enzyme in cilia of *Tetrahymena* (Schultz *et al.*, 1983) and *Paramecium* (Klumpp and Schultz, 1982) would seen to extend the potential functions of this important regulatory enzyme.

Of the various glycosyltransferases previously reported in flagellar surfaces of *Chlamydomonas* and *Euglena*, only one in *E. gracilis* has been partially characterized (Chen and Bouck, 1984; Bouck and Chen, 1984): (1) Addition of UDP-[³H]glucose to isolated flagella resulted in incorporation into a fraction soluble in chloroform/methanol. (2) Autoradiograms of SDS gels of flagella incubated in UDP-[³H]glucose showed that most of the label migrated to the dye front—a region to which glycolipids characteristically migrate. The same region of the gel was strongly PAS-positive. (3) When UDP-[³H]glucose was added to a detergent (CHAPS) extract of flagella in the presence of

Table II. Enzymes Associated with the Flagellar and/or Ciliary Membrane

Enzyme	Substrate	Organism	Reference
Guanylate cyclase	GTP	*Arbacia*	Ward *et al.* (1985, 1986)
	GTP	*Chaetopteris*	Gray and Drummond (1976)
	GTP	*Hemicentrotus*	Sano (1976)
	GTP	*Lytechinus*	Gray and Drummond (1976)
	GTP	*Paramecium*	Schultz and Klumpp (1980), Klumpp and Schultz (1982), Klumpp *et al.* (1984)
	GTP	*Strongylocentrotus*	Gray and Drummond (1976)
	GTP	*Tetrahymena*	Schultz *et al.* (1983)
Adenylate cyclase	ATP	*Chlamydomonas*	Pasquale and Goodenough (1987)
	ATP	*Paramecium*	Schultz and Klumpp (1983), Klumpp *et al.* (1984), Gustin and Nelson (1987), Schultz *et al.* (1987)
Cyclic nucleotide phosphodiesterase	cGMP	*Hemicentrotus*	Sano (1976)
Mg²⁺/Ca²⁺-dependent ATPase	ATP	*Paramecium*	Doughty (1978), Travis and Nelson (1986)
	ATP	*Tetrahymena*	Dentler (1977, 1988)
Ca²⁺-dependent ATPase	ATP	*Chlamydomonas*	Bessen *et al.* (1980)
	ATP	*Paramecium*	Brugerolle *et al.* (1980), Andrivon *et al.* (1983), Doughty and Kaneshiro (1985)
Protein kinase	Protein	*Paramecium*	Lewis and Nelson (1980)
Glycosyltransferases	Protein	*Chlamydomonas*	McLean and Bosmann (1975), Köhle *et al.* (1980)
Xylosyltransferases	Lipids	*Euglena*	Geetha-Habib and Bouck (1982)
Glucosyltransferase	Lipids	*Euglena*	Chen and Bouck (1984)

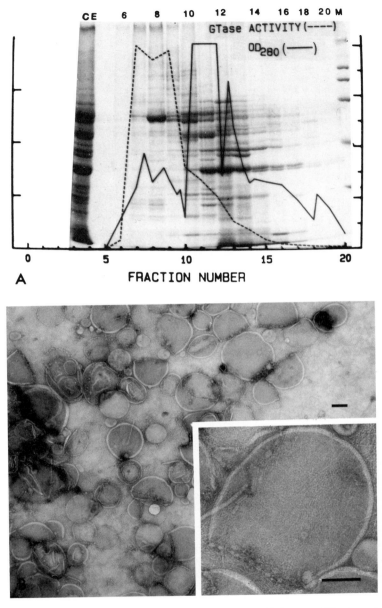

Figure 8. (A) CHAPS extract (CE) of *Euglena* flagella separated by FPLC on a gel filtration column. Column fractions were resolved on an SDS acrylamide gel and tested for glycosyltransferase activity using cholesterol as a substrate. The activity profile (dotted line) showed most of the activity to be in fractions 7, 8, and 9, but the bulk of the proteins (OD_{280}, solid line), were collected in later fractions. Lane M shows molecular weight standards: 205, 116, 97, 66, 45, 29 kDa. (B) Flagellar membrane vesicles prepared by alkali extraction of isolated flagella of *E. gracilis*. These preparations retained lipid glycosyltransferase activity. The helmetlike profiles (inset) are common. Bar = 100 nm.

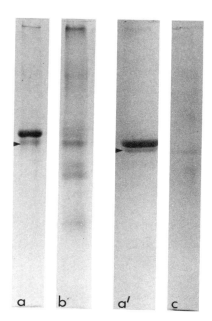

Figure 9. Comparison of the polypeptides from flagellar fractions with lipid glycosyltransferase activity. Lanes a and a' are the active fraction from the gel filtration column; lane b is a CHAPS extract from isolated flagellar membrane vesicles; lane c is the active fraction eluted from a phenyl Sepharose (hydrophobic) affinity column. Arrowheads indicate the common band in all three active fractions.

an endogenous lipid substrate, a lipophilic fraction became radiolabeled. (4) The *in vivo* (whole flagella) or the *in vitro* (CHAPS extract) labeled product migrated on silica gel plates in appropriate solvents to the region expected of a steryl glucoside. (5) When cholesterol was added as an enzyme substrate, incorporation of UDP-[^{3}H]glucose into a lipophilic product was greatly enhanced (Rosiere and Bouck, unpublished). All these data taken together strongly suggest that *Euglena* flagella have an endogenous membrane-associated glycosyltransferase which is capable of catalyzing the transfer of glucose to a sterol-like membrane lipid.

Isolating this enzyme has proven to be difficult. Earlier results (Chen and Bouck, 1984) based on dial-[^{3}H]-UDP binding as an affinity label implicated a 32-kDa protein, but an antibody specific to this polypeptide did not immunoprecipitate or affect enzyme activity (Rosiere and Bouck, unpublished). More recently, activity has been subfractionated from CHAPS extracts of flagella using a variety of chromatographic methods. Especially useful has been a sizing column in conjunction with FPLC which separates the bulk of the CHAPS-soluble proteins from enzyme activity (Fig. 8A). SDS gel analysis of the active fractions yields a prominent polypeptide band at ∼ 68 kDa and a slightly faster migrating minor band (Fig. 9a,a'). When a similar CHAPS extract containing enzyme activity was loaded on a phenyl Sepharose (hydrophobic) affinity column, the bulk of the proteins passed through the column. All the activity, however, remained bound and could be eluted (albeit with reduced activity) with 0.32% NP-40. Acrylamide gels of the eluate revealed an enrichment of a number of relatively minor polypeptides (Fig. 9b), one of which coincided with the minor band from the sizing column. In an effort to reduce the number of initial CHAPS-soluble proteins, attempts were made to isolate membrane vesicles prior to extraction. In these experiments, isolated flagella were first treated with

NaOH to remove all peripheral (axonemal) and soluble proteins, and then the NaOH-resistant membranes were separated from debris and soluble proteins on a sucrose gradient. As seen in Fig. 8B, these membranes were still contaminated with mastigonemes, but surprisingly they retained substantial enzyme activity as did a CHAPS extract from such membrane preparations. By comparing the polypeptides remaining in the active fraction obtained from gel filtration with the polypeptides separated by hydrophobic affinity chromatography and those present in a CHAPS extract of isolated flagellar membrane vesicles, it is evident that there is only one band in common (Fig. 9). We tentatively conclude that this 64-kDa polypeptide is the monomeric form of a flagellar steryl glucosyltransferase.

An important question remains as to whether the flagellar membrane enzyme is a unique flagellar component or a leftover from a cell-surface enzyme inadvertently or passively trapped in the developing flagellum. Several lines of evidence suggest that the enzyme is flagellum-specific: (1) Although enzymes with similar substrate and acceptor requirements are present on cell surfaces, they have different cation sensitivities for stimulation and inhibition (see Bouck and Chen, 1984). (2) If acquisition of a flagellar glucosyltransferase was the result of random trapping, presumably other kinds of surface enzymes should also be present in the flagellar membrane. Plasma membrane glycosyltransferases glucosylate endogenous proteins as readily as they glucosylate lipids—probably indicating a greater variety of glucose-utilizing enzymes on the cell surface. Yet under conditions in which 50% of the UDP-[^{3}H]glucose is incorporated into TCA-precipitable labeled products (proteins, glycoproteins) in plasma membranes, virtually no TCA-precipitable products were labeled in isolated flagella (Chen and Bouck, 1984). (3) Several different radiolabeled lipophilic products have been isolated from both flagella and cell surfaces after labeling with UDP-[^{3}H]glucose. One product from each source chromatographs similarly, but another product separates in a different fraction of a lipophilic LH-60 column and has a different M_r on TLC plates (Chen and Bouck, 1984). The sum of these observations suggests that at least some of the lipid glucosyltransferases of flagella are probably different from those of the plasma membrane, and are not therefore remnants of imperfect membrane segregation.

4. Assembly of the Flagellar Surface

How are surface molecules transported to and assembled on the flagellar surface? A general model for flagellar surface assembly will first be presented and then specific data supporting this model will be discussed. The supporting data are derived from different kinds of cells, since no one system illustrates every facet of the model. Mastigoneme and scale ontogeny are combined only to demonstrate how flagellar surfaces *can* be generated from Golgi-derived materials and delivered to surface pools for subsequent redistribution.

The study of flagellar assembly has been facilitated by the discovery that in many organisms, flagellar amputation is followed by flagellar regeneration under a variety of experimental conditions (reviewed in Lefebvre and Rosenbaum, 1986). Assuming that the kinetics and regulation of assembly in these artificially deflagellated cells reflect normal development (see, however, Coggin and Kochert, 1986), a composite model of flagellar development would involve three different assembly patterns: (1) axonemal microtubule doublet assembly at the distal (tip), fast-growing end (reviewed in Lefebvre and Rosenbaum, 1986); (2) central pair microtubule assembly at the proximal transition region of the flagellum (reviewed in Dentler, 1981); (3) flagellar surface recruitment from the proximal flagellar region. We will consider here only the last process.

Several lines of evidence support the hypothesis that the flagellar surface is assembled at least in part from material available near the flagellar base. Perhaps the most convincing observation is the finding of assembled mastigonemes and scales at or near the flagellar base. In pulse–chase experiments in *Euglena* using either radiolabeled surface moieties (Geetha-Habib and Bouck, 1982) or immunologically tagged surface components (Rogalski and Bouck, 1982), it is evident that there is a transfer of surface material from the basal (proximal) to the more distal flagellar regions. A second supporting argument for basal accretion of the flagellar surface is less direct and is based on the numerous observations showing that in many cells, mastigonemes and flagellar scales are assembled within the Golgi complex (reviewed in Moestrup, 1982). This observation eliminates other potential models of assembly in these cells. For example, Golgi-associated mastigonemes and scales are clearly not assembled *in situ* on the flagellar surface, and therefore do not follow the pattern of tip assembly which characterizes growth of the axonemal doublet microtubules. It is also unlikely that scales and mastigonemes are carried to the flagellar tip inside the growing flagellum since this would pose a formidable transport problem within the narrow confines of the axoneme and, equally importantly, such transport has not been observed.

Flagellar regeneration in *Euglena* following amputation is somewhat different than flagellar assembly during cell division. Mechanical, pH, or cold shock-induced amputation always occurs at a site about 6 μm distal to the basal body/flagellum junction. By contrast, during cell division the proximal flagellar region which also harbors the crystalline paraflagellar body (the presumptive photoreceptor) must be fully assembled. This proximal region appears to lack the organized arrays of mastigoneme units found elsewhere on the flagellum, and the distribution of intramembrane particles likewise does not match the pattern found in the distal region (Melkonian *et al.*, 1982). An appealing working model of surface assembly of the *Euglena* flagellum proposes that mastigonemes are first deposited or assembled on the reservoir membrane adjacent to the flagellum, and then are rearranged on the flagellum using membrane "templates" identified as intramembrane particles on the EF (external face) of the proximal region of the flagellum (Melkonian *et al.*, 1982) to form the two half-sheaths which are displaced distally as the axonemal microtubules elongate. The process of distributing mastigonemes and flagellar scales could follow the same general scheme in other organisms. Departures from the *Euglena* model would include obvious differences in mastigoneme and scale organization and attachment, and the presence or absence of reservoirlike surface pools. The evidence for this or a similar type of model will be discussed in more detail below.

4.1. Origin of Flagellar Scales, Mastigonemes, and Membranes

Evidence for the presence of flagellar scales and mastigonemes within the compartments of the Golgi apparatus and other intramembrane compartments has been reviewed (Moestrup, 1982). Not uncommonly, a *single* Golgi cisternum may contain as many as four different scale types as well as mastigonemes (Moestrup and Walne, 1979). Within the more proximal (*cis*) regions of the Golgi stack, an individual cisternum may contain partially formed scales, leading to the unmistakeable impression that there is a gradient of scale maturation from proximal (*cis*) to distal (*trans*) regions. This form of cisternal progression as marked by their contents of mastigonemes or scales is difficult to reconcile with models of Golgi apparatus function based on vesicular transport from cisternum to cisternum in a more or less stationary stack (for detailed discussions, see Robinson and Kristen, 1982; Farquhar, 1985; McFadden and Melkonian, 1986). Whether scale and mastigoneme production by cisternal progression (membrane flow) is a "rare formula" or a particularly clear case of a general phenonemon remains to be seen.

Although the Golgi apparatus has been implicated in the processing of scales and tubular mastigonemes in numerous studies, the origin of nontubular mastigonemes (defined in Bouck, 1972) such as those found in *E. gracilis* has remained uncertain. Either these mastigonemes are more labile to preparative methods or mastigonemes are not fully formed before their release to the reservoir surface. Equally uncertain is the origin of the flagellar membrane itself which has been presumed to be at least partially derived from the Golgi membrane vesicles which contain the mastigonemes and/or scales. Estimates of the amount of cisternal membrane which potentially can be released from the Golgi complex during scale production are at first glance quite staggering. In the scaly green flagellate *Scherffelia dubia*, the two Golgi complexes produce a total of 54,400 scales which are distributed among four flagella. Assuming an average of 31 scales/cisternum, it has been calculated that during scale release there would be a sufficient cisternal membrane available to cover 127.5 flagella (McFadden and Melkonian, 1986). Thus, at best only 3.1% of the Golgi derived membrane could be incorporated into the flagellar membrane in these cells. McFadden and Melkonian (1986) have proposed, however, that perhaps small vesicles containing the scales and/or mastigonemes are blebbed from the *trans* (mature face) cisternae and transported to the cell surface, a mechanism similar to the *trans*-Golgi network hypothesis of Griffiths and Simons (1986), whereas the bulk of the cisternal membrane is recycled. Thus, the amount of membrane actually available for flagellar incorporation may be closer to the amount that can be utilized.

4.2. Composition of Scales and Mastigonemes

Flagellar scales are said to be "unmineralized" apparently on the basis of extrapolation of data obtained from surface theca (Manton *et al.*, 1973), and by the transparency of scales to the electron beam (Moestrup, 1982). Rod-shaped scales from the flagella of the scaly green flagellate *Tetraselmis striata* were resistent to protease and had Con A-binding sites; pentagonal scales from flagella of the same organism were protease-sensitive and also bound Con A. A family of 10–12 glycopeptides (50–70 kDa) were extracted as a minor component of pentagonal scales (Melkonian *et al.*, 1985). The scale-derived theca of *Tetraselmis* was recently analyzed by mass spectrometry and gas–liquid chromatography (Becker *et al.*, 1989).*

Mastigonemes of *Chlamydomonas* migrate in SDS gels as a single glycoprotein with

---

*The main constituent (42%) of the theca was the acidic compound 3-deoxy-*manne*-2-oculosonic acid (KDO). A methylated KDO was also identified as a minor component (7%).

an estimated molecular mass of 170 kDa (Witman *et al.*, 1972) or 230 kDa (Monk *et al.*, 1983). In the chromophyte *Ochromonas* (see Fig. 1), tubular mastigonemes have been separated on SDS gels into six well-defined glycoproteins of about 51, 64, 100, 115, 176, and 255 kDa. Antibodies raised against these glycopolypeptides in individual gel bands were found to extensively cross-react, but peptide maps of each band using either chymotrypsin or *Staphylococcus aureus* V-8 protease digestion showed no similarities in one-dimensional polyacrylamide gels (Kawano and Bouck, 1984). Immunolocalization with ferritin-conjugated antibodies suggested that the 100- and 115-kDa proteins were associated with the mastigoneme shaft. In *Euglena*, the mastigoneme fraction has been separated on SDS gels into a large number of glycopeptides and polypeptides (Rogalski and Bouck, 1980) which have not yet been correlated with the long or short mastigonemes or with specific parts of the mastigoneme units. These polypeptides range in molecular mass from 20 kDa to proteins which barely enter the acrylamide gels. It can be concluded that in the three organisms for which there are some data (*Chlamydomonas, Ochromonas, Euglena*), the composition of these mastigonemes differs significantly as does mastigoneme structure. Their commonality lies in the fact that they are all firmly bound to the axoneme and yet are positioned mostly outside the flagellar membrane, and in the perhaps trivial fact that they are all glycosylated.

4.3. Release of Mastigonemes and Flagellar Scales at the Cell Surface

In *Euglena* and some scaly green flagellates, the flagella originate within a pocket-like depression of the cell surface. This pocket (reservoir) in *Euglena* is often lined with filaments resembling mastigonemes. Since there is a gap in our knowledge of the events immediately preceding the appearance of mastigonemes in the reservoir, it is postulated that some assembly step which renders mastigonemes visible or less labile to preparative methods must be taking place at the reservoir membrane. These "intermediate" stage mastigonemes constitute a surface pool which can be incorporated into the regenerating flagellum, as judged by the results of pulse–chase experiments (Geetha-Habib and Bouck, 1982; Rogalski and Bouck, 1982). In some scaly green flagellates, the Golgi-derived scales and mastigonemes collect in a "scale reservoir" which connects by a duct to the flagellar pocket (McFadden and Wetherbee, 1985). It is not clear whether the scale reservoir is totally depleted during flagellar assembly or whether a pool remains. Cycloheximide totally blocked regeneration in the scaly green flagellate *Pyramimonas gelidicola*, whereas tunicamycin only inhibited regeneration by 50% (McFadden and Wetherbee, 1985). This could be interpreted to indicate that there was a pool of glycoproteins (scales?) sufficient only to regenerate a half-length flagellum. In similar experiments with *Tetraselmis striata*, cycloheximide (0.35 µM) completely inhibited flagellar regeneration, and it was concluded that there was no pool of flagellar precursors in these cells (Reize and Melkonian, 1987). Tunicamycin, however, had relatively little effect until a second regeneration, indicating that there may indeed be sufficient glycoproteins available for one complete flagellar assembly.

In most cells the release of Golgi system-derived materials appears to be closely linked to flagellar assembly so that scales and mastigonemes are rarely observed in transition at the surface. Therefore, *Euglena* and scale-bearing flagellates that have scale reservoirs provide important experimental systems for following the transition stages from an essentially disordered collection of surface components to the genus-specific pattern characteristic of the mature flagellar surface. Further work with these systems is certainly needed. A striking and singular case of mastigonemes and mastigoneme-like "somatonemes" persisting on the flagellar and cell surfaces, respectively, has recently been reported (Brugerolle and Bardele, 1988) in the parasitic chromophyte *Proteromonas lacertae*. These cells may provide a unique system for examining the factors directing mastigonemes to the flagellar surface and (misdirecting them?) to the cell surface. Interestingly, as with tubular mastigonemes of chromophytes, the surface somatonemes are bound by a detergent-resistant connection to microtubules underlying the plasma membrane.

4.4. Organizing the Flagellar Surface

Following their release to the cell surface, mastigonemes and scales are first targeted to the flagella, and then distributed in some programmed manner to their ultimate position on the flagellar surface. In modeling surface development, we rely on two observations: (1) In most cases the distribution of mastigonemes and scales is linked to flagellar elongation (for possible exception in brown flagellates, see discussion in Moestrup, 1982), i.e., at no stage in flagellar growth does there appear to be an absence of mastigonemes or scales on the flagellar surface that is filled in subsequently. (2) The usual arrangement of tubular mastigonemes is in one or more longitudinal files which thus parallels the direction of flagellar growth and the direction of bidirectional particle transport demonstrated in other flagellates (Bloodgood, 1977, 1987). It is significant that the 240-kDa surface glycoprotein associated with particle transport in the green flagellate *Chlamydomonas moewusii* has recently been shown to remain associated with the axoneme after removal of the flagellar membrane with detergent (Reinhart and Bloodgood, 1988). Similar detergent-resistant binding to the axoneme has been demonstrated both for tubular mastigonemes of chromophytes (Markey and Bouck, 1977) and for the nontubular mastigonemes of *Euglena* (Bouck *et al.,* 1978). Therefore, the motive force of particle transport could then in theory also be harnessed for the unidirectional transport of mastigonemes during flagellar growth. Recall that the axonemal doublet microtubules grow at their distal tip, whereas mastigonemes and scales are produced at the flagellar base, necessitating a continual repositioning of the mastigonemes relative to the axoneme during flagellar elongation. The notion that at least tubular mastigonemes might track along the axonemal microtubules in a manner similar to particles is attractive. It cannot be ruled out, however, that mastigoneme transport might be the result of passive displacement as the flagellar membrane extends distally during axonemal growth (the finger in the glove model).

Positioning scales on a flagellar surface is more difficult to explain, but the process could follow the same general scheme. For example, 24 rows of small pentagonal scales are positioned around the circumference of some scaly green flagellates (Melkonian, 1982; McFadden and Wetherbee, 1985; Hori and Moestrup, 1987). Adjacent rows are staggered, thus giving the appearance of helical winding. Interestingly, four rows of rectangular scales (two each on opposite sides of the flagellum) are arranged exactly parallel to the long axis of the flagellum; mastigonemes are attached in the same region. Scales do not extend through the flagellar membrane but are superficially bound via electrostatic interactions (Melkonian, 1982). A possible sequence for scale orientation during flagellar growth might include: (1) a primary positioning by the mastigonemes with their axonemal attachment; (2) association of the rows of rectangular scales with the mastigonemes which could direct the distal flow of scales; (3) the remainder of the pentagonal scales would accompany the rows of rectangular scales by ionic interaction and a ''best-fit configuration would be produced by a process of auto-regulation'' (McFadden and Wetherbee, 1985). Yet to be explained is the positioning of the larger, outer layer of flagellar scales in their nine imbricating rows. In *Pyramimonas*, there is some evidence for preassembly of scale complexes in the scale reservoir before their appearance on the flagellum (McFadden and Wetherbee, 1985).

To extend these ideas to the *Euglena* flagellum requires an explanation of the origin of its exceptional arrangement of mastigonemes, and an explanation of how mastigonemes are translocated since most are not directly attached to axonemal doublet microtubules. It seems clear that the tiers of overlapping mastigoneme units must be organized from the apparently random collection attached to the reservoir membrane. Melkonian *et al.* (1982) have suggested that this organization may be taking place in the proximal portion of the flagellum, and they have called attention to the surprisingly different intramembrane particle distribution in that region as contrasted with the distal region of the flagellum. In particular, the particles which are helically ordered in the distal region are more or less longitudinally positioned in the proximal region, and these particles partition to the PF in the distal region and to the EF in the proximal region. The mechanisms and forces involved in the generation of helical tiers of mastigonemes from longitudinal arrays of particles are not yet evident, but future studies focusing particularly on the proximal/distal junction may help resolve this problem.

How then do the multiple mastigoneme units move distally during elongation when few are directly attached to microtubule doublets? Distal movement of mastigonemes during flagellar elongation must involve the paraxial ribbon and the PR to which they are attached. Displacement of the *entire flagellar rod and ribbon* relative to the doublet microtubules is a hypothetical possibility which could help account for apparent coordination between the mastigoneme units of the two half-sheaths. Such bulk movement of rod and ribbon would also obviate the need for a separate motor for each of the 7000 mastigoneme units estimated to be present in each flagellum (Bouck *et al.*, 1978). Moreover, the demonstrated attachment of rod and ribbon to axonemal microtubules (Fig. 3) provides a rationale for a microtubule-based motility system rather than a paraxial ribbon- and PR-based system. Displacement of the PR and ribbon relative to the axonemal doublet microtubules could be assessed by pulse labeling flagella and determining by autoradiography whether there is proximal assembly during flagellar regeneration. Proximal assembly (if demonstrated) would *require* that the paraflagellar rod and ribbon be translocated along the axonemal microtubules during flagellar elongation.

5. The Control of Flagellar Surface Assembly in *Euglena*

Flagellar regeneration kinetics in *Euglena* are similar to those of other cells and therefore presumably axonemal assembly is governed by the same principals (see review in Lefebvre and Rosenbaum, 1986). Flagellar surfaces in some cells seem to be in a state

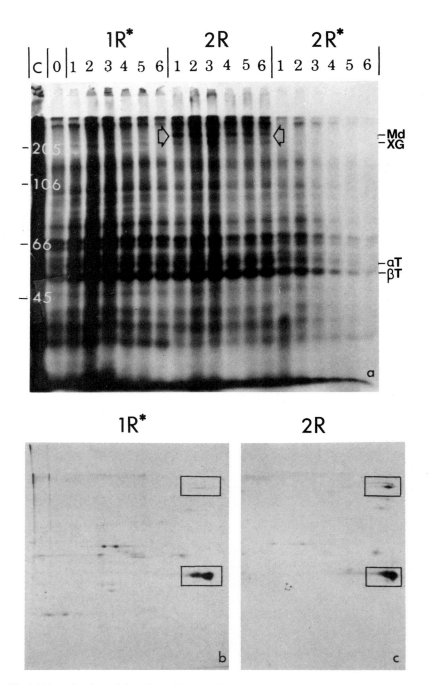

Figure 10. A high-molecular-weight polypeptide is synthesized during one regeneration cycle but not incorporated into the flagellum until the next round of regeneration. Cells of *E. gracilis* were deflagellated and allowed to regenerate flagella. Radiolabeled methionine was added in 30-min pulses during the regeneration period. After 240 min the cells were deflagellated and flagella from each pulse period were solubilized. The radiolabeled flagellar polypeptides were separated on an acrylamide gel (lanes 1R*). The deflagellated cells from the pulse labeling experiment were then allowed to regenerate flagella a second time without additional label, and after

of continual flux as evidenced by membrane loss and protein turnover (see Bloodgood, 1987). Mastigonemes may be more stable, perhaps because of their attachment to the axoneme, and in *Euglena* membranes and mastigonemes both appear to be sufficiently long-lived so that they can be tagged and followed during regeneration. Experiments with inhibitors of protein synthesis (cycloheximide) and N-linked glycosylation (tunicamycin) as well as precursor uptake studies indicate that there is a complex interplay between pools and *de novo*-synthesized flagellar surface components.

Tunicamycin (2 μg/ml) administered to deflagellated cells of *E. gracilis* had relatively little effect on flagellar regeneration although 96% of the control levels of [^{14}C]-xylose incorporation was inhibited (Geetha-Habib and Bouck, 1982). Interestingly, such treatment profoundly affected regeneration of flagella if the same treated cells were deflagellated a second time in the absence of tunicamycin. These data suggested that (1) some flagellar glycoprotein essential for flagellar growth was available in sufficient quantity to supply one complete flagellum, but depletion of this pool prevented further growth, and (2) some glycoprotein(s) synthesized during regeneration was not available for assembly during that same round of regeneration. Similar tunicamycin effects have also been observed on the regeneration of flagella of *Tetraselmis striata* (Reize and Melkonian, 1987).

To directly identify a protein or glycoprotein which might be synthesized during one round of regeneration but not utilized until a second regeneration cycle, cells of *E. gracilis* were pulse labeled with [^{35}S]methionine at 30-min intervals after flagellar amputation. After a total of 240 min of regeneration, cells were deflagallated a second time and allowed to regenerate a new flagellum in the absence of label. After an additional 240 min of regeneration, these chased flagella as well as those from the first regeneration were separately collected, solubilized, and the proteins resolved on SDS gels. Fluorograms of these gels (Fig. 10a) shows a labeled high-molecular-weight species present in the second-round regenerated flagella which was absent from the first-round regenerated flagella (Levasseur and Bouck, 1986). The period of labeling during the first regeneration seemed to make little difference on the intensity of the high-molecular-weight band. The experiment was repeated with continuous labeling during the first regeneration cycle. Fluorograms of the two-dimensional gels of the labeled flagella from the latter experiment (Fig. 10b,c) revealed a fairly narrow, high-molecular-weight streak with an isoelectric range in this system which approximated that of tubulin. It is not yet known if this band is glycosylated or tunicamycin-sensitive, but the experiments do support the hypothesis that at least one protein is not utilized, or not transported to the flagellum, during the regeneration cycle in which it is synthesized. The presence of this (these) protein(s) seems to be sufficiently critical that the flagellum cannot regenerate in its (their) absence. These results concur with the experiments showing a redistribution (and depletion) of putative glycoprotein

---

240 min these cells were redeflagellated. The flagella were solubilized, separated on an acrylamide gel, and fluorographed (lanes 2R). Note that a new high-molecular-weight polypeptide appears in the fluorogram (arrows) even though no label was added during this regeneration period. In panels b and c, cells were labeled continuously only during the first of two regeneration cycles and the flagellar proteins were separated on two-dimensional gels. The fluorogram of the first flagella regenerated shows little label in the upper boxed region (b), but a similar gel of flagella from a second regeneration cycle (c) shows a substantial increase of label in the upper boxed area. The lower box marks the position of tubulins.

pools from the *Euglena* reservoir to the flagellar surface (see above), i.e., there is a utilization and exhaustion of the surface pool (Rogalski and Bouck, 1982). These experiments also suggest that at least some newly synthesized glycoprotein is not accessible to the elongating flagellum until the primary pool is first depleted.

6. Summary and Prospectus

The remarkable complexity of the flagellar surface of *Euglena* indicates that the flagellar surface is not simply an extension of the cell surface, but is intricately constructed from at least some material specifically directed to or modified on the flagellum. In addition to their intrinsic interest, these flagellar surface-specific moieties and their underlying axonemal attachments can provide markers useful for tracing the origin, targeting, and positioning of flagellar surface components. At least two general concepts have emerged from the study of mastigonemes alone: (1) basal addition of flagellar surface elements and (2) anchorage of glycoproteins through the membrane to axonemes. These findings have invoked models of flagellar growth which require active displacement of surface and possibly even axonemal components relative to the peripheral doublet microtubules. Perhaps the need for such displacement provides a specific function for the surface motility identified in flagella of other cells (Bloodgood, 1987).

The emphasis in this chapter has been primarily on *E. gracilis* with its impressive surface architecture, but it should be noted that other euglenoids may differ in flagellar number and organization. Mastigoneme units in *Peranema,* for example, do not appear to be like those of *Euglena,* although their arrangement on the flagellum in half-sheaths is similar (Hilenski and Walne, 1985). Too little is known about the composition of mastigonemes from euglenoids to make useful biochemical comparisons among different genera.

Thus far, no experiments have been carried out on *Euglena* flagella at the molecular level. A genomic library has been constructed (Curtis and Rawson, 1981), and used for other purposes. More useful at present may be a λgtll expression library (Levasseur and Bouck, unpublished) which is being used to identify flagellum-specific cDNAs by screening with appropriate antiflagellar antibodies. Since there are several flagellum-specific proteins (e.g., PR and mastigonemal proteins) in *Euglena,* the regulation of flagellar assembly can be approached by sequencing these flagellar genes and their upstream regions. Comparative sequence analysis of DNA coding for the proteins of PRs of *Euglena* and trypanosomes may also clarify their potential homologies. The ability to carry out reverse genetics with the lambda expression library together with cell transformation might well circumvent some of the difficulties which previously have precluded direct genetic studies of *Euglena.*

ACKNOWLEDGMENTS. Previously unpublished work reported herein was supported by NSF Grant DCB-8602793 to G.B.B., a University of Illinois fellowship to P.J.L., and a fellowship to T.K.R. from the Laboratory for Cell, Molecular and Developmental Biology of the University of Illinois at Chicago.

References

Andrivon, C., Brugerolle, G., and Delachambre, D., 1983, Specific Ca^{2+}-ATPase of the ciliary membrane of *Paramecium tetraurelia, Biol. Cell.* **47**:351–364.

Becker, B., Hard, K., Melkonian, M., Kamerling, J. P., and Vliegenthart, J. F. G., 1989, Identification of 3-deoxy-*manno*-2-octulosonic acid, 3-deoxy-5-0-methyl-*manno*-2-octulosonic acid and 3-deoxy-lyxo-2-heptolosaric acid in the cell wall (theca) of the green alga *Tetraselmis striata* Butcher (prasinophyceae), *Eur. J. Biochem.* **182**:153–160.

Bessen, M., Fay, R. B., and Witman, G. B., 1980, Calcium control of waveform in isolated flagellar axonemes of *Chlamydomonas, J. Cell Biol.* **86**:446–455.

Bloodgood, R. A., 1977, Motility occurring in association with the surface of the *Chlamydomonas* flagellum, *J. Cell Biol.* **75**:983–989.

Bloodgood, R. A., 1987, Glycoprotein dynamics in the *Chlamydomonas* flagellar membrane, *Adv. Cell Biol.* **1**: 97–130.

Bouck, G. B., 1972, Architecture and assembly of mastigonemes, in: *Advances in Cell and Molecular Biology,* Volume 2 (E. J. Dupraw, ed.), Academic Press, New York, pp. 237–271.

Bouck, G. B., 1982, Flagella and the cell surface, in: *The Biology of Euglena,* Volume 3 (D. E. Buetow, ed.), Academic Press, New York, pp. 29–51.

Bouck, G. B., and Chen, S. J., 1984, Synthesis and assembly of the flagellar surface, *J. Protozool.* **31**:21–24.

Bouck, G. B., and Rogalski, A. A., 1982, Surface properties of the euglenoid flagellum, *Symp. Soc. Exp. Biol.* **35**:381–397.

Bouck, G. B., Rogalski, A., and Valaitis, A., 1978, Surface organization and composition of *Euglena.* II. Flagellar mastigonemes, *J. Cell Biol.* **77**:805–826.

Brugerolle, G., and Bardele, C. F., 1988, Cortical cytoskeleton of the flagellate *Proteromonas lacertae:* Interrelation between microtubules, membrane and somatonemes, *Protoplasma* **142**:46–54.

Brugerolle, G., Andrivon, C., and Bohatier, J., 1980, Isolation, protein pattern and enzymatic characterization of the ciliary membrane of *Paramecium tetraurelia, Biol. Cell.* **37**:251–260.

Cavalier-Smith, T., 1982, The evolutionary origin and phylogeny of eukaryotic flagella, *Symp. Soc. Exp. Biol.* **35**:465–493.

Chen, S. J., and Bouck, G. B., 1984, Endogenous glycosyltransferases glucosylate lipids in flagella of *Euglena, J. Cell Biol.* **98**:1825–1835.

Coggin, S. J., and Kochert, G., 1986, Flagellar development and regeneration in *Volvox carteri* (Chlorophyta), *J. Phycol.* **22**:370–381.

Corliss, J. O., 1984, The kingdom Protista and its 45 phyla, *Biosystems* **17**:87–126.

Cunha, N. L., De Souza, W., and Hasson-Voloch, A., 1984, Isolation of the flagellum and characterization of the paraxial structure of *Herpetomonas megaseliae, J. Submicrosc. Cytol.* **16**:705–713.

Curtis, S. E., and Rawson, J. R. Y., 1981, Characterization of the nuclear ribosomal DNA of *Euglena gracilis, Gene* **15**:237–247.

Dentler, W. L., 1977, Fine structural localization of phosphatases in cilia and basal bodies of *Tetrahymena pyriformis, Tissue Cell* **9**:209–222.

Dentler, W. L., 1981, Microtubule–membrane interactions in cilia and flagella, *Int. Rev. Cytol.* **72**:1–47.

Dentler, W. L., 1988, Fractionation of *Tetrahymena* ciliary membranes with Triton X-114 and the identification of a ciliary membrane ATPase, *J. Cell Biol.* **107**:2679–2688.

Doughty, M. J., 1978, Ciliary Ca^{2+}-ATPase from the excitable membrane of *Paramecium.* Some properties and purification by affinity chromatography, *Comp. Biochem. Physiol.* **60B**:339–345.

Doughty, M. J., and Kaneshiro, E. S., 1985, Divalent cation-dependent ATPase activities in ciliary membranes and other surface structures in *Paramecium tetraurelia:* Comparative *in vitro* studies, *Arch. Biochem. Biophys.* **238**:118–128.

Dubreuil, R. R., and Bouck, G. B., 1985, The membrane skeleton of a unicellular organism consists of bridged, articulating strips, *J. Cell Biol.* **101**:1884–1896.

Dubreuil, R. R., and Bouck, G. B., 1988, Interrelationships among the plasma membrane, the membrane skeleton and surface form in a unicellular flagellate, *Protoplasma* **143**:150–164.

Dubreuil, R. R., Rosiere, T. K., Rosner, M. C., and Bouck, G. B., 1988, Properties and topography of the major integral plasma membrane protein of a unicellular organism, *J. Cell Biol.* **107**:191–200.

Farquhar, M. G., 1985, Progress in unraveling pathways of Golgi traffic, *Annu. Rev. Cell Biol.* **1**:447–488.

Gallo, J. M., and Schrével, J., 1985, Homologies between paraflagellar rod proteins from trypanosomes and euglenoids revealed by a monoclonal antibody, *Eur. J. Cell Biol.* **36**:163–168.

Gealt, M. A., Adler, J. H., and Nes, W. R., 1981, The sterols and fatty acids from purified flagella of *Chlamydomonas reinhardi, Lipids* **16**:133–136.

Geetha-Habib, M., and Bouck, G. B., 1982, Synthesis and mobilization of flagellar glycoproteins during regeneration in *Euglena, J. Cell Biol.* **93**:432–441.

Gibbs, S. P., 1978, The chloroplasts of *Euglena* may have evolved from symbiotic green algae, *Can. J. Bot.* **56**: 2883–2889.

Gray, J. P., and Drummond, G. I., 1976, Guanylate cyclase of sea urchin sperm: Subcellular localization, *Arch. Biochem. Biophys.* **172**:31–38.

Griffiths, G., and Simons, K., 1986, The *trans* Golgi network: Sorting at the exit site of the Golgi complex, *Science* **234**:438–443.

Gunderson, J. H., Elwood, H., Ingold, A., Kindle, K., and Sogin, M. L., 1987, Phylogenetic relationships between chlorophytes, chrysophytes, and oomycetes, *Proc. Natl. Acad. Sci. USA* **84**:5823–5827.

Gustin, M. C., and Nelson, D. L., 1987, Regulation of ciliary adenylate cyclase by Ca^{2+} in *Paramecium, Biochem. J.* **246**:337–345.

Hilenski, L. L., and Walne, P. L., 1985, Ultrastructure of the flagella of the colorless phagotroph *Peranema trichophorum* (Euglenophyceae). 1. Flagellar mastigonemes, *J. Phycol.* **21**:114–125.

Hofmann, C., and Bouck, G. B., 1976, Immunological and structural evidence for patterned, intussusceptive growth in a unicellular organism, *J. Cell Biol.* **69**:693–715.

Hoops, H. J., and Witman, G. B., 1983, Outer doublet heterogeneity reveals structural polarity related to beat direction in *Chlamydomonas* flagella, *J. Cell Biol.* **97**:902–908.

Hori, T., and Moestrup, Ø., 1987, Ultrastructure of the flagellar apparatus in *Pyramimonas octopus* (Prasino-phyceae). I. Axoneme structure and numbering of peripheral doublets/triplets, *Protoplasma* **138**:137–148.

Huang, B. P. H., 1986, *Chlamydomonas reinhardtii:* A model system for the genetic analysis of flagella structure and motility, *Int. Rev. Cytol.* **99**:181–215.

Hyams, J. S., 1982, The *Euglena* paraflagellar rod: Structure relationship to other flagellar components and preliminary biochemical characterization, *J. Cell Sci.* **55**:199–210.

Kawano, L. S., and Bouck, G. B., 1984, CER, cell surface–flagellum relationship during flagellar develop-ment, in: *Compartments in Algal Cells and Their Interaction* (W. Wiessner, D. Robinson, and R. C. Starr, eds.), Springer-Verlag, Berlin, pp. 76–87.

Kivic, P. A., and Walne, P. L., 1984, An evaluation of a possible phylogenetic relationship between the Euglenophyta and Kinetoplastida, *Origins Life* **13**:269–288.

Klumpp, S., and Schultz, J. E., 1982, Characterization of a Ca^{2+}-dependent guanylate cyclase in the excitable membrane from *Paramecium, Eur. J. Biochem.* **124**:317–324.

Klumpp, S., and Schultz, J. E., 1982, Characterization of a Ca^{2+}-dependent guanylate cyclase in the excitable ciliary membrane from *Paramecium, Eur. J. Biochem.* **124**:317–324.

Köhle, D., Lang, W., and Kauss, H., 1980, Agglutination and glycosyltransferase activity of isolated gametic flagella of *Chlamydomonas reinhardii, Arch. Microbiol.* **127**:239–243.

Leedale, G. F., 1978, Phylogenetic criteria in euglenoid flagellates, *Biosystems* **10**:183–187.

Lefebvre, P. A., and Rosenbaum, J. L., 1986, Regulation of the synthesis and assembly of ciliary and flagellar proteins during regeneration, *Annu. Rev. Cell Biol.* **2**:517–546.

Lefort-Tran, M., Bré, M. H., Ranck, J., and Pouphile, M., 1980, *Euglena* plasma membrane during normal and vitamin B_{12} starvation growth, *J. Cell Sci.* **41**:245–261.

Levasseur, P. J., and Bouck, G. B., 1986, Early synthesis and later redistribution of a high molecular weight flagellar protein in *Euglena, J. Cell Biol.* **103**:278a.

Lewis, R. M., and Nelson, D. L., 1980, Biochemical studies of the excitable membrane of *Paramecium*. IV. Protein kinase activities of cilia and ciliary membrane, *Biochim. Biophys. Acta* **615**:341–353.

McFadden, G. I., and Melkonian, M., 1986, Golgi apparatus activity and membrane flow during scale bio-genesis in the green flagellate *Scherffelia dubia* (Prasinophyceae). I: Flagellar regeneration, *Protoplasma* **130**:186–198.

McFadden, G. I., and Wetherbee, R., 1985, Flagellar regeneration and associated scale deposition in *Pyrami-monas gelidicola* (Prasinophyceae, Chlorophyta), *Protoplasma* **128**:31–37.

McLean, R. J., and Bosmann, B. H., 1975, Cell–cell interactions: Enhancement of glycosyl transferase ectoenzyme systems during *Chlamydomonas* gametic contact. *Proc. Natl. Acad. Sci. USA* **72**:310–313.

Manton, I., Oates, K., and Gooday, G., 1973, Further observations on the chemical composition of thecae of *Platymonas tetrathele* West (Prasinophyceae) by means of the x-ray microanalyzer electron microscope (EMMA), *J. Exp. Bot.* **24**:223–229.

Markey, D. R., and Bouck, G. B., 1977, Mastigoneme attachment in *Ochromonas, J. Ultrastruct. Res.* **59**: 173–177.

Maruta, H., Greer, K., and Rosenbaum, J. L., 1986, The acetylation of alpha tubulin and its relationship to the assembly and disassembly of microtubules, *J. Cell Biol.* **103**:571–579.

Melkonian, M., 1982, Effect of divalent cations on flagellar scales in the green flagellate *Tetraselmis cordiformis, Protoplasma* **111**:221–233.

Melkonian, M., Robenek, H., and Rassat, J., 1982, Flagellar membrane specializations and their relationship to mastigonemes and microtubules in *Euglena gracilis, J. Cell Sci.* **55**:115–135.

Melkonian, M., Reize, I. B., and McFadden, G., 1985, Flagellar scales in the green flagellate *Tetraselmis striata:* Isolation, characterization and biogenesis, *Eur. J. Cell Biol.* **36** (suppl 7):44.

Mignot, J. P., 1966, Structure et ultrastructure de quelques euglénomonadines, *Protistologica* **2**:51–117.

Miller, K. R., and Miller, G. J., 1978, Organization of the cell membrane in *Euglena, Protoplasma* **95**:11–24.

Moestrup, Ø., 1982, Flagellar structure in algae: A review, with new observations particularly on the Chrysophyceae, Phaeophyceae (Fucophyceae), Euglenophyceae and *Reckertia, Phycologia* **21**:427–528.

Moestrup, Ø., and Walne, P., 1979, Studies on scale morphogenesis in the Golgi apparatus of *Pyramimonas tetrarhynchus* (Prasinophyceae), *J. Cell Sci.* **36**:437–459.

Monk, B. C., Adair, W. S., Cohen, R. A., and Goodenough, U. W., 1983, Topography of *Chlamydomonas:* Fine structure and polypeptide components of the gametic flagellar membrane surface and the cell wall, *Planta* **158**:517–533.

Murray, J. M., 1984, Three-dimensional structure of a membrane microtubule complex, *J. Cell Biol.* **98**:283–295.

Mylnikov, A. P., 1986, Ultrastructure of a colorless flagellate *Phyllomites apiculatus* Kinetoplastida, *Arch. Protistenkd.* **132**:1–10.

Nakano, Y., Urade, Y., Urade, R., and Kitaoka, S., 1987, Isolation, purification, and characterization of the pellicle of *Euglena gracilis* Z, *J. Biochem.* **102**:1053–1063.

Pasquale, S. M., and Goodenough, U. W., 1987, Cyclic AMP functions as a primary sexual signal in gametes of *Chlamydomonas reinhardtii, J. Cell Biol.* **105**:2279–2292.

Reinhart, F. D., and Bloodgood, R. A., 1988, Membrane–cytoskeleton interactions in the flagellum: A 240,000 M_r surface-exposed glycoprotein is tightly associated with the axoneme in *Chlamydomonas moewusii, J. Cell Sci.* **89**:521–531.

Reize, I. B., and Melkonian, M., 1987, Flagellar regeneration in the scaly green flagellate *Tetraselmis striata* Prasinophyceae. Regeneration kinetics and effect of inhibitors, *Helgol. Wiss. Meeresunters.* **41**:149–164.

Robinson, D. G., and Kristen, U., 1982, Membrane flow via the Golgi apparatus of higher plant cells, *Int. Rev. Cytol.* **77**:89–127.

Rogalski, A. A., 1981, Flagellar surface complex of *Euglena gracilis.* Ph.D. thesis, University of Chicago.

Rogalski, A. A., and Bouck, G. B., 1980, Characterization and localization of a flagellar-specific membrane glycoprotein in *Euglena, J. Cell Biol.* **86**:424–435.

Rogalski, A. A., and Bouck, G. B., 1982, Flagellar surface antigens in *Euglena.* Immunological evidence for an external glycoprotein pool and its transfer to the regenerating flagellum, *J. Cell Biol.* **93**:758–766.

Rosiere, T. K., and Bouck, G. B., 1987, Flagellar membranes of *Euglena* lack the major cell surface integral protein and fail to bind membrane skeletal proteins, *J. Cell Biol.* **105**:131a.

Russell, D. G., Newsam, R. J., Palmer, G. C. N., and Gull, K., 1983, Structural and biochemical characterization of the paraflagellar rod of *Crithidia fasciculata, Eur. J. Cell Biol.* **30**:137–143.

Russell, D. G., Miller, D., and Gull, K., 1984, Tubulin heterogeneity in the trypanosome *Crithidia fasciculata, Mol. Cell. Biol.* **4**:779–790.

Sano, M., 1976, Subcellular localizations of guanylate cyclase and 3',5'-cyclic nucleotide phosphodiesterase in sea urchin sperm, *Biochim. Biophys. Acta* **428**:525–531.

Schultz, J. E., and Klumpp, S., 1980, Guanylate cyclase in the excitable ciliary membrane of *Paramecium, FEBS Lett.* **122**:64–66.

Schultz, J. E., and Klumpp, S., 1983, Adenylate cyclase in cilia from *Paramecium*. Localization and partial characterization, *FEBS Lett.* **154:**347–350.

Schultz, J. E., Schönefeld, U., and Klumpp, S., 1983, Calcium/calmodulin-regulated guanylate cyclase and calcium-permeability in the ciliary membrane from *Tetrahymena, Eur. J. Biochem.* **137:**89–94.

Schultz, J. E., Uhl, D. G., and Klumpp, S. E., 1987, Ionic regulation of adenylate cyclase from the cilia of *Paramecium tetraurelia, Biochem. J.* **246:**187–192.

Sogin, M. L., Elwood, H. J., and Gunderson, J. H., 1986, Evolutionary diversity of eukaryotic small-subunit rRNA genes, *Proc. Natl. Acad. Sci. USA* **83:**1383–1387.

Travis, S. M., and Nelson, D. L., 1986, Characterization of Ca^{2+}- or Mg^{2+}-ATPase of the excitable ciliary membrane from *Paramecium tetraurelia:* Comparison with a soluble Ca^{2+}-dependent ATPase, *Biochim. Biophys. Acta* **862:**39–48.

Walne, P. L., Moestrup, Ø., Norris, R. E., and Ettl, H., 1986, Light and electron microscopical studies of *Eutreptiella eupharyngea* new species (Euglenophyceae) from Danish and American waters, *Phycologia* **25:**109–126.

Ward, G. E., Garbers, D. L., and Vacquier, V. D., 1985, Effects of extracellular EEG factors on sperm guanylate cyclase, *Science* **227:**768–770.

Ward, G. E., Moy, G. W., and Vacquier, V. D., 1986, Phosphorylation of membrane-bound guanylate cyclase of sea urchin spermatozoa, *J. Cell Biol.* **103:**95–101.

Willey, R. L., Walne, P. L., and Kivic, P., 1988, Phagotrophy and the origins of the euglenoid flagellates, *Crit. Rev. Plant Sci.* **7:**303–340.

Witman, G. B., Carlson, K., Berliner, J., and Rosenbaum, J. L., 1972, *Chlamydomonas* flagella. I. Isolation and electrophoretic analysis of microtubules, matrix, membranes, and mastigonemes, *J. Cell Biol.* **54:**507–539.

Wolters, J., and Erdmann, V. A., 1986, Cladistic analysis of 5S rRNA and 16S rRNA secondary and primary structures. The evolution of eukaryotes and their relation to Archaebacteria, *J. Mol. Evol.* **24:**152–166.

Gliding Motility and Flagellar Glycoprotein Dynamics in *Chlamydomonas*

Robert A. Bloodgood

1. Introduction

Although there are many forms of motility and contractility within eukaryotic cells, most cases of whole cell locomotion can be conveniently divided into one of two classes: (1) movement of cells through a liquid medium due to the propagation of bends along cilia and flagella or (2) movement of cells while in adhesive contact with a solid or semisolid surface. In the first case (swimming of ciliated and flagellated cells), locomotion results from a viscous coupling between the ciliary or flagellar surface and the liquid medium. In the second case (exemplified by amoeboid and fibroblastic movements), cell-surface molecules with specific binding properties mediate the transfer of energy between the cell and the extracellular matrix or solid substrate (Buck and Horwitz, 1987; Burridge *et al.*, 1988; Lackie, 1986).

The eukaryotic, unicellular biflagellate green alga, *Chlamydomonas,* exhibits both of these forms of whole cell locomotion. When sufficient liquid medium is available, *Chlamydomonas* swims through the medium by the coordinated initiation and propagation of bends along its two flagella (Ringo, 1967). However, *Chlamydomonas* is a soil alga which often finds itself without sufficient liquid for swimming motility and yet presumably still needs the ability to move in order to maintain itself in an optimal level of light for photosynthesis. In this case, it is able to utilize a form of whole cell locomotion referred to as gliding motility.

Chlamydomonas possesses three different cell-surface domains (Fig. 1); (1) cell body plasma membrane, (2) flagellar membrane, and (3) cell wall. While the flagellar membrane is a specialized region of the cell's plasma membrane and is in continuity with the rest of the plasma membrane, these two plasma membrane domains differ in a number of properties (Bloodgood, 1987). Recent studies using *C. eugametos* (Musgrave *et al.,*

Robert A. Bloodgood • Department of Anatomy and Cell Biology, University of Virginia School of Medicine, Charlottesville, Virginia 22908.

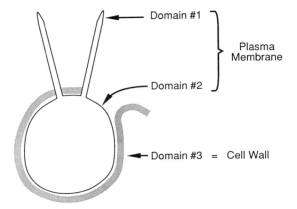

Figure 1. Diagram illustrating the three major cell surface domains of *Chlamydomonas*: (1) flagellar membrane, (2) the rest of the plasma membrane, and (3) the multilayered cell wall composed of glycoproteins.

1986) and *C. reinhardtii* (Bloodgood, 1988b) suggest that there are indeed two separate domains of the plasma membrane characterized by a barrier to lateral diffusion of plasma membrane proteins. The multilayered cell wall, composed of hydroxyproline-rich glycoproteins, normally covers the cell body plasma membrane but not the flagellar membrane (Roberts *et al.*, 1985a). Unlike many other types of eukaryotic cells (and most mammalian cells), the only domain of the *Chlamydomonas* plasma membrane normally available for cell–cell or cell–substrate interactions is the flagellar membrane (Figs. 1, 2). The initial interaction of gametes during the sexual aspect of the cell cycle occurs via the gametic flagellar surfaces, which possess specialized sexual agglutinin molecules, which are hydroxyproline-rich glycoproteins (see van den Ende *et al.*, this volume).

　　Both forms of whole cell locomotion exhibited by *Chlamydomonas* result from the activity of the flagella. The coordinated flagellar bending motility that allows swimming

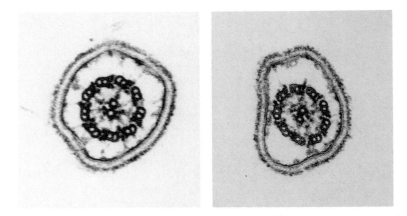

Figure 2. Transmission electron micrographs of cross sections through flagella of *Chlamydomonas reinhardtii*. The flagellar membranes have swollen artifactually allowing a better visualization of the slender electron-dense links between the flagellar membrane and the axoneme. A prominent glycocalyx is visible on the external surface of the flagellar membrane. Magnification 90,000×. Micrographs provided by Dr. Gregory S. May.

is primarily a property of the axoneme and can occur in the absence of the flagellar membrane (see Witman, this volume). The nature of the flagellar waveform is regulated by the intraaxonemal free calcium concentration, which is controlled by the flagellar membrane (see Preston and Saimi, this volume). On the other hand, whole cell gliding motility is a manifestation of the flagellar surface and results from an interaction of the flagellar membrane with both the cytoskeletal components within the flagellum (see Dentler, this volume, for a discussion of membrane–cytoskeletal interactions in cilia and flagella) and an extracellular substrate.

This chapter will review what is known about gliding motility, describe the dynamic properties of the *Chlamydomonas* flagellar surface, and examine the relationship of flagellar membrane glycoprotein dynamics to whole cell gliding motility. Except where indicated, the information to be discussed has come from studies of *C. reinhardtii*. However, some information about and comparison with *C. moewusii/C. eugametos* will be included. For reasons that will become clear, any discussion of flagellar surface dynamics in *Chlamydomonas* will inevitably include some discussion of the flagellar surface events involved in mating (see also van den Ende *et al.*, this volume).

2. Gliding Motility

2.1. Gliding Motility—An Overview

Gliding motility (Jarosch, 1962; Halfen, 1979; King, 1988) represents a special case of eukaryotic whole cell locomotion involving cell–substrate interactions. Gliding motility is generally associated with free-living or parasitic protists; it has been described in diatoms, desmids, euglenoid flagellates, sporozoan protozoa (*Eimeria* and *Gregarina*), *Plasmodium* sporozoites, and Labyrinthulales. In contrast to fibroblastic and amoeboid movements observed in vertebrate cells on the one hand and ciliary and flagellar swimming motility on the other, gliding motility is characterized by both (1) the absence of any detectable cell deformation or cell shape change during locomotion and (2) the absence of any visible organelle that might be responsible for the movement. Gliding locomotion is also found associated with many prokaryotic organisms (especially the cyanobacteria, the Cytophagales, and the Myxobacteriales) (Burchard, 1981; Castenholz, 1982; Pate, 1985); interestingly enough, the overall phenomenology of gliding motility in prokaryotes and lower eukaryotes can exhibit striking similarities. Most attention has been devoted to understanding gliding motility in prokaryotic gliding organisms, at least in part due to the availability of nongliding mutants (Glaser and Pate, 1973; Hodgkin and Kaiser, 1977); despite considerable effort, little understanding of the mechanism underlying gliding motility has emerged for any prokaryotic or eukaryotic system. Nevertheless, numerous hypotheses to explain prokaryotic and eukaryotic gliding motility have been put forward (reviewed in Burchard, 1981; Castenholz, 1982; Lapidus and Berg, 1982; King, 1988). These include (1) rotary motors similar to those that underlie bacterial flagellar motility, (2) slime production and expulsion, (3) surface tension, (4) contraction of cytoplasmic filaments, and (5) the linear movement of cell-surface membrane components. One particularly interesting feature of gliding motility has emerged from these studies: many organisms that glide (both prokaryotic and eukaryotic) can move inert marker objects (such as polystyrene microspheres) along their surfaces, and these particle movements are coupled to gliding motility (King, 1981, 1988; Lapidus and Berg, 1982; Bloodgood, 1981, 1988a). Mutant cell lines that do not exhibit gliding motility do not exhibit microsphere movement (Pate and Chang, 1979; Lewin, 1982; Reinhart and Bloodgood, 1988b); physiological conditions or drugs that prevent the expression of gliding motility prevent microsphere movement (Pate and Chang, 1979; King and Lee, 1982; C. A. King, unpublished results). In the case of *Cytophaga,* Lapidus and Berg (1982) have suggested that their observations are "consistent with a model for gliding in which sites to which glass and polystyrene strongly adsorb move within the fluid outer membrane along tracks fixed to a rigid peptidoglycan framework."

With few exceptions (Tamm, 1967), little attention has been paid to gliding motility in flagellated organisms. Indeed, as recently as 1979, Halfen in a review on gliding motility stated that "no one in this century has seriously suggested that flagella or cilia are implicated in gliding. . . ." On the contrary, *Chlamydomonas* does indeed exhibit gliding motility and it is their flagella that are clearly responsible for the gliding motility. Although there were a few early reports of gliding in Chlamydomonads (see Bloodgood, 1981, for references), it was Lewin (1952, 1954) who first described gliding in *Chlamydomonas* in any detail; several recent reports have further added to our general description of this phenomenon (Bloodgood, 1981; Lewin, 1982; Reinhart and Bloodgood, 1988b).

2.2. Gliding Motility in *Chlamydomonas*

When *Chlamydomonas* flagella contact a solid substrate such as a glass slide or cover glass, they will often adhere by their tips, which will begin to migrate along the substrate in random directions (Fig. 3). The cell body will then become "strung out" between the migrating flagella; this will result in the establishment of a characteristic gliding configuration in which the flagella are oriented 180° from one another and appear to lay flat against the substrate (Fig. 4). In this configuration, the cell will glide along the solid substrate at a velocity of 1.5–2.0 μm/sec, occasionally stopping and starting and reversing the direction of movement. Flagella do not appear to exhibit any shape changes or bend propagation along their length during gliding. It is clear that normal flagellar motility (bend initiation and propagation) is not involved in gliding because a wide range of nonmotile mutants of *Chlamydomonas* defective in swimming motility due to the absence of various axonemal components (central pair microtubules, radial spokes, outer dynein arms, inner dynein arms) exhibit normal gliding behavior. A number of observations suggest that it is the leading flagellum that is providing the force for cell locomotion during gliding. When using a wild-type strain of *Chlamydomonas,* a gliding cell will sometimes briefly lift one or the other of its flagella off the substrate. If the leading flagellum loses contact with the substrate, the cell ceases moving until the flagellum renews its contact with the substrate; if the trailing flagellum detaches from the substrate, the cell continues to glide. Both flagella can play the role of active flagellum, as the cell can reverse the direction of gliding without losing contact with the substrate. One can observe cells in the gliding configuration that do not exhibit any movement for long periods of time; this could reflect a situation in which both flagella are functionally interacting with the substrate resulting in balanced forces acting upon the cell body. Uniflagellate cells can be generated by applying light pressure to the coverslip; such cells

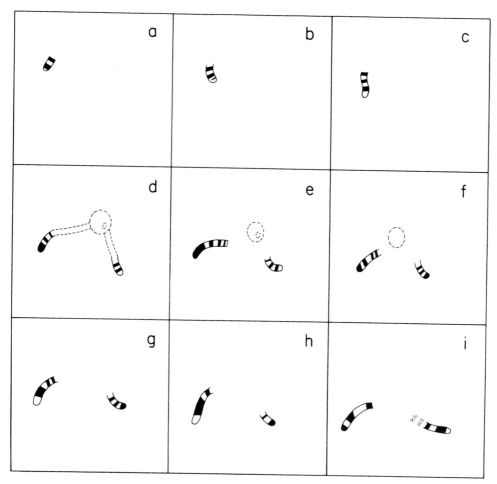

Figure 3. Series of tracings from a videotape of a *Chlamydomonas* cell interacting with a solid substrate by its flagellar surfaces as it assumes the characteristic gliding configuration. The videotape was made using interference reflection microscopy and the dark areas are presumed to represent areas of close contact between the flagellar membrane and the glass substrate. First one flagellum and then the other make adhesive contact with the substrate; migration of the flagellar tips along the substrate results in pulling the cell body closer to the substrate and increasing the extent of the flagellum that is apposed to the substrate. Migration of the flagellar surfaces will eventually result in the cell body being "strung out" between the flagella that are then oriented 180° from one another (see Fig. 5). Redrawn from the unpublished data of Barbara Daniel and Conrad King, University College London.

glide continuously in one direction with the cell body trailing the flagellum. These observations (Bloodgood, 1981) suggest that it is always the leading flagellum that is propelling the cell during gliding motility. If this is true, gliding forces in the flagellum are unidirectional such that the motor moves toward the base of the flagellum (microtubule minus ends) during gliding, propelling the leading flagellum forward (away from the cell body) relative to the substrate. This makes the flagellar gliding motor equivalent

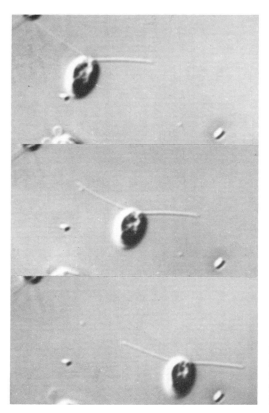

Figure 4. Three frames from a videotape of a *C. reinhardtii* strain PF-18 cell gliding with the left flagellum in contact with the glass substrate; the time sequence is from bottom to top. Differential interference contrast optics. Magnification 1250×.

to dynein (rather than kinesin) in its directionality (see Section 11.1). This conclusion conflicts with the clear bidirectionality of movement of polystyrene microspheres along the flagellar surface (Section 3) and the rather clear genetic and drug inhibition data (Bloodgood and Salomonsky, 1989; Lewin, 1982; Reinhart and Bloodgood, 1988b) indicating that microsphere movement is a reflection of the same motor responsible for whole cell gliding motility. During the course of gliding, the tip of the leading flagellum will occasionally wiggle; on some occasions, the course of whole cell movement will follow a curved path due to a bend in the tip of the flagellum (Bloodgood, 1981); this may reflect a mechanism by which the cell can "steer" its course during gliding motility, perhaps in response to external stimuli such as light.

In order to learn more about the interaction of the flagellar surface with the solid substrate during gliding motility, Conrad King and Barbara Daniel at University College London observed gliding motility in *Chlamydomonas* using interference reflection microscopy (Figs. 3, 5); this technique provides information about the distance between the lower cell surface and the underlying substrate. Although one might have predicted there would be either a single localized site of close contact on the leading flagellum or a continuous zone of close contact with the substrate running the entire length of the leading flagellum, this is not what was observed. King and Daniel observed localized regions of

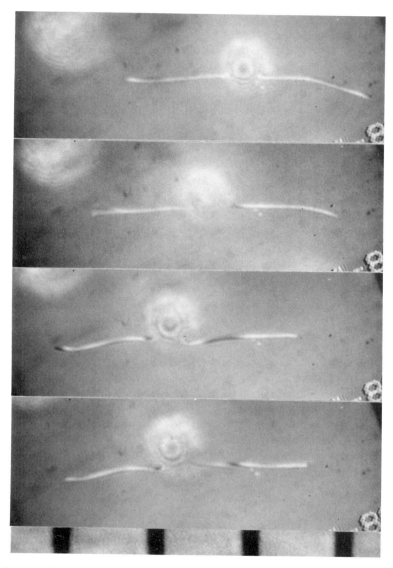

Figure 5. Sequence of frames (10 sec apart) from a videotape of a *C. reinhardtii* cell exhibiting gliding motility as viewed using interference reflection microscopy. In the top two frames, the cell is gliding to the left. In the third frame it has ceased movement in order to reverse the direction of gliding; in the fourth (bottom) frame, it is gliding to the right. The dark patches on the flagella indicate the areas of flagellar surface that are in closest contact with the glass substrate. The scale at the bottom is 10 μm per division. These micrographs have been provided by Dr. Conrad King, University College London.

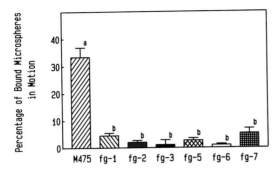

Figure 6. Histogram showing the level of flagellar surface motility (polystyrene microsphere movement) exhibited by several gliding-defective mutant cell lines of *C. moewusii* and the parent cell strain (M475) from which the mutants were selected. Error bars are S.E.M. a, b: Cell line bars sharing the same letter are not significantly different (*p* < 0.01). Reproduced from Reinhart and Bloodgood (1988b) with permission.

close contact on both the leading and trailing flagella; more than one site of close contact was observed simultaneously on the same flagellum and the positions of these contact sites changed during gliding motility and during reversal of the direction of gliding (Figs. 3, 5).

Lewin (1982) isolated a number of nongliding mutant cell lines from *C. moewusii*. Although all of these nongliding mutant cell lines fail to exhibit any microsphere movements (confirming that microsphere movement is an expression of the motor underlying gliding motility) (Fig. 6; Reinhart and Bloodgood, 1988b), the different nongliding mutant cell lines show very different levels of flagellar surface adhesiveness, as judged by the binding of polystyrene microspheres (Fig. 7; Reinhart and Bloodgood, 1988b). Examination of flagellar proteins from the parent cell line and the various nongliding mutant cell lines of *C. moewusii* suggests a correlation between the level of flagellar surface adhesiveness (as judged by binding of polystyrene microspheres) and the amount of a 240-kDa flagellar membrane glycoprotein (Reinhart and Bloodgood, 1988a, b). Since the expression of gliding motility requires a mechanical connection between the flagellar surface and a solid substrate, it is to be expected that mutant cell lines (such as fg-3) in which the flagellar surface adhesiveness is greatly reduced will exhibit a defect in gliding

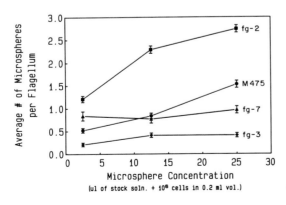

Figure 7. Graph showing the level of flagellar surface adhesiveness (as judged by the binding of polystyrene microspheres) exhibited by the M475 parent cell line of *C. moewusii* and three of the nongliding mutant cell lines (fg-2, fg-7, and fg-3) as determined for three different concentrations of polystyrene microspheres. Error bars are S.E.M. Reproduced from Reinhart and Bloodgood (1988b) with permission.

motility. It is intriguing that one of the nongliding mutants (fg-2) exhibits greatly increased flagellar surface adhesiveness; it is conceivable that too tight an interaction with the substrate can impede gliding locomotion. This is suggested by experiments with cells gliding on poly-L-lysine-coated glass surfaces (Bloodgood, 1981); some poly-L-lysine appears to promote gliding while a higher level of poly-L-lysine prevents gliding altogether. In the case of the nongliding mutant cell lines (such as fg-1) in which the properties of the flagellar surface appear to be normal, it is likely that the defect in gliding motility stems from a defect in the intraflagellar force-transducing machinery. Further analysis of these nongliding mutant cell lines may shed additional light on the mechanism of gliding motility.

3. Polystyrene Microsphere Movements

Polystyrene microspheres and other inert marker particles have been used to visualize the movement of plasma membrane domains in a variety of cell types (reviewed by Bloodgood, 1988a). Polystyrene microspheres will bind to and translocate at velocities of 1.5–2.0 μm/sec along the surface of flagella from both vegetative and gametic *Chlamydomonas* (Bloodgood, 1977; Bloodgood *et al.*, 1979; Hoffman and Goodenough, 1980). Bound microspheres exhibit saltatory motility; they can stop and start and reverse their direction of movement at any point along the flagellar surface. Microsphere movements are always rectilinear occurring in the direction of the long axis of the flagellum; microspheres are never observed to spiral around the flagellum. Freeze-fracture studies of the *Chlamydomonas* flagellar membrane reveal rows of large intramembrane particles oriented in the long axis of the flagellum (Bergman *et al.*, 1975; Snell, 1976); these particle arrays may reflect functional tracks within the membrane that dictate the direction of microsphere movement (and gliding motility). Treatment of cells with cycloheximide results in a slow loss of the ability to bind and translocate polystyrene microspheres (Bloodgood *et al.*, 1979) along with a disruption in the regular rows of flagellar intramembrane particles (Bloodgood, 1981), suggesting that the intramembrane particle arrays play some role in the microsphere movement. A number of observations have demonstrated that microsphere movements are a manifestation of the same motor that is responsible for whole cell gliding motility. Lewin (1982) isolated nongliding mutants of *C. moewusii* and these mutants failed to exhibit polystyrene microsphere movements (Reinhart and Bloodgood, 1988b). A number of experimental treatments can inhibit polystyrene microsphere movements (Bloodgood, 1987; Bloodgood and Salomonsky, 1989); in all cases, cells that are unable to express microsphere movement are unable to exhibit gliding motility (see Table III). In the parasitic protozoan cell *Gregarina*, all drug treatments (including trifluoperazine) that prevent gliding motility also inhibit polystyrene microsphere movements along the cell surface (King and Lee, 1982; King, 1988). In at least one prokaryotic gliding organism (*Cytophaga*), it has also been shown that polystyrene microsphere movements are the expression of the motor for gliding motility, utilizing both nongliding mutants and physiological conditions that do not allow the expression of gliding motility in wild-type cells (Pate and Chang, 1979).

Calcium appears to play some role in polystyrene microsphere movements (and presumably gliding motility). Lowering the free calcium concentration in the medium beneath 10^{-6} M results in a reversible decrease in polystyrene microsphere movements (Bloodgood *et al.*, 1979). Calmodulin inhibitors (TFP and W-7) inhibit polystyrene microsphere movements along gametic flagella (Detmers and Condeelis, 1986). Lidocaine, a local anesthetic reported to interfere with the movement of calcium across membranes, reversibly inhibits polystyrene microsphere movements, and this effect can be overcome by addition of excess calcium (Snell *et al.*, 1982).

Polystyrene microsphere movements provide a convenient method for visualizing and quantitating the cell surface force transduction responsible for gliding motility. In addition, observations made using microspheres have provided a number of additional pieces of information about force transduction at the flagellar surface. All parts of the flagellar surface are capable of adhering to and exerting force on an external substrate (be it a microsphere, a glass surface, or another flagellar surface). Individual small domains on the flagellar surface can operate independently of one another. This has been shown in two ways: (1) When a number of individual polystyrene microspheres are attached to and moving along the surface of the same flagellum, the movements of each microsphere are independent of the movements of the others. (2) A gliding cell can exhibit polystyrene microsphere movements on the free surface of both the leading and trailing flagella; the direction of movement and the reversals in direction of the microspheres are uncoupled from the direction of gliding of the cell. Whatever flagellar membrane proteins are involved in the force transduction at the flagellar surface, they must be widely distributed over the surface of the flagellum; similarly, whatever intraflagellar contractile machinery is responsible for the actual energy transduction, it, too, must be widely distributed along the inner surface of the flagellar membrane. If the machinery responsible for force transduction at the flagellar surface is anchored on the outer doublet microtubules, then both the relevant flagellar membrane glycoproteins and the relevant energy-transducing molecules would be expected to be localized to nine "tracks" along the flagellar surface, one track lying over each of the nine outer doublet microtubules.

In *Chlamydomonas,* the normal manifestations of the force-transducing system operating at the flagellar surface are gliding motility (see Section 2.2) and possibly the migration of gametic flagellar contact sites during the early events in mating (see Section 5). An intriguing observation is that polystyrene microsphere movements also occur on the surface of sea urchin blastula cilia (Bloodgood, 1980) where neither gliding motility nor mating interactions occur. In this case, it is possible that ciliary surface motility is a reflection of a mechanism by which ciliary axonemal proteins or protein assemblages (such as tubulin, dyneins, radial spoke proteins) are transported along the length of the organelle, in association with the inner surface of the ciliary membrane, during normal assembly and turnover. This kind of facilitated transport system would be particularly useful for axonemal precursors since most axonemal assembly occurs at the distal end of the organelle (Rosenbaum and Child, 1967; Witman, 1975), at least in *Ochromonas* and *Chlamydomonas* flagella. Connections between the ciliary/flagellar membrane and the distal end of the axoneme have been extensively documented (Dentler, 1981, 1987, this volume) and might provide a route for subunits transported along the flagellar membrane to be routed onto the growing ends of the outer doublet microtubules during flagellar growth.

4. Flagellar Surface Motility

Collectively, whole cell gliding motility and polystyrene microsphere movements in *Chlamydomonas* are referred to as flagellar surface motility; these events are common to the flagella of both vegetative and gametic cells. The important scientific questions about flagellar surface motility that need to be addressed are:

1. Which flagellar surface components are responsible for the adhesive interaction and the mechanical coupling of the force (presumably transduced within the flagellum) to the external substrate?
2. What is the nature of the motor that is responsible for transducing the force that is applied by the flagellar surface to the external substrate? Is the energy-transducing system anchored on the flagellar axoneme?
3. What is the pathway of transmembrane signaling in the flagellum? What is the nature of the signal (cAMP, calcium) generated within the flagellum in response to events occurring at the flagellar surface? Are there common flagellar signaling mechanisms that are utilized in both vegetative and gametic cells?
4. How is the flagellar surface motility system regulated locally within the plane of the flagellum? In other words, how can local domains within the flagellar membrane operate independently? How large are these local domains and what is their relationship to the flagellar cytoskeleton (in particular, the outer doublet microtubules)?

Efforts to obtain answers to these important questions will be discussed in the subsequent sections of this review. Due to the easy accessibility of the flagellar surface and the availability of probes for the flagellar surface, the most progress has been made in answering the first question. More recently, the use of specific inhibitors has provided considerable new information about signaling mechanisms operating in the flagellum (see Section 8).

5. Mating-Associated Dynamic Flagellar Surface Events

One additional class of dynamic flagellar surface-associated events remains to be discussed; this occurs during the early steps in the sexual interactions between *Chlamydomonas* gametes (see van den Ende *et al.*, this volume). The initial adhesive interaction between plus and minus gametes occurs by the action of highly specific sexual agglutinin molecules (hydroxyproline-rich glycoproteins) located on the flagellar surfaces (Adair, 1985; van den Ende, 1985). This flagellar adhesion event, which may involve cross-linking and patching of flagellar agglutinin molecules, activates a transmembrane signaling pathway that alters the level of one or more intracellular second messengers (see Section 8). Following the initial adhesive interaction of the gametic flagellar surfaces, the sites of flagellar contact migrate so that the flagella become aligned with the tips in tight contact with one another (see Fig. 3 in Bloodgood, 1982; Mesland, 1976). This flagellar alignment is accompanied by a process of flagellar tip activation in which the flagellar tips swell and are observed to contain an accumulation of densely staining material (Mesland *et al.*, 1980). The process of flagellar alignment involves an active movement of one flagellar surface along another flagellar surface; it is possible that this process utilizes the same motor that is responsible for whole cell gliding motility, a possibility first suggested by Lewin (1952).

Supporting this suggestion are the observations that some of the early events in

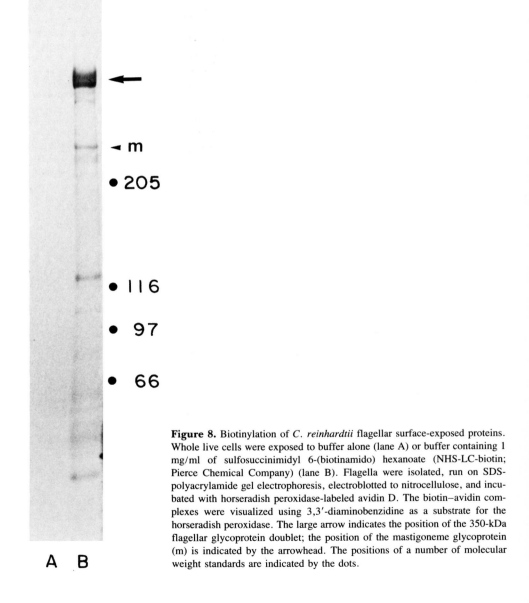

Figure 8. Biotinylation of *C. reinhardtii* flagellar surface-exposed proteins. Whole live cells were exposed to buffer alone (lane A) or buffer containing 1 mg/ml of sulfosuccinimidyl 6-(biotinamido) hexanoate (NHS-LC-biotin; Pierce Chemical Company) (lane B). Flagella were isolated, run on SDS-polyacrylamide gel electrophoresis, electroblotted to nitrocellulose, and incubated with horseradish peroxidase-labeled avidin D. The biotin–avidin complexes were visualized using 3,3′-diaminobenzidine as a substrate for the horseradish peroxidase. The large arrow indicates the position of the 350-kDa flagellar glycoprotein doublet; the position of the mastigoneme glycoprotein (m) is indicated by the arrowhead. The positions of a number of molecular weight standards are indicated by the dots.

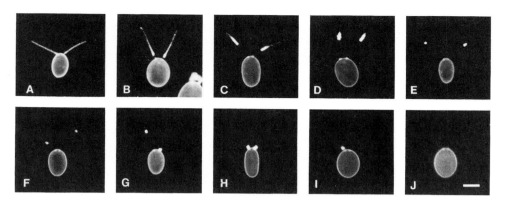

Figure 9. Redistribution of the FMG-1 monoclonal antibody along the flagellar surface of live strain pf-18 cells. Live cells were incubated with the monoclonal antibody on ice and then the unbound antibody was removed. The cells were fixed with glutaraldehyde at various times after warming the cells to 23°C. After fixation, the cells were labeled with the FITC-conjugated second antibody. The cell in A was fixed while the cells were still cold. The other cells were fixed at 5 min (B–F), 10 min (G, H), and 15 min (I, J) after warming. Bar = 5 μm. Reproduced from Bloodgood *et al.* (1986) with permission.

Chlamydomonas mating are inhibited by the same drug treatments (lidocaine, TFP, W-7) that inhibit polystyrene microsphere movements (Snell *et al.*, 1982; Detmers and Condeelis, 1986). These agents have no effect on flagellar agglutination but appear to inhibit the flagellar signal responsible for cell wall release and mating structure activation. Flagellar tip activation is inhibited by TFP (Detmers and Condeelis, 1986) but not by lidocaine (Snell *et al.*, 1982). This is interesting because recent results (Section 8) suggest that TFP may be exerting its effect through an alteration in intracellular cAMP level while lidocaine may be acting through an alteration in calcium fluxes.

An alternative (although not mutually exclusive) possibility is that the dynamic flagellar surface events associated with flagellar alignment and flagellar tip activation may reflect or be triggered by the slow redistribution (tipping) of flagellar membrane proteins (Fig. 9) described in Section 7 that occurs in vegetative (Bloodgood *et al.*, 1986) and gametic (Homan *et al.*, 1988) cells in response to cross-linking of the surface with certain antibodies and lectins but has also been shown to accompany the early events in mating (Homan *et al.*, 1987; Musgrave and van den Ende, 1987). Drug treatments that inhibit glycoprotein redistribution in vegetative cells (the calmodulin antagonists TFP and W-7 and the cyclic nucleotide-dependent protein kinase inhibitor H-8) (Figs. 10–12) also inhibit mating in gametes (Detmers and Condeelis, 1986; Pasquale and Goodenough, 1987). Looked at in a different perspective, it may be that flagellar membrane protein dynamics (visualized as tipping) underlie both gliding motility (visualized also as polystyrene microsphere movements) and mating events and that is why there are drug treatments (such as TFP and W-7) that interfere with all three events.

The possible nature of the flagellar signals utilized during mating in *Chlamydomonas* is discussed in Section 8.1.

Figures 10–12. Effect of calmodulin antagonists [trifluoperazine (TFP), W-7] and an inhibitor of cyclic nucleotide-dependent protein kinase (H-8) on redistribution of the FMG-1 mouse monoclonal antibody (recognizing the 350-kDa flagellar membrane glycoproteins) along the flagellar surface of *C. reinhardtii*. In each case, cells were incubated at 4°C with the antibody and the drug, unbound antibody was removed, and the cells were warmed to 25°C. After 10 min, the cells were fixed and scored in a fluorescence microscope. The graphs show the percentage of cells in a population exhibiting each of the patterns of redistribution illustrated in the sketches along the abscissa. Although these data were obtained using vegetatively grown cells, similar results have been obtained using gametic cells. Similar results were also obtained using Con A instead of the FMG-1 monoclonal antibody.

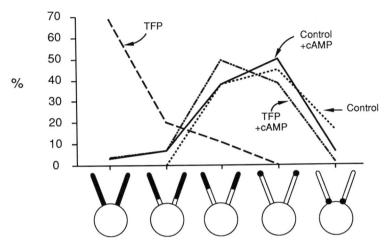

Figure 10. TFP (16 μM) inhibits the redistribution of the 350-kDa glycoproteins induced by FMG-1 antibody cross-linking; this inhibition is reversed by the presence of 20 mM dibutyryl cAMP or dibutyryl cGMP. TFP is known to be an effective calmodulin antagonist at this concentration.

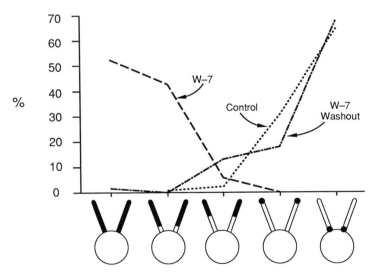

Figure 11. W-7 (18 μM) inhibits the redistribution of the 350-kDa glycoproteins induced by FMG-1 antibody cross-linking and this inhibition is reversed either by washing out the drug or by addition of dibutyryl cAMP (data not shown). W-7 is known to be an effective calmodulin antagonist at this concentration.

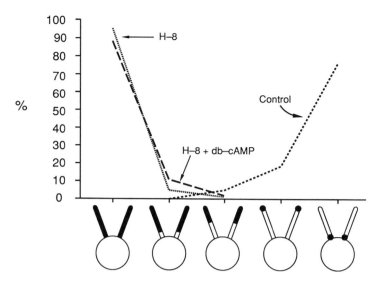

Figure 12. H-8 (2 mM), a known inhibitor of cyclic nucleotide-dependent protein kinases (Hidaka *et al.*, 1984), completely inhibits the redistribution of the 350-kDa flagellar membrane glycoproteins induced by FMG-1 antibody cross-linking. This inhibition was not reversed by dibutyryl cAMP but was reversed after washing the H-8 out of the cells (data not shown).

6. Characterization of the Major Flagellar Glycoproteins in *C. reinhardtii*

The flagellar membrane is a specialized domain of the *Chlamydomonas* plasma membrane possessing a limited number of functions and a fairly simple composition (Bloodgood, 1987). The flagellar membrane has a very prominent "fuzzy coat" or glycocalyx (Fig. 2) and is linked to the underlying cytoskeleton (axoneme) by slender connections (Fig. 2). Flagella can be released from the cell bodies and subsequently purified using rather straightforward procedures (Witman, 1986). Because the flagellum contains no internal membranes and because the cell body is stabilized by the presence of a thick cell wall, flagellar membranes can be obtained with minimal contamination by other cellular membranes. The most likely source of contamination of flagellar preparations is from cell wall glycoproteins and this does occur to a certain extent in both *C. reinhardtii* (Monk *et al.*, 1983) and *C. eugametos* (Musgrave *et al.*, 1983).

Nonionic detergent extracts of whole flagella or intact flagellar membrane vesicles isolated from either vegetative or gametic cells of *C. reinhardtii* are enriched in a closely migrating pair of high-molecular-weight glycoproteins (Bergman *et al.*, 1975; Bloodgood and May, 1982; Bloodgood, 1984; Monk *et al.*, 1983; Snell, 1976; Witman *et al.*, 1972; see Fig. 8). Commonly referred to as the 350-kDa glycoproteins, they are the major glycoproteins (Snell, 1976; Bloodgood, 1982) and the major Con A-binding species (Monk *et al.*, 1983) in the *C. reinhardtii* flagellum. Although we continue to use the 350-kDa terminology to refer to these proteins, this is likely to be a very inaccurate figure for their molecular mass as they are heavily glycosylated. Moreover, they migrate more slowly in SDS-polyacrylamide gels than does the *Chlamydomonas* axonemal dynein heavy chains, which have molecular masses of 415–480 kDa (Witman, 1989). Monk *et al.*, (1983) have referred to these two flagellar surface proteins as F2 and F3 and assigned them molecular masses of 375 and 350 kDa, respectively.

The 350-kDa glycoproteins are the principal proteins labeled by vectorial labeling of the flagellar surface by biotinylation (Fig. 8) or iodination (Monk *et al.*, 1983; Bloodgood and Workman, 1984). Using two different types of substrate immobilized iodination systems, it has been demonstrated that at least one (and perhaps both) of these two high-molecular-weight glycoproteins contacts the substrate during the expression of gliding motility (Bloodgood and Workman, 1984); no other flagellar proteins were significantly labeled by these procedures. Pronase treatment of whole cells results in the partial digestion of the faster migrating of the two 350-kDa glycoproteins concomitant with a dramatic decrease in the level of binding of polystyrene microspheres (Bloodgood and May, 1982). This suggests that the smaller of the two 350-kDa glycoproteins may be the component that provides the adhesiveness for the vegetative flagellar surface. Adair *et al.* (1983) have estimated that there are approximately 90,000 copies of the major flagellar membrane glycoproteins per gametic flagellum. These two high-molecular-weight glycoproteins are not present in the *C. reinhardtii* cell wall (Monk *et al.*, 1983; Bloodgood, unpublished observations) and are barely detectable in the deflagellated cell body by Western blot analysis (Bloodgood, unpublished observations). It is still not entirely clear whether the 350-kDa glycoproteins are unique to the flagellar domain of the plasma membrane or if they are present, albeit in very reduced concentration, in the cell body plasma membrane. Gel filtration chromatography (using Bio-Gel A15m and A5m) of nonionic detergent extracts of purified flagella suggests that there is a large native aggregate containing multiple (at least two) copies of each of the two major species (Bloodgood, unpublished results). In contrast to the major cell wall glycoproteins and the sexual agglutinin glycoproteins specific for the gametic flagellar membrane (neither of which are integral membrane proteins), the 350-kDa flagellar membrane glycoproteins are integral membrane proteins and do not contain any appreciable amount of hydroxyproline (Table I). This observation makes the 350-kDa glycoproteins the first *Chlamydomonas* cell-surface proteins to be characterized that are not in the class of hydroxyproline-rich glycoproteins. In the full-length flagellum, the 350-kDa glycoproteins turn over rapidly relative to axonemal proteins (Bloodgood, 1984); this explains the ability of cells to recover fairly rapidly from the effects of proteolytic modification of the flagellar surface (Bloodgood and May, 1982).

The sugar composition of the 350-kDa membrane glycoproteins purified from vegetative flagella by Con A affinity chromatography and Bio-Gel A15m chromatography is given in Table II, which shows the sugar composition for glycoproteins from the flagella of strain PF-18 cells and cells of a mutant cell line (L-23) selected from mutagenized PF-18 cells for alterations in cell-surface carbohydrate epitopes (see Sections 9 and 10). Although not entirely clear, these sugar compositions are suggestive of a mixture of *N*-linked and *O*-linked carbohydrates. This sugar composition is somewhat similar to that reported for the major cell wall glycoprotein of *C. reinhardtii* (2BII or volvin) by Roberts *et al.* (1985a). However, it is unlikely that the 350-kDa flagellar glycoprotein preparation is significantly contaminated with cell wall glycoproteins (most of which are hydroxyproline-rich glycoproteins) because of the absence of any detectable hydroxyproline in our purified flagellar glycoprotein preparation. Table I shows a comparison of the amino

Table I. *Chlamydomonas reinhardtii* **Cell Surface Glycoproteins**

Amino acid	Amino acid composition (no. of residues per 1000)			
	(+) Agglutinin[a]	(−) Agglutinin[a]	Wall protein[b]	350 kDa[c]
Lys	22	41	39	31
His	31	20	6	0
Arg	36	22	35	24
Asx	95	104	112	82
Thr	59	67	75	138
Ser	103	113	78	110
Glx	88	83	64	30
Pro	43	56	67	50
Gly	88	88	71	79
Ala	77	78	86	161
Val	46	62	68	88
Cys	50.6	ND[d]	ND	23
Met	10	8	12	11
Ile	27	35	38	40
Leu	51	36	69	94
Trp	ND	ND	ND	ND
Tyr	15	11	31	23
Phe	28	31	38	26
Hyp	123	120	112	8

[a]Data from Adair (1985).
[b]Data from Catt *et al.* (1976) for a major cell wall glycoprotein (2BI).
[c]Data from six runs (Bloodgood, unpublished results).
[d]ND, not determined.

acid composition of the 2BII cell wall glycoprotein, the vegetative 350-kDa flagellar membrane glycoproteins, and the sexual agglutinin molecules from plus and minus gametic flagella. One class of anticarbohydrate monoclonal antibodies prepared to the 2BII wall protein (Class III exemplified by the MAC-3 antibody) (Smith *et al.*, 1984) and a polyclonal antibody made to a 2BII glycopeptide rich in hydroxyproline, arabinose, and galactose (Roberts *et al.*, 1985b) both cross-react with the 350-kDa flagellar membrane

**Table II. The Molar Ratios
of the Constituent Monosaccharides
in 350-kDa Glycoproteins**

Monosaccharide[a]	L-23	PF-18
Arabinose	12	5
Rhamnose	0.3	0.2
Xylose	3	2
Mannose	10	2
Galactose	8	5
Glucose	0	1
N-Acetylglucosamine	1	1

[a]Values were normalized to GlcNAc.

glycoproteins. These observations suggest that the major cell wall protein (2BII) shares carbohydrate antigenic sites with the 350-kDa flagellar glycoproteins and explain why most of the anticarbohydrate monoclonal antibodies we and others have obtained to *C. reinhardtii* flagellar glycoproteins cross-react with the cell wall (see below).

The *C. reinhardtii* flagellar membrane contains several other proteins in addition to the 350-kDa glycoproteins, albeit in lower amounts (Monk *et al.*, 1983). Using surface iodination techniques, Bloodgood and Workman (1984) found that the principal flagellar protein available for surface iodination, besides the 350-kDa glycoproteins, is a protein of about 65 kDa, which has been further characterized by Remillard and Witman (1982) and Bloodgood (1984). It is unclear which of the iodinated flagellar surface components described by Monk *et al.* (1983), if any, correspond to this 65-kDa protein. Monk *et al.* (1983) have identified a number of other iodinatable flagellar surface components in gametic cells.

Associated with the *C. reinhardtii* flagellar surface are 0.9-μm-long structures referred to as mastigonemes; these structures have been purified (Witman *et al.*, 1972; Snell, 1976) and contain a single glycoprotein of 230 kDa (Monk *et al.*, 1983). The mastigoneme protein is the second most prominent protein (after the 350-kDa flagellar membrane glycoproteins) associated with the *C. reinhardtii* flagellar surface. This protein was not found to be significantly iodinated by Bloodgood and Workman (1984); however, it is labeled by surface biotinylation ("m" in Fig. 8).

In contrast to the great deal that is known about the synthesis, processing, and assembly of *Chlamydomonas* flagellar axonemal proteins (Silflow *et al.*, 1981; Lefebvre and Rosenbaum, 1986), relatively little is known about the synthesis, processing, insertion, and turnover of *Chlamydomonas* flagellar membrane proteins (Bloodgood, 1987), or any other plasma membrane or cell wall protein in this organism for that matter. There is no known endocytic pathway in *Chlamydomonas;* flagellar membrane turnover occurs by the budding of membrane vesicles into the medium. This process of flagellar membrane blebbing occurs in both vegetative and gametic cells; in gametic cells, this is the origin of the "gamone" which is known to isoagglutinate gametes of the opposite mating type (Wiese, 1965; McLean *et al.*, 1974). This flagellar blebbing phenomenon demands that there be turnover of flagellar membrane lipids and proteins. It is known that the 350- and 65-kDa flagellar membrane proteins turn over faster than any other flagellar proteins (Bloodgood, 1984; Remillard and Witman, 1982) in the vegetative cells.

Recent evidence from *C. eugametos* (Musgrave *et al.*, 1986) and *C. reinhardtii* (Bloodgood, 1988b) suggests that a barrier to protein diffusion exists between the flagellar membrane domain and the rest of the plasma membrane. This would necessitate that proteins destined for the flagellar membrane be inserted into the cell surface at the flagellar membrane domain. There is no evidence that membrane vesicles can enter the flagellum; indeed, the base of the flagellum is blocked by a basal body complex that is tightly adherent to the surrounding plasma membrane (Ringo, 1967). This plasma membrane targeting problem raises interesting mechanistic questions similar to those being approached in other systems, such as polarized epithelial cells and mammalian sperm (Gumbiner and Louvard, 1985; Wolf, 1987; Cardullo and Wolf, this volume). A recent electron microscopic immunolocalization study (Grief and Shaw, 1987) utilizing a monoclonal antibody to the *C. reinhardtii* cell wall 2BII glycoprotein that cross-reacts with flagellar membrane proteins suggests an unusual secretory pathway in *Chlamydomonas*. This study suggests that the contractile vacuole may be the route by which these glycoproteins (after transit through the Golgi) reach the cell surface.

In addition to those proteins common to the vegetative and gametic flagellar membrane, there is one class of very-high-molecular-weight glycoproteins that is specific to the gametic flagellar membrane. These are the species-specific and mating type-specific sexual agglutinin molecules; they are very-high-molecular-weight, hydroxyproline-rich glycoproteins that possess a very characteristic morphology when viewed in the electron microscope (Adair, 1985; van den Ende, 1985). It is interesting to note that many monoclonal antibodies prepared to the purified sexual agglutinin glycoproteins from *C. reinhardtii* flagella also cross-react with the *C. reinhardtii* cell wall glycoproteins and with nonagglutinin flagellar glycoproteins, including the 350-kDa flagellar membrane glycoproteins (Adair, 1985). Gametes possess a pool of sexual agglutinin molecules within the cell body; flagellar agglutination between plus and minus gametes greatly speeds up the turnover of the flagellar agglutinin molecules (Saito *et al.*, 1985; Snell and Moore, 1980). Demets *et al.* (1988) have demonstrated that sexual agglutination in *C. eugametos* is a self-enhancing process; agglutination results in a change in the flagellar surface that increases both agglutinability and binding of a monoclonal antibody (66.3) specific for the sexual agglutinin. Goodenough (unpublished results) has recently observed that raising the intracellular cAMP level in unmated gametic cells results in a tenfold increase in the amount of sexual agglutinin on the flagellar membrane of *C. reinhardtii*.

7. Dynamics of Flagellar Membrane Glycoproteins

Two classes of carbohydrate-binding probes have been utilized to follow the distribution of surface-exposed glycoproteins in the *C. reinhardtii* flagellar membrane: (1) anticarbohydrate mouse monoclonal antibodies and (2) the lectin Con A. In particular, one monoclonal antibody, designated FMG-1, primarily recognizes the 350-kDa glycoproteins in the flagellar membrane, in addition to cross-reacting with cell wall polypeptides containing the same or a similar carbohydrate epitope (Bloodgood *et al.*, 1986). Based on lectin affinity chromatography and lectin blots, the 350-kDa glycoproteins are the primary, although not exclusive, proteins of the flagellar membrane recognized by Con A (see Section 6).

When fixed cells or live cells at 4°C are incubated with fluorescein-labeled Con A or FMG-1 monoclonal antibody, both are observed to label the flagellar surface and the cell wall uniformly. If live gametic or vegetative cells are labeled in the cold with either of these reagents and then are warmed to room temperature, the label redistributes within the plane of the flagellar membrane in a characteristic pattern on a time scale of minutes (Fig. 9), reminiscent of the capping of receptors observed on lymphocytes (Schreiner and Unanue, 1976) and many other mammalian cell types. The redistribution is temperature-sensitive; it does not occur at 4°C and proceeds progressively more rapidly as the temperature is raised (Bloodgood *et al.*, 1986). This process (reviewed in Bloodgood, 1987) occurs on both vegetative and gametic cells and was named *tipping* by Goodenough and

Jurivich (1978), who first observed it on *Chlamydomonas* flagella. This relatively slow redistribution of flagellar membrane proteins is phenomenologically different from the faster (time scale of seconds), local and bidirectional movements of flagellar surface domains that are visualized as microsphere movements (Section 3) and gliding motility (Section 2.2). Gametes of *C. eugametos* will exhibit tipping of a wheat germ agglutinin-binding protein and an antibody to the sexual agglutinin glycoprotein (Homan *et al.*, 1987, 1988; Musgrave and van den Ende, 1987; Kooijman, personal communication); in this species, it has been demonstrated that redistribution of one flagellar protein can occur independent of redistribution of other flagellar proteins. As is the case with capping of surface receptors in lymphocytes (Taylor *et al.*, 1971; Schreiner and Unanue, 1976), cross-linking of the flagellar surface-exposed proteins is necessary to induce their redistribution; Fab antibody fragments and succinyl Con A are ineffective. A similar process of membrane protein redistribution and shedding induced by antibodies has been reported for the cilia of *Paramecium* (Barnett and Steers, 1984), the flagella of the parasitic protozoan *Leishmania* (Dwyer, 1976) and sea urchin sperm tail flagella (Trimmer and Vacquier, 1988).

Several observations suggest that tipping of flagellar membrane proteins may be involved in mating in *Chlamydomonas*. Using labeled monoclonal antibody (66.3) to the sexual agglutinin of *C. eugametos*, van den Ende's laboratory (Homan *et al.*, 1987; Musgrave and van den Ende, 1987) showed that during the normal process of mating in *C. eugametos*, the sexual agglutinin molecules on the flagellar surface redistribute to the flagellar tips, while other flagellar surface proteins do not redistribute. This process may be reflected in the accumulation of fuzzy material at the gametic flagellar tips reported by Mesland *et al.* (1980) and may be involved in the processes of flagellar tip activation, flagellar tip locking, and flagellar signaling (see Section 5). After cell fusion and the deadhesion of the flagellar surfaces, the labeled monoclonal antibodies are lost from the flagellar surfaces. All of the multivalent ligands (lectins, antibodies) that induce the early events of mating in gametes of a single mating type [polyclonal antisera to the *C. reinhardtii* gametic or vegetative flagellum (Goodenough and Jurivich, 1978; Claes, 1977), monoclonal antibody 66.3 to the sexual agglutinin of *C. eugametos* (Homan *et al.*, 1988), Con A in the case of *C. reinhardtii* (Claes, 1975, 1977), wheat germ agglutinin in the case of *C. eugametos* (Kooijman, personal communication)] also redistribute (exhibit tipping) along the flagellar surface (Goodenough and Jurivich, 1978; Bloodgood *et al.*, 1986; Homan *et al.*, 1987, 1988; Musgrave and van den Ende, 1987).

Mesland *et al.* (1980) reported that high concentrations of colchicine (10 mg/ml) and vinblastine (120 μg/ml) inhibit flagellar tip alignment, flagellar tip activation, and quadriflagellate pair formation during mating in *C. reinhardtii*. Homan *et al.* (1988) observed that a similar concentration of colchicine inhibited glycoprotein redistribution in the gametic flagella of *C. eugametos*. Taken together, these observations suggest that glycoprotein redistribution in the flagellum is necessary for successful mating. Further, the observations of Homan and van den Ende suggest that glycoprotein redistribution along the flagellar surface may be dependent upon tubulin in some form that is sensitive to colchicine, perhaps a membrane-associated tubulin lattice similar to that described by Stephens *et al.* (1987) for mollusk gill cilia. Bloodgood and Salomonsky (unpublished results) found that redistribution of a monoclonal antibody along the vegetative flagellar surface of *C. reinhardtii* was rather insensitive to colchicine; a significant reduction in the rate of redistribution was only seen at 20 mg/ml or higher. Observations showing an effect of antimitotic drugs on *Chlamydomonas* flagella are somewhat puzzling because intact flagellar microtubules are insensitive to these drugs and a previous report of membrane tubulin in *C. reinhardtii* flagella (Adair and Goodenough, 1978) was later retracted (Monk *et al.*, 1983). However, flagellar tip activation is accompanied by a certain amount of tubulin assembly onto the A subfibers of the outer doublets (Mesland *et al.*, 1980), which one would certainly expect to be inhibited by the concentration of colchicine utilized (Rosenbaum *et al.*, 1969); this could explain the colchicine effect on flagellar tip activation. Vinblastine has been reported to bind to calmodulin in a calcium-dependent manner (Watanabe and West, 1982) and to inhibit the activation of the erythrocyte calmodulin-dependent Ca^{2+}-transport ATPase (Gietzen *et al.*, 1982) and brain calmodulin-dependent phosphodiesterase activity (Watanabe and West, 1982). Since calmodulin antagonists inhibit flagellar tip activation (Detmers and Condeelis, 1986), the effect of vinblastine on this *Chlamydomonas* mating event could be due to an effect on calmodulin. High concentrations of colchicine have been reported to have a direct effect on a number of plasma membrane functions (Cheng and Katsoyannis, 1975; Mizel and Wilson, 1972; Rembold and Langenbach, 1978; Schellenberg and Gillespie, 1977).

In summary, little is currently known about the actual mechanism involved in the active movement of flagellar membrane proteins in the plane of the vegetative and gametic flagellar membrane but growing evidence suggests that these movements have important implications both for mating in gametes and for whole cell gliding motility (see Section 10). We are, however, beginning to learn more about the transmembrane signaling mechanisms whereby an alteration of the flagellar surface (interaction of sexual agglutinin molecules or cross-linking with antibodies or lectins) activates the machinery for membrane protein redistribution.

8. Flagellar Signaling in *Chlamydomonas*

Events occurring at the cell surface are communicated to the interior of the cell by a process referred to as transmembrane signaling (Hollenberg, 1986). In particular, the binding of ligands to plasma membrane proteins can result, often through the action of G proteins, in the activation of membrane-associated enzymes such as adenylate cyclase and phospholipase C. This results in changes in the levels of intracellular second messengers, such as calcium, cAMP, cGMP, IP_3, IP_4, and diacyl glycerol. These low-molecular-weight second messengers can then alter the activity or function of cytoplasmic proteins and enzymes, in particular protein kinases. Binding of antibodies or lectins to cell-surface proteins (usually receptors for specific ligands) can result in signals being sent to the cytoplasm of the cell resulting in activation of a variety of cellular processes (Kahn *et al.*, 1981; Schreiber *et al.*, 1983; Lis and Sharon, 1986), including redistribution (patching and capping) of cell-surface receptors (Taylor *et al.*, 1971; Schreiner and Unanue, 1976).

One of the most exciting recent developments in the study of cilia and flagella is the growing awareness that transmembrane signaling events are critically involved in the dynamic events associated with cilia and flagella. Olfactory cilia are enriched in a G_s

regulatory GTP-binding protein as well as an odorant and GTP-activated adenylate cyclase (Anholt, 1987; Lancet and Pace, 1987). In addition, olfactory cilia have a membrane conductance that is gated directly by cAMP or cGMP (Nakamura and Gold, 1987). Calcium-regulated guanylate and adenylate cyclases and cyclic nucleotide-activated protein kinases have been described in *Paramecium* cilia (Klumpp and Schultz, 1982; Schultz *et al.*, 1987; Gustin and Nelson, 1987; Schultz and Jantzen, 1980; Preston and Saimi, this volume) and mammalian spermatozoa (Gross *et al.*, 1987; Garbers *et al.*, 1973) and are clearly involved in the regulation of ciliary and flagellar motility in these systems (Tash *et al.*, 1986; Preston and Saimi, this volume). Much recent data have been obtained concerning the signaling mechanisms that operate in vegetative and gametic cells of *Chlamydomonas*, as will be discussed in the following two sections.

8.1. Signaling Related to Gametic Interactions

As already mentioned, mating in *Chlamydomonas* involves a flagellar signaling event in which contact of the surface of plus and minus gametic flagella results in FTA, cell wall release, and mating structure activation (Snell, 1985; van den Ende, 1985). It is known that, under certain circumstances, antibodies to flagellar membrane proteins and certain lectins (Con A in *C. reinhardtii* and wheat germ agglutinin in *C. eugametos*) can induce mating events in a population of unmated gametes (Claes, 1975, 1977; Goodenough and Jurivich, 1978; Homan *et al.*, 1988; van den Ende *et al.*, this volume). Cross-linking of flagellar membrane proteins appears to be an essential step in this process because monovalent Fab fragments are not effective (Goodenough and Jurivich, 1978; Homan *et al.*, 1988). This suggests that cross-linking of the flagellar surface may induce transmembrane signaling and an alteration in the level of a second messenger within the flagellum or cell body. What is the second messenger utilized for signal transduction in the *Chlamydomonas* flagellum?

Calcium was the first candidate to be considered as a second messenger in gametic flagellar signaling. A number of observations indirectly suggest an involvement of calcium: (1) trifluoperazine (TFP) and W-7 (both inhibitors of calmodulin) do not affect gametic flagellar agglutination but inhibit all subsequent steps in mating (Detmers and Condeelis, 1986); (2) lidocaine (xylocaine) does not affect gametic flagellar agglutination but prevents two of the major consequences of signaling (cell wall release and mating structure activation) (Snell *et al.*, 1982); (3) a transient increase in efflux of calcium occurs upon initiation of mating (Bloodgood and Levin, 1983); and (4) mating is accompanied by the redistribution of calcium within the cell body (Kaska *et al.*, 1985). On the other hand, mating does not require the presence of calcium in the medium (Bloodgood and Levin, 1983); this suggests that intracellular free calcium levels, if important to events in mating, are being regulated using an intracellular source of calcium (Kaska *et al.*, 1985).

The presence of calcium-regulated adenylate cyclase activity in *Paramecium* cilia (Gustin and Nelson, 1987; Schultz *et al.*, 1987), *Chlamydomonas* flagella (Pasquale and Goodenough, 1987), and mammalian sperm (Gross *et al.*, 1987) suggests that there may be integration of both calcium and cAMP signaling systems in cilia and flagella (Stephens and Stommel, 1989; Tash and Means, 1983).

Mating in both *C. eugametos* (Pijst *et al.*, 1984) and *C. reinhardtii* (Pasquale and Goodenough, 1987) is accompanied by a transient rise in intracellular cAMP. In *C. reinhardtii*, unmated gametes can be sexually activated by exposing them to a combination of exogenous dibutyryl cAMP and isobutyl methylxanthine (IBMX; an inhibitor of cyclic nucleotide phosphodiesterase) (Pasquale and Goodenough, 1987). Inhibitors of calmodulin (TFP, W-7) interfere with mating in *C. reinhardtii* (Detmers and Condeelis, 1986) and in *C. eugametos* (Pijst, 1985). In *C. reinhardtii*, TFP interferes with the mating-induced rise in intracellular cAMP; TFP inhibition can be overcome by a combination of exogenous dibutyryl cAMP and IBMX (Pasquale and Goodenough, 1987). Calmodulin is found in association with the membrane fraction of *C. reinhardtii* flagella (Gitelman and Witman, 1980) and the flagellar membrane contains the highest specific activity of adenylate cyclase in the cell (Pasquale and Goodenough, 1987); flagellar membrane adenylate cyclase activity in inhibited by TFP *in vitro*, albeit at very high concentrations of TFP (Pasquale and Goodenough, 1987). Although TFP inhibition studies should be interpreted with some caution (Roufogalis, 1982), the observations cited above justifiably led Pasquale and Goodenough (1987) to propose that the mating-associated rise in intracellular cAMP may be due to an activation of a flagellar membrane-associated, calcium–calmodulin-regulated adenylate cyclase, with or without some modulation in activity of a cAMP-dependent phosphodiesterase activity. This suggests that a mating-induced rise in free calcium concentration might be expected in order to activate this calmodulin-dependent adenylate cyclase activity. Some evidence for calcium fluxes during mating (Snell *et al.*, 1982; Bloodgood and Levin, 1983; Kaska *et al.*, 1985) has already been discussed; however, a number of workers, including Pasquale and Goodenough (1987), have failed to confirm the original reports by Claes (1980) and Goodenough (1980) that gametes could be activated using the divalent cation ionophore A23187 in the presence of external calcium.

An alternative mechanism for activation of the membrane-associated adenylate cyclase could be via a G_s regulatory protein (Gilman, 1987). Although Pasquale and Goodenough (1987) argued against a role for G proteins in regulation of the *Chlamydomonas* flagellar membrane adenylate cyclase, these same workers now claim (personal communication) that GTPτS stimulates this adenylate cyclase activity.

What is the mechanism by which a rise in intracellular cAMP activates later events in mating in *Chlamydomonas*? One obvious possibility is that the rise in intracellular cAMP activates a cyclic nucleotide-dependent protein kinase. Pijst *et al.* (1984) demonstrated that *C. eugametos* gametes contain a cAMP-dependent protein kinase and Hasegawa *et al.* (1987) have shown that *C. reinhardtii* vegetative flagella have a cAMP-dependent protein kinase activity. Pasquale and Goodenough (1987) have shown that postagglutination mating events in *C. reinhardtii* are inhibited by H-8, an inhibitor of cyclic nucleotide-dependent protein kinases (Hidaka *et al.*, 1984).

8.2. Signaling Related to Glycoprotein Redistribution in Vegetative Flagella

Because redistribution of flagellar membrane proteins accompanies the early events of mating (Homan *et al.*, 1987; Musgrave and van den Ende, 1987; van den Ende *et al.*, this volume) and because antibodies and lectins can induce sexual signaling in gametes (Section 8.1), the redistribution of the 350-kDa glycoproteins along the vegetative and gametic flagellar surface induced by antibodies and Con A (Bloodgood *et al.*, 1986) might be expected to involve signaling events similar to those utilized during mating. Recent results from the author's laboratory (Bloodgood and Salomonsky, 1988) suggest that this may be the case.

Redistribution of the FMG-1 mouse monoclonal antibody and Con A along the flagella of both vegetative and gametic cells of *C. reinhardtii* is inhibited by the calmodulin antagonists TFP (Fig. 10) and W-7 (Fig. 11), and this inhibition can be reversed by providing exogenous dibutyryl cAMP (Fig. 10) or dibutyryl cGMP or by washing out the drug (Fig. 11). The relatively inactive analogue, W-5, is without effect. These observations are similar to those of Pasquale and Goodenough (1987) that TFP can inhibit sexual signaling and that this inhibition can be reversed by raising intracellular cAMP levels. Our observations suggest that cAMP may be used as a signal for activating the machinery necessary for redistribution of glycoproteins in the plane of the flagellar membrane. We postulate that antibody cross-linking of the flagellar surface results in a calcium–calmodulin activation of flagellar membrane-associated adenylate cyclase which results in an increase in the cAMP level within the flagellum and/or the cell body. Since both *C. reinhardtii* vegetative cells (Hasegawa *et al.*, 1987) and *C. eugametos* gametic cells (Pijst *et al.*, 1984) have been reported to contain a cAMP-dependent protein kinase activity, it is likely that the rise in intracellular cAMP concentration is activating a cyclic nucleotide-dependent protein kinase. Support for this comes from the observation that H-8, an inhibitor of cyclic nucleotide-dependent protein kinases (Hidaka et al., 1984), completely inhibits redistribution of the FMG-1 antibody along the vegetative flagellar surface of *C. reinhardtii* (Fig. 12). Pasquale and Goodenough (1987) demonstrated that H-8 completely inhibited the signaling events that are essential for successful mating of *Chlamydomonas* gametes. As would be expected, neither effect of H-8 was reversed by exogenous dibutyryl cAMP (Fig. 12; Pasquale and Goodenough, 1987).

Assuming that protein phosphorylation is a step in the signal transduction pathway in both vegetative and gametic flagella, it will be of interest to learn what the flagellar substrates are for the cAMP-dependent protein kinase and to determine how protein phosphorylation is regulating the machinery responsible for flagellar membrane protein redistribution. A large number of phosphorylated proteins have been identified in the *C. reinhardtii* flagellum (Huang *et al.*, 1981) including flagellar proteins whose phosphorylation is stimulated by calcium (Segal and Luck, 1985) or cAMP (Hasegawa *et al.*, 1987).

If a flagellar membrane adenylate cyclase is regulated by calcium and calmodulin and if the activation of this enzyme is necessary for the induction of flagellar membrane protein redistribution and for the postagglutination events in gamete mating, then it would be expected that changes in intracellular calcium levels would result from sexual agglutination and cross-linking of the flagellar surface with antibodies or lectins. Evidence for a role of calcium in mating has already been cited. Contrary to what was originally reported (Claes, 1980; Goodenough, 1980), it appears that the ionophore A23187 in the presence of excess calcium will not activate gametes (Pasquale and Goodenough, 1987). Flagellar glycoprotein redistribution in vegetative cells is dependent upon a critical concentration of free calcium in the extracellular medium (Bloodgood and Salomonsky, unpublished observations). At a free calcium concentration of 10^{-7} M or lower, no redistribution of the FMG-1 monoclonal antibody is observed; at a free calcium concentration of 2×10^{-5} M, maximal redistribution is observed. At intermediate free calcium concentrations, partial redistribution is observed. It is interesting to note that the calcium dependence of antibody redistribution closely mirrors that for polystyrene microsphere movements (Bloodgood *et al.*, 1979).

At least two other flagellar functions in *Chlamydomonas* can be regulated by alterations in the levels of both cAMP and calcium. Flagellar beating motility is regulated by intracellular calcium (Bessen *et al.*, 1980; Hyams and Borisy, 1978) and cAMP levels (Rubin and Filner, 1973; Hartfiel and Amrhein, 1976; Hasegawa *et al.*, 1987). Hasegawa *et al.* (1987) identified two axonemal proteins that are phosphorylated in response to cAMP-induced inhibition of axonemal motility. Flagellar stability appears to be regulated by calcium and cAMP levels in an antagonistic manner; lowering the calcium (Lefebvre *et al.*, 1978; Quader *et al.*, 1978) and raising the cAMP (Rubin and Filner, 1973; Lefebvre *et al.*, 1980) levels both lead to flagellar resorption and prevent regeneration. May and Rosenbaum (1983) showed that a set of axonemal proteins (including several radial spoke proteins) were dephosphorylated during flagellar resorption and phosphorylated during flagellar regeneration.

The flagellar membrane protein redistribution (tipping) phenomenon demonstrates that flagellar membrane glycoproteins can exhibit dramatic movements in the plane of the flagellar membrane in both gametes and vegetative cells. The inhibitor studies using TFP, W-7, and H-8 cited above suggest that a similar transmembrane signaling system (utilizing cAMP and calcium as second messengers) may operate in both vegetative and gametic flagella (Pasquale and Goodenough, 1987; Bloodgood and Salomonsky, 1988). The active redistribution of membrane proteins in the plane of the flagellar membrane is likely to be an important event in mating interactions of gametic cells (see Section 7) and may also be an essential part of the mechanism underlying gliding motility (see Section 10).

9. Use of Carbohydrate Probes in Conjunction with FACS to Isolate Mutant Cell Lines with Carbohydrate Defects

In order to understand the mechanism of gliding motility, it is of interest to ask whether the movement of flagellar membrane proteins within the flagellar membrane is essential for the expression of gliding motility (and microsphere movement). This is suggested by the observation that TFP and W-7 inhibit both flagellar glycoprotein redistribution (Figs. 10, 11) and polystyrene microsphere movements (Detmers and Condeelis, 1986), the latter clearly being a manifestation of the same motor used for gliding motility. We sought a more direct way to prevent the lateral movement of flagellar membrane glycoproteins in *C. reinhardtii* in order to test the hypothesis that the lateral movements of flagellar membrane glycoproteins are a prerequisite for gliding motility. This was accomplished by obtaining a unique mutant cell line with altered cell-surface properties (Bloodgood and Salomonsky, 1989).

We have utilized fluorescence-activated cell sorting (FACS) to select mutagenized *C. reinhardtii* strain PF-18 cells that express an elevated or reduced level of binding of

Con A or an anticarbohydrate mouse monoclonal antibody (FMG-1) which recognizes primarily the 350-kDa flagellar glycoproteins (Bloodgood *et al.*, 1986). Two of these strategies were successful: cell lines were obtained that expressed (1) significantly elevated levels of Con A binding and (2) a total absence of binding of the FMG-1 monoclonal antibody (Bloodgood *et al.*, 1987). Interestingly, some of the cell lines that were selected for the absence of FMG-1 monoclonal antibody binding also expressed elevated levels of Con A binding and all of the cell lines that were selected for elevated binding of Con A failed to exhibit binding of the FMG-1 monoclonal antibody. One mutant cell line expressing both of these properties, designated L-23 (Figs. 13, 14), was studied in detail (Bloodgood *et al.*, 1987). Table II shows that the high-molecular-weight flagellar glycoproteins from this mutant exhibit an increased abundance of mannose residues, suggesting a defect in *N*-linked oligosaccharide processing.

Under the conditions normally utilized to study the redistribution of Con A (100 μg/ml) along the PF-18 flagellar surface, the L-23 cell line exhibited no redistribution (tipping), presumably because of extensive cross-linking of the Con A-binding glycoproteins within the plane of the flagellar membrane preventing their lateral movement. This hypothesis was confirmed by reducing the level of Con A (to approximately 10 μg/ml) until the degree of binding of the mutant flagellar surface was approximately equivalent to

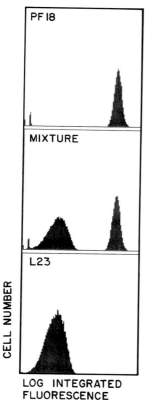

Figure 13. Flow cytometer histogram of integrated fluorescence intensity versus cell number for the PF-18 parent cell line (top panel), the L-23 mutant cell line derived from mutagenized PF-18 cells (bottom panel), and a 1 : 1 mixture of the two cell lines (middle panel). The cells were labeled with fluorescein-conjugated FMG-1 monoclonal antibody. Log fluorescence on the abscissa increases from left to right. Reproduced from Bloodgood *et al.* (1987) with permission.

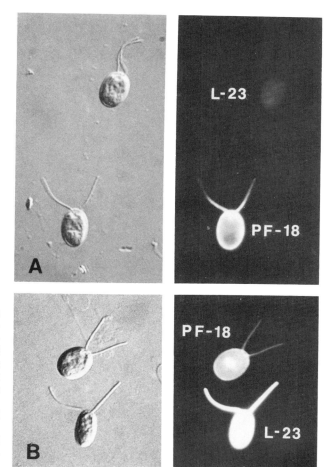

Figure 14. Differential interference contrast (left panels) and immunofluorescence (right panels) microscopy of a mixture of PF-18 and L-23 cells stained with the FMG-1 mouse monoclonal antibody (A) or with FITC–Con A (B). The L-23 cells stain very poorly with the FMG-1 monoclonal antibody and very well with Con A relative to the PF-18 cells. Magnification 1100×. Reproduced from Bloodgood (1988b) with permission.

what was seen with the parent cell strain (PF-18) at 100 μg/ml. Under these conditions, the L-23 mutant cell line exhibited Con A redistribution, albeit at a rate somewhat slower than the PF-18 parent cell line (see Table III and Bloodgood and Salomonsky, 1989).

10. Use of the L-23 Mutant Cell Line to Demonstrate That Flagellar Membrane Glycoprotein Movements Are Essential for Gliding Motility

The observation that polystyrene microspheres move along the flagellar surface of *Chlamydomonas* flagella suggests that some protein or protein complex to which the microsphere adheres is moving within the lipid bilayer of the flagellar membrane during the course of microsphere motility and during gliding motility. In fact, one can observe the movement of irregularly shaped microsphere aggregates along the flagellar surface, and the orientation of the aggregate relative to the long axis of the flagellum remains fixed

**Table III. Effect of Con A on Glycoprotein Redistribution
and Microsphere Movement**

Cell strain	Con A concentration	Con A redistribution	Microsphere movement	Gliding motility
PF-18	10 µg/ml	Yes	Yes	Yes
	50 µg/ml	Yes	Yes	Yes
	100 µg/ml	Yes	Yes	Yes
L-23	10 µg/ml	Yes	Yes	Yes
	50 µg/ml	No	No	No
	100 µg/ml	No	No	No

during transit. This observation suggests a continuous attachment of the surface marker to some component(s) of the flagellar membrane surface during microsphere movement.

The L-23 mutant cell line (described in Section 9) provided an opportunity to determine whether the expression of microsphere movement and gliding motility depends on the ability of the major Con A-binding glycoproteins of the flagellar membrane to move within the plane of the flagellar membrane (Bloodgood and Salomonsky, 1989). PF-18 and L-23 cells lines were assayed for their ability to express glycoprotein redistribution, microsphere movement, and gliding motility after treatment with various concentrations of Con A (Table III). It was observed that, under all conditions where cells were able to exhibit glycoprotein redistribution, they were able to move polystyrene microspheres and to exhibit whole cell gliding motility. Under all conditions where cells were unable to redistribute the Con A-binding glycoproteins, they were also unable to move polystyrene microspheres or to express gliding motility. This correlation strongly argues that the expression of gliding motility requires the movement, within the plane of the flagellar membrane, of one or more Con A-binding glycoproteins (Bloodgood and Salomonsky, 1989). This is probably the first demonstration, in any cell type, that whole cell locomotion along a solid substrate requires glycoprotein movement within the plane of the plasma membrane. Based upon Western blots of whole flagella probed with Con A and upon Con A Sepharose affinity chromatography of nonionic detergent extracts of flagella, the principal Con A-binding glycoproteins of the flagellar membrane are the 350-kDa glycoproteins (Monk *et al.*, 1983; Bloodgood, unpublished observations). The major flagellar surface-exposed proteins that contact the substrate during the expression of gliding motility and polystyrene microsphere movement are the 350-kDa flagellar glycoproteins (Bloodgood and Workman, 1984). Taken together, these observations suggest, but do not prove, that the 350-kDa glycoproteins move within the flagellar membrane during the expression of gliding motility in *C. reinhardtii*.

11. Mechanisms and Motors

11.1. Candidates for the Motor Responsible for Flagellar Membrane Protein Redistribution and Gliding Motility

Because of the easy accessibility of the flagellar surface to probes such as antibodies and lectins, much attention has been paid to characterizing the flagellar surface proteins that may be involved in the dynamic events associated with gliding motility and the early events of mating in *Chlamydomonas*. Currently, much less is known about the energy-transducing system (motor) responsible for the cell surface applying force to the substrate, presumably through the active translocation of integral membrane proteins in the plane of the flagellar membrane. Properties expected of the motor are that it have ATPase activity and be capable of interacting with a population of flagellar membrane proteins, presumably the 350-kDa Con A-binding glycoproteins. In addition, it would not be surprising if the motor were associated with the outer doublet microtubules. Based on the preliminary data utilizing H-8 to inhibit cyclic nucleotide-dependent protein kinases (Section 8.2), it is possible that some component of the motor is a substrate for phosphorylation by a member of this class of protein kinases.

Until recently, most cases of eukaryotic cell motility were thought to be associated with one of two energy transducers: myosin (operating in association with f-actin filaments) and dynein (operating in association with microtubules). Recently, a number of additional microtubule-associated translocator molecules have been discovered. These include: kinesin, which has been found in squid axoplasm (Vale *et al.*, 1985a), mammalian brain (Kuznetsov and Gelfand, 1986), sea urchin eggs (Porter *et al.*, 1987), and *Drosophila* (Saxton *et al.*, 1988); MAP1C, a dynein analogue from mammalian brain (Paschal *et al.*, 1987); a 292-kDa protein from squid nervous tissue (Gilbert and Sloboda, 1986); a 400-kDa protein isolated from the nematode *Caenorhabditis* (Lye *et al.*, 1987); and a 440-kDa protein from the giant amoeba *Reticulomyxa* (Euteneuer *et al.*, 1988). Only the translocator molecule isolated from *Reticulomyxa* (Euteneuer *et al.*, 1988) has been shown to support bidirectional movement of polystyrene microspheres along microtubules *in vitro*. All of the other known translocator molecules exhibit unidirectional movement; kinesin and kinesinlike molecules exhibit microtubule plus-end directed movements (equivalent to movement of microspheres toward the tip of the flagellum) while dynein and dyneinlike molecules exhibit microtubule minus-end directed movements (equivalent to movement of microspheres toward the base of the flagellum). Since polystyrene microsphere movements along the flagellar surface of *Chlamydomonas* are clearly bidirectional (Section 3), there are two possibilities: (1) more than one translocator molecule is involved in flagellar surface dynamics or (2) the flagellar surface is associated with a bidirectional motor, such as appears to operate in *Reticulomyxa* (Euteneuer *et al.*, 1988).

Particularly relevant to the present discussion are those translocator molecules that have been found associated with the *Chlamydomonas* flagellum. The principal energy-transducing molecule in all cilia and flagella is dynein; *Chlamydomonas* flagellar axonemes contain six dynein ATPases localized to the inner and outer arms (Piperno, 1988; Witman, 1989). A number of *C. reinhardtii* dynein outer arm mutants (including PF-13, PF-22, PF-28, and ODA-38) have been examined in terms of their surface properties and found to exhibit normal gliding motility, polystyrene microsphere movement, and antibody redistribution (Bloodgood, unpublished results). This suggests that outer arm dynein ATPases are not the motors underlying any of the flagellar surface properties in *C. reinhardtii*.

With the exception of quail cilia (Sandoz *et al.*, 1982), myosin has never been reported in cilia or flagella. Actin has been reported in both quail oviduct cilia (Sandoz *et al.*, 1982) and *Chlamydomonas* flagella (Piperno and Luck, 1979; Detmers *et al.*, 1985). The actinlike protein from the *C. reinhardtii* flagellum appears to be a component of the inner dynein arm (Piperno and Luck, 1979; Piperno, 1988) and hence is unlikely to be in a position to affect events at the flagellar surface.

Watanabe and Flavin (1976) identified a 3 S calcium-specific ATPase in the *C. reinhardtii* flagellum; this enzyme activity has been localized to the membrane–matrix compartment (Bessen *et al.*, 1980). Because this enzyme does not appear to be an integral membrane protein, it is unlikely to be a calcium pump for the flagellum, leaving it as a good candidate for a translocator molecule.

No efforts have been made to use the *in vitro* motility assays (movement of microspheres coated with protein along microtubules or microtubule gliding on a glass surface coated with the protein) developed for identifying translocator molecules in other systems (Porter *et al.*, 1987; Paschal and Vallee, 1987) to assay flagellar extracts or partially purified flagellar proteins for translocator activity. This should be a fruitful avenue for future investigation, especially if used in combination with mutant cell lines having defects in gliding motility.

11.2. A Proposed Mechanism for Gliding Motility

The preceding sections have summarized what is known about gliding motility and about the dynamics of flagellar membrane proteins in vegetative and gametic cells of *Chlamydomonas*. This information has been used as the basis for formulating a working model for the mechanism of whole cell gliding motility. In the model (Fig. 15), a large transmembrane glycoprotein has an adhesive site on the outside surface of the flagellum and a site for connection to the motor on the cytoplasmic surface of the flagellar membrane. Alternatively, there may be a protein complex in which the two functions are associated with two different polypeptides. Cross-linking (and perhaps local clustering) of glycoproteins on the flagellar surface by contact with a solid substrate (glass surface or polystyrene microsphere) induces a signal (rise in intracellular cAMP level due to an activation of a calcium–calmodulin-dependent flagellar membrane-associated adenylate cyclase). This rise in cAMP then activates a cAMP-dependent protein kinase resulting in phosphorylation of one or more flagellar proteins. Protein phosphorylation directly or indirectly activates an ATPase-containing cross-bridge located between the outer doublet microtubules and the cytoplasmic domains of the large transmembrane flagellar glycoproteins. In analogy with dynein bridge action, ATP hydrolysis is coupled to conformational changes in the bridge structure that result in force being applied to the flagellar membrane glycoproteins, causing their lateral movement within the flagellar membrane in the direction defined by the outer doublet microtubule. The movement of the 350-kDa transmembrane glycoprotein in the plane of the flagellar membrane applies force to the substrate resulting in either the movement of a microsphere along the flagellar surface or the movement of the entire cell along the substrate in the opposite direction.

The illustration in Fig. 15 suggests two alternative possibilities for the action of the flagellar membrane–cytoskeleton cross-bridge structure. In one case (the "carrying the torch" model), the cross-bridge remains permanently linked to the membrane protein complex while "walking" along the lattice of an outer doublet microtubule. This may be equivalent to what occurs when membranous organelles move along a flagellar axoneme (Gilbert *et al.*, 1985), when polystyrene microspheres coated with known translocator molecules move along brain microtubules (Vale *et al.*, 1985a,b), or when flagellar ax-

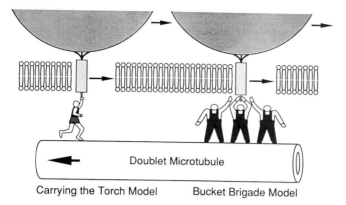

Figure 15. Diagram illustrating two models for the mechanism of gliding motility and polystyrene microsphere movement associated with the *Chlamydomonas* flagellum. It is postulated that a translocator molecule mechanically interconnects the outer doublet microtubule with 350-kDa glycoprotein complexes in the flagellar membrane. Both models postulate that conformational changes of the translocator molecule associated with ATP hydrolysis result in lateral movement of the 350-kDa glycoprotein complex within the plane of the flagellar membrane. In one model ("carrying the torch" model), the translocator molecule is permanently associated with the flagellar membrane glycoprotein complex and translocates along the surface of the outer doublet microtubule. In the second model ("bucket brigade" model), the translocator molecules are permanently associated with the surface lattice of the outer doublet microtubules and exhibit transient associations with the flagellar membrane glycoprotein complex. In this case, a membrane glycoprotein complex is passed along a series of translocator molecules. The net effect of either mechanism is that the translocation of the glycoproteins through the flagellar membrane results in the movement of a polystyrene microsphere along the flagellar surface or the movement of the entire flagellum (and attached cell) relative to the substrate.

onemes glide along a glass surface coated with known translocator molecules (Porter *et al.*, 1987; Paschal and Vallee, 1987). In the other scenario suggested in Fig. 15 (the "bucket brigade" model), the cross-bridges remain tightly associated with the outer doublet microtubule while transient interactions of neighboring cross-bridges with the membrane glycoproteins result in the transfer of the particular membrane protein complex from one cross-bridge to the next. These two models carry different predictions as to the relative affinities of the putative cross-bridge molecules for the membrane glycoproteins versus the outer doublet microtubules. Moreover, they help to define the approaches that will be appropriate for identifying and purifying the components of the cross-bridge. Consider the analogy with a dynein arm cross-bridge interconnecting the A subfiber of one microtubule doublet with the B subfiber of an adjacent doublet microtubule. In the case of the "torch" model, the microtubule would be functionally equivalent to the B subfiber; in the "bucket brigade" model, the microtubule would be functionally equivalent to the A subfiber.

The conclusion that directed flagellar glycoprotein movements, driven by an intraflagellar motor, underlie whole cell gliding motility is appealing. On the other hand, there are problems in visualizing the global redistribution of flagellar membrane glycoproteins observed in response to antibodies or lectins as the underlying basis for microsphere movement and gliding motility, which are more local and probably more highly controlled processes. One way to deal with this apparent discrepancy is to recognize that both the binding and movement of a polystyrene microsphere and the gliding of a single cell only involve limited portions of the flagellar surface. If we assume that the solid substrate contacts a limited region of the flagellar surface and locally cross-links a limited number of flagellar membrane glycoproteins or glycoprotein complexes, local patching of these glycoproteins may induce only local signal transduction and/or local activation of the mechanism responsible for movement of the flagellar membrane glycoproteins in the plane of the membrane. Indeed, Peng *et al.* (1981) demonstrated that polylysine-coated polystyrene microspheres bound to the surface of muscle cells in culture resulted in recruiting acetylcholine receptors to the points of contact of the microspheres with the muscle cell plasma membrane. The global capping-like phenomenon observed when the entire flagellar surface is coated with antibody or lectin is probably not relevant to what is occurring during gliding motility (or microsphere movement) although it may be more relevant to events occurring during mating (discussed in Sections 5, 7, and 8.1).

It is obvious that a number of regulatory features of gliding motility remain unaddressed by the proposed model. In particular, the ability of gliding cells and moving polystyrene microspheres to reverse their direction of movement remains to be explained.

The model presented here has certain similarities to those proposed to explain gliding motility and microsphere movement in the protozoan *Gregarina* (King, 1988) and the gliding bacterium *Cytophaga* (Lapidus and Berg, 1982). This is not to imply that these systems are likely to utilize similar molecular mechanisms for achieving gliding motility.

12. Conclusions

The flagellar membrane of *Chlamydomonas* is very dynamic, in a number of interesting ways. As an experimental system, it possesses a number of advantages for understanding cell–substrate interactions, transmembrane signaling, plasma membrane dynamics, and the force transduction responsible for whole cell locomotion. In this regard, the ultimate value of *Chlamydomonas* lies in the ability to combine the power of genetics and the specificity of immunological probes with the more traditional cell biological approaches. Recent data resulting from this integrated approach suggest that whole cell gliding motility in *Chlamydomonas* results from the directed movement of flagellar membrane glycoproteins within the plane of the flagellar membrane.

ACKNOWLEDGMENTS. The author's research discussed herein has been supported by research grants from the National Institutes of Health (GM 28766) and the National Science Foundation (DCB-8905530). Ms. Nancy Salomonsky has provided excellent technical assistance in the author's research. Dr. Herman van Halbeek provided assistance in determining the sugar compositions of flagellar membrane glycoproteins. Dr. Jim Sullivan assisted in the preparation of Figs. 1, 4, 10–12, and 15. Ms. Barbara Daniel and Dr. Conrad King (University College London) have kindly provided the illustrations for Figs. 3 and 5 along with access to their unpublished manuscripts. Comments on a draft version of the manuscript provided by Drs. H. van den Ende, W. Snell, and R. D. Sloboda resulted in a much improved final product.

References

Adair, W. S., 1985, Characterization of *Chlamydomonas* sexual agglutinins, *J. Cell Sci. Suppl.* **2**:233–260.

Adair, W. S., and Goodenough, U. W., 1978, Identification of a membrane tubulin in *Chlamydomonas* flagella, *J. Cell Biol.* **79**:54a.

Adair, W. S., Hwang, C., and Goodenough, U. W., 1983, Identification and visualization of the sexual agglutinin from the mating-type plus flagellar membrane of *Chlamydomonas, Cell* **33**:183–193.

Anholt, R. R. H., 1987, Primary events in olfactory reception, *Trends Biochem. Sci.* **12**:58–62.

Barnett, A., and Steers, E., Jr., 1984, Antibody-induced membrane fusion in *Paramecium, J. Cell Sci.* **65**:153–162.

Bergman, K., Goodenough, U. W., Goodenough, D. A., Jawitz, J., and Martin, H., 1975, Gametic differentiation in *Chlamydomonas reinhardtii.* II. Flagellar membranes and the agglutination reaction, *J. Cell Biol.* **67**:606–622.

Bessen, M., Fay, R. B., and Witman, G. B., 1980, Calcium control of waveform in isolated flagellar axonemes of *Chlamydomonas, J. Cell Biol.* **86**:446–455.

Bloodgood, R. A., 1977, Rapid motility occurring in association with the *Chlamydomonas* flagellar membrane, *J. Cell Biol.* **75**:983–989.

Bloodgood, R. A., 1980, Direct visualization of dynamic membrane events in cilia, *J. Exp. Zool.* **213**:293–295.

Bloodgood, R. A., 1981, Flagella-dependent gliding motility in *Chlamydomonas, Protoplasma* **106**:183–192.

Bloodgood, R. A., 1982, Dynamic properties of the flagellar surface, *Symp. Soc. Exp. Biol.* **35**:353–380.

Bloodgood, R. A., 1984, Protein turnover in the *Chlamydomonas* flagellum, *Exp. Cell Res.* **150**:488–493.

Bloodgood, R. A., 1987, Glycoprotein dynamics in the *Chlamydomonas* flagellar membrane, *Adv. Cell Biol.* **1**:97–130.

Bloodgood, R. A., 1988a, The use of microspheres in the study of cell motility, in: *Microspheres: Medical and Biological Applications* (A. Rembaum and Z. A. Tokes, eds.), CRC Press, Boca Raton, pp. 165–192.

Bloodgood, R. A., 1988b, Gliding motility and the dynamics of flagellar membrane glycoproteins in *Chlamydomonas reinhardtii, J. Protozool.* **35**:552–558.

Bloodgood, R. A., and Levin, E. N., 1983, Transient increase in calcium efflux accompanies fertilization in *Chlamydomonas, J. Cell Biol.* **97**:397–404.

Bloodgood, R. A., and May, G. S., 1982, Functional modification of the *Chlamydomonas* flagellar surface, *J. Cell Biol.* **93**:88–96.

Bloodgood, R. A., and Salomonsky, N. L., 1988, Transmembrane signaling in *Chlamydomonas, J. Cell Biol.* **107**:776a.

Bloodgood, R. A., and Salomonsky, N. L., 1989, Expression of gliding motility in *Chlamydomonas* requires that a concanavalin A binding glycoprotein be free to move within the plane of the flagellar membrane, *Cell Motil. Cytoskeleton* **13**:1–8.

Bloodgood, R. A., and Workman, L. J., 1984, A flagellar surface glycoprotein mediating cell–substrate interaction in *Chlamydomonas, Cell Motil.* **4**:77–87.

Bloodgood, R. A., Leffler, E. M., and Bojczuk, A. T., 1979, Reversible inhibition of *Chlamydomonas* flagellar surface motility, *J. Cell Biol.* **82**:664–674.

Bloodgood, R. A., Woodward, M. P., and Salomonsky, N. L., 1986, Redistribution and shedding of flagellar membrane glycoproteins visualized using an anticarbohydrate monoclonal antibody and concanavalin A, *J. Cell Biol.* **102**:1797–1812.

Bloodgood, R. A., Salomonsky, N. L., and Reinhart, F. D., 1987, Use of carbohydrate probes in conjunction with fluorescence activated cell sorting to select mutant cell lines of *Chalmydomonas* with defects in cell surface glycoproteins, *Exp. Cell Res.* **173**:572–585.

Buck, C. A., and Horwitz, A. F., 1987, Cell surface receptors for extracellular matrix molecules, *Annu. Rev. Cell Biol.* **3**:179–205.

Burchard, R. P., 1981, Gliding motility of prokaryotes: Ultrastructure, physiology, and genetics, *Annu. Rev. Microbiol.* **35**:497–529.

Burridge, K., Fath, K., Kelley, T., Nuckolls, G., and Turner, C., 1988, Focal adhesions: Transmembrane junctions between the extracellular matrix and the cytoskeleton, *Annu. Rev. Cell Biol.* **4**:487–525.

Castenholz, R. W., 1982, Motility and taxes, *Bot. Monogr.* **19**:413–439.

Catt, J. W., Hills, G. J., and Roberts, K., 1976, A structural glycoprotein, containing hydroxyproline, isolated from the cell wall of *Chlamydomonas reinhardtii, Planta* **131**:165–171.

Cheng, K., and Katsoyannis, P. G., 1975, The inhibition of sugar transport and oxidation in fat cell ghosts by colchicine, *Biochem. Biophys. Res. Commun.* **64**:1069–1075.

Claes, H., 1975, Influence of concanavalin A on autolysis of gametes from *Chlamydomonas reinhardtii, Arch. Microbiol.* **103**:225–230.

Claes, H., 1977, Non-specific stimulation of the autolytic system in gametes from *Chlamydomonas reinhardtii, Exp. Cell Res.* **108**:221–229.

Claes, H., 1980, Calcium ionophore-induced stimulation of secretory activity in *Chlamydomonas reinhardtii, Arch. Microbiol.* **124**:81–86.

Demets, R., Tomson, A. M., Homan, W. L., Stegwee, D., and van den Ende, H., 1988, Cell–cell adhesion in conjugating *Chlamydomonas* gametes: A self-enhancing process, *Protoplasma* **145**:27–36.

Dentler, W. L., 1981, Microtubule–membrane interactions in cilia and flagella, *Int. Rev. Cytol.* **72**:1–47.

Dentler, W. L., 1987, Cilia and flagella, *Int. Rev. Cytol. Suppl.* **17**:391–456.

Detmers, P. A., and Condeelis, J. S., 1986, Trifluoperazine and W-7 inhibit mating in *Chlamydomonas* at an early stage of gametic interaction, *Exp. Cell Res.* **163**:317–326.

Detmers, P. A., Carboni, J. M., and Condeelis, J., 1985, Localization of actin in *Chlamydomonas* using antiactin and NBD-phallacidin, *Cell Motil.* **5**:415–430.

Dwyer, D. M., 1976, Antibody-induced modulation of *Leishmania donovani* surface membrane antigens, *J. Immunol.* **117**:2081–2091.

Euteneuer, U., Koonce, M. P., Pfister, K. K., and Schliwa, M., 1988, An ATPase with properties expected for the organelle motor of the giant amoeba, *Reticulomyxa, Nature* **332**:176–178.

Garbers, D. L., First, N. L., and Lardy, H. A., 1973, Properties of adenosine 3′,5′-monophosphate dependent protein kinases isolated from bovine epididymal spermatozoa, *J. Biol. Chem.* **248**:875–879.

Gietzen, K., Wuthrich, A., and Bader, H., 1982, Effects of microtubular inhibitors on plasma membrane calmodulin dependent Ca^{2+}-transport ATPase, *Mol. Pharmacol.* **22**:413–420.

Gilbert, S. P., and Sloboda, R. D., 1986, Identification of a MAP 2-like ATP-binding protein associated with axoplasmic vesicles that translocate on isolated microtubules, *J. Cell Biol.* **103**:947–956.

Gilbert, S. P., Allen, R. D., and Sloboda, R. D., 1985, Translocation of vesicles from squid axoplasm on flagellar microtubules, *Nature* **315**:245–248.

Gilman, A. G., 1987, G proteins: Transducers of receptor-generated signals, *Annu. Rev. Biochem.* **56**:615–649.

Gitelman, S. E., and Witman, G. B., 1980, Purification of calmodulin from *Chlamydomonas:* Calmodulin occurs in cell bodies and flagella, *J. Cell Biol.* **98**:764–770.

Glaser, J., and Pate, J. L., 1973, Isolation and characterization of gliding motility mutants of *Cytophaga columnaris, Arch. Mikrobiol.* **93**:295–309.

Goodenough, U. W., 1980, Ionophore stimulation of mating signals in *Chlamydomonas, J. Cell Biol.* **87**:37a.

Goodenough, U. W., and Jurivich, D., 1978, Tipping and mating-structure activation induced in *Chlamydomonas* by flagellar membrane antisera, *J. Cell Biol.* **79**:680–693.

Grief, C., and Shaw, P. J., 1987, Assembly of cell-wall glycoproteins of *Chlamydomonas reinhardtii:* Oligosaccharides are added in medial and trans Golgi compartments, *Planta* **171**:302–312.

Gross, M. K., Toscano, D. G., and Toscano, W. A., Jr., 1987, Calmodulin-mediated adenylate cyclase from mammalian sperm, *J. Biol. Chem.* **262**:8672–8676.

Gumbiner, B., and Louvard, D., 1985, Localized barriers in the plasma membrane: A common way to form domains, *Trends Biochem. Sci.* **10**:435–438.

Gustin, M. C., and Nelson, D. L., 1987, Regulation of ciliary adenylate cyclase by Ca^{2+} in *Paramecium, Biochem. J.* **246**:337–345.

Halfen, L. N., 1979, Gliding movements, *Encycl. Plant Physiol.* **7**:250–267.

Hartfiel, G., and Amrhein, N., 1976, The action of methylxanthines on motility and growth of *Chlamydomonas reinhardtii* and other flagellated algae. Is cyclic AMP involved? *Biochem. Physiol. Pflanz.* **169**:531–556.

Hasegawa, E., Hayashi, H., Asakura, S., and Kamiya, R., 1987, Stimulation of in vitro motility of *Chlamydomonas* axonemes by inhibition of cAMP-dependent phosphorylation, *Cell Motil. Cytoskeleton* **8**:302–311.

Hidaka, H., Inagaki, M., Kawamoto, S., and Sasaki, Y., 1984, Isoquinolinesulfanamides, novel and potent inhibitors of cyclic nucleotide dependent protein kinase, *Biochemistry* **23**:5036–5041.

Hodgkin, J., and Kaiser, D., 1977, Cell-to-cell stimulation of movement in nonmotile mutants of *Myxococcus,* *Proc. Natl. Acad. Sci. USA* **74:**2938–2942.

Hoffman, J. L., and Goodenough, U. W., 1980, Experimental dissection of flagellar surface motility in *Chlamydomonas, J. Cell Biol.* **86:**656–665.

Hollenberg, M. D., 1986, Mechanisms of receptor-mediated transmembrane signaling, *Experientia* **42:**718–727.

Homan, W., Sigon, C., van den Briel, W., Wagter, R., de Nobel, H., Mesland, D., Musgrave, A., and van den Ende, H., 1987, Transport of membrane receptors and the mechanics of sexual cell fusion in *Chlamydomonas eugametos, FEBS Lett.* **215:**323–326.

Homan, W., Musgrave, A., de Nobel, H., Wagter, R., de Wit, D., Kolk, A., and van den Ende, H., 1988, Monoclonal antibodies directed against the sexual binding site of *Chlamydomonas eugametos* gametes, *J. Cell Biol.* **107:**177–189.

Huang, B., Piperno, G., Ramanis, Z., and Luck, D. J. L., 1981, Radial spokes of *Chlamydomonas* flagella: Genetic analysis of assembly and function, *J. Cell Biol.* **88:**80–88.

Hyams, J. S., and Borisy, G. G., 1978, Isolated flagellar apparatus of *Chlamydomonas:* Characterization of forward swimming and alteration of waveform and reversal of motion by calcium ions in vivo, *J. Cell Sci.* **33:**235–253.

Jarosch, R., 1962, Gliding, in: *Physiology and Biochemistry of Algae* (R. Lewin, ed.), Academic Press, New York, pp. 573–581.

Kahn, C. R., Baird, K. L., Flier, J. S., Grunfield, C., Harmon, J. T., Harrison, L. C., Karlsson, F. A., Kasuga, M., King, G. L., Lang, U. C., Podskalny, J. M., and van Obberghen, E., Insulin receptors, receptor antibodies, and the mechanism of insulin action, 1981, *Recent Prog. Horm. Res.* **37:**477–533.

Kaska, D. D., Piscopo, I. C., and Gibor, A., 1985, Intracellular calcium redistribution during mating in *Chlamydomonas reinhardtii, Exp. Cell Res.* **160:**371–379.

King, C. A., 1981, Cell surface interaction of the protozoan *Gregarina* with concanavalin A beads—Implications of models of gregarine gliding, *Cell Biol. Int. Rep.* **5:**297–305.

King, C. A., 1988, Cell motility of sporozoan protozoa, *Parasitol. Today* **4:**315–319.

King, C. A., and Lee, K., 1982, Effect of trifluoperazine and calcium ions on gregarine gliding, *Experientia* **38:**1051–1052.

Klumpp, S., and Schultz, J. E., 1982, Characterization of a Ca^{2+}-dependent guanylate cyclase in the excitable ciliary membrane from *Paramecium, Eur. J. Biochem.* **124:**317–324.

Kuznetsov, S. A., and Gelfand, V. I., 1986, Bovine brain kinesin is a microtubule-activated ATPase, *Proc. Natl. Acad. Sci. USA* **83:**8530–8534.

Lackie, J. M., 1986, *Cell Movement and Cell Behavior,* Allen & Unwin, London.

Lancet, D., and Pace U., 1987, The molecular basis of odor recognition, *Trends Biochem. Sci.* **12:**63–66.

Lapidus, I. R., and Berg, H. C., 1982, Gliding motility of *Cytophaga* sp, strain U67, *J. Bacteriol.* **151:**384–398.

Lefebvre, P. A., and Rosenbaum, J. L., 1986, Regulation of the synthesis and assembly of ciliary and flagellar proteins during regeneration, *Annu. Rev. Cell Biol.* **2:**517–546.

Lefebvre, P. A., Nordstrom, S. A., Moulder, J. E., and Rosenbaum, J. L., 1978, Flagellar elongation and shortening in *Chlamydomonas.* IV. Effects of flagellar detachment, regeneration, and resorption on the induction of flagellar protein synthesis, *J. Cell Biol.* **78:**8–27.

Lefebvre, P. A., Silflow, C. D., Wieben, E. D., and Rosenbaum, J. L., 1980, Increased levels of mRNAs for tubulin and other flagellar proteins after amputation or shortening of *Chlamydomonas* flagella, *Cell* **20:**469–477.

Lewin, R. A., 1952, Studies of the flagella of algae, I. General observations of *Chlamydomonas moewusii* Gerloff, *Biol. Bull. (Woods Hole, Mass.)* **103:**74–79.

Lewin, R. A., 1954, Mutants of *Chlamydomonas moewusii* with impaired motility, *J. Gen. Microbiol.* **11:**358–368.

Lewin, R. A., 1982, A new kind of motility mutant (non-gliding) in *Chlamydomonas, Experientia* **38:**348–349.

Lis, H., and Sharon, N., 1986, Biological properties of lectins, in: *The Lectins: Properties, Functions, and Applications in Biology and Medicine* (I. E. Liener, N. Sharon, and I. J. Goldstein, eds.), Academic Press, New York, pp. 265–291.

Lye, R. J., Porter, M. E., Scholey, J. M., and McIntosh, J. R., 1987, Identification of a microtubule-based cytoplasmic motor in the nematode *C. elegans, Cell* **51:**309–318.

McLean, R. J., Laurendi, C. J., and Brown, R. M., 1974, The relationship of gamone to the mating reaction in *Chlamydomonas moewusii, Proc. Natl. Acad. Sci. USA* **71:**2610–2613.

May, G. S., and Rosenbaum, J. L., 1983, Flagellar protein phosphorylation during flagellar regeneration and resorption in *Chlamydomonas reinhardtii, J. Cell Biol.* **97:**195a.

Mesland, D. A. M., 1976, Mating in *Chlamydomonas eugametos,* a scanning electron microscopical study, *Arch. Mikrobiol.* **109:**31–35.

Mesland, D. A. M., Hoffman, J. L., Caligor, E., and Goodenough, U. W., 1980, Flagellar tip activation stimulated by membrane adhesions in *Chlamydomonas* gametes, *J. Cell Biol.* **84:**599–617.

Mizel, S. B., and Wilson, L., 1972, Nucleoside transport in mammalian cells. Inhibition of colchicine. *Biochemistry* **11:**2573–2578.

Monk, B. C., Adair, W. S., Cohen, R. A., and Goodenough, U. W., 1983, Topography of *Chlamydomonas:* Fine structure and polypeptide components of the gametic flagellar membrane surface and the cell wall, *Planta* **158:**517–533.

Musgrave, A., and ven den Ende, H., 1987, How *Chlamydomonas* court their partners, *Trends Biochem. Sci.* **12:**470–473.

Musgrave, A., de Wildt, P., Broekman, R., and van den Ende, H., 1983, The cell wall of *Chlamydomonas eugametos.* Immunological aspects, *Planta* **158:**82–89.

Musgrave, A., de Wildt, P., van Etten, I., Pijst, H., Scholma, C., Kooyman, R., Homan, W., and van den Ende, H., 1986, Evidence for a functional membrane barrier between the flagellum and cell body of *Chlamydomonas eugametos* gametes, *Planta* **167:**544–553.

Nakamura, T., and Gold, G. H., 1987, A cyclic nucleotide-gated conductance in olfactory receptor cilia, *Nature* **325:**442–444.

Paschal, B. M., and Vallee, R. B., 1987, Retrograde transport by the microtubule-associated protein MAP 1C, *Nature* **330:**181–183.

Paschal, B. M., Shpetner, H. S., and Vallee, R. B., 1987, MAP 1C is a microtubule-activated ATPase which translocates microtubules in vitro and has dynein-like properties, *J. Cell Biol.* **105:**1273–1282.

Pasquale, S. M., and Goodenough, U. W., 1987, Cyclic AMP functions as a primary sexual signal in gametes of *Chlamydomonas reinhardtii, J. Cell Biol.* **105:**2279–2292.

Pate, J. L., 1985, Gliding motility in *Cytophaga, Microbiol. Sci.* **2:**289–295.

Pate, J. L., and Chang, L.-Y. E., 1979, Evidence that gliding motility in prokaryotic cells is driven by rotary assemblies in the cell envelopes, *Curr. Microbiol.* **2:**59–64.

Peng, H. B., Cheng, P.-C., and Luther, P. W., 1981, Formation of ACh receptor clusters induced by positively charged latex beads, *Nature* **292:**831–834.

Pijst, H. L. A., 1985, Sexual agglutination of the green alga *Chlamydomonas eugametos:* Symptoms and signals, Doctoral dissertation, University of Amsterdam, Chapter 6.

Pijst, H. L. A., van Driel, R., Janssens, P. M. W., Musgrave, A., and van den Ende, H., 1984, Cyclic AMP is involved in sexual reproduction of *Chlamydomonas eugametos, FEBS Lett.* **174:**132–136.

Piperno, G., 1988, Isolation of a sixth dynein subunit adenosine triphosphatase of *Chlamydomonas* axonemes, *J. Cell Biol.* **106:**133–140.

Piperno, G., and Luck, D. J. L., 1979, An actin-like protein is a component of axonemes from *Chlamydomonas* flagella, *J. Biol. Chem.* **254:**2187–2190.

Porter, M. E., Scholey, J. M., Stemple, D. L., Vigers, G. P. A., Vale, R. D., Sheetz, M. P., and McIntosh, J. R., 1987, Characterization of the microtubule movement produced by sea urchin egg kinesin, *J. Biol. Chem.* **262:**2794–2802.

Quader, H., Cherniack, J., and Filner, P., 1978, Participation of calcium in flagellar shortening and regeneration in *Chlamydomonas reinhardtii, Exp. Cell Res.* **113:**295–301.

Reinhart, F. D., and Bloodgood, R. A., 1988a, Membrane–cytoskeleton interactions in the flagellum: A 240 kDa surface-exposed glycoprotein is tightly associated with the axoneme in *Chlamydomonas moewusii, J. Cell Sci.* **89:**521–530.

Reinhart, F. D., and Bloodgood, R. A., 1988b, Gliding defective mutant cell lines of *Chlamydomonas moewusii* exhibit alterations in a 240 kDa surface-exposed flagellar glycoprotein, *Protoplasma* **144:**110–118.

Rembold, H., and Langenbach, T., 1978, Effect of colchicine on cell membrane and on biopterin transport in *Crithidia fasciculata, J. Protozool.* **25:**404–408.

Remillard, S. P., and Witman, G. B., 1982, Synthesis, transport, and utilization of specific flagellar proteins during flagellar regeneration in *Chlamydomonas, J. Cell Biol.* **93:**615–631.

Ringo, D. L., 1967, Flagellar motion and fine structure of the flagellar apparatus in *Chlamydomonas*, *J. Cell Biol.* **33**:543–571.

Roberts, K., Grief, C., Hills, G. J., and Shaw, P. J., 1985a, Cell wall glycoproteins: Structure and function, *J. Cell Sci. Suppl.* **2**:105–127.

Roberts, K., Phillips, J., Shaw, P., Grief, C., and Smith, E., 1985b, An immunological approach to the plant cell wall, in: *Biochemistry of Plant Cell Walls* (C. T. Brett and J. R. Hillman, eds.), Cambridge University Press, London, pp. 125–154.

Rosenbaum, J. L., and Child, F. M., 1967, Flagellar regeneration in protozoan flagellates, *J. Cell Biol.* **54**:507–539.

Rosenbaum, J. L., Moulder, J. E., and Ringo, D. L., 1969, Flagellar elongation and shortening in *Chlamydomonas*. The use of cycloheximide and colchicine to study the synthesis and assembly of flagellar proteins, *J. Cell Biol.* **41**:600–619.

Roufogalis, B. D., 1982, Specificity of trifluoperazine and related phenothiazines for calcium-binding proteins, in: *Calcium and Cell Function*, Volume 3 (W. Y. Cheung, ed.), Academic Press, New York, pp. 129–159.

Rubin, R. W., and Filner, P., 1973, Adenosine 3′,5′-cyclic monophosphate in *Chlamydomonas reinhardtii*. Influence on flagellar function and regeneration. *J. Cell Biol.* **56**:628–635.

Saito, T., Tsubo, Y., and Matsuda, Y., 1985, Synthesis and turnover of cell body-agglutinin as a pool of flagellar surface-agglutinin in *Chlamydomonas reinhardtii* gamete, *Arch. Microbiol.* **142**:205–210.

Sandoz, D., Gounon, P., Karsenti, E., and Sauron, M.-E., 1982, Immunochemical localization of tubulin, actin, and myosin in axonemes of ciliated cells from quail oviduct, *Proc. Natl. Acad. Sci. USA* **79**:3198–3202.

Saxton, W. M., Porter, M. E., Cohn, S. A., Scholey, J. M., Raff, E. C., and McIntosh, J. R., 1988, *Drosophila* kinesin: Characterization of microtubule motility and ATPase, *Proc. Natl. Acad. Sci. USA* **85**: 1109–1113.

Schellenberg, R. R., and Gillespie, E., 1977, Colchicine inhibits phosphatidylinositol turnover induced in lymphocytes by concanavalin A, *Nature* **265**:741–742.

Schreiber, A. B., Libermann, T. A., Lax, I., Yarden, Y., and Schlessinger, J., 1983, Biological role of epidermal growth factor-receptor clustering, *J. Biol. Chem.* **258**:846–853.

Schreiner, G. F., and Unanue, E. R., 1976, Membrane and cytoplasmic changes in B lymphocytes induced by ligand–surface immunoglobulin interaction, *Adv. Immunol.* **24**:37–165.

Schultz, J. E., and Jantzen, H. M., 1980, Cyclic nucleotide-dependent protein kinases from cilia of *Paramecium tetraurelia*, *FEBS Lett.* **116**:75–78.

Schultz, J. E., Uhl, D. G., and Klumpp, S., 1987, Ionic regulation of adenylate cyclase from the cilia of *Paramecium tetraaurelia*, *Biochem. J.* **246**:187–192.

Segal, R. A., and Luck, D. J., 1985, Phosphorylation in isolated *Chlamydomonas* axonemes: A phosphoprotein may mediate the Ca^{2+}-dependent photophobic response, *J. Cell Biol.* **101**:1702–1712.

Silflow, C. D., Lefebvre, P. A., McKeithan, T. W., Schloss, J. A., Keller, L. R. and Rosenbaum, J. L., 1981, Expression of flagellar protein genes during flagellar regeneration in *Chlamydomonas*, *Cold Spring Harbor Symp. Quant. Biol.* **46**:157–169.

Smith, E., Roberts, K., Hutchings, A., and Galfre, G., 1984, Monoclonal antibodies to the major structural glycoprotein of the *Chlamydomonas* cell wall, *Planta* **161**:330–338.

Snell, W. J., 1976, Mating in *Chlamydomonas*: A system for the study of specific cell adhesion. I. Ultrastructural and electrophoretic analysis of flagellar surface components involved in adhesion, *J. Cell Biol.* **68**:48–69.

Snell, W. J., 1985, Cell–cell interactions in *Chlamydomonas*, *Annu. Rev. Plant Physiol.* **36**:287–315.

Snell, W. J., and Moore, W. S., 1980, Aggregation-dependent turnover of flagellar adhesion molecules in *Chlamydomonas* gametes, *J. Cell Biol.* **84**:203–210.

Snell, W. J., Buchanan, M., and Clausell, A., 1982, Lidocaine reversibly inhibits fertilization in *Chlamydomonas*: A possible role in sexual signaling, *J. Cell Biol.* **94**:607–612.

Stephens, R. E., and Stommel, E. W., 1989, Role of cyclic adenosine monophosphate in ciliary and flagellar motility, in: *Cell Movement*, Volume 1: *The Dynein ATPases* (F. D. Warner, P. Satir, and I. R. Gibbons, eds.), Alan R. Liss, New York, pp. 299–316.

Stephens, R. E., Oleszki-Szuts, S., and Good, M. J., 1987, Evidence that tubulin forms an integral membrane skeleton in molluscan gill cilia, *J. Cell Sci.* **88**:527–535.

Tamm, S. L., 1967, Flagellar development in the protozoan *Peranema trichophorum, J. Exp. Zool.* **164:**163–186.

Tash, J. S., and Means, A. R., 1983, Cyclic adenosine 3',5'monophosphate, calcium, and protein phosphorylation in flagellar motility, *Biol. Reprod.* **28:**75–104.

Tash, J. S., Hidaka, H., and Means, A. R., 1986, Axokinin phosphorylation by cAMP-dependent protein kinase is sufficient for activation of sperm flagellar motility, *J. Cell Biol.* **103:**649–655.

Taylor, R. B., Duffus, W., Raff, M., and de Petris, S., 1971, Redistribution and pinocytosis of lymphocyte surface immunoglobulin molecules induced by anti-immunoglobulin antibody, *Nature New Biol.* **233:** 225–229.

Trimmer, J. S., and Vacquier, V. D., 1988, Monoclonal antibodies induce the translocation, patching, and shedding of surface antigens of sea urchin spermatozoa, *Exp. Cell Res.* **175:**37–51.

Vale, R. D., Reese, T. S., and Sheetz, M. P., 1985a, Identification of a novel force generating protein (kinesin) involved in microtubule-based motility, *Cell* **41:**39–50.

Vale, R. D., Schnapp, B. J. Mitchison, T., Steuer, E., Reese, T. S., and Sheetz, M. P., 1985b, Different axoplasmic proteins generate movement in opposite directions along microtubules in vitro, *Cell* **43:**623–632.

van den Ende, H., 1985, Sexual agglutination in Chlamydomonads, *Adv. Microb. Physiol.* **26:**89–123.

Watanabe, K., and West, W. L., 1982, Calmodulin, activated cyclic nucleotide phosphodiesterase, microtubules, and vinca alkaloids, *Fed. Proc.* **41:**2292–2299.

Wantanabe, T., and Flavin, M., 1976, Nucleotide-metabolizing enzymes in *Chlamydomonas* flagella, *J. Biol. Chem.* **251:**182–192.

Wiese, L., 1965, On sexual agglutination and mating type substances (gamones) in isogamous heterothallic Chlamydomonads. I. Evidence of the identity of the gamones with the surface components responsible for sexual flagellar contact, *J. Phycol.* **1:**46–54.

Witman, G. B., 1975, The site of in vivo assembly of flagellar microtubules, *Ann. N.Y. Acad. Sci.* **253:**178–191.

Witman, G. B., 1986, Isolation of *Chlamydomonas* flagella and flagellar axonemes, *Methods Enzymol.* **134:** 280–290.

Witman, G. B., 1989, Perspective: Composition and molecular organization of the dyneins, in: *Cell Movement,* Volume 1: *The Dynein ATPases* (F. D. Warner, P. Satir, and I. R. Gibbons, eds.), Alan R. Liss, Inc., New York, pp. 25–35.

Witman, G. B., Carlson, K., Berliner, J., and Rosenbaum, J. L., 1972, *Chlamydomonas* flagella. I. Isolation and electrophoretic analysis of microtubules, matrix, membranes and mastigonemes, *J. Cell Biol.* **54:** 507–539.

Wolf, D. E., 1987, Overcoming random diffusion in polarized cells—Corralling the drunken begger, *BioEssays* **6:**116–121.

The Role of Flagella in the Sexual Reproduction of *Chlamydomonas* Gametes

H. van den Ende, A. Musgrave, and F. M. Klis

1. Introduction

A striking characteristic of *Chlamydomonas* flagella is that they transform into sexual organelles during gametogenesis. They mediate a species-specific adhesion or agglutination reaction between cells of opposite mating type, due to the presence of molecules called agglutinins which are located on their surface. This interaction between flagella serves not only to maintain sexually compatible cells in close proximity, but also to generate an intracellular signal which prepares the cells for fusion. The most important response is the protrusion of a specialized zone of the cell surface called the mating structure, which is located at the anterior end of the cell near the flagellar bases. Activation of the mating structure is accompanied by the local hydrolysis of the cell wall.

   In this chapter, recent work on two species, the heterothallic and isogamous *C. reinhardtii* and *C. eugametos,* will be reviewed. The mating processes in these two species of *Chlamydomonas* differ in some significant aspects. In *C. reinhardtii,* the cell wall is released completely during sexual agglutination and the activation of the mating structure results in the formation of a long fertilization tubule in mt^+ cells and a smaller protuberance in mt^- cells, by which means the naked cells fuse rapidly (Goodenough and Weiss, 1975; Goodenough *et al.,* 1982; Detmers *et al.,* 1983). In *C. eugametos,* however, activation of the mating structure leads to the formation of a very small papilla which just penetrates the cell wall in both cell types (Brown *et al.,* 1968; Mesland, 1976). The two cells fuse by means of these papillae, which results in a plasma bridge, while the cells remain surrounded by their cell walls. Thus, a tandem of cells is formed, the so-called vis-à-vis pair. In *C. eugametos,* vis-à-vis pairs are relatively long-lived and motile. Only many hours later, do the cells settle and fuse completely to form a diploid zygote. An intricate zygote cell wall is formed under the gamete walls, which only very slowly disappear.

H. van den Ende, A. Musgrave, and F. M. Klis • Department of Molecular Cell Biology, University of Amsterdam, 1098 SM Amsterdam, The Netherlands.

2. The Agglutination Process

Sexual agglutination has a number of features by which it is distinguished from any other aggregate-forming process; they are particularly apparent in *C. eugametos*. When two gamete suspensions of opposite mating type are mixed, the cells form clumps within seconds. The flagella of agglutinating cells change their movement from the regular swimming beat to a vehement twitching movement, which gives the clumps a vibrating appearance (Homan *et al.*, 1980). Remarkably, at high cell densities, cell clusters tend to grow into large cell aggregates, comprising hundreds of cells, because free swimming cells join existing clumps rather than forming new ones (Mesland, 1978). At low cell densities, cells aggregate in small clusters or simply in pairs.

Another conspicuous feature of sexual agglutination is that the flagella always tend to associate tip-to-tip (Wiese, 1969). This is typical for both species. It is the result of a tip-oriented movement of the contact sites to the respective tips (see Section 5). Additionally, in *C. eugametos* the flagella tend to become associated over their whole length, which undoubtedly is functional in establishing contact between the two very small plasma papillae. In this species, the formation of pairs is further promoted by a mating-type-specific reorientation of the flagella (Musgrave *et al.*, 1985). Agglutinating mt^- flagella become reflexed around their own cell body; in contrast, the mt^+ flagella are held forward in a complementary manner (Fig. 1). This behavior accounts for the vis-à-vis orientation and favors the apposition of the papillae so that the plasma bridge can be established with high efficiency. Together with the distally oriented movement of the flagellar contact sites, this results in a sorting of cells, so that mt^+ gametes are predominantly concentrated at the periphery and the mt^- gametes at the center of the clump. This can be easily demonstrated by labeling one of the mating strains with fluorescein isothiocyanate, without the viability of the cells being affected. In *C. reinhardtii*, such an orderly behavior of

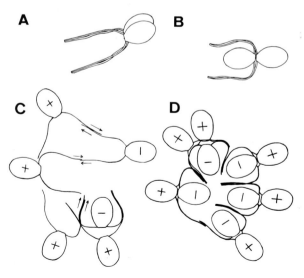

Figure 1. Diagram illustrating how tipping helps gametes of *C. eugametos* to sort themselves into pairs. The most stable contacts are realized by the alignment and the specific orientation of mt^+ and mt^- flagella. From Homan *et al.* (1987a) with permission.

agglutinating gametes does not seem to be required, since the available contact area on the mating structures is much larger (e.g., Goodenough *et al.*, 1982).

The flagella lose their sexual agglutinability within a few minutes after cell fusion and resume their swimming beat. In *C. reinhardtii*, all four flagella of a quadriflagellate zygote propel the cell. In *C. eugametos*, it is only the mt^+ pair flagella that resume the swimming motion. The mt^- flagella do not beat during the lifetime of the vis-à-vis pairs (Lewin, 1952; Crabbendam *et al.*, 1984). Settling vis-à-vis pairs weakly aggregate together in rows with the cell bodies lying side by side. Again this is due to mating-type-specific adhesiveness of the flagella (Musgrave *et al.*, 1985). Presumably the same flagellar surface components are involved in gamete agglutination and vis-à-vis pair aggregation. Only later do they differentiate into zygotes.

3. The Agglutinins

The adhesiveness of gamete flagella is due to the presence of specific molecules, the agglutinins, on the flagellar surface. They have been characterized for *C. reinhardtii* (Adair *et al.*, 1983; Cooper *et al.*, 1983; Collin-Osdoby and Adair, 1985; Saito and Matsuda, 1984b; Goodenough *et al.*, 1985; Adair, 1985) as well as for *C. eugametos* and the closely related *C. moewusii* (Musgrave *et al.*, 1981; Klis *et al.*, 1985; Samson *et al.*, 1987a,b; Crabbendam *et al.*, 1986). In all species they are large linear glycoproteins. From the Stokes radii and sedimentation coefficients, their molecular mass has been determined to be in the range of 10^6 Da. The carbohydrate content is approximately 50%, which leads to a molecular mass of the protein core on the order of 5×10^5 Da. The molecular mass of the completely deglycosylated mt^+ agglutinin of *C. reinhardtii* is 480 kDa (Adair *et al.*, 1983). The overall amino acid composition of the agglutinins of *C. reinhardtii* and *C. eugametos* is remarkably similar (Table I). Considering the obvious differences between the two species in the restriction patterns of their chloroplast DNA (Lemieux and Lemieux, 1985; Jupe *et al.*, 1988), this suggests that their agglutinin structures have been highly conserved. The agglutinins are rich in hydroxyproline and serine (Collin-Osdoby and Adair, 1985; Cooper *et al.*, 1983; Samson *et al.*, 1987a,b), two amino acids that function as attachment sites for *O*-glycosidically linked carbohydrate chains. The predominant sugars are arabinose and galactose. There is also evidence for the presence of *N*-acetylglucosamine (Samson *et al.*, 1987a,b). This correlates with the sensitivity of flagellar adhesiveness to tunicamycin that is observed in many strains (Ray and Gibor, 1982; Matsuda *et al.*, 1981; Wiese and Mayer, 1982). Table II summarizes some of the characteristics of the agglutinins.

Electron microscopic studies of three different species, *C. reinhardtii*, *C. eugametos*, and the syngens I and II of *C. moewusii* (the latter also called *C. moewusii yapensis;* Harris, 1989), have shown that the agglutinins are highly asymmetric, rodlike molecules. In *C. reinhardtii*, the mt^+ and mt^- agglutinins are very similar, with subtle differences in conformation and dimensions (Fig. 2). The molecules consist of a shaft, with one or two distinct bends at specific points. At one end this shaft terminates in a globular head, and at the other end in a hooklike structure (Collin-Osdoby and Adair, 1985; Goodenough *et al.*, 1985). The globular domain might comprise the binding site,

Table I. Amino Acid Composition
of the Agglutinins of *C. reinhardtii*
and *C. eugametos*[a]

Amino acid	Amino acid composition (no. of residues per 1000)			
	C. reinhardtii		*C. eugametos*	
	mt^+	mt^-	mt^+	mt^-
Lys	22	41	80	30
His	31	20	20	10
Arg	36	22	60	30
Asx	95	104	70	70
Thr	59	67	40	60
Ser	103	113	120	130
Glx	88	83	90	100
Pro	43	56	40	30
Gly	88	88	110	120
Ala	77	78	60	90
Val	46	62	40	60
Lys	50	ND	ND	10
Met	10	8	ND	10
Ile	27	35	30	40
Leu	51	36	60	60
Trp	ND	ND	ND	ND
Tyr	15	11	10	20
Phe	28	31	20	30
Hyp	123	120	120	100
SerP	ND	ND	3	ND

[a]Data from Adair (1985) and Samson *et al.* (1987b).

Table II. Characteristics of the Agglutinins
of *Chlamydomonas* Species[a]

	C. reinhardtii		*C. eugametos*		*C. moewusii yapensis*	
	mt^+	mt^-	mt^+	mt^-	mt^+	mt^-
Terminal head	Yes	Yes	Yes	No	Yes	No
Length of shaft (nm)	218	218	200	345	245	349
Terminal hook	Yes	Yes	Yes	Yes	?	?
M_r (mDa)			1.2	1.3	1.0	1.2

[a]Data from Goodenough *et al.* (1985), Adair *et al.* (1983), Samson *et al.* (1987a,b), and Crabbendam *et al.* (1986).

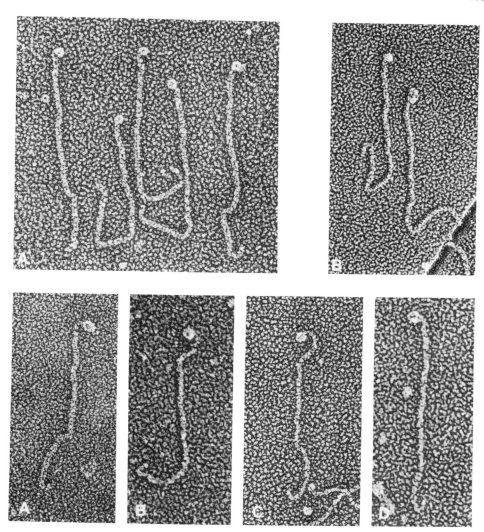

Figure 2. Morphology of the agglutinins of *C. reinhardtii, mt*⁺ (top) and *mt*⁻ (bottom). 250,000×. Micrographs by Dr. J. E. Heuser. From Goodenough *et al.* (1985) with permission.

since thermolysin digestion or reduction followed by alkylation of the *C. reinhardtii mt*⁺ agglutinin (Collin-Osdoby *et al.*, 1984), or freeze-thawing of the similar *mt*⁺ agglutinin from *C. eugametos* (Crabbendam *et al.*, 1986), inactivates the molecules, resulting in a modified head morphology or clustering of the molecules via their globular heads.

In *C. moewusii* and *C. eugametos*, it is only the *mt*⁺ agglutinin that shows a distal globular domain (Crabbendam *et al.*, 1986; Samson *et al.*, 1987a,b). The *mt*⁻ agglutinins are longer, stringy, more flexible molecules (Fig. 3). That of *C. eugametos mt*⁻ is

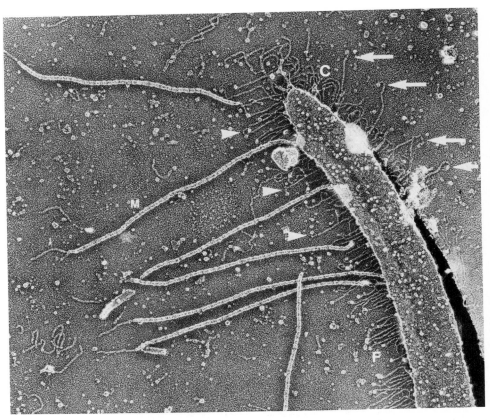

Figure 3. Flagellum of an *mt⁻* gamete of *C. reinhardtii*. Arrows indicate molecules with the morphology of *mt⁻* agglutinin. M, mastigonemes; C, crescent fibrils. 115,000×; reproduced at 70%. Micrograph by Dr. J. E. Heuser. From Goodenough *et al.* (1985) with permission.

segmented, consisting of five rigid domains bound together by flexible joints that are sometimes obvious as pronounced kinks. The distribution of these joints bestows an asymmetry on the molecule. That of *C. moewusii yapensis mt⁻* has a corkscrew appearance, while the agglutinins of *C. moewusii* syngen I are very similar to those of *C. eugametos*. The putative binding site on these agglutinins that lack a globular head is unknown. Since the shaft varies considerably in length between the compatible species *C. eugametos* and *C. moewusii* syngen I, we presume it acts simply as a ''spacer,'' presenting the active site at its extremity. The hook structure that is frequently seen at the proximal end of the agglutinin in these species, probably contains the domain by which the molecule is attached to the membrane. This is suggested by electron micrographs of the flagellar surface of gametes of *C. reinhardtii*, obtained with the rapid-freeze deep-etch procedure (Heuser, 1983), in which molecules are seen on the surface that have a form that unambiguously identifies them as agglutinins (Goodenough *et al.*, 1985). They appear to be anchored to the flagellar surface by their hook end, with their globular heads situated distally (Fig. 4). A particularly interesting observation is that these putative

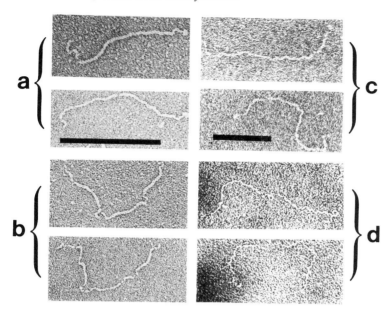

Figure 4. Agglutinins of *C. eugametos* and *C. moewusii yapensis*, visualized by negative staining. (a) *C. eugametos, mt⁺*; (B) *C. eugametos, mt⁻*; bar = 100 nm; (c) *C. moewusii yapensis, mt⁺*; (D) *C. moewusii yapensis, mt⁻*; bar = 200 nm. From Crabbendam *et al.* (1986) and Samson *et al.* (1987a) with permission.

agglutinin molecules are sometimes seen to associate at their proximal ends to form fibrils. Their constant diameter suggests that they are not just random aggregates.

Adair (1985) has emphasized that the agglutinins also show a striking resemblance to another class of hydroxyproline-rich proteins, the major cell wall components of *Chlamydomonas*. Many of these glycoproteins are rodlike molecules, which display molecular recognition properties, manifested by their self-assembly into wall-like crystalline aggregates *in vitro* (Roberts, 1974; Roberts *et al.*, 1985; Goodenough *et al.*, 1986b; Adair *et al.*, 1987). These molecules contain a high proportion of polypeptide in the left-handed polyproline II helix conformation (Homer and Roberts, 1979), which is reinforced by short-chain arabinosides linked to hydroxyproline residues. The relatively high hydroxyproline and arabinose content of the agglutinins strengthens the idea that the rodlike domains of agglutinins also have the polyproline II conformation. Cooper *et al.* (1983) presented evidence that proline hydroxylation is required for the biosynthesis of functional agglutinin.

At least in *C. eugametos,* it seems that terminal mannose is important for the activity of the *mt⁺* agglutinin. Wiese and Shoemaker (1970) and Musgrave *et al.*(1979) showed that Con A interferes with the agglutinability of *mt⁺* gametes. This is consistent with the fact that the *mt⁺* agglutinin (but not the *mt⁻* agglutinin) is bound by Con A. Wiese and Wiese (1975) also demonstrated that α-mannosidase specifically inactivates *mt⁺* gametes, whereas *mt⁻* gametes are resistant to this treatment. Again, this is reflected by the effect of α mannosidase on the isolated agglutinins (Samson *et al.*, 1987b), indicating that terminal α-mannosyl residues are required for biological activity. However, the mating reaction cannot be inhibited by mannose or α-methyl-D-mannose (Wiese, 1974). On the other hand, the mt^- agglutinin of *C. eugametos* is susceptible to α-galactosidase (Wiese and Mayer, 1982; Samson *et al.*, 1987b). While the terminal α-mannosyl or α-galactosyl residues may be needed just to stabilize the tertiary structure, it is attractive to think that they are part of the sexual binding site. A comparison of the susceptibilities of the agglutinins of noncompatible species (Table III) suggests that terminal mannose and galactose residues play a role in determining the binding specificity of these molecules.

The mt^- agglutinin of *C. eugametos* is very susceptible to β-elimination by alkali, by which a mixture of mono-, di-, and trisaccharides is released (Lens *et al.*, 1983). Finally, it should be mentioned that Wiese and Wiese (1978) reported that mt^- activity in *C. eugametos* is lost by treatment with sulfatase, which suggests that sulfated sugars are essential for biological activity. Samson *et al.* (1987a) found that only the combined action of sulfatase and periodate oxidation can destroy the biological activity of isolated mt^- agglutinin from *C. moewusii* syngen II. The presence of sulfated sugars in the *Volvocales* is well documented (Roberts *et al.*, 1985; Wenzl *et al.*, 1984). Taken together, the evidence indicates a important role for the carbohydrate portion of the agglutinins.

A useful tool in the structural and functional analysis of large molecules like the agglutinins are antibodies. Polyclonal antibodies raised against gamete glycoproteins bind to a variety of glycoproteins on all cell types (Wiese, 1965; Claes, 1977; Goodenough and Jurivich, 1978; Lens *et al.*, 1980, 1983), and consequently monoclonal antibodies (mAbs) have been selected against the agglutinins, in the expectation that some of them may be agglutinin-specific or at least may recognize specific domains on the molecules. The results so far have been of mixed success. Adair *et al.* (1983) and Adair (1985) characterized mAbs against the mt^+ agglutinin of *C. reinhardtii* that had been selected for their ability to isoagglutinate mt^+ gametes. None proved to be agglutinin-specific. However, two of them appeared to be gamete-specific, and only cross-reacted with one or two other high-molecular-weight glycoproteins. They were very useful for mapping different epitopes, because they revealed two periodate-sensitive epitopes common to the head and hook ends of the molecule that were not present in the shaft region, indicating that the head and hook possess at least two specific carbohydrate determinants restricted to these

Table III. Inactivation of Isolated Agglutinins of *Chlamydomonas* Species by Modification of Their Carbohydrate Moieties[a]

	Inactivation					
	C. reinhardtii		*C. eugametos*		*C. moewusii yapensis*	
Treatment	mt^+	mt^-	mt^+	mt^-	mt^+	mt^-
Periodate (10 mM)	Yes	Yes	Yes	Yes	Yes	Yes[b]
α-Galactosidase	No	No	No	Yes	No	No
α-Mannosidase	No		Yes	No	No	Yes

[a]Data from Collin-Osdoby and Adair (1985) and van den Ende *et al.* (1988).
[b]After preincubation with either phosphatase or sulfatase; this preincubation did not affect the agglutinative activity.

regions. Also, the shaft was shown to possess two unique repeating epitopes, one being periodate-sensitive and the other periodate-insensitive.

mAbs have also been selected for their ability to block agglutinin activity. Snell *et al.* (1986) characterized a mAb that specifically blocked the activity of the mt^+ agglutinin in *C. reinhardtii*, both *in vivo* and *in vitro*. However, it did not bind any components in immunoblots of flagellar proteins, so its specificity for the mt^+ agglutinin remains to be proven. Homan *et al.* (1988) selected several mAbs against the mt^- agglutinin of *C. eugametos*, which both in their native form and as Fab fragments inhibited mt^- ag-glutinability *in vivo* and *in vitro*. They also bound specifically to the mt^- agglutinin in immunoblots of mt^- flagellar extracts. This indicates that the antibody binds close to or in the binding site(s) of the mt^- agglutinin. These mAbs also bound to two mt^+ glycopro-teins, one of them being the mt^+ agglutinin, but they did not inhibit its binding activity. Such mAbs have proved to be important as specific agglutinin labels in studying their distribution during the mating process (see below).

There is evidence that the agglutinins are extrinsically bound to the flagellar mem-brane. They can be solubilized not only by detergents, but also by pH and osmotic shock (Klis *et al.*, 1985), EDTA (Adair *et al.*, 1982; Saito and Matsuda, 1984a), or by ex-traction with chaotropic agents such as guanidine thiocyanate (Musgrave *et al.*, 1981) that do not dissolve the lipid membrane. In addition, Homan *et al.* (1982) presented evidence for the agglutinins being anchored to an intrinsic membrane protein. They showed that agglutinin-containing membrane vesicles ("isoagglutinins," released into the culture liquid by blebbing of the flagellar membrane; Förster *et al.*, 1956; Wiese, 1984) could be inactivated by mild sonication and then reactivated by incubating them with isolated agglutinin. Reactivation did not occur after trypsinizing the inactivated vesicles, presum-ably because the anchor protein had been destroyed. Further evidence for the presence of an anchor protein comes from the observation that living gametes are more sensitive to enzymes such as trypsin and chymotrypsin than are the isolated agglutinins (Wiese and Hayward, 1972; Saito and Matsuda, 1984b; Collin-Osdoby and Adair, 1985). In this case, we assume that these enzymes digest the anchor protein, thus releasing the agglutinin from the membrane surface without inactivating the agglutinin per se.

Although only flagellum-associated agglutinins are functional in sexual interaction in both *C. eugametos* and *C. reinhardtii*, a large part of the total pool of agglutinin mole-cules is present in the cell body (Pijst *et al.*, 1983; Saito *et al.*, 1985). The latter authors argue that the cell body-associated agglutinin in *C. reinhardtii* is internal, because when naked protoplasts are isolated, their body surfaces are nonagglutinable. However, if the agglutinins are internal, one must explain why they should be sensitive to externally added trypsin or be so readily released by EDTA (Saito *et al.*, 1985). A possibility is that the trypsin or EDTA inactivates or removes the flagellar surface agglutinin molecules, there-by initiating a flow of other agglutinin molecules from an intracellular pool to the flagellar surface. This process continues until all of the intracellular agglutinin has been processed through the flagellar surface. In fact, Saito *et al.* (1985) propose that the cell body agglutinin functions as a pool for the flagella, for when mt^- gametes agglutinate with a nonfusing mt^+ mutant, the amount of cell body agglutinin initially drops sharply and then recovers to its original level. This recovery is blocked by cycloheximide. In *C. eu-gametos*, only a small part of the cell body agglutinin is in the form of an available pool (e.g., Tomson *et al.*, 1988), for the major fraction is extracellular, either associated with the plasma membrane or free from it in the periplasmatic space (Pijst *et al.*, 1983). In contrast to *C. reinhardtii*, this material is restricted to the cell body surface by a barrier at the base of each flagellum separating the cell membrane into two independent regions, the flagellar and the cell body domains (Musgrave *et al.*, 1986; see also Bloodgood, this volume). This separation is most clearly illustrated by an experiment in which flagella were regenerated in 4-hr-old vis-à-vis pairs of *C. eugametos*. By this time the plasma membrane constituents of both participating cells had merged. Using antibodies that recognized mating-type-specific antigens, it was shown that the regenerated flagella maintained their original specific antigenicity while the cell body surface exhibited mixed antigenicity (Musgrave *et al.*, 1985).

4. Mode of Action of the Agglutinins

Little is known about how the agglutinins interact, although some work has been done on isolated flagella. Goodenough (1986) confirmed earlier results of Köhle *et al.* (1980) who found that isolated mt^+ and mt^- flagella when mixed clump together in a sex-specific but random fashion. The interaction is sensitive to the ionic strength of the medium, and also occurs between flagella fixed in glutaraldehyde. It seems that the interaction is not disturbed by cross-linking the peptide part of either molecule.

The question of whether agglutinins of different mating type interact with each other or with unidentified receptors on the flagella cannot be answered with certainty. Isolated agglutinins do not mask each other's activity when mixed (Collin-Osdoby and Adair, 1985; Samson *et al.*, 1987b), but since such experiments have been carried out at nanomolar levels or lower, this might simply mean that the dissociation constant of the agglutinin–agglutinin complex is too high to obtain a measurable association under these conditions. In *C. eugametos*, there is evidence for a direct interaction between the agglutinins. As mentioned above, Homan *et al.* (1988) described mAbs (particularly mAb 66.3) that inhibit the adhesiveness of isolated mt^- agglutinin. They also inhibit *in vivo* agglutination of mt^- gametes; therefore, it is unlikely that components other than agglutinins are involved because they would still be free to bind the mt^+ flagella. What is more, in binding the mt^- agglutinin on the gamete flagella, mAbs stimulate all the sexual responses such as papillar formation, so there is no need to invoke other receptors. Finally, deep-etch images of adhering flagella suggest that the complementary agglutinins associate laterally into a meshwork of cablelike structures during the adhesion process (Goodenough *et al.*, 1986a). So for the time being, we must assume that *Chlamydomonas* possess a unipolar recognition system in which only the agglutinins are involved.

5. Longitudinal Redistribution of Agglutinins

Isolated flagella adhere at random, but the interaction does not evolve as it does in living cells. *In vivo*, the initial random contacts normally become tip-to-tip associations due to migration of the contact sites. Another important difference is that in living cells there is a considerable increase in the turnover rate of agglutinins, due to the inactivation of the molecules on the flagellar surface (Snell and Moore, 1980; Pijst *et al.*, 1984b). Snell *et al.* (1986) have recently shed some light on the nature of the inactivation. They used their mAb that binds the *mt*⁺ agglutinin of *C. reinhardtii*. When *mt*⁺ gametes were added to beads coated with the antibody, they bound the beads, reacted as if agglutinating but then de-adhered because they had inactivated them. In a similar manner, live gametes can agglutinate and inactivate fixed partners. What Snell and co-workers have now shown is that while the beads are being inactivated by *mt*⁺ gametes, they acquire an affinity for *mt*⁻ gametes. This suggests that during inactivation, flagellar membrane material is lost from the gametes onto the beads (or its sexual partner), thus sterically blocking the antibodies or agglutinins.

Before mating, the agglutinins are evenly distributed over the gamete flagella, and any part of the flagellar surface can be involved in the initial adhesion. With the selection of mAb 66.3, directed against the *mt*⁻ agglutinin of *C. eugametos*, this distribution could be visualized using immunolabeling. However, the agglutinin distribution may not be so uniform, for using the rapid-freeze deep-etch technique, Goodenough *et al.* (1985) showed that in *C. reinhardtii*, the agglutinins seem to occur in rows along the length of the flagella, as if they are bound to components of the axoneme. Interestingly, there are pictures of similar linear arrays of intramembrane particles along the length of the flagellar membrane (Bergman *et al.*, 1975; Snell, 1976). However, they are observed in gametes as well as in vegetative cells, so it is not likely that they represent the intrinsic membrane protein that anchors the agglutinin to the membrane, or other agglutinin-associated proteins.

During sexual agglutination, there is an extensive longitudinal redistribution of the agglutinins. This was first reported by Goodenough and Jurivich (1978), on the basis of their observation that antibodies directed against flagellar surface components in *C. reinhardtii* agglutinated gametes via their tips whereas vegetative cells were agglutinated at any point on their flagella. This was corroborated by the observation in paralyzed mutants that the contact sites between flagella migrated to the tips (Goodenough *et al.*, 1980). By labeling the agglutinins with mAb 66.3, their accumulation at the flagellar tips could be demonstrated directly in *C. eugametos* (Homan *et al.*, 1987a,b; Musgrave and van den Ende, 1987). In contrast to what has been reported about *C. reinhardtii* (Bloodgood *et al.*, 1986; Bloodgood, 1987), in *C. eugametos* this redistribution is specific for agglutinins and proteins associated with the agglutinins, in particular the wheat germ agglutinin-binding protein (Kooijman *et al.*, 1988). Flagellar surface components that are not associated with the agglutinins are not redistributed to the tips. The same phenomenon can be evoked by presenting single gametes with mAb 66.3. Fab fragments of this antibody are ineffective, which suggests that tipping is due to agglutinin cross-linking. This raises the problem of how such cross-linking occurs during sexual adhesion. Isolated agglutinins of all species studied are unable to cause isoagglutination of gametes of the opposite mating type, suggesting that they are functionally monovalent. However, Goodenough *et al.* (1985) observed that agglutinin molecules tend to associate longitudinally at their proximal ends into fibrils, from which the distal ends (considered to contain the binding site) project. One could easily imagine that the interaction of such aggregates might lead to large agglutinin clusters, which subsequently would be transported to the tips of the flagella. The mechanism of this transport is unknown but since it is inhibited by colchicine (Mesland *et al.*, 1980; Homan *et al.*, 1988), it may well be tubulin-mediated (but see Bloodgood, this volume).

It seems that adhesion sites that have migrated to the tips become fixed there and are less likely to migrate back onto the rest of the flagellar surface. Similarly, flagellum-bound polystyrene beads that have migrated into the tip region on an agglutinating flagellum also become temporarily fixed there (Hoffman and Goodenough, 1980). So there seems to be some connection between the tip-oriented movement of flagellar membrane constituents and particle movement (see Bloodgood, this volume, for a detailed discussion).

In *C. reinhardtii*, the consequences of tipping is clearly seen by the more or less exclusive adhesion of the flagella at their distal ends, the bases obviously being less involved. In *C. eugametos*, in contrast, the whole flagellum is agglutinable, and partner flagella adhere from their bases to their aligned tips. Consequently, the cells are closely appressed, so that contact of the papillae and thus cell fusion is favored.

6. The Signaling Action of Sexual Agglutination

As discussed above, sexual agglutination serves not only to bring the mating cells in juxtaposition, but also to induce cellular responses by which the cells become fusion-competent. Apart from the activation of the mating structure, accompanied by the action of an autolysin to dissolve the cell wall, there are a number of other, possibly related responses, including a typical change in the behavior of the flagella, called twitching, and a change in the ultrastructure of the flagellar tips, called flagellar tip activation (Mesland *et al.*, 1980). It seems evident that the agglutinins are involved in generating the signals by which these phenomena are evoked. We do not know what the primary reactions are that occur at the surface of interacting flagella, but the effects of antiagglutinin antibodies strongly suggest that the cross-linking of agglutinins is an important early step in the signal sequence. Treating *Chlamydomonas* gametes with antibodies results in the same cellular responses as observed during sexual agglutination (Claes, 1977; Goodenough and Jurivich, 1978; Snell *et al.*, 1986; Homan *et al.*, 1988) but treatment with monovalent Fab fragments that cannot cross-link the agglutinins does not induce a response (Goodenough and Jurivich, 1978; Homan *et al.*, 1988). Homan *et al.* (1988) have shown that the amount of antibody needed to induce mating structure activation is about 40 times higher than that required to induce twitching or tipping, and is similar to that needed to block the agglutinability of live mt^- gametes. This implies that most of the agglutinin molecules at the flagellar surface have to be occupied before the induction signal is strong enough to trigger mating structure activation. This reaction is also obtained with mAbs that bind the agglutinins outside their binding site, as long as the antibody is polyvalent. Thus, receptor clustering is important for signaling and for the induction of tipping. It has also been involved in other similar systems, such as the action of epidermal growth factor (Schlessinger, 1986), platelet fibrinogen receptor (Isenberg *et al.*, 1987), and the complement receptor on human neutrophils (Detmers *et al.*, 1987).

While we do not know how receptor clustering elicits cellular responses, there is now strong evidence that cAMP metabolism is a major constituent of the signal transduction pathway. When *C. eugametos* gametes are mixed, there is a rapid transient rise in the intracellular cAMP level (Pijst *et al.*, 1984a). A similar increase in cAMP content can be evoked by presenting single mating type cells with isolated flagella of the other mating type. This indicates that it is the direct consequence of flagellar interaction. A cAMP-dependent protein kinase was found in cell homogenates, suggesting that signal transduction proceeds along the same lines as in animal transduction systems. A more extensive study was made in *C. reinhardtii* by Pasquale and Goodenough (1987). They also found a rapid rise in cAMP content after mixing mt^+ and mt^- gametes, but the kinetics were slightly different. While in *C. eugametos* the transient increase was very rapid and terminated in less than a minute, in *C. reinhardtii* the return of the cAMP concentration to its original level coincided with the completion of the mating process by cell fusion. Using a nonfusing mt^+ mutant (*imp-1*) and a wild-type mt^- strain, the agglutination process was prolonged considerably, and the cAMP concentration did not return to its original level. The importance of cAMP was demonstrated by showing that single mt^+ and mt^- gametes respond to exogenous dibutyryl cAMP by activating their mating structures. Thus, flagellaless or nonagglutinating mutants could be induced to fuse by a simple cAMP treatment, bypassing the agglutination process. This leads to the conclusion that cAMP is an important second messenger in the signaling pathway.

It may well be that the agglutination reaction results in the activation of adenylate cyclase via guanidine nucleotide-binding G proteins (e.g., Levitsky, 1987), but an intermediary role for Ca^{2+} is not excluded. A transient increase in efflux of Ca^{2+} occurs at the start of the mating reaction (Bloodgood and Levin, 1983) and mating is accompanied by the redistribution of Ca^{2+} in the cell body (Kaska *et al.*, 1985). Calmodulin blockers, such as trifluoperazine and W-7, inhibit the mating response in agglutinating cells (Snell *et al.*, 1982; Detmers and Condeelis, 1986). This inhibition can be reversed by the addition of cAMP (Pasquale and Goodenough, 1987). Calmodulin and a calmodulin-regulated protein kinase are known to be present in the flagella of *Chlamydomonas* (Gitelman and Witman, 1980; Schleicher *et al.*, 1984). In animal signaling systems the intracellular Ca^{2+} concentration is generally regulated by phosphoinositides which are derived from the phospholipase C-catalyzed breakdown of phosphatidylinositol bis- and trisphosphates (Berridge and Irvine, 1984). The occupied receptors again activate the appropriate enzymes via membrane-associated G proteins (e.g., Neer and Clapham, 1988). Whether a similar system is operative in *Chlamydomonas* remains to be elucidated.

7. Modulation of Sexual Agglutinability

Flagellar agglutinability is a transitory property that disappears when cell fusion is completed. It is obvious therefore that adhesion between mating cells is not a passive process, but is subject to regulatory influences from the cell. There are several examples showing that flagellar adhesiveness can be modulated, e.g., by altering the agglutinin content, the distribution in the plane of the flagellar membrane, or their covalent structure.

In a light–dark regimen, gametes of *C. eugametos* show a diurnal periodicity in mating competence. At the beginning of the light period, this mating competence is high, but it declines during the course of the light period and increases again during the next dark period (Demets *et al.*, 1987). This rhythm persists in the dark, but it is rapidly damped out in continuous light (Tomson *et al.*, 1988). The fluctuations are paralleled by periodic changes in agglutinability, which confirms the notion that it is the sexual adhesiveness of the flagella which determines the mating competence. Extracts of isolated flagella also showed changes in activity that corresponded with the agglutinability *in vivo*. Thus, the evidence suggests that it is the agglutinin content of the flagellar surface that determines the degree of agglutinability.

A similar mechanism might be envisaged with respect to the decrease in agglutinability after gamete fusion. As already mentioned, there is a considerable turnover of flagellar agglutinins during sexual agglutination (Snell and Moore, 1980; Pijst *et al.*, 1984b). To maintain agglutinability, new agglutinin has to be continuously supplied to the flagellar membrane from a cellular pool. If this supply is arrested on fusion, the agglutinin content of the flagella will rapidly decline, as has in fact been demonstrated by Musgrave *et al.* (1985).

An important aspect of sexual agglutinability in *C. eugametos* and *C. reinhardtii* is that it increases upon sexual contact between the gametes (Tomson *et al.*, 1986; Demets *et al.*, 1988; U. W. Goodenough, personal communication). A severalfold increase in sexual agglutinability was observed shortly after the onset of sexual agglutination. If the agglutination reaction was disturbed by vigorous shaking, the cells de-adhered and the agglutinability of both mating types dropped within minutes to the original level. This phenomenon appeared to be fully reversible. The increase in adhesiveness could also be realized by presenting gametes of a single mating type with flagella or isolated agglutinin of the opposite mating type. The increase is never much more than eightfold, and is independent of the initial level of agglutinability (R. Demets, personal communication).

The most obvious supposition would be that the agglutinin content of the flagella increases on contact with its partner, but, at least for *C. eugametos*, that seems not to be a sufficient explanation. By analyzing flagella amputated before and after sexual stimulation, it was found that their adhesiveness had increased approximately 8-fold, in accordance with the *in vivo* situation. However, the rise in agglutinin content was only 1.4-fold which suggests that the contact-induced rise in agglutinability is not so much due to an increase in agglutinin content, but rather to a higher avidity of the agglutinins already present. Again the lateral mobility of the agglutinins may be involved in this phenomenon. When gametes are treated with wheat germ agglutinin, their flagellar adhesiveness is rapidly enhanced, as if they were agglutinating (R. Kooijman, personal communication; van den Ende *et al.*, 1989). Wheat germ agglutinin binds to a membrane component which is associated with agglutinin (it is codistributed with the agglutinins during tipping), and because it is multivalent, it must cluster the agglutinin molecules by cross-linking the associated protein. This leads to an increase in the binding competence of the agglutinins, either by a cooperative effect (a cluster of receptors often has a much higher avidity than the affinity of each receptor; see, e.g., Detmers *et al.*, 1987) or by an aggregation-induced allosteric change whereby the affinity of each receptor is increased.

Whichever explanation is appropriate, it is clear that agglutinin molecules are subject to *in situ* conformational changes. An illuminating example is the effect of light on some strains of *C. eugametos* and *C. moewusii* (Förster and Wiese, 1954; Lewin, 1956; Kooijman *et al.*, 1986, 1988). In the dark, such strains are absolutely nonagglutinable, but they acquire this ability when illuminated. This agglutinability is rapidly lost when the cells are returned to the dark. One must conclude that it is the affinity or the concentration of the agglutinins at the flagellar surface that is modulated by light. Studies by Kooijman *et al.* (1988) have shown that light activates a structural modification of the agglutinins themselves that is presumably mediated by an ectoenzyme, such as a protein kinase, phosphatase, or sulfatase. The modification results in a change in the conformation of the molecule by which its binding constant with the complementary agglutinin is dramatically altered. The fact that the light requirement for agglutinability occurs in mt^+ as well as in mt^- strains suggests that the structural modification occurs in a domain that is common to both agglutinins. Unfortunately, the nature and location of the photoreceptor are unknown, although action spectra have been described (Lewin, 1956; Förster, 1957).

8. Conclusions

From the considerations presented above, we suggest that the following sequence of events occurs during sexual agglutination in *Chlamydomonas*. Flagellar contact between compatible gametes results in agglutinin–agglutinin interaction that rapidly leads to a patching of these molecules in the plane of the flagellar membrane. This patching may be mediated by a submembranous cytoskeleton. Agglutinin patching leads to an increased binding competence, consolidating the contact between the flagella. This interaction progresses into a more extensive cross-linking of the agglutinins and associated proteins, culminating in tip-oriented transport of the complexes. At the same time, agglutinin cross-linking activates adenylate cyclase, located internally at the flagellar membrane, which produces a rapid increase in the cAMP level that initiates a signaling cascade leading ultimately to the activation of the mating structure.

This picture is far from complete. The mechanism underlying the longitudinal redistribution of agglutinins, resulting in enhanced agglutinability and tipping, is unknown. Also, the role of the agglutinins in the signal-generating process is unclear. While agglutinin cross-linking seems to be an essential step in the process, this can also be achieved by agents that cross-link agglutinin-associated proteins (e.g., wheat germ agglutinin in *C. eugametos*). So it seems that the agglutinins only serve to endow specificity on the agglutination reaction. If so, it should be possible to mate incompatible species using antibodies cross-reacting with flagellar proteins on both partners. The chances are slight that such an approach would result in cell fusion, since we know that cell fusion is also a very specific process (Mesland and van den Ende, 1978; Forest, 1983).

These and many other questions will undoubtedly keep this area of research lively for years to come.

References

Adair, W. S., 1985, Characterization of *Chlamydomonas* sexual agglutinins, *J. Cell Sci. Suppl.* **2:**233–260.

Adair, W. S., Monk, B. C., Cohen, R., and Goodenough, U. W., 1982, Sexual agglutinins from the *Chlamydomonas* flagellar membrane. Partial purification and characterization, *J. Biol. Chem.* **257:**4593–4602.

Adair, W. S., Hwang, C. J., and Goodenough, U. W., 1983, Identification and visualization of the sexual agglutinin from mating type plus flagellar membranes of *Chlamydomonas*, *Cell* **33:**183–193.

Adair, W. S., Steinmetz, S. A., Mattson, D. M., Goodenough, U. W., and Heuser, E. J., 1987, Nucleated assembly of *Chlamydomonas* and *Volvox* cell walls, *J. Cell Biol.* **105:**2373–2382.

Bergman, K., Goodenough, U. W., Goodenough, D. A., Jawitz, J., and Martin, H., 1975, Gametic differentiation in *Chlamydomonas reinhardtii.* II. Flagellar membranes and the agglutination reaction, *J. Cell Biol.* **67:**606–622.

Berridge, M. J., and Irvine, R. F., 1984, Inositol trisphosphate, a novel second messenger in cellular signal transduction, *Nature* **312:**315–321.

Bloodgood, R. A., 1987, Glycoprotein dynamics in *Chlamydomonas* flagellar membrane, *Adv. Cell Biol.* **1:**97–130.

Bloodgood, R. A., and Levin, E. N., 1983, Transient increase in calcium efflux accompanies fertilization in *Chlamydomonas, J. Cell Biol.* **97:**397–404.

Bloodgood, R. A., Woodward, M. P., and Salomonsky, N. L., 1986, Redistribution and shedding of flagellar membrane glycoproteins visualized using an anti-carbohydrate monoclonal antibody and concanavalin A, *J. Cell Biol.* **102:**1797–1812.

Brown, R. M., Johnson, C., and Bold, H. C., 1968, Electron and phase-contrast microscopy of sexual reproduction in *Chlamydomonas moewusii, J. Phycol.* **4:**100–120.

Claes, H., 1977, Non-specific stimulation of the autolytic system in gametes from *Chlamydomonas reinhardii, Exp. Cell Res.* **108:**221–229.

Collin-Osdoby, P., and Adair, W. S., 1985, Characterization of the purified *Chlamydomonas* minus agglutinin, *J. Cell Biol.* **101:**1144–1152.

Collin-Osdoby, P., Adair, W. S., and Goodenough, U. W., 1984, *Chlamydomonas* agglutinin conjugated to agarose beads as an in vitro probe of adhesion, *Exp. Cell Res.* **150:**282–291.

Cooper, J. B., Adair, W. S., Mecham, R. P., Heuser, J. E., and Goodenough, U. W., 1983, *Chlamydomonas* agglutinin is a hydroxyproline-rich glycoprotein, *Proc. Natl. Acad. Sci. USA* **80:**5898–5901.

Crabbendam, K. J., Nanninga, N., Musgrave, A., and van den Ende, H., 1984, Flagellar tip activation in vis-a-vis pairs of *Chlamydomonas eugametos, Arch. Microbiol.* **138:**220–223.

Crabbendam, K. J., Klis, F. M., Musgrave, A., and van den Ende, H., 1986, Ultrastructure of the plus and minus sexual agglutinins of *Chlamydomonas eugametos,* as visualized by negative staining, *J. Ultrastruct. Mol. Struct. Res.* **96:**151–159.

Demets, R., Tomson, A. M., Stegwee, D., and van den Ende, H., 1987, Control of the mating competence rhythm in *Chlamydomonas eugametos, J. Gen. Microbiol.* **133:**1081–1088.

Demets, R., Tomson, A. M., Homan, W. L., Stegwee, D., and van den Ende, H., 1988, Cell–cell adhesion in conjugating *Chlamydomonas* gametes: A self-enhancing process, *Protoplasma* **145:**27–36.

Detmers, P. A., and Condeelis, J. S., 1986, Trifluoperazine and W-7 inhibit mating in *Chlamydomonas* at an early stage of gametic interaction, *Exp. Cell Res.* **163:**317–326.

Detmers, P. A., Goodenough, U. W., and Condeelis, J., 1983, Elongation of the fertilization tubule in *Chlamydomonas:* New observations on the core microfilaments and the effect of transient intracellular signals on their structural integrity, *J. Cell Biol.* **97:**522–532.

Detmers, P. A., Wright, S. D., Olsen, E. B., Kimball, B., and Kohn, Z. A., 1987, Aggregation of complement receptors on human neutrophils in the absence of ligand, *J. Cell Biol.* **105:**1137–1145.

Forest, C. L., 1983, Specific contact between mating structure membranes observed in conditional fusion-defective *Chlamydomonas* mutants, *Exp. Cell Res.* **148:**143–154.

Förster, H., 1957, Das Wirkungsspektrum der Kopulation von *Chlamydomonas eugametos, Z. Naturforsch.* **12b:**765–770.

Förster, H., and Wiese, L., 1954, Untersuchungen zur Kopulationsfähigkeit von *Chlamydomonas eugametos, Z. Naturforsch.* **9b:**470–471.

Förster, H., Wiese, L., and Braunitzer, G., 1956, Ueber das agglutinierend wirkende Gynogamon von *Chlamydomonas eugametos, Z. Naturforsch.* **11b:**315–317.

Gitelman, S. E., and Witman, G. B., 1980, Purification of calmodulin from *Chlamydomonas:* Calmodulin occurs in cell bodies and flagella, *J. Cell Biol.* **98:**764–770.

Goodenough, U. W., 1986, Experimental analysis of the adhesion reaction between isolated *Chlamydomonas* flagella, *Exp. Cell Res.* **166:**237–246.

Goodenough, U. W., and Jurivich, D., 1978, Tipping and mating-structure activation induced in *Chlamydomonas* gametes by flagellar membrane antisera, *J. Cell Biol.* **79:**680–691.

Goodenough, U. W., and Weiss, R. L., 1975, Gametic differentiation in *Chlamydomonas reinhardtii.* III. Cell wall lysis and microfilament-associated mating structure activation in wild-type and mutant strains, *J. Cell Biol.* **67:**623–637.

Goodenough, U. W., Adair, W. S., Caligor, E., Forest, C. L., Hoffman, J. L., Mesland, D. A. M., and Spath, S., 1980, Membrane–membrane and membrane–ligand interactions in *Chlamydomonas* mating, in: *Membrane–Membrane Interactions* (B. Gilula, ed.), Raven Press, New York, pp. 131–153.

Goodenough, U. W., Detmers, P. A., and Hwang, C., 1982, Activation for cell fusion in *Chlamydomonas:* Analysis of wild-type gametes and nonfusing mutants, *J. Cell Biol.* **92:**378–386.

Goodenough, U. W., Adair, W. S., Collin-Osdoby, P., and Heuser, J. E., 1985, Structure of the *Chlamydomonas* agglutinin and related flagellar surface proteins in vitro and in situ, *J. Cell Biol.* **101:**924–941.

Goodenough, U. W., Adair, W. S., Collin-Osdoby, P., and Heuser, J. E., 1986a, *Chlamydomonas* cells in contact, in: *Cells in Contact* (E. Gall and G. M. Edelman, eds.), Wiley, New York, pp. 111–135.

Goodenough, U. W., Gebhart, B., Mecham, R. P., and Heuser, J. E., 1986b, Crystals of the *Chlamydomonas reinhardtii* cell wall: Polymerization, depolymerization and purification of glycoprotein monomers, *J. Cell Biol.* **103:**405–417.

Harris, E. H., 1989, *The Chlamydomonas Sourcebook,* Academic Press, New York.

Heuser, J. E., 1983, A method for freeze-drying molecules absorbed to mica flakes, *J. Mol. Biol.* **169:** 155–196.

Hoffman, J. L., and Goodenough, U. W., 1980, Experimental dissection of flagellar surface motility in *Chlamydomonas, J. Cell Biol.* **86:**656–665.

Homan, W. L., Musgrave, A., Molenaar, E. M., and van den Ende, H., 1980, Isolation of monovalent sexual binding components from *Chlamydomonas eugametos* flagellar membranes, *Arch. Microbiol.* **128:**120–125.

Homan, W. L., Gijsberti Hodenpijl, P., Musgrave, A., and van den Ende, H., 1982, Reconstitution of biological activity on isoagglutinins from *Chlamydomonas eugametos, Planta* **155:**529–535.

Homan, W. L., Sigon, C., van den Briel, W., Wagter, R., de Nobel, H., Mesland, D. A. M., Musgrave, A., and van den Ende, H., 1987a, Transport of membrane receptors and the mechanics of sexual cell fusion in *Chlamydomonas eugametos, FEBS Lett.* **215:**323–326.

Homan, W. L., Kalshoven, H., Kolk, A. H. J., Musgrave, A., Schuring, F., and van den Ende, H., 1987b, Monoclonal antibodies to surface glycoconjugates in *Chlamydomonas eugametos* recognize strain-specific O-methyl sugars, *Planta* **170:**328–335.

Homan, W. L., Musgrave, A., de Nobel, H., Wagter, R., de Wit, D., Kolk, A. H. J., and van den Ende, H., 1988, Monoclonal antibodies directed against the binding site of *Chlamydomonas eugametos* gametes, *J. Cell Biol.* **107:**177–189.

Homer, R. B., and Roberts, K., 1979, Glycoprotein conformation in plant cell walls. Circular dichroism reveals a polyproline II structure, *Planta* **146:**217–222.

Isenberg, W. M., McEver, R. P., Phillips, D. R., and Shuman, M. A., 1987, The platelet fibrinogen receptor: An immunogold-surface replica study of agonist-induced ligand binding and receptor clustering, *J. Cell Biol.* **104:**1655–1663.

Jupe, E. R., Chapman, R. L., and Zimmer, E. A., 1988, Nuclear ribosomal RNA genes and algal phylogeny. The *Chlamydomonas* example. *BioSystems* **21:**223–230.

Kaska, D. D., Piscopo, I. C., and Gibor, A., 1985, Intracellular calcium redistribution during mating in *Chlamydomonas reinhardtii, Exp. Cell Res.* **160:**371–379.

Klis, F. M., Samson, M. R., Touw, E., Musgrave, A., and van den Ende, H., 1985, Sexual agglutination in the unicellular green alga *Chlamydomonas eugametos.* Identification and properties of the mating type plus agglutination factor, *Plant Physiol.* **79:**740–745.

Köhle, D., Lang, W., and Kauss, H., 1980, Agglutination and glycosyltransferase activity of isolated gametic flagella from *Chlamydomonas reinhardii, Arch. Microbiol.* **127:**239–243.

Kooijman, R., Elzenga, T. J. M., de Wildt, P., Musgrave, A., Schuring, F., and van den Ende, H., 1986, Light dependence of sexual agglutinability in *Chlamydomonas eugametos, Planta* **169:**370–378.

Kooijman, R., de Wildt, P., Homan, W. L., Musgrave, A., and van den Ende, H., 1988, Light affects flagellar agglutinability in *Chlamydomonas eugametos* by modification of the agglutinin molecules, *Plant Physiol.* **86:**216–223.

Lemieux, B., and Lemieux, C., 1985, Extensive sequence rearrangements in the chloroplast genomes of the green algae *Chlamydomonas eugametos* and *Chlamydomonas reinhardtii, Curr. Genet.* **10:**213–219.

Lens, P. F., van den Briel, W., Musgrave, A., and van den Ende, H., 1980, Sex-specific glycoproteins in *Chlamydomonas* flagella. An immunological study, *Arch. Microbiol.* **126:**77–81.

Lens, P. F., Olofsen, F., van Egmond, P., Musgrave, A., and van den Ende, H., 1983, Isolation of an antigenic determinant from flagellar glycoproteins of *Chlamydomonas eugametos*, *Arch. Microbiol.* **135**:311–314.

Levitsky, A., 1987, Regulation of adenylate cyclase by hormones and G proteins, *FEBS Lett.* **211**:113–118.

Lewin, R. A., 1952, Studies on the flagella of algae. I. General observations on *Chlamydomonas moewusii* Gerloff, *Biol. Bull. (Woods Hole, Mass.)* **102**:74–79.

Lewin, R. A., 1956, Control of sexual activity in *Chlamydomonas* by light, *J. Gen. Microbiol.* **15**:170–185.

Matsuda, Y., Sakamoto, K., Mizuochi, T., Kobata, A., Tamura, G., and Tsubo, Y., 1981, Mating type specific inhibition of gametic differentiation of *Chlamydomonas reinhardtii* by tunicamycin, *Plant Cell Physiol.* **22**:1607–1611.

Mesland, D. A. M., 1976, Mating in *Chlamydomonas eugametos*. A scanning electron microscopical study, *Arch. Microbiol.* **109**:31–35.

Mesland, D. A. M., 1978, Sexual cell interaction in *Chlamydomonas eugametos*, Thesis, University of Amsterdam.

Mesland, D. A. M., and van den Ende, H., 1978, An inhibitor of cell fusion in *Chlamydomonas* gametes, *Arch. Microbiol.* **117**:131–134.

Mesland, D. A. M., Hoffman, J. L., Caligor, E., and Goodenough, U. W., 1980, Flagellar tip activation stimulated by membrane adhesions in *Chlamydomonas* gametes, *J. Cell Biol.* **84**:599–617.

Musgrave, A., and van den Ende, H., 1987, How *Chlamydomonas* court their partners, *Trends Biochem. Sci.* **12**:470–473.

Musgrave, A., van der Steuyt, P., and Ero, L., 1979, Concanavalin A binding to *Chlamydomonas eugametos* flagellar proteins and its effect on sexual reproduction, *Planta* **147**:51–56.

Musgrave, A., van Eyk, E., te Welscher, R., Broekman, R., Lens, P. F., Homan, W. L., and van den Ende, H., 1981, Sexual agglutination factor from *Chlamydomonas eugametos*, *Planta* **153**:362–369.

Musgrave, A., de Wildt, P., Schuring, F., Crabbendam, K., and van den Ende, H., 1985, Sexual agglutination in *Chlamydomonas eugametos* before and after cell fusion, *Planta* **166**:234–243.

Musgrave, A., de Wildt, P., van Etten, Y., Pijst, H., Kooijman, R., Homan, W., and van den Ende, H., 1986, Evidence for a functional membrane barrier in the transition zone between the flagellum and the cell body of *Chlamydomonas eugametos* gametes, *Planta* **167**:544–553.

Neer, E. J., and Clapham, D. E., 1988, Roles of G protein subunits in transmembrane signalling, *Nature* **333**:129–134.

Pasquale, S. M., and Goodenough, U. W., 1987, Cyclic AMP functions as a primary sexual signal in gametes of *Chlamydomonas reinhardtii*, *J. Cell Biol.* **105**:2279–2292.

Pijst, H. L. A., Zilver, R. J., Musgrave, A., and van den Ende, H., 1983, Agglutination factor in the cell body of *Chlamydomonas eugametos*, *Planta* **158**:403–409.

Pijst, H. L. A., van Driel, R., Janssens, P. M. W., Musgrave, A., and van den Ende, H., 1984a, Cyclic AMP is involved in sexual reproduction of *Chlamydomonas eugametos*, *FEBS Lett.* **174**:132–136.

Pijst, H. L. A., Ossendorp, F. A., van Egmond, P., Kamps, A., Musgrave, A., and van den Ende, H., 1984b, Sex specific binding and inactivation of agglutination factor in *Chlamydomonas eugametos*, *Planta* **160**:529–536.

Ray, D. A., and Gibor, A., 1982, Tunicamycin-sensitive glycoproteins involved in the mating of *Chlamydomonas reinhardi*, *Exp. Cell Res.* **141**:245–252.

Roberts, K., 1974, Crystalline glycoproteins of algae: Their structure, composition and assembly, *Philos. Trans. R. Soc. London Ser. B* **268**:129–146.

Roberts, K., Grief, C., Hills, G., and Shaw, P. J., 1985, Cell wall glycoproteins: Structure and function, *J. Cell Sci. Suppl.* **2**:105–127.

Saito, T., and Matsuda, Y., 1984a, Sexual agglutinin of mating-type minus gametes in *Chlamydomonas reinhardtii*. I. Loss and recovery of agglutinability of gametes treated with EDTA, *Exp. Cell Res.* **152**:322–330.

Saito, T., and Matsuda, Y., 1984b, Sexual agglutinin of mating-type minus gametes in *Chlamydomonas reinhardtii*. II. Purification and characterization of minus agglutinin and comparison with plus agglutinin, *Arch. Microbiol.* **139**:95–99.

Saito, T., Tsubo, Y., and Matsuda, Y., 1985, Synthesis and turnover of cell body-agglutinin as a pool for flagellar surface-agglutinin in *Chlamydomonas reinhardii*, *Arch. Microbiol.* **142**:207–210.

Samson, M. R., Klis, F. M., Crabbendam, K. J., van Egmond, P., and van den Ende, H., 1987a, Purification, visualization and characterization of the sexual agglutinins of green alga *Chlamydomonas moewusii yapensis, J. Gen. Microbiol.* **133**:3183–3191.

Samson, M. R., Klis, F. M., Homan, W. L., van Egmond, P., Musgrave, A., and van den Ende, H., 1987b, Composition and properties of the sexual agglutinins of the flagellated green alga *Chlamydomonas eugametos, Planta* **170**:314–321.

Schleicher, M., Lukas, T. J., and Watterson, D. M., 1984, Isolation and characterization of calmodulin from the motile green alga *Chlamydomonas reinhardtii, Arch. Biochem. Biophys.* **229**:33–42.

Schlessinger, J., 1986, Allosteric regulation of the epidermal growth factor receptor kinase, *J. Cell Biol.* **103**:2067–2072.

Snell, W. J., 1976, Mating in *Chlamydomonas:* A system for the study of specific cell adhesion. I. Ultrastructural and electrophoretic analysis of flagellar surface components involved in adhesion, *J. Cell Biol.* **68**:48–69.

Snell, W. J., and Moore, W. S., 1980, Aggregation-dependent turnover of flagellar adhesion molecules in *Chlamydomonas* gametes, *J. Cell Biol.* **84**:203–210.

Snell, W. J., Buchanan, M., and Clausell, A., 1982, Lidocaine reversibly inhibits fertilization in *Chlamydomonas.* A possible role in sexual signaling, *J. Cell Biol.* **94**:607–612.

Snell, W. J., Kosfiszer, M. G., Clausell, A., Perillo, N., Imam, S., and Hunnicutt, G., 1986, A monoclonal antibody that blocks adhesion of *Chlamydomonas* mt$^+$ gametes, *J. Cell Biol.* **103**:2449–2456.

Tomson, A. M., Demets, R., Sigon, C. A. M., Stegwee, D., and van den Ende, H., 1986, Cellular interactions during the mating process in *Chlamydomonas eugametos, Plant Physiol.* **81**:522–526.

Tomson, A. M., Demets, R., Homan, W. L., Stegwee, D., and van den Ende, H., 1988, Endogenous oscillator controls surface density of mating receptors in *Chlamydomonas eugametos, Sex. Plant Reprod.* **1**:46–50.

van den Ende, H., Klis, F. M., and Musgrave, A., 1988, The role of flagella in sexual reproduction of *Chlamydomonas eugametos, Acta Bot. Neerl.* **37**:327–350.

van den Ende, H., Tomson, Z. M., Demets, R., and Kooijman, R., 1989, Modulation of sexual agglutinability in *Chlamydomonas eugametos,* in: *Algae as Experimental Systems* (A. W. Coleman, L. J. Goff, and J. R. Stein-Taylor, eds.), Alan R. Liss, New York, pp. 187–200.

Wenzl, S., Thym, D., and Sumper, M., 1984, Development-dependent modification of the extracellular matrix by sulphated glycoproteins in *Volvox carteri, EMBO J.* **73**:739–744.

Wiese, L., 1965, On sexual agglutination and mating-type substances (gamones) in isogamous heterothallic *Chlamydomonas.* I. Evidence of the identity of the gamones with the surface components responsible for sexual flagellar contact, *J. Phycol.* **1**:46–54.

Wiese, L., 1969, Algae, in: *Fertilization: Comparative Morphology, Biochemistry and Immunology,* Volume 2 (C. B. Metz and A. Monroy, eds.), Academic Press, New York, pp. 135–188.

Wiese, L., 1974, Nature of sex specific glycoprotein agglutinins in *Chlamydomonas, Ann. N.Y. Acad. Sci.* **234**:383–395.

Wiese, L., 1984, Mating systems in unicellular algae, *Encycl. Plant Physiol. N. Ser.* **17**:238–260.

Wiese, L., and Hayward, P. C., 1972, On sexual agglutination and mating-type substances in isogamous dioecious *chlamydomonads.* III. The sensitivity of sex cell contact to various enzymes, *Am. J. Bot.* **59**:530–536.

Wiese, L., and Mayer, R. A., 1982, Unilateral tunicamycin sensitivity of gametogenesis in dioecious isogamous *chlamydomonas* species, *Gamete Res.* **5**:1–9.

Wiese, L., and Shoemaker, D., 1970, On sexual agglutination and mating type substances in isogamous heterothallic chlamydomonads, *Biol. Bull. (Woods Hole, Mass.)* **138**:88–95.

Wiese, L., and Wiese, W., 1975, On sexual agglutination and mating type substances in isogamous dioecious *Chlamydomonads.* IV. Unilateral inactivation of the sex contact capacity in compatible and incompatible taxa by alpha-mannosidase and snake venom protease, *Dev. Biol.* **43**:264–276.

Wiese, L., and Wiese, W., 1978, Sex cell contact in *Chlamydomonas,* a model for cell recognition, *Symp. Soc. Exp. Biol.* **32**:83–104.

The Role of Ciliary Surfaces in Mating in *Paramecium*

Tsuyoshi Watanabe

1. Introduction

The cell surface of the unicellular organism *Paramecium* is covered by thousands of cilia. The cilia are projections of the cell surface and are covered with membranes which are continuous with the rest of the cell surface membrane. The cilia and basal bodies of *Paramecium* have a conventional ultrastructure which does not differ basically from that in other protozoa or higher animals (Jurand and Selman, 1969; Allen, 1971). Cilia have important functions in the life of *Paramecium,* such as locomotion through the surrounding water and ingestion of food into the cytostome (see Wichterman, 1985). The cilia responsible for the ingestion of food are mainly localized in the gullet, which is a funnel-shaped depression of the cell surface. Probably all other cilia, except the caudal ones, are responsible for locomotion. The caudal cilia are somewhat longer than somatic cilia and are known to be immotile. In addition to these functions, cilia play an essential role in the initial step of the mating reaction in *Paramecium*. Mating in *Paramecium* is referred to as conjugation, as is the case in other unicellular organisms. Cilia that can take part in conjugation are found only at a particular location on the cell surface and occur only at certain times in the life cycle and culture phase of paramecia (Hiwatashi, 1969). In this chapter, the role of the ciliary surface in the conjugation of *Paramecium* and some features of temporal and spatial differentiation of cilia as mating organelles will be described.

2. The Events Occurring during Conjugation in *Paramecium*

In 1937, Sonneborn discovered the mating types of *Paramecium* and observed the events of conjugation induced by mixing two different mating types. When cultures derived from single individuals are mixed together, in some combinations of the cultures the individuals immediately clump together in clusters. This agglutination reaction is

Tsuyoshi Watanabe • Department of Biological Science, Tohoku University, Kawauchi, Sendai 980, Japan.

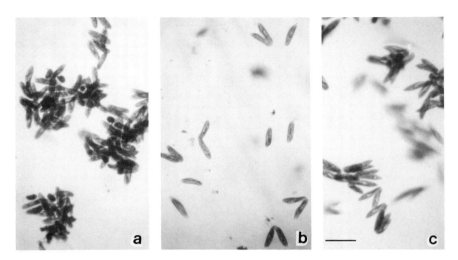

Figure 1. Events of conjugation in *P. caudatum*. (a) Mating reaction—10 min; (b) holdfast union—60 min; (c) beginning of the paroral union—120 min after mixing cells of the complementary mating types. Bar = 200 μm.

called the *mating reaction*. The agglutination process is known to be mating type specific, because it occurs only between cells of complementary mating types. Typical conjugating pairs emerge from the clusters afterwards. The duration of the mating reaction is dependent on the temperature and on the species of *Paramecium*. Generally speaking, however, the mating reaction continues for about 1 hr in many species of *Paramecium* at ordinary room temperature (Fig. 1).

At the end of the mating reaction, the clusters dissociate into smaller ones and many pairs emerge from them. The cells in these pairs contact each other at their anterior regions. This type of cell adhesion is called the *holdfast union*. When the contact area of the pair is extended to the post-oral region, the cell adhesion is called the *paroral union*. The paroral union is formed at approximately 2 hr after the onset of the mating reaction, and continues for about 5 hr in *P. tetraurelia* (Jurand and Selman, 1969) and 15 hr in *P. caudatum* (Mikami and Hiwatashi, 1975). Then the members of the pair separate into two exconjugants. During the paroral union, an ordered series of nuclear events, including meiosis, pronuclear exchange, and karyogamy, takes place (Mikami and Hiwatashi, 1975). Breakdown of the old macronucleus occur after pair separation (see Hiwatashi and Mikami, 1989). Thus, only partial and temporary cell fusion occurs during the conjugation of *Paramecium*.

3. Role of Ciliary Surfaces in the Mating Reaction

3.1. Mating Substances in the Ciliary Membrane

The sexual cell recognition molecules utilized in the mating reaction are designated as the mating substances (see Hiwatashi and Kitamura, 1985). When paramecia of complementary mating types are mixed, cells adhere to their partners using the mating substances. The mating-competent cells are referred to as "mating-reactive" cells and these cells exhibit the mating substances on their ciliary surfaces. Interactions between these mating substances initiate activation processes of conjugation.

The chemical nature of the mating substance was first investigated by means of inactivation of the mating reactivity of living cells or mating-reactive killed cells. Metz (1954) summarized the results of his research on this subject and concluded that the mating reactivity is dependent upon protein integrity of ciliary surfaces. Further studies on the chemical properties of the mating substances have been performed using isolated cilia because the cilia are known to be responsible for the cell adhesion that occurs during the mating reaction. Cilia have been identified as the structural components of the cell that carry the mating substances by many investigators utilizing various methods. Cohen and Siegel (1963) first obtained mating-reactive cilia from *P. bursaria* by disruption of the cells followed by differential centrifugation. Their preparation agglutinated with the cilia of the complementary mating type cells but was contaminated with other particulates (Cohen and Siegel, 1963). Ciliary preparations which induce clumping of the cells of the complementary mating type have been obtained from *Paramecium* using a variety of methods including $K_2Cr_2O_7$ (Miyake, 1964), $MnCl_2$ (Fukushi and Hiwatashi, 1970), and Triton X-100 + $CaCl_2$ (Takahashi *et al.*, 1974). Watanabe (1977a) obtained mating-reactive detached cilia by a modified STEEP + $CaCl_2$ method which was originally developed for isolation of *Chlamydomonas* flagella by Witman *et al.* (1972). These preparations of cilia from one mating type can induce not only agglutination but also selfing pairs as well as successive processes of conjugation in cells of the opposite mating type. On the other hand, deciliated cell bodies never showed mating reactivity of any kind. Hence, investigators agree that the mating substances reside in or on the cilia (see Hiwatashi, 1981; Kitamura, 1988). Watanabe (1977a) and Kitamura and Hiwatashi (1978) investigated the chemical nature of the mating substances by means of inactivation of the mating reactivity of detached cilia. Among the enzymes tested for inactivation, only proteolyic enzymes effectively destroyed the mating reactivity of the cilia, whereas glycosidases and lipases showed no effect at all. These results indicate that the mating substances or at least their active site must involve proteins of the ciliary membrane (Watanabe, 1977a; Kitamura and Hiwatashi, 1978). Watanabe (1977b) treated the detached cilia with Triton X-100 and ascertained by electron microscopy that the extent of loss of the mating reactivity correlates with the extent of removal of the ciliary membrane. Furthermore, employing the procedures for chemical dissection of cilia designed by Gibbons (1965), Watanabe (1977b) confirmed that a pure ciliary membrane fraction retains strong mating reactivity, while the reactivity cannot be detected in any of the axonemal components (Fig. 2a–d).

Kitamura and Hiwatashi (1976) treated the detached cilia with 2 M urea and 0.1 mM Na_2-EDTA and then centrifuged the preparation at 9000g for 10 min. The supernatant was percolated through a 0.45-μm membrane filter, dialyzed against 10 mM Tris-HCl (pH 7.3), and then centrifuged at 105,000g for 60 min. The mating reactivity was detected in the pellet but not in the supernatant of the final centrifugation. Electron microscopy of the pellet revealed membrane vesicles of 100–150 nm in diameter (Fig. 2d). Using these membrane vesicles as the starting material, Kitamura and Hiwatashi (1980) succeeded in first dissolving and then reconstituting the membrane vesicles. They dissolved the membrane vesicles isolated by the urea–EDTA method in 9 mM lithium diiodosalicylate (LIS)

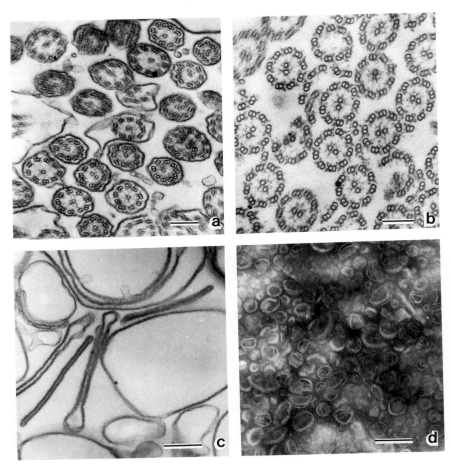

Figure 2. (a) Mating-reactive isolated cilia. (b) Non-mating-reactive demembranated axonemal fraction; (c) Mating-reactive ciliary membrane fraction obtained using the method of Gibbons (1965). (d) Negatively stained ciliary membrane vesicles prepared by the urea—EDTA method of A. Kitamura. This fraction also showed mating reactivity. Bars = 0.2 μm.

and then centrifuged the sample at 105,000g for 60 min. The supernatant was dialyzed against 10 mM Tris-HCl (pH 7.0) to remove LIS. The dialysate was centrifuged at 105,000g for 60 min, and the pellet was obtained. This pellet consisted of membrane vesicles with diameters of 50–100 nm. When these reconstituted membrane vesicles were added to mating-reactive cells of the opposite mating type, no mating agglutination of the latter was observed. In about 1 hr, however, conjugating pairs appeared. As will be described later, formation of conjugating pairs in cells of one mating type without the mating reaction is termed the chemical induction of conjugation (see Hiwatashi, 1969; and Section 5). However, in the case of the reconstituted membrane vesicles, induction of the pairs was mating-type-specific, not like the chemical induction where pairs are formed in a non-mating-type-specific fashion. In a recent review on the mating substances of *Paramecium,* Kitamura (1988) concluded that the substances are intrinsic proteins of ciliary membranes and their active site is made up of simple proteins.

3.2. Localization of Mating-Reactive Cilia

During the mating reaction, cells form large clumps. Until 1961, it had been believed that all of the cilia on the surface of paramecia exhibited adhesiveness. Hiwatashi (1961) claimed that almost all of the adhesiveness was restricted to the ventral surface of the cell. He confirmed this by splitting the killed cells into ventral and dorsal halves. Cohen and Siegel (1963) and Cohen (1964) reported similar regional differences in the distribution of mating-reactive cilia in *P. bursaria.* In this species, the mating reactivity appears with a diurnal rhythmicity (Ehret, 1953). During the mating cycle of *P. bursaria,* the area of mating reactivity on the cell surface spreads anteriorly from the posterior ventrolateral region until at the time of peak mating reactivity the entire ventrolateral surface is involved (Cohen, 1964). These observations were made using phase-contrast microscopy of the interaction between living cells and detached cilia of the opposite mating type. Miyake (1964) also reported that the detached cilia of *P. multimicronucleatum* did not adhere evenly to the whole surface of the paramecia but rather adhered mainly to the ventral side.

3.3. Attempts to Isolate Pure Mating-Reactive Cilia

Because the reactive cilia are localized to a restricted area of the cell surface, ciliary preparations must contain both mating-reactive cilia and large amounts of nonreactive cilia. Hiwatashi (1981) postulated that the proportion of the reactive cilia is probably as small as one-tenth of the total cilia. Since mating substances are present only on the mating-reactive cilia, it would be very useful for biochemical and immunological studies of the mating substances if we could separate reactive cilia from nonreactive ones.

Attempts have been made to separate mating-reactive cilia from nonreactive ones (see Hiwatashi, 1981). The idea is based on a bioaffinity principle: when detached whole cilia are mixed with living cells of the complementary mating type, only the reactive cilia adhere to cells of the opposite mating type, and the latter clump to form large clusters. Then the clusters of the reacting cells can be separated from the remaining free cilia and unclumped single cells by slow centrifugation or by allowing the clusters to settle. Finally, the reactive cilia attached to cells can be separated by centrifugation after killing the cells.

In all cases, however, the recovered cilia did not show stronger mating reactivity than the original preparation (Hiwatashi, 1981). A quantitative measurement of the reactivity of the cilia is possible if we determine the lowest concentration (mg/ml) of ciliary protein that can agglutinate a definite percentage of cells in a fixed volume of culture with a fixed cell density (Kitamura and Hiwatashi, 1976). The results suggest that inactivation of mating reactivity occurred during the isolation procedure. As mentioned previously, mating substances are sensitive to proteases. Therefore, the possible inactivator was thought to be a protease which might be secreted from living cells. Kitamura (unpublished, cited by Hiwatashi, 1981) used TLCK (*N*-α-*p*-tosyl-L-lysine chloromethyl ketone HC1), a protease inhibitor, during the purification but failed to recover cilia with a stronger reactivity than the original ones.

Kitamura (1983) examined the effect of 11 different protease inhibitors on the early events of conjugation between living cells of complementary mating types. All inhibitors had no detectable effect on the mating reaction or pair formation. The author concluded that no active proteases were secreted from cells during the mating reaction. Metz (1954) reported that loss of cellular mating reactivity occurred during the mating reaction. One possible explanation of such inactivation is that the mating interaction between the isolated cilia and the cilia on the whole cells may itself be inactivating the isolated cilia. This is what appears to happen in *Chlamydomonas* mating using whole cells of one mating type and isolated flagella of the opposite mating type (see van den Ende, 1985). If this is also the case for *Paramecium*, the cilia once reacted with the cells and separated from them should contain a reduced amount of the active substances, and thus the bioaffinity procedure may not be useful for isolation of the pure mating-reactive cilia and other methods for separation should be sought (Hiwatashi, 1981).

3.4. Nature of the Ciliary Interactions

The mating reaction is a specific and instantaneous cell agglutination. In the mating reaction, cells of the complementary mating types stick together by their cilia in a manner of tip-to-tip contact. When the mating agglutination in living cells of one mating type is induced by detached cilia (or ciliary membrane fractions) of the opposite mating type, the cells stick to each other by ciliary tip-to-tip contact mediated by the detached cilia. These observations suggest that mating substances reside only on the tips of the cilia. However, we cannot exclude the possibility that the substance is distributed on the entire surface of the ventral cilia. So far, determination of the exact location of the mating reactivity on the ciliary surface using detached cilia has been difficult, because these cilia often curl or swell when they are mixed with living cells (Watanabe, unpublished observation).

Presence of the mating substance in the ciliary membranes is recognized only by the agglutination reaction between living cells or between living cells and killed cells or detached cilia. On the other hand, when mating-reactive killed cells of complementary mating types are mixed together, no agglutination takes place. This is also the case for *Chlamydomonas* gametes (Goodenough, 1977). Therefore, living cells of one of the two mating types are necessary in order to assay for mating substance activity of cells or ciliary preparations. Hence, involvement of a soluble cofactor(s) secreted by living cells has been hypothesized (Hiwatashi, 1981). Takahashi *et al.* (1974) prepared detached cilia from paramecia of both mating types using the Triton X-100 + $CaCl_2$ method which was developed for obtaining Triton X-100-extracted models for studying swimming mechanisms (Naitoh and Kaneko, 1973). When these cilia were mixed together, agglutination between cilia occurred. This result indicates that no soluble cofactor secreted by living cells is necessary for the agglutination of cilia. This result has also been observed in flagella isolated from *Chlamydomonas* gametes (Köhle *et al.*, 1980; Goodenough, 1986). Takahashi *et al.* (1974) used 10 mM $CaCl_2$ in 10 mM Tris-maleate buffer (pH 7.0) to prepare a ciliary suspension. In order to obtain mating-type-specific agglutination between detached cilia, the concentration of Ca^{2+} appears to be critical, because nonspecific agglutination is often observed at lower Ca^{2+} concentrations (Watanabe, unpublished observations).

The Triton-extracted models are dead cells but they can swim if ATP is added (Naitoh and Kaneko, 1973). When the Triton-extracted models are prepared from complementary mating types and reactivated by ATP, they not only swim but also adhere to each other as do living cells (Takahashi *et al.*, unpublished observations; Watanabe, unpublished observations). Thus, the mating reaction can take place in the absence of living cells. In spite of the work of Takahashi *et al.* (1974), the distribution of the mating substances on the ciliary membrane remains uncertain.

3.5. Hydrophobic Interactions between Cilia and Polystyrene Surfaces

Kitamura (1982) found that only mating-reactive cells of *P. caudatum* attached to the polystyrene surface of petri dishes used for bacterial culture. The attachment of cells occurred exclusively at the tip of ventrally located cilia and appeared to involve a hydrophobic interaction. Although the ability to attach to polystyrene is closely correlated with mating reactivity, the attachment sites seem to be distinct from the mating recognition sites, as suggested by the following observations: (1) Cells that lose their mating reactivity by treatment with trypsin still attach to the polystyrene surface; (2) the attachment of cells is temperature-dependent within a range where mating reactivity remains unchanged; (3) when mating-reactive cells are applied to the dish, they remain attached but their mating reactivity is quickly lost. Attachment to polystyrene induces the early micronuclear migration (EMM) (see Section 5.2) but does not lead to further nuclear changes.

Kitamura and Steers (1983) reported a method that induces attachment to polystyrene in cells which are not mating-reactive. If sexually immature cells as well as log-phase cells are treated with purified immunoglobulin G or some other hydrophobic reagent such as phenethylamine, benzylamine, amphetamine, or phenylethylamine, cell–polystyrene adhesion is induced even though the cells are not mating-reactive. In these cases, however, the EMM does not occur. Kitamura (1984) showed that the mating reaction and the treatment with the conjugation-inducing chemicals, both of which activate conjugation, increase the hydrophobicity of the cell surface of *P. caudatum*. This increase of hydrophobicity occurred within 5 min after initiation of the mating reaction, suggesting a rapid change in the properties of ciliary membrane surfaces. The earliest change so far observed microscopically is inactivation of ciliary movement (Kitamura *et al.*, 1980; see Section 5.1). This ciliary inactivation is not the result of a change in membrane potential but probably is correlated with some other change in the properties of the ciliary membrane, such as membrane fluidity (Kitamura and Hiwatashi, 1984a). Kitamura and Hiwatashi (1984a) proposed possible mechanisms for the increase in hydrophobicity; the increase may be a result of the accumulation of hydrophobic proteins or of the exposure of free lipid surfaces at the tip of the ventral cilia (Fig. 3).

Kitamura (1986) extended this line of work to other species of *Paramecium*. Mating reactivity-dependent attachment to polystyrene surfaces was seen not only in *P. caudatum* but also in *P. multimicronucleatum*, *P. tetraurelia*, and *P. trichium*. In *P. bursaria*, however, the attachment was rarely seen even if the cells showed strong mating reactivity. On the contrary, cells of *P. duboscqui* showed strong affinity for polystyrene surfaces even when they had no mating reactivity or when applied to polystyrene dishes with

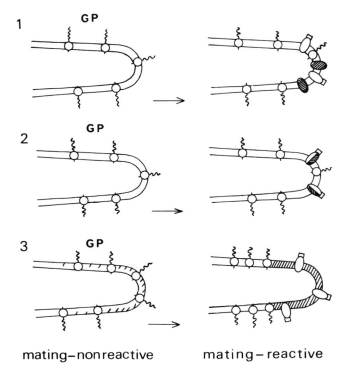

mating–nonreactive   mating–reactive

Figure 3. Schematic presentation of three possible relationships between mating substances and attachment sites for polystyrene in the ciliary membrane. Open squares: active sites of the mating substances. Striped portions: attachment sites for polystyrene. GP: glycoproteins of the surface coat (Section 6). From Kitamura and Hiwatashi (1984a) with permission of the Zoological Society of Japan.

reduced hydrophobicity. The early micronuclear migration occurs in attached cells of *P. caudatum,* but not in those of other species.

4. Regulation of the Expression of Mating Reactivity

4.1. Genetic Control of the Mating Type Specificity

Mating type determination and its genetic control have been extensively studied and reviewed by many authors (Sonneborn, 1974; Butzel, 1974; Nanney, 1977; Hiwatashi, 1981; Tsukii and Hiwatashi, 1983; Hiwatashi and Kitamura, 1985; Tsukii, 1988). The genetic systems used to control mating types are different from species to species. Here, I will describe the system used by *P. caudatum,* because in this species, genes controlling the mating type specificity have been identified (Tsukii and Hiwatashi, 1983). The species *P. caudatum* is subdivided into many mating groups called *syngen* (Sonneborn, 1957). Each syngen of *P. caudatum* is composed of two complementary mating types. The two types of each syngen in *P. caudatum* are designated O (odd-numbered) and E (even-numbered) types according to their numerical designation. The O mating type is expressed

Table I. Mating Types and Corresponding Genotypes of Strains in Each Syngen and Intersyngen Hybrids of *P. caudatum*[a]

Phenotypes	Genotypes
Strains in each syngen	
E^1	$Mt^1/Mt^1; MA^1/MA^1; MB^1/MB^1$ $MT^1/mt; MA^1/MA^1; MB^1/MB^1$
O^1	$mt/mt; MA^1/MA^1; MB^1/MB^1$
E_3	$MT^3/MT^3; MA^3/MA^3; MB^3/MB^3$ $MT^3/mt; MA^3/MA^3; MB^3/MB^3$
O^3	$mt/mt; MA^3/MA^3; MB^3/MB^3$
Intersyngen hybrids	
E^1E^3	$MT^1/MT^3; MA^b/MA^b; MB^b/MB^b$
O^1O^3	$mt/mt; MA^1/MA^3; MB^1/MB^3$
O^{0c}	$mt/mt; MA^1/MA^1; MB^3/MB^3$

[a] From Tsukii (1988).
[b] MT^1 and Mt^3 can suppress the expression of *MA* and *MB* regardless of the latter's syngen specificity.
[c] O^0: mating-typeless strains.

when cells are homozygous for the recessive gene *mt* (*mt/mt*). Cells containing the dominant alleles (*Mt/Mt* or *Mt/mt*) express the E type (Hiwatashi, 1981). Recent work of Tsukii and Hiwatashi (1983) using artificially induced intersyngenic crosses showed that the syngenic specificities of the mating type are controlled by at least three loci: *Mt*, *MA*, and *MB* (Table I). Codominant multiple alleles at the *Mt* locus control the syngenic specificity of E mating types, and those at the two other loci, *MA* and *MB*, control the specificity of O mating types. Since *Mt* is epistatic to *MA* and *MB*, the O mating type can be expressed only in the recessive homozygote (*mt/mt*) of the *Mt* locus. In addition, at least one allele each at the *MA* and *MB* loci must have a common syngen specificity for the expression of O types. Therefore, if *MA* and *MB* are both homozygous for different syngens, no mating type is expressed (Tsukii and Hiwatashi, 1983). These results suggest, although there is as yet no direct evidence, that the genes are most probably coding for the mating substances (Tsukii, 1988; Hiwatashi, 1988) (Fig. 4).

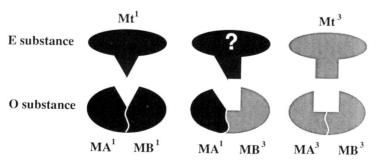

Figure 4. Model for the mating substances of type E and type O. Active O substances are composed of the product of *MA* and *MB* genes. If the O substance is a hybrid of the MA^1 and MB^3 (or MA^3 and MB) products, it will not react with E1 or E3 substances, resulting in mating-typeless cells. Courtesy of Dr. Tsukii; modified from Tsukii (1988) with permission of Springer-Verlag.

4.2. Temporal Differentiation of Mating-Reactive Cilia

4.2.1. Sexual Immaturity and Maturity. The life cycle of *Paramecium* consists of sequential periods of immaturity, maturity, and senility. During the immaturity period which follows conjugation, the exconjugant clones never show mating reactivity even under appropriate conditions. The paramecia exhibit strong mating reactivity in their period of maturity and weak or no reactivity in the senility period. The length of the immaturity period is measured in terms of the number of fissions after conjugation (Sonneborn, 1957; Miwa and Hiwatashi, 1970). In *P. caudatum,* the immaturity period was measured as about 50–60 fissions (Miwa and Hiwatashi, 1970).

The difference between sexually immature and mature cells was studied by transplantation of cytoplasm from immature to mature cells (Miwa *et al.,* 1975). When mature cytoplasm was injected into immature cells, the recipient immature cells showed no change. In contrast, when immature cytoplasm was injected into mature cells, the recipient mature cells lost their mating reactivity. Moreover, suppression of the mating reactivity in the injected mature cells lasted for several subsequent cell generations (Miwa *et al.,* 1975). The authors assumed that some factors in the cytoplasm of immature cells inhibit expression of the mating reactivity in mature cells. Haga and Hiwatashi (1981) isolated from immature cells a protein of about 10 kDa which controls sexual immaturity; they named this protein *immaturin*. Miwa (1979a, b) tested its species specificity and found that the protein from *P. primaurelia* is effective on cells of *P. tetraurelia, P. caudatum,* and *P. multimicronucleatum*. Therefore, the immaturin molecules may be common in these species. However, immaturin from *P. primaurelia* is not effective on cells of *P. bursaria*. Miwa (1979a) found that immature cells of *P. bursaria* have their own immaturin.

Myohara and Hiwatashi (1978) obtained early maturity mutants which became mature (20–25 fissions earlier than the wild-type clones when they were homozygous for the *Emt A* gene. Miwa (1984) injected the cytoplasm of the *Emt A* homozygote immediately before maturation into late immature cells of wild type and showed that about 10% of the recipient cells expressed mating reactivity while control cells showed no reactivity. This result suggests that the cytoplasm of the early mature mutant immediately before maturation contains some material which accelerates maturation when injected into wild-type cells in the late stage of immaturity. Therefore, Miwa concluded that the function of the early mature gene product might be the destruction of immaturin activity. Although the mechanism of action of immaturin is unknown, a repressorlike action seems most probable (Hiwatashi and Kitamura, 1985).

4.2.2. Culture Phase and Mating Reactivity. Cells of *Paramecium* exhibit mating reactivity when they are moderately starved. Thus, in typical cultures of paramecia, cells in the stationary phase of growth are competent to mate. The mating-competent cells are in the G_1 phase of the cell cycle (Fujishima, 1983). Matsuda and Hiwatashi (1979) isolated dividing cells from a *P. caudatum* culture into buffer solution and then tested for the mating ability of the cells at various times after isolation. The cells express mating ability more than 12 hr after isolation. Interestingly, mating-reactive cells divided once before exhibiting mating reactivity. Measurement of DNA content in the macronuclei of the once-divided (reactive) and the nondivided (nonreactive) cells revealed that the latter contained about two times as much DNA as the former. This suggested that the nonreactive cells are in the S or G_2 phase of growth and that the mating reactivity appears in the G_1 phase (Matsuda and Hiwatashi, 1979). Miwa and Umehara (1983) also studied the relationship between the cell cycle and mating reactivity expression using highly mating-reactive mutants of *P. tetraurelia* but obtained different results from those of Matsuda and Hiwatashi (1979). They cultured a single cell in a microcapillary containing 1 μl of fresh culture medium. Under the conditions used, the cells divide three times and express mating reactivity after the second division (four-cell stage). The reactivity continued through the four-cell to the eight-cell stage except for the cells just dividing. From these results, together with the measurement of nuclear DNA content by microspectro-photometry, they concluded that the cells express mating reactivity not only in the G_1 phase but also in other phases (S and G_2) of the cell cycle except for the M phase (Miwa and Umehara, 1983). Thus, the appearance of the mating substance on the cilia is correlated neither with the cell cycle nor with cessation of cell division during the stationary phase of growth. Some metabolic shift is probably necessary for the changes of cellular physiological conditions from vegetative growth to sexual phase (Hiwatashi, 1981).

When certain protein and RNA synthesis inhibitors are applied before cells express mating reactivity, no reactivity appears in the cells (Nobili, 1963; Cohen, 1965; Clark, 1972). In *P. caudatum,* the cell's mating reactivity usually lasts for several days in the stationary phase. If the inhibitors are applied to the reactive cells, the cells gradually reduce or totally lose their reactivity (for review, see Miyake, 1981; Kitamura, 1988). These results suggest that *de novo* synthesis of RNAs and proteins is necessary for the appearance as well as the maintenance of the mating reactivity. Watanabe and Fujishima (unpublished results) studied the loss of mating reactivity in enucleated cells of *P. caudatum.* When macronuclei were removed from mating-reactive cells, the ability to clump with cells of the opposite mating type was maintained for 1 hr in all cells. Approximately 2–3 hr after enucleation, 30% of the enucleated cells lost their reactivity (Table II). Even if the enucleated cells clumped with cells of the opposite mating type, no conjugating pairs were formed. This observation is consistent with the results of experiments in which RNA and protein synthesis inhibitors were applied during the mating reaction. Thus, *de novo* synthesis of RNA and protein is necessary not only for the

Table II. The Mating Reactivity of Enucleated Cells[a,b]

Time after enucleation	No. of cells enucleated	No. of mating-reactive cells[c]	Percent of reactive cells
1 hr	8	8	100
1–2	23	16	69.6
2–4.5	14	4	28.6
4.5	9	0	0

[a]From Watanabe and Fujishima (unpublished).
[b]Enucleation was performed in 1-day-starved cells of stock 27aG3 in *P. caudatum,* syngen 3, mating type V.
[c]The reactivity was assayed by mixing the cell with mating-reactive cells of stock cys13-9 of mating type VI.

appearance and maintenance of mating substances but also for the formation of conjugant pairs, because inhibitors block pair formation if they are added during the mating reaction (Miyake, 1981). On the other hand, when moderately starved cells were refed, mating reactivity was lost rather quickly. Mishima (1978) found that when 1-day-starved reactive cells were fed, they lost their reactivity within 2 hr and when cells with no reactivity were transferred into bacteria-free medium, mating reactivity reappeared in 2 hr. Shimomura and Takagi (1985) reported that the complete loss of mating reactivity of 1-day-starved cells occurred in about 1.5 hr after refeeding bacteria. These observations show that the cellular physiological conditions necessary for the expression of mating ability are correlated with the nutritional state of the cells.

5. Results of Ciliary Interactions

5.1. Decrease in Swimming Velocity

The earliest change detected in the cells after the onset of the mating reaction is an inactivation of ciliary movement. Cells in the agglutinates do not swim. This is not only due to a physical capture by adhesion of the ciliary surface but also due to the presence of some mechanism for inactivation of ciliary movement (see Hiwatashi, 1981). Kitamura *et al.* (1979, 1980) and Kitamura and Hiwatashi (1984b) studied the relationship between the reception of the sexual signal and the decrease in swimming velocity. In these studies, small membrane vesicles obtained by treating detached cilia with 4 mM LIS were used. The membrane vesicles thus obtained were designated as LIS-membrane vesicles (LIS-MV) and showed a property similar to that of reconstituted membrane vesicles (Kitamura and Hiwatashi, 1980; Section 3.1). Addition of LIS-MV to cells of the opposite mating type did not cause agglutination in the cells but resulted in an immediate decrease in swimming velocity and later the formation of conjugating pairs. Addition of about 50 μg/ml of LIS-MV decreased the swimming velocity of the cells to 50% of normal in 1 min and to 15% in 10 min. If the cells that had decreased their swimming velocity were washed free of the LIS-MV, they recovered their normal swimming speed. Microscopic observation of the cells revealed not only immobilization of the cilia on the ventral surface but also loss of typical metachronal coordination of the dorsal cilia. Ciliary movement is controlled by membrane potential (Naitoh and Eckert, 1974). A positive correlation between membrane depolarization and change in ciliary beat frequency or shift of the beat direction has also been found to exist (Machemer, 1974; Preston and Saimi, this volume). Kitamura *et al.* (1979, 1980) measured the membrane potential of paramecia before and after the addition of LIS-MV but were unable to detect any change in membrane potential or resistance.

Induction of conjugating pairs by chemical treatment of cells of one mating type has been referred to as the chemical induction of conjugation (Miyake, 1958, 1968a). During the chemical induction of conjugation, the same decrease in swimming velocity of cells has been observed. When mating-reactive cells of one mating type were transferred into conjugation-inducing chemicals (e.g., 6 mM KCl–50 mM methylurea–calcium-poor conditions for *P. caudatum,* 20 mM KCl–1 wt% acriflavine–calcium-poor conditions for *P. tetraurelia),* a marked change in swimming behavior was observed (Miyake, 1958; Tsukii and Hiwatashi, 1978; Cronkite, 1979). About 1 hr later, conjugating pairs or sometimes chains of united cells were formed (Miyake, 1958, 1968a). For the chemical induction of conjugation of *Paramecium,* the reader is referred to other reviews (Cronkite, 1979; Kitamura and Hiwatashi, 1984a; Hiwatashi and Kitamura, 1985). When cells whose swimming was stopped in the conjugation-inducing medium were observed by phase-contrast microscopy, perturbation of the ciliary metachronal waves was seen not only in the ventral cilia but also over the dorsal cilia (Kitamura and Hiwatashi, 1984b). Thus, the observed change in swimming behavior after immersion into the conjugation-inducing medium was almost the same as that observed in the LIS-MV treatment.

5.2. Early Micronuclear Migration

In cells of *P. caudatum* in the stationary phase of growth, the micronucleus is located in the concavity of the macronucleus. When the cells are stimulated by the mating reaction or by conjugation-inducing chemicals, the micronucleus moves out of the macronuclear concavity (Fujishima and Hiwatashi, 1977). This phenomenon is called *early micro-nuclear migration* (EMM). EMM takes place within 30 min in more than 90% of the cells involved in mating agglutination. This early nuclear event seems to be a prerequisite for the successive nuclear changes such as premeiotic DNA synthesis, meiosis, and pronucleus formation (Fujishima, 1983). Since EMM occurs only in mating-reactive cells, it is used as an indicator of nuclear activation (Cronkite, 1976, 1979; Kitamura, 1982). The migration of the micronucleus which is induced by interaction of the mating substances is inhibited by colchicine, suggesting the involvement of microtubules in EMM (Fujishima and Hiwatashi, 1977). Although possible mechanisms for EMM induction have been discussed in relation to the change in the intracellular free Ca^{2+} concentration (Cronkite, 1979; Hiwatashi, 1981; Kitamura and Hiwatashi, 1984a), the actual mechanism is still uncertain. Fujishima and Hiwatashi (unpublished, cited by Fujishima, 1988) found that the velocity of cytoplasmic streaming increased remarkably within 5 min after the beginning of the mating reaction. This phenomenon occurs much earlier than does EMM. It is generally known that an increase in the intracellular free Ca^{2+} concentration often produces vigorous protoplasmic streaming (Alberts *et al.,* 1983). Further studies to clarify the relationship among EMM, Ca^{2+}, and cytoplasmic streaming are under way (Fujishima, 1988).

5.3. Local Degeneration of Cilia and Pair Formation

A much later effect of the interaction of the mating substances is the formation of conjugating pairs, which begins about 1 hr after onset of the mating reaction. Wichterman (1946) reported a degeneration of cilia as well as trichocysts along the fused oral surfaces of the living conjugants. The degeneration of cilia proceeds posteriorly beyond the region where cells are united and frequently even to the extreme posterior ends of the conjugants. Hiwatashi (1955) and Vivier and André (1961) confirmed this observation using light microscopy and electron microscopy, respectively. These observations suggest that the holdfast union and successive paroral union take place on the cilia-free naked surface and that deciliation is a prerequisite for the formation of conjugating pairs. Miyake (1966) observed that cilia in the holdfast region disappeared at a time that was approximately halfway between the beginning of the mating reaction and the appearance of a holdfast union, which occurred in about 60 min.

Watanabe (1978) studied the processes, extent, and location of deciliation using scanning electron microscopy. His observations revealed that the degeneration of cilia begins at the anterior end of the ventral surface 30 min after onset of the mating reaction. The degeneration of cilia then extends posteriorly along the suture. On the anterior part of the cell, deciliation occurs on both sides of the suture whereas on the posterior part, it occurs only on the right side of the suture (Fig. 5a). Watanabe (1982) showed that the location of deciliation was correlated with the location of the anterior suture and the oral apparatus, although the function of these structures in the deciliation was uncertain. The extent of spreading of the deciliated area seems to correlate with the strength of the mating reactivity. When cells showed a very strong agglutination reaction, holdfast unions consisting of more than two cells were frequently formed. Such holdfast unions imply the presence of extra deciliated free surface area with which other deciliated cells then make contact.

The maximum extent of deciliation is attained by 3 hr after the beginning of the mating reaction or chemical treatment for the induction of conjugation. At this time, the

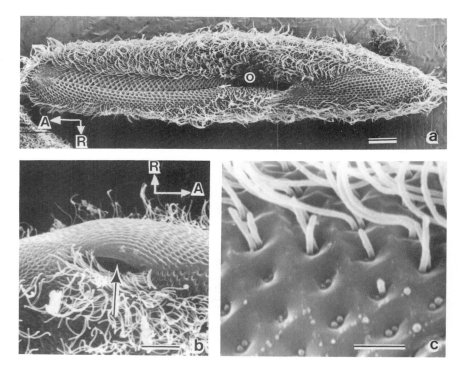

Figure 5. Scanning electron micrographs of the local degeneration of cilia. (a) Ninety minutes after onset of the mating reaction. At the anterior portion, deciliation occurs on both sides of the anterior suture (arrow). (b) Five hours after beginning the chemical induction of autogamy. Deciliation has extended to the right wall of the oral cavity (arrow). (c) Shortened cilia and stubs of degenerated cilia. O: opening of the mouth; A, R: direction of anterior and right side, respectively. Bars = 10 μm (a, b), 2 μm (c).

cilia of the cytopharynx also degenerate (Fig. 5b). This explains the cessation of ingestion which is a visible sign of the cell's irreversible entrance into the sexual process (Fujishima, 1983; Takagi *et al.*, 1981; Shimomura and Takagi, 1985). In usual conjugants, the deciliation area remains cilia-free until the time of separation of the pair. Regeneration of cilia begins before complete separation of conjugants and is complete within 60 min after pair separation (Watanabe, 1983). Pair formation during the chemical induction of conjugation can be inhibited by proteases, such as ficin, papain, or trypsin. In spite of the inhibition of pairing, a series of nuclear changes proceeds and results in artificial induction of autogamy (Miyake, 1968b,c; Mikami and Koizumi, 1979; Tsukii and Hiwatashi, 1979). Despite this inhibition of pairing by the proteases, deciliation occurs normally and continues until the time of pair separation during normal conjugation (Watanabe, 1983). Thus, degeneration and regeneration of ventral cilia are regulated by the same mechanisms of *activation* as nuclear events and the cessation of ingestion.

Miyake (1981) found that if puromycin (500 µg/ml) and actinomycin S_3 (50 µg/ml) were added 1 hr before the onset of chemical induction of conjugation, the degeneration of cilia and subsequent pair formation did not occur. On the other hand, proteases or lipase did not inhibit local deciliation but did inhibit pair formation during the chemical induction of conjugation. Hence, Miyake (1969, 1978, 1981) suspected that the holdfast substance, a hypothetical substance for cell adhesion of the holdfast union, may be lipoprotein.

Little is known concerning the mechanism of the degeneration of cilia. Scanning electron microscopic observation of cells undergoing ciliary degeneration revealed short cilia of various lengths together with normal ones (Watanabe, 1978) (Fig. 5c). The basal bodies of the completely degenerated cilia remained as stubs. The partially degenerated cilia remained as stubs. The partially degenerated cilia did not show the typical tapering ends. Therefore, the degeneration is most probably due to resorption of axoneme from the distal ends. Miyake (1981) pointed out the possibility that the membranes of the resorbed cilia may still maintain mating agglutinability during the early stage of the holdfast union.

As described previously, the mating reaction is a mating-type-specific cell contact formed between cells of complementary mating types. Contrary to this, the holdfast union pairs are formed even in cells of the same mating type as shown in the artificial induction of conjugation by chemical treatment or by detached cilia. This indicates that the holdfast union is not mating type specific. Hiwatashi (1951) examined whether the pair always consist of two individuals of different mating types in the natural conjugation using a differential vital staining method. He found that most of the pairs were composed of two mating types but a few, about 5% of the total pairs, were homotypic pairs. This result suggests that mating-type-specific cell contact by cilia is still working in the process of the holdfast union formation, even though the *holdfast substance* is nonspecific.

Jennings (1941) reported that in normal conjugation of *P. bursaria*, conjugating pairs always consisted of two different mating types. On the other hand, chemical induction of conjugation has been reported to be unsuccessful in this species (Miyake, 1968a). It seems that in *P. bursaria,* cell adhesion in the holdfast union is mating type specific. Otherwise, there should be other mechanisms preventing selfing conjugation. Endoh *et al.* (1987) and Endoh (1987) studied whether selfing conjugation can be induced artificially. Results of these studies indicate that selfing pairs can be induced by conjugation-inducing chemicals or by detached cilia derived from cells of complementary mating type. Thus, the holdfast substance of *P. bursaria* is also mating type nonspecific. When the holdfast union pairs are separated by centrifugation, cells can adhere again to different partners. The second holdfast union pairs were expected to be both homotypic and heterotypic ones. Endoh *et al.* (1987) showed that all pairs of the second holdfast union were heterotypic and that this may be due to residual mating-reactive cilia on the ventral surface. They showed that more than 80% of split cells in the holdfast union retained mating reactivity even at 4 hr after beginning the mating reaction. Thus, in *P. bursaria*, remaining reactive cilia contribute to heterotypic pair formation.

6. Biochemical and Morphological Approaches to Characterizing the Mating Substances

Biochemical studies designed to find differences between mating-reactive and nonreactive cilia as well as between complementary mating types have been performed, though the number of such investigations is limited (see Kitamura, 1988). Every effort to isolate the mating substances from cilia, or to identify them on the ciliary surfaces, have been unsuccessful. As stated above, the mating substances are located on the ventral cilia of sexually mature paramecia in the stationary phase. Comparisons of the protein composition of cilia or ciliary membranes between immature and mature cells, between log- and stationary-phase cells, or between cells of complementary mating types have been performed using SDS-polyacrylamide gel electrophoresis. Adoutte *et al.* (1980) analyzed proteins of ciliary membranes in *P. tetraurelia* using one- and two-dimensional gel

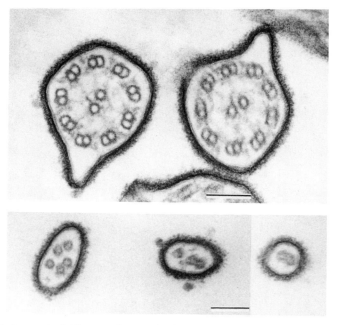

Figure 6. Surface coat of ciliary membrane stained with ruthenium red. Bars = 0.1 μm.

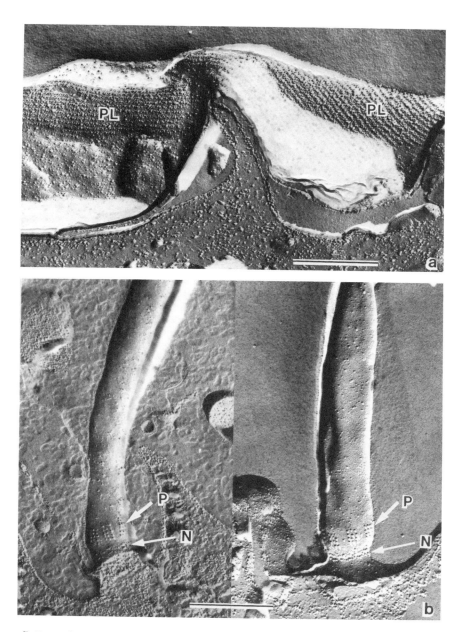

Figure 7. Freeze-fracture electron micrographs. (a) Linearly arranged membrane particle plates (PL) on the ventral surface. (B) Ciliary necklace (N) and patches (P) are visible at the base of all cilia. Bars = 0.1 μm.

electrophoresis. They found a peptide of 31 kDa which was absent in ciliary membranes from cells of log phase and present in those of early stationary phase. Thus, this peptide seems to correlate with the presence of cellular mating reactivity, although the function of the peptide is unknown. Kitamura and Hiwatashi (1980) and Kitamura (unpublished, cited in Kitamura, 1988) also tried to find differences in the protein constitution of ciliary membrane vesicles from log- and stationary-phase *P. caudatum,* but no differences were observed.

Morphological approaches to finding differences between mating-reactive and non-reactive ciliary surfaces or between complementary mating types have been utilized. Watanabe (1981) compared the thickness of the surface coat revealed by ruthenium red staining (Wyroba and Przelecka, 1973) in relation to cellular mating reactivity in *P. caudatum* (Fig. 6). No difference in morphology was observed between reactive and nonreactive cell surfaces, although cells of mating type VI have a thicker layer than those of mating type V (Watanabe, 1981).

Freeze-fracture studies of the cell membrane of *Paramecium* have revealed specialized arrangements of particles in the surface membrane. Such particle arrays have been correlated with specific organelles, membranes, or macromolecular structures lying close to the cytoplasmic face of this membrane (Allen, 1978a,b). One of the most fascinating arrangements of membrane particles found in our study consisted of large rectilinearly arranged particle plates in the cell membrane (Fig. 7a). These plates are found almost exclusively on the anterior ventral surface of the cell along both sides of the anterior suture (Allen, 1978a). As discussed by Allen (1978a), the location of these unique particle plates allows us to speculate on several functions of this structure, such as (1) localized receptors for the mechanical stimuli which induce membrane depolarization followed by ciliary reversal (Naitoh and Eckert, 1969; Ogura and Machemer, 1980), (2) cell fusion sites between conjugants, and (3) sensory sites for chemicals. To validate these possibilities, further studies are necessary. Freeze-fracture electron microscopy of the ciliary membranes showed the ciliary necklace and ciliary patch at the base of all cilia (for review see Allen, 1978b), but specialized features of ciliary membrane particles specifically correlated with mating reactivity were not observed (Fig. 7b).

7. General Discussion and Conclusions

As in the fertilization of multicellular organisms, the essential purpose of conjugation in *Paramecium* is the production of a new generation which has a new combination of genome sets. The principal role of the mating substances is to ensure the occurrence of crossbreeding between cells of complementary mating types. Mating substances act in the first step of conjugation not only as adhesion molecules but also as initiators of a series of cytoplasmic and nuclear activation events. Certain ventral cilia then degenerate in order to provide a free surface for sites of holdfast union and paroral union and for subsequent membrane fusion. At the beginning of pair formation, those reactive cilia still remaining contribute to the formation of the conjugating pair consisting of two different mating types.

Interactions of the mating substance molecules probably induce a change in the intracellular concentration of free Ca^{2+} (Cronkite, 1979) and a reduction in swimming velocity (Hiwatashi and Kitamura, 1985). On the other hand, mating phenomena are also initiated by treatment with conjugation-inducing chemicals. A transient increase in internal Ca^{2+} concentration or a removal of Ca^{2+} from surface membranes is postulated to be necessary for initiation of the chemical induction of conjugation, whereas these are not required for normal mating (Cronkite, 1976; Kitamura and Hiwatashi, 1984c). The mating reaction must continue for about 60 min before pair formation occurs; the duration of the chemical treatment necessary for artificial induction of the holdfast union is also about 60 min. Miyake (1969) showed that chemical treatment can substitute for the mating reaction in terms of effecting pair formation, suggesting that the mating reaction and the chemical treatment give the same stimulus to the cell.

Cells in which the mating substances are not active because of trypsinization or genetic mismatching of the *MA* and *MB* genes (Section 4.1) cannot conjugate through the mating reaction but are capable of chemical induction of conjugation (Miyake, 1979; Tsukii and Hiwatashi, 1983). Thus, the mating substances may not be a receptor for this chemical induction of conjugation. It is, however, highly probable that the receptor for the chemical induction resides in the mating-reactive cilia, regardless of the integrity of the mating substance molecules.

Temporal differentiation of the mating-reactive cilia has provided an interesting problem in specific gene expression during clonal life history and during different nutritional states of the culture phase. Little experimental analysis has been done on the mechanism of the spatial differentiation of the mating-reactive cilia. The differential distribution of mechanoreceptor potential along the longitudinal axis of the cell (Ogura and Machemer, 1980) may have no connection to the localized distribution of mating-reactive cilia. Regional differences observed in the mating processes may be based on the morphological differentiation of the cortex of *Paramecium*. Unfortunately, it has not yet been possible to isolate the mating substances or to obtain specific antibodies to the mating substances. Future studies should increase our understanding of the initiation mechanisms of mating and of the local differentiation of ciliary functions in *Paramecium*.

ACKNOWLEDGMENTS. I thank Dr. K. Hiwatashi for his encouragement and help during the preparation of the manuscript. I am also indebted to Dr. R. A. Bloodgood for kindly reviewing the manuscript and making valuable comments.

References

Adoutte, A., Ramanathan, R., Lewis, R. M., Dute, R. R., Ling, K.-Y., Kung, C., and Nelson, D. L., 1980, Biochemical studies of the excitable membrane of *Paramecium tetraurelia*. III. Proteins of cilia and ciliary membranes, *J. Cell Biol.* **84**:717–738.

Alberts, B., Bray, D., Lewis, J., Raff, M., Roberts, K., and Watson, J. D., 1983, *Molecular Biology of the Cell,* Garland Publishing, New York.

Allen, R. D., 1971, Fine structure of membranous and microfibrillar systems in the cortex of *Paramecium caudatum, J. Cell Biol.* **49**:1–20.

Allen, R. D., 1978a, Particle arrays in the surface membrane of *Paramecium:* Junctional and possible sensory sites, *J. Ultrastruct. Res.* **63**:64–78.

Allen, R. D., 1978b, Membrane of ciliates: Ultrastructure, biochemistry and fusion, in: *Membrane Fusion* (G. Poste and G. L. Nicolson, eds.), Elsevier/North-Holland, Amsterdam, pp. 657–763.

Butzel, H. M., 1974, Mating type determination and development in *Paramecium aurelia,* in: *Paramecium: A Current Survey* (W. J. Van Wagtendonk, ed.), Elsevier, Amsterdam, pp. 91–130.

Clark, M. A., 1972, Control of mating type expression in *Paramecium multimicronucleatum,* syngen 2, *J. Cell. Physiol.* **79:**1–14.

Cohen, L. W., 1964, Diurnal intracellular differentiation in *Paramecium bursaria, Exp. Cell Res.* **36:** 398–406.

Cohen, L. W., 1965, The basis for the circadian rhythm of mating in *Paramecium bursaria, Exp. Cell Res.* **37:** 360–367.

Cohen, L. W., and Siegel, R. W., 1963, The mating-type substances of *Paramecium bursaria, Genet. Res.* **4:** 143–150.

Cronkite, D. L., 1976, A role of calcium ions in chemical induction of mating in *Paramecium tetraurelia, J. Protozool.* **23:**431–433.

Cronkite, D. L., 1979, The genetics of swimming and mating behavior in *Paramecium,* in: *Biochemistry and Physiology of Protozoa,* 2nd ed., Volume 2 (M. Levandowsky and S. H. Hutner, eds.), Academic Press, New York, pp. 221–273.

Ehret, C. F., 1953, An analysis of the role of electromagnetic radiations in the mating reaction of *P. bursaria, Physiol. Zool.* **26:**274–300.

Endoh, H., 1987, Genetic control of chemical induction of conjugation in *Paramecium bursaria, Heredity* **59:** 397–403.

Endoh, H., Watanabe, T., and Hiwatashi, K., 1987, Artificial induction of selfing conjugation in *Paramecium bursaria, J. Exp. Zool.* **241:**333–338.

Fujishima, M., 1983, Microspectrophotometric and autoradiographic study of the timing and duration of premeiotic DNA synthesis in *Paramecium caudatum, J. Cell Sci.* **60:**51–65.

Fujishima, M., 1988, Conjugation, in: *Paramecium* (H.-D. Gortz, ed.), Springer-Verlag, Berlin, pp. 70–84.

Fujishima, M., and Hiwatashi, K., 1977, An early step in initiation of fertilization in *Paramecium:* Early micronuclear migration, *J. Exp. Zool.* **201:**127–134.

Fukushi, T., and Hiwatashi, K., 1970, Preparation of mating reactive cilia from *Paramecium caudatum* by $MnCl_2$, *J. Protozool.* **17:**s21.

Gibbons, I. R., 1965, Chemical dissection of cilia, *Arch. Biol.* **76:**317–352.

Goodenough, U. W., 1977, Mating interaction in *Chlamydomonas,* in: *Receptors and Recognition, Series B,* Volume 3 (J. L. Reisseg, ed.), Chapman & Hall, London, pp. 323–350.

Goodenough, U. W., 1986, Experimental analysis of the adhesion reaction between isolated *Chlamydomonas* flagella, *Exp. Cell Res.* **166:**237–246.

Haga, N., and Hiwatashi, K., 1981, A protein called immaturin controlling sexual immaturity in *Paramecium caudatum, Nature* **289:**177–179.

Hiwatashi, K., 1951, Studies on the conjugation of *Paramecium caudatum.* IV. Conjugating behavior of individuals of two mating types marked by a vital staining method, *Sci. Rep. Tohoku Univ. Ser. 4* **19:**95–99.

Hiwatashi, K., 1955, Studies on the conjugation of *Paramecium caudatum.* VI. On the nature of the union of conjugation, *Sci. Rep. Tohoku Univ. Ser. 4* **21:**207–218.

Hiwatashi, K., 1961, Locality of mating reactivity on the surface of *Paramecium caudatum, Sci. Rep. Tohoku Univ. Ser. 4* **27:**93–99.

Hiwatashi, K., 1969, Paramecium, in: *Fertilization,* Volume 2 (C. B. Metz and A. Monroy, eds.), Academic Press, New York, pp. 255–293.

Hiwatashi, K., 1981, Sexual interaction of the cell surface in *Paramecium,* in: *Sexual Interaction in Eukaryotic Microbes* (D. H. O'Day and P. A. Hagen, eds.), Academic Press, New York, pp. 351–378.

Hiwatashi, K., 1988, Sexual recognition in *Paramecium,* in: *Eukaryote Cell Recognition—Concepts and Model Systems* (G. P. Chapman, ed.), Cambridge University Press, London, pp. 77–91.

Hiwatashi, K., and Kitamura, A., 1985, Fertilization in *Paramecium,* in: *Biology of Fertilization,* Volume 1 (C. B. Metz and A. Monroy, eds.), Academic Press, New York, pp. 57–85.

Hiwatashi, K., and Mikami, K., 1989, Fertilization in *Paramecium:* Processes of the nuclear reorganization, *Int. Rev. Cytol.* **114:**1–9.

Jennings, H. S., 1941, *Paramecium bursaria.* II. Self-differentiation and self-fertilization of clones, *Proc. Am. Philos. Soc.* **85:**25–48.

Jurand, A., and Selman, G. G., 1969, *The Anatomy of Paramecium aurelia,* Macmillan & Co., London.

Kitamura, A., 1982, Attachment of *Paramecium* to polystyrene surfaces: A model system for the analysis of sexual cell recognition and nuclear activation, *J. Cell Sci.* **58**:185–199.

Kitamura, A., 1983, Effects of protease inhibitors on early events in the conjugation of *Paramecium caudatum*, *J. Exp. Zool.* **225**:501–503.

Kitamura, A., 1984, Evidence for an increase in the hydrophobicity of the cell surface during sexual interaction of *Paramecium*, *Cell Struct. Funct.* **9**:91–95.

Kitamura, A., 1986, Attachment of mating reactive *Paramecium* to polystyrene surfaces. IV. Comparison of adhesiveness among six species of the genus *Paramecium*, *Biol. Bull. (Woods Hole, Mass.)* **171**:350–359.

Kitamura, A., 1988, Mating-type substances, in: *Paramecium* (H. D. Gortz, ed.), Springer-Verlag, Berlin, pp. 85–96.

Kitamura, A., and Hiwatashi, K., 1976, Mating-reactive membrane vesicles from cilia of *Paramecium caudatum*, *J. Cell Biol.* **69**:736–740.

Kitamura, A., and Hiwatashi, K., 1978, Are sugar residues involved in specific cell recognition of mating in *Paramecium? J. Exp. Zool.* **203**:99–108.

Kitamura, A., and Hiwatashi, K., 1980, Recognition of mating active membrane vesicles in *Paramecium*, *Exp. Cell Res.* **125**:486–489.

Kitamura, A., and Hiwatashi, K., 1984a, Cell contact and the activation of conjugation in *Paramecium*, *Zool. Sci.* **1**:161–168.

Kitamura, A., and Hiwatashi, K., 1984b, Inactivation of cell movement following sexual cell recognition in *Paramecium caudatum*, *Biol. Bull. (Woods Hole, Mass.)* **166**:156–166.

Kitamura, A., and Hiwatashi, K., 1984c, A possible mechanism of chemical induction of conjugation in *Paramecium:* Importance of cationic exchange on the cell surface, *J. Exp. Zool.* **231**:303–307.

Kitamura, A., and Steers, E., Jr., 1983, Attachment of *Paramecium* to polystyrene surfaces. II. Induction of the attachment by hydrophobic reagents or immune immunoglobulin G, *J. Cell Sci.* **62**:209–222.

Kitamura, A., Onimaru, H., Naitoh, Y., and Hiwatashi, K., 1979, Relation between sexual cell recognition and swimming behavior in *Paramecium caudatum*, *Doubutsugaku Zasshi* **88**:528 (abstract in Japanese).

Kitamura, A., Onimaru, H., Naitoh, Y., and Hiwatashi, K., 1980, Ciliary inactivation following sexual cell recognition in *Paramecium*, *Dev. Growth Differ.* **22**:722 (Abst.).

Köhle, D., Lang, W., and Kauss, H., 1980, Agglutination and glycosyltransferase activity of isolated gametic flagella from *Chlamydomonas reinhardii*, *Arch. Microbiol.* **127**:239–243.

Machemer, H., 1974, Frequency and directional responses of cilia to membrane potential changes in *Paramecium*, *J. Comp. Physiol.* **92**:293–316.

Matsuda, J., and Hiwatashi, K., 1979, Expression of mating reactivity and cell cycle in *Paramecium caudatum*, *Jpn. J. Protozool.* **12**:26–27 (abstract in Japanese).

Metz, C. B., 1954, Mating substances and the physiology of fertilization in ciliates, in: *Sex in Microorganisms* (D. H. Wenrich, ed.), Am. Assoc. Adv. Sci, Washington, D.C., pp. 284–334.

Mikami, K., and Hiwatashi, K., 1975, Macronuclear regeneration and cell division in *Paramecium caudatum*, *J. Protozool.* **22**:536–540.

Mikami, K., and Koizumi, S., 1979, Induction of autogamy by treatment with trypsin in *Paramecium caudatum*, *J. Cell Sci.* **35**:177–184.

Mishima, S., 1978, Feeding and mating reactivity in *Paramecium multimicronucleatum*, *J. Protozool.* **25**:75–76.

Miwa, I., 1979a, Specificity of the immaturity substances in *Paramecium*, *J. Cell Sci.* **36**:253–260.

Miwa, I., 1979b, Immaturity substances in *Paramecium primaurelia* and their specificity, *J. Cell Sci.* **38**:193–200.

Miwa, I., 1984, Destruction of immaturin activity in early mature mutants of *Paramecium caudatum*, *J. Cell Sci.* **72**:111–120.

Miwa, I., and Hiwatashi, K., 1970, Effect of mitomycin C on the expression of mating ability in *Paramecium caudatum*, *Jpn. J. Genet.* **45**:269–275.

Miwa, I., and Umehara, N., 1983, Conjugation between G_1 and G_2 phase cells in *Paramecium tetraurelia*, *J. Protozool.* **30**:271–274.

Miwa, I., Haga, N., and Hiwatashi, K., 1975, Immaturity substances: Material basis for immaturity in *Paramecium*, *J. Cell Sci.* **19**:369–378.

Miyake, A., 1958, Induction of conjugation by chemical agents in *Paramecium caudatum*, *J. Inst. Polytech. Osaka City Univ. Ser. D* **9**:251–296.

Miyake, A., 1964, Induction of conjugation by cell-free preparations in *Paramecium multimicronucleatum*, *Science* **146**:1583–1585.

Miyake, A., 1966, Local disappearance of cilia before the formation of holdfast union in conjugation of *Paramecium multimicronucleatum*, *J. Protozool. Suppl.* **13**:28.

Miyake, A., 1968a, Induction of conjugation by chemical agents in *Paramecium*, *J. Exp. Zool.* **167**:359–380.

Miyake, A., 1968b, Artificial induction of autogamy by chemical agents in *Paramecium multimicronucleatum*, syngen 2, *Jpn. J. Dev. Biol.* **22**:62–63.

Miyake, A., 1968c, Chemical induction of nuclear reorganization without conjugating union in *Paramecium multimicronucleatum*, syngen 2, *Proc. 12th Int. Congr. Genet. Tokyo* **1**:72.

Miyake, A., 1969, Mechanism of initiation of sexual reproduction in *Paramecium multimicronucleatum*, *Jpn. J. Genet.* **44** (Suppl):388–395.

Miyake, A., 1978, Cell communication, cell union, and initiation of meiosis in ciliate conjugation, *Curr. Top. Dev. Biol.* **12**:37–82.

Miyake, A., 1981, Physiology and biochemistry of conjugation in ciliates, in: *Biochemistry and Physiology of Protozoa*, 2nd ed., Volume 4, (M. Levandowsky and S. H. Hutner, eds.), Academic Press, New York, pp. 125–198.

Myohara, K., and Hiwatashi, K., 1978, Mutants of sexual maturity in *Paramecium caudatum* selected by erythromycin resistance, *Genetics* **90**:227–241.

Naitoh, Y., and Eckert, R., 1969, Ionic mechanisms controlling behavioral responses of *Paramecium* to mechanical stimulation, *Science* **164**:963–965.

Naitoh, Y., and Eckert, R., 1974, The control of ciliary activity in protozoa, in: *Cilia and Flagella* (M. A. Sleigh, ed.), Academic Press, New York, pp. 305–352.

Naitoh, Y., and Kaneko, H., 1973, Reactivated Triton-extracted models of *Paramecium:* Modification of ciliary movement by calcium ions, *Science,* **176**:523–524.

Nanney, D. L., 1977, Cell–cell interaction in ciliates: Evolutionary and genetic constraints, in: *Microbial Interactions* (J. L. Reissig, ed.), Chapman & Hall, London, pp. 351–397.

Nobili, R., 1963, Effects of antibiotics, base- and amino acid-analogues on mating reactivity of *Paramecium aurelia*, *J. Protozool.* **10**:s24.

Ogura, A., and Machemer, H., 1980, Distribution of mechanoreceptor channels in the *Paramecium* surface membrane, *J. Comp. Physiol.* **135**:233–242.

Shimomura, F., and Takagi, Y., 1985, Commitment to sexual and asexual phases in *Paramecium multimicronucleatum*, *J. Exp. Zool.* **233**:317–326.

Sonneborn, T. M., 1937, Sex, sex inheritance and sex determination in *Paramecium aurelia*, *Proc. Natl. Acad. Sci. USA* **23**:378–385.

Sonneborn, T. M., 1957, Breeding systems, reproductive methods, and species problems in protozoa, in: *The Species Problem* (E. Mayer, ed.), Am. Assoc. Adv. Sci., Washington, D.C., pp. 155–324.

Sonneborn, T. M., 1974, *Paramecium aurelia*, in: *Handbook of Genetics*, Volume 2 (R. C. King, ed.), Plenum Press, New York, pp. 496–594.

Takagi, Y., Yoshiki, Y., Baba, C., and Yamada, S., 1981, Determination of macronuclear fragmentation, stabilization of cell union, and cessation of feeding activity in triplet conjugants of *Paramecium*, *J. Protozool.* **28**:99–102.

Takahashi, M., Takeuchi, N., and Hiwatashi, K., 1974, Mating agglutination of cilia detached from complementary mating types of *Paramecium*, *Exp. Cell Res.* **87**:415–417.

Tsukii, Y., 1988, Mating-type inheritance, in: *Paramecium* (H.-D. Gortz, ed.), Springer-Verlag, Berlin, pp. 59–69.

Tsukii, Y., and Hiwatashi, K., 1978, Inhibition of early events of sexual processes in *Paramecium* by concanavalin A., *J. Exp. Zool.* **205**:439–446.

Tsukii, Y., and Hiwatashi, K., 1979, Artificial induction of autogamy in *Paramecium caudatum*, *Genet. Res.* **34**:163–172.

Tsukii, Y., and Hiwatashi, K., 1983, Genes controlling mating-type specificity in *Paramecium caudatum:* Three loci revealed by intersyngenic crosses, *Genetics* **104**:41–62.

van den Ende, H., 1985, Sexual agglutination in *Chlamydomonas*, *Adv. Microb. Physiol.* **26**:89–123.

Vivier, E., and André, J., 1961, Donnees structurales et ultrastructurales nouvelles sur la conjugaison de *Paramecium caudatum*, *J. Protozool.* **8**:416–426.

Watanabe, T., 1977a, Chemical properties of mating substances in *Paramecium caudatum:* Effect of various agents on mating reactivity of detached cilia, *Cell Struct. Funct.* **2:**241–247.

Watanabe, T., 1977b, Ciliary membranes and mating substances in *Paramecium caudatum, J. Protozool.* **24:** 426–429.

Watanabe, T., 1978, A scanning electron-microscopic study of the local degeneration of cilia during sexual reproduction in *Paramecium, J. Cell Sci.* **32:**55–66.

Watanabe, T., 1981, Electron microscopy of cell surfaces of *Paramecium caudatum* stained with ruthenium red, *Tissue Cell* **13:**1–7.

Watanabe, T., 1982, Correlation between ventral surface structures and local degeneration of cilia during conjugation in *Paramecium, J. Embryol. Exp. Morphol.* **70:**19–28.

Watanabe, T., 1983, Local degeneration of cilia and nuclear activation during sexual reproduction in *Paramecium caudatum, Dev. Growth Differ.* **25:**113–120.

Wichterman, R., 1946, The behavior of cytoplasmic structures in living conjugants of *Paramecium bursaria, Anat. Rec.* **94:**93–94.

Wichterman, R., 1985, *The Biology of Paramecium,* 2nd ed., Plenum Press, New York.

Witman, G. K., Carlson, K., Berliner, J., and Rosenbaum, J. L., 1972, *Chlamydomonas* flagella. I. Isolation and electrophoretic analysis of microtubules, matrix, membranes and mastigonemes, *J. Cell Biol.* **54:**507–539.

Wyroba, E., and Przelecka, A., 1973, Studies on the surface coat of *Paramecium aurelia.* I. Ruthenium red staining and enzyme treatment, *Z. Zellforsch. Mikrosk. Anat.* **143:**343–353.

Calcium Ions and the Regulation of Motility in *Paramecium*

Robin R. Preston and Yoshiro Saimi

1. Introduction

Paramecium is a large unicell: of the most commonly studied species, *P. tetraurelia* can reach 150–200 μm in length, while *P. caudatum* may exceed 300 μm. The cell body of both species supports 5000–6000 cilia, which are individually covered by a membrane that collectively accounts for about half of the total cell surface area. The ciliary and somatic membranes are physically contiguous, but their biochemical compositions are quite distinct. How this segregation is achieved is an interesting problem in itself (for more details, the reader is referred to chapters in this volume by Bouck *et al.*, Williams, Chailley *et al.*, and Kaneshiro), but it ensures a functional specialization of the two surfaces. A primary role of the ciliary membrane is in providing a barrier between the locomotory apparatus of the cilium (the axoneme) and the external environment, but it is also involved in establishing physical contact between mating-competent cells. This cell agglutination reaction is discussed in some detail by Watanabe (this volume). Also, there is evidence to suggest that cilia may be sensitive to chemicals in the environment (Preston and Usherwood, 1988). However, it is perhaps in its role as an ion-selective filter that the ciliary membrane is most important to the day-to-day survival of the unicell.

Paramecium lives in fresh water, where the ion concentration is typically low. For convenience, this is usually replaced in the laboratory by a buffered solution containing 4 mM K^+ and 1 mM Ca^{2+}, resulting in a "resting" membrane potential of -35 to -40 mV, inside negative. When left undisturbed in this solution, *Paramecium* swims forward in a left-handed helical path at speeds of 1–2 mm/sec. However, if the cell collides with an object, or encounters a noxious chemical or ion concentration, it stops, backs up for a short distance, and then swims off in a new direction. This simple behavioral response, dubbed an "avoiding reaction" (Jennings, 1906), has provoked considerable interest.

Robin R. Preston and Yoshiro Saimi • Laboratory of Molecular Biology, University of Wisconsin, Madison, Wisconsin 53706.

From a number of studies (reviewed by Eckert and Brehm, 1979; Kung and Saimi, 1982), it is apparent that touching the cell anteriorly elicits a depolarizing mechanoreceptor potential. This in turn triggers the rapid opening of voltage-sensitive Ca^{2+} channels in the ciliary membrane, causing Ca^{2+} to flood into the cilium. The concomitant regenerative depolarization is graded with stimulus intensity, and has an upstroke that is about ten times slower than firing in neurons, but it is often referred to as an action potential or spike. The spike is terminated by a delayed K^+ efflux through voltage-sensitive K^+ channels, but the presence of Ca^{2+} within the cilium triggers a series of events that culminate in the cilia reversing their beat direction. The cell swims backward as a consequence (Fig. 1). Extreme environmental conditions can occasionally lead to prolonged depolarization and lengthy bouts of backward swimming. This results in the activation of Ca^{2+}-dependent K^+ channels that may serve as an "emergency backup system" for membrane repolarization.

Paramecium also demonstrates a response to being touched posteriorly, but in this example, the cell hyperpolarizes and darts forward (the "escape response"; Naitoh, 1974). The response perhaps enables the cell to escape from predators, or from a confined space. The ionic events that underlie the hyperpolarization are less well understood than those mediating the avoiding reaction, but probably involve a Ca^{2+} flux through a set of hyperpolarization-dependent channels distinct from those of the ciliary membrane (Hennessey, 1987). Hyperpolarization also elicits voltage-dependent and Ca^{2+}-dependent K^+ fluxes (Oertel *et al.*, 1978; Richard *et al.*, 1986). A biochemical consequence of the membrane potential change is a rise in the intraciliary concentration of cAMP, which in the role of second messenger can increase the frequency of posteriorly directed ciliary beat (Bonini *et al.*, 1986; Fig. 1).

While the survival value of the escape response is open to question, the avoiding reaction is clearly central to *Paramecium* behavior, since it enables the cell to avoid

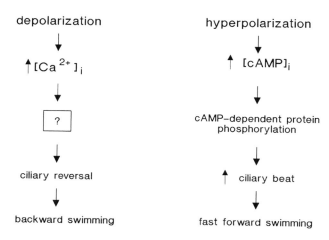

Figure 1. Sequence of events that follow membrane depolarization or hyperpolarization in *Paramecium*. (Left) Depolarizing stimuli, such as touching the anterior end of the cell, activate voltage-sensitive Ca^{2+} channels in the ciliary membrane. Ca^{2+} flows into the cilium and effects ciliary reorientation, so that the effective power stroke becomes directed toward the anterior of the cell. The cell swims backward as a consequence. The molecular nature of the Ca^{2+} sensor and/or associated ciliary reversal mechanism (indicated by "?") is unknown. (Right) Membrane hyperpolarization, caused by, for example, posterior mechanostimulation, is accompanied by activation of adenylate cyclase and an increase in intraciliary cAMP concentration. cAMP stimulates cAMP-dependent protein kinase(s) to phosphorylate a number of ciliary proteins. As a result, the cilia beat with increased frequency in a more posterior direction, and the cell swims faster forward.

entering areas that contain potentially dangerous physical or chemical extremes, and permits the cell to follow chemical gradients that lead to a food source (Van Houten and Van Houten, 1982). The Ca^{2+} channels that mediate the avoiding reaction are thus an important constituent of the ciliary membrane. This chapter examines how these channels activate and inactivate during depolarization, where they may be located in the ciliary membrane, their possible mechanisms of regulation, and discusses the sequence of events that occur after Ca^{2+} is permitted access to the intraciliary space. Although attention is focused on *Paramecium*, a number of parallel studies on the related holotrich *Tetrahymena* are also described. Preliminary membrane potential recordings (Onimaru *et al.*, 1980; Connolly and Kerkut, 1981, 1983, 1984) suggest that membrane excitation in *Tetrahymena* may be similar to that in *Paramecium*, so these observations may be of help in a more general understanding of ciliary membrane function in unicellular organisms. For a review of electrical control of ciliary activity in hypotrichous ciliates, interested readers are referred to Machemer and Deitmer (1987).

2. Voltage-Dependent Calcium Channels

2.1. Voltage-Clamp Analysis of Calcium Currents

The action potential represents summed electrical effects of one or more ion species passing across the cell membrane. In dissecting this event, it is convenient to clamp the membrane potential at a constant voltage and then monitor current flow with time in response to an imposed voltage step. Like most cells, *Paramecium* maintains an intracellular free Ca^{2+} concentration of about 0.1 μM (Naitoh and Kaneko, 1972, 1973; Browning and Nelson, 1976; Martinac and Hildebrand, 1981). Since the external Ca^{2+} concentration is 1 mM, the electrochemical driving forces for Ca^{2+} entry into the cell are considerable. Thus, when the membrane potential of *Paramecium* is clamped at -40 mV and then depolarized, the resulting Ca^{2+} flow through the ciliary membrane is observed as a rapidly activating, inward current. The current peaks about 1.8 msec after the onset of the voltage step and decays exponentially thereafter (Fig. 2A). A maximum current of 7– 8 nA is observed when *P. tetraurelia* is depolarized to -10 mV, whereas the Ca^{2+} current of *P. caudatum* is somewhat larger (10–12 nA). The current decays during the depolarization because Ca^{2+} ions entering the cilium subsequently inactivate the Ca^{2+} channels (Brehm and Eckert, 1978b; Brehm *et al.*, 1980). Thus, if free Ca^{2+} is removed as soon as it enters the cilium (by injecting a chelator such as EGTA), the inactivation is weakened or removed (Brehm *et al.*, 1980; Fig. 2B). The mechanism of inactivation is unknown, but in the light of studies of Ca^{2+} channel inactivation in higher organisms (see Chad *et al.*, 1987), phosphorylation/dephosphorylation reactions may be involved. The channels are also subject to a voltage-dependent inactivation during depolarizations of 1 sec or more (Hennessey and Kung, 1985). This process is not affected by EGTA, but may also involve covalent modification of the channel protein. Both types of inactivation are common to Ca^{2+} channels in a variety of animal cells (Eckert *et al.*, 1981; Fox, 1981; Kass and Scheuer, 1982; Chad *et al.*, 1984; Eckert and Chad, 1984).

The Ca^{2+} channel normally passes only Ca^{2+}, but if Ba^{2+} or Sr^{2+} is added to the bathing medium, these cations will also carry current through the channel (Brehm and

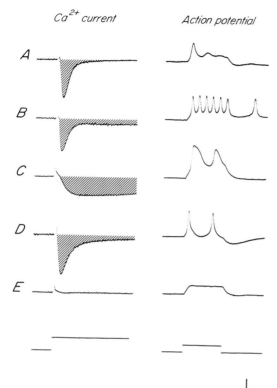

Ca²⁺ current   *Action potential*

A

B

C

D

E

Figure 2. Ca²⁺ currents and action potentials of wild-type and mutant *Paramecium*. Ca²⁺ currents and action potentials were recorded from *Paramecium* using methods similar to those described by Hinrichsen and Saimi (1984). Isolated Ca²⁺ currents under voltage-clamp were recorded using 4 M CsCl electrodes from cells bathed in a solution containing 10 mM tetraethylammonium, 1 mM Ca²⁺, 0.01 mM EDTA, 1 mM HEPES, pH 7.2. The cells were held at a membrane potential of −40 mV and currents elicited by depolarization to −10 mV (lower trace). Action potentials evoked by current injection were recorded using 0.5 M KCl electrodes from cells bathed in a solution containing 4 mM K⁺, 1 mM Ca²⁺, 0.01 mM EDTA, 1 mM HEPES, pH 7.2. 0.2-nA, 200-msec current pulses were used to elicit the membrane potential responses.

(A) The Ca²⁺ current from a wild-type cell (left) activates rapidly in response to depolarization under voltage-clamp, peaks at ca. 1.8 msec, and then inactivates. This inactivation is Ca²⁺-dependent. The action potential in the wild-type cell (right) is graded with stimulus intensity and is not recurrent, due in part to inactivation of the current. This inactivation can be weakened by an iontophoretic injection of EGTA (7.5 nA, 20 sec), causing a sustained current (B, left), and repeated, all-or-nothing action potentials in response to depolarization (right). Ba²⁺ does not substitute for Ca²⁺ in its ability to inactivate the Ca²⁺ channels, so that when Ca²⁺ in the bathing medium is replaced by 4 mM Ba²⁺, the resulting Ba²⁺ current is sustained (C, left) and the action potential is all-or-nothing (right). (D) *Dancer* is a mutant with a defective Ca²⁺ channel that does not inactivate fully in response to Ca²⁺ (left; compare with the wild-type cell). This causes the mutant cell to respond to depolarizing stimuli with repeated, all-or-nothing action potentials (right). Conversely, *pwB* (E) is a mutant that lacks most of its Ca²⁺ current (left), and consequently is unable to respond to membrane depolarization with an action potential (right).

The vertical calibration bar represents 5 nA or 25 mV; the horizontal bar represents 10 or 80 msec for the current traces and voltage traces, respectively.

Eckert, 1978b). Ba²⁺ and Sr²⁺ are not as effective as Ca²⁺ in their ability to inactivate the channel, however, so that a Ba²⁺ or Sr²⁺ current is usually prolonged compared with that carried by Ca²⁺ (Fig. 2C). In addition, Ba²⁺ and Sr²⁺ currents activate and deactivate (channel closure when the voltage step is terminated) more slowly than the Ca²⁺ current, suggesting that Ca²⁺ may also be involved in these processes (Saimi and Kung, 1982).

The *Paramecium* Ca²⁺ channel is similar to those of higher organisms with respect to its inactivation properties and ion selectivity, but, like other ion channels of *Paramecium*, its pharmacology is distinct. The Ca²⁺ current is insensitive to many of the well-known Ca²⁺ channel inhibitors, including verapamil and D600, diltiazem, and many dihydropyridines including nifedipine, a potent antiarrhythmic agent. To date, the only drug shown to have specific effects on the current is W-7, an anticalmodulin agent that reversibly inhibits the Ca^{2+} influx at micromolar concentrations (Hennessey and Kung, 1984). These dissimilarities are not unexpected. Ca^{2+} channels appear to be membrane components of most, if not all, cell types, suggesting that they arose from a common ancestor early in evolutionary history. Protozoa separated from the metazoan stock early in evolution, so while the molecular mechanisms of ion channel gating may be similar to those in higher animals, it is likely that protozoan Ca^{2+} channels were acquired before the channels became kinetically and pharmacologically diversified.

2.2. Calcium Channel Mutants

One of the advantages of using *Paramecium* as a model system for the study of membrane excitability is that because membrane potential changes affect the direction and frequency of ciliary beat, it is possible to "observe" these electrical events, simply by watching a cell swim. This relationship enables a visual selection of mutant cells with specific defects in membrane excitability, since cells with an abnormally large Ca^{2+} current might be expected to overreact to depolarizing stimuli. Conversely, cells lacking functional Ca^{2+} channels should be unable to swim backward when stimulated. This method of mutant screening has been exploited fully by Kung and his colleagues (Kung, 1971b, 1979; Kung *et al.*, 1975), who have isolated hundreds of cell lines with defects in four of the nine known ion currents of *P. tetraurelia*. These include the mutants alluded to above. *Dancer,* as its name suggests, gives exaggerated, jerking, responses to depolarizing stimuli (Hinrichsen *et al.*, 1984b). A record of *Dancer*'s membrane potential during depolarization shows that its action potentials are all-or-nothing (Fig. 2D), as opposed to the graded spikes of the wild-type. Voltage-clamp analyses show that while the peak amplitude of the Ca^{2+} current in *Dancer* is comparable with that of the wild-type, its inactivation and deactivation properties are slower in the mutant, allowing excessive amounts of Ca^{2+} to enter the cilium (Hinrichsen and Saimi, 1984). *Dancer* is particularly interesting because its phenotype suggests that the defect may be in the structure of the channel itself, rather than in a secondary, regulatory mechanism.

Perhaps the best known of the *Paramecium* mutants are the *pawns* (*pw*), named after the chess piece for their inability to move backwards (Kung, 1971a,b). *Pawns* have been used extensively as controls for studies of Ca^{2+} channel function in wild-type *Paramecium*, so it may be useful to examine them in some detail. Whereas the first *pw* mutants were isolated for their inability to undergo ciliary reversal, others have since been selected for their resistance to K^{+} (Shusterman, 1981) or Ba^{2+} (Schein, 1976b). Ba^{2+} normally passes through the wild-type Ca^{2+} channel and poisons the cell. All *pawns* are characterized by their inability to generate action potentials when stimulated, and by their complete or near-complete lack of a Ca^{2+} current under voltage-clamp (Oertel *et al.*, 1977; Satow and Kung, 1980; Fig. 2E). There are several hundred lines of *pawns*, which fall into four complementation groups: *pwA, pwB, pwC,* and *pwD* (Kung, 1971b; Saimi and Kung, 1987). All are recessive to the wild-type. The *pwB* mutants are generally less leaky than the other three classes, while the severity of the loss of excitability caused by the *pwA* mutation is wide-ranging. All *pwC* mutants and a number of *pwA*'s are temperature-sensitive, so that their phenotypes are only expressed following growth at restrictive temperatures. This characteristic was exploited by Satow *et al.* (1974) in early studies of Ca^{2+} channel stability. When temperature-sensitive *pwA* mutants are transferred to growth media at restrictive temperatures, there is a gradual loss of reversal behavior, and by inference, Ca^{2+} channel function, over 4–12 hr. Schein (1976a) also investigated Ca^{2+} channel stability, but in this example, the loss of excitability occurred as a result of inducing *pwA/pwA* homozygosity in *pwA/+* heterozygotes (achieved through autogamy; see below). The *pawn A* homozygotes retain the ability to undergo ciliary reversal for several days, since they still possess functional, wild-type Ca^{2+} channels. However, there is a gradual loss of excitability as the channels are turned over, which enabled Schein (1976a) to estimate a mean channel lifetime of 5–8 days in nongrowing cells.

P. tetraurelia is commonly used in mutageneses because it naturally undergoes a process of self-fertilization, called autogamy. This in effect ensures a 50% probability that an induced recessive mutation will be expressed in the ex-autogamous daughter cells. Autogamy does not occur naturally in the related species, *P. caudatum,* but self-fertilization can be induced under appropriate circumstances, which has facilitated the isolation of *pawn* equivalents (Takahashi and Naitoh, 1978; Takahashi, 1979; Takahashi *et al.,* 1985). These mutants, named *CNR*s (for *c*audatum *n*on *r*eversal), genetically comprise four unlinked complementation groups, *cnrA, cnrB, cnrC,* and *cnrD,* and like the *pw* mutants, they are mostly recessive. A single exception is the K^+-agitated mutant (*cnrB^{Kg}*, formerly K^+-sensitive), so named for its exaggerated ciliary reversal response to high concentrations of K^+ (Takahashi, 1979, 1988). Complementation testing has shown this mutant to be a dominant allele of *cnrB,* however. Thus, mutations in the *cnrB* gene can result in either a loss of Ca^{2+} channel function or, in this example, a phenotype more akin to the semidominant *Dancer* mutant of *P. tetraurelia.* Discerning the nature of the gene product of *cnrB* will perhaps shed considerable light on Ca^{2+} channel function in *Paramecium.*

In addition to membrane excitation mutants, behavioral screenings for *pawns* net individuals that are defective in the mechanics of ciliary reversal. Such a phenotype may result from a structural defect in the axoneme, or from an inability of the reversal mechanism to sense Ca^{2+} when it enters the cilium. These mutants, called *atalantas* (Kung *et al.,* 1975; Hinrichsen and Kung, 1984; Hinrichsen *et al.,* 1984a), can be distinguished from true *pawns* by testing the ability of their axonemes to respond to Ca^{2+} after their membranes have been disrupted with a detergent. Detergent-treated cells are to all intents dead since they lack membranes, but their axonemal structures can be functionally reactivated if provided with Mg^{2+} and ATP. This causes the dead cells (often referred to as "models") to swim forward (Naitoh and Kaneko, 1972). When Ca^{2+} is added to the reactivation medium, it has direct access to the axoneme, causing the cilia of the wild-type and true *pawns* to reverse their beating direction. Cells with structural defects in their locomotory apparatus remain incapable of a normal response to Ca^{2+}, however, so that they either spin in place or continue to swim forward.

2.3. Location of Calcium Channels

Incubation in chloral hydrate or dibucaine, or mechanical agitation in ethanol, causes deciliation and a concomitant loss of the Ca^{2+} action potential in *Paramecium* (Ogura and

Takahashi, 1976; Dunlap, 1977; Machemer and Ogura, 1979). This treatment is not lethal, however, and paramecia will subsequently regenerate their cilia if removed from the deciliation medium. Reciliation is accompanied by the reappearance of the action potential, which strongly suggests that the Ca^{2+} channels are exclusive to the ciliary membrane of *Paramecium*. The exact location of the channels within the ciliary membrane has been the subject of some controversy, however.

Freeze-fracture electron micrographs of the *Paramecium* ciliary membrane reveal the presence of two intramembranous specializations. The "ciliary necklace" consists of two bands of particles encircling the ciliary base and is common to a number of organisms (see Dentler, this volume; Chailley *et al.*, this volume). A second specialization, called "cili-

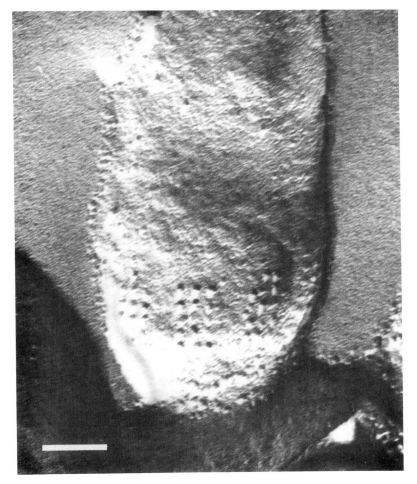

Figure 3. Ciliary plaques. Ciliary plaques are arrays of intramembranous particles observed in freeze-fracture replicas of the ciliary base. The arrays comprise three particle rows, each row containing three to seven particles. The cell was fixed in 2.5% glutaraldehyde and cryoprotected for 1 hr with 30% (v/v) glycerol prior to fracturing, using methods described by Preston and Newman (1986). The calibration bar represents 0.1 μm.

ary plaques,'' comprises three rows of three to seven particles arranged in nine groups above the necklace (Plattner, 1975; Fig. 3). Since the occurrence of a particle in membrane replicas generally reflects the presence of an underlying protein or protein cluster, it is conceivable that the plaques could be proteins involved in the regulation of ion fluxes (Byrne and Byrne, 1978b), perhaps representing Ca^{2+} pumps or the Ca^{2+} channels themselves. This contention is supported by the facts that the plaques lie adjacent to, and appear to be associated with, each of the outer nine microtubule doublets (Dute and Kung, 1978), and that the Ca^{2+} sensor for ciliary reversal appears to be located at the ciliary base (Hamasaki and Naitoh, 1985). It is unlikely that the ciliary plaque particles represent Ca^{2+} channels, however, for the following reasons. First, freeze-fracture electron micrographs of *pawn* mutants show superficially normal plaques (Bardele, 1981). The only cell line with gross abnormalties or deficiencies in the plaque region is a Na^+ channel mutant, *paranoiac* (*PaA;* Byrne and Byrne, 1978a; Byrne *et al.,* 1988). Second, the plaques are restricted to a select group of ciliated protozoa (Bardele, 1981); if they represented Ca^{2+} channels, they might be expected to be more widely distributed, as is the necklace. Third, a French-pressed preparation of ciliary membranes from *P. tetraurelia* yields two populations of vesicles with different densities (Thiele *et al.,* 1982). When loaded with Arsenazo III, a Ca^{2+} indicator, a heavy, multilamellar vesicle population is shown to be relatively impermeable to Ca^{2+}, while a low-density population readily admits Ca^{2+}. Since low-density vesicles prepared from a *pwA–pwB* double mutant are less permeable to Ca^{2+} than those from the wild-type, it is assumed that these vesicles contain the Ca^{2+} channel (Thiele *et al.,* 1983). The ciliary plaques are estimated to cover 3–5% of the ciliary surface, yet the low-density vesicles comprise 80% of total recovered ciliary membrane surface area, suggesting that the channels are spread along the length of the cilium. Finally, Moss and Tamm (1987) have recorded propagation of a Ca^{2+} action potential using extracellular recording electrodes placed at various positions along giant cilia of ctenophore comb plates. The comb plates are highly specialized structures, but these observations suggest that at least in this example, Ca^{2+} channels and the axonemal target(s) for Ca^{2+} are distributed along the length of the cilium.

2.4. Calcium Channel Activity in Isolated Ciliary Membranes

While spectrophotometric assays of Ca^{2+} fluxes across ciliary membrane vesicles suggest that Ca^{2+} channels are active and functional *in vitro* (Thiele and Schultz, 1981; Thiele *et al.,* 1982), such assays are inevitably crude and they measure channel activity indirectly. There are at least two electrophysiological methods that are routinely used in direct studies of channel activity, either in the native membrane, or following their incorporation into artificial membranes. The patch-clamp technique of single channel recording is currently being used to describe the activity of two Ca^{2+}-dependent K^+ channels in the somatic membrane of *P. tetraurelia* (Martinac *et al.,* 1986; Saimi and Martinac, 1989; see also Martinac *et al.,* 1988); this technique has yet to be applied to a study of the ciliary Ca^{2+} channels. The second method involves incorporating functional channels into an artificial membrane. This is effected by fusing ciliary membrane vesicles with an artificial lipid bilayer. When a voltage is imposed across the bilayer, channel activity is apparent as steps of current flow as the channels fluctuate between open and closed states. Boheim and collaborators (Hanke *et al.,* 1981; Boheim *et al.,* 1982; Schultz

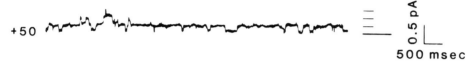

Figure 4. Single channel currents from *Paramecium* ciliary membrane vesicles incorporated into a lipid bilayer. An artificial lipid bilayer was formed across a small pore (50–100 μm) separating two solution-filled compartments. Purified ciliary membrane vesicles, prepared according to the method of Adoutte *et al.* (1980), were added to the *cis* compartment for fusion with the bilayer. If the membrane vesicles insert correctly, the *cis* compartment corresponds to the extracellular side of the membrane. The *cis* compartment was filled with a solution containing 50 mM $BaCl_2$; the *trans* compartment contained 50 mM $MgCl_2$ and 0.1 mM $BaCl_2$. A voltage was then applied across the bilayer to induce channel opening and current flow: +50 indicates the holding potential (in mV) on the *cis* compartment of the chamber relative to the *trans* compartment. The long horizontal line to the right of the current trace indicates the zero current level, while the short lines indicate the number of single channels open. Upward deflections from the zero current level represent flow of positive (Ba^{2+}) current from the *cis* to the *trans* compartment. This putative Ca^{2+} channel shows voltage-dependence and has a conductance of 1.6 pS. From Ehrlich *et al.* (1984), with permission. Copyright 1984 by the AAAS.

et al., 1984a) described both a cation-selective channel with a conductance of 16 pS and a voltage-, Ca^{2+}-dependent channel with a conductance of 45 pS. Ehrlich *et al.* (1984) also described two channels from *Paramecium* ciliary membranes, both of which are divalent cation selective. The first channel, conductance 1.5–2 pS (Fig. 4), is strongly voltage-dependent and has a divalent cation selectivity that mirrors that of the Ca^{2+} current in intact cells. This channel is also blocked by W-7 (Ehrlich *et al.,* 1988), making it a likely candidate for the ciliary Ca^{2+} channel. The second, larger-conductance channel (30 pS) does not discriminate between Mg^{2+} and Ba^{2+}, suggesting that it may represent the channel that mediates the inward mechanoreceptor current in *Paramecium*. Finally, Oosawa and Sokabe (1985; Oosawa *et al.,* 1988) reported a 211-pS, cation-selective channel in ciliary membrane vesicles from *Tetrahymena*. The channel may have a role in maintaining the "resting" membrane potential of this ciliate.

2.5. Curing of *pawns* and *CNRs*

One of the more intriguing aspects of the study of membrane excitability in *Paramecium* is that certain mutant phenotypes can be "cured," albeit temporarily, by injecting wild-type cytoplasm. The importance of this observation is that identification of the active factors may provide information as to how Ca^{2+} channels are synthesized, assembled, maintained, and regulated.

Curing was first observed in *P. tetraurelia* as a natural phenomenon that occurred when a *pwA* mutant was mated to a wild-type cell (Berger, 1976). During conjugation, two mating cells establish cytoplasmic contact for the exchange of nuclei, but the ability to swim backward is conferred on the mutant cell prior to nuclear exchange, so it is likely that a soluble cytoplasmic factor is responsible. The phenomenon has since been investigated systematically by Hiwatashi and his co-workers, who established that microinjection of ca. 50 pl of wild-type cytoplasm (about 10% of the cell's volume) into *CNR C* is sufficient to restore Ca^{2+} channel function (Hiwatashi *et al.,* 1980). The mutants become capable of backward swimming within 1–2 hr of the injection, with curing reaching a maximum after 8 hr and lasting 2–3 days in the absence of cell growth and protein synthesis (Haga *et al.*, 1983). Similar curings have been demonstrated in *CNR A, B,* and *D,* and in *pawn A, B,* and *C* (Haga *et al.*, 1983; Takahashi *et al.*, 1985). The donor cytoplasm does not necessarily have to originate from a wild-type cell, or even from the same species of *Paramecium: pawn A, B,* and *C* of *P. tetraurelia* are able to restore some degree of excitability to the four *CNR*s of *P. caudatum,* and vice versa (Haga *et al.*, 1983). The only restriction on the curing effect is that the cytoplasmic donor must not be from the same complementation group as the recipient, even though the two cells may be mutated in different parts of the same gene. This study implies that the *cnr* and *pw* mutations map to different genes, so that at least seven genes are necessary for wild-type Ca^{2+} channel function (Hiwatashi *et al.*, 1980). Since the factors responsible for curing *pawn A, B,* and *C* are proteins (Haga *et al.*,, 1982, 1984c), Nock *et al.* (1982) reasoned that it should be possible to restore excitability to these mutants by injecting wild-type cytoplasmic RNA. This expectation has been fully realized: *pwC* cells are able to swim backward in response to depolarizing stimuli 14 hr after injecting ca. 150 pg of RNA from the wild-type. This technique may be valuable in future studies of Ca^{2+} channel processing and assembly. The *pawn* curing factors all sediment with microsomal cell fractions (Haga *et al.*, 1984c), but the protein that confers excitability on *CNR C* (*cnrC* factor) is soluble and cytoplasmic (Haga and Hiwatashi, 1982). Recently, this factor has become a focus of attention, in the hopes of discerning its molecular nature (Haga *et al.*, 1984a,b). It has not yet been purified to homogeneity, but it is tentatively identified as an acidic (pI 4.5–5), heat-labile protein with a molecular mass in the range of 17–30 kDa. The factor is not calmodulin (CaM), nor can CaM substitute for the *cnrC* factor in curing *CNR C* (Haga *et al.*, 1984a). It is possible that the *cnrC* factor comprises one or more of three phosphoproteins, and perhaps an abnormal bias in the factor's phosphorylation state prevents Ca^{2+} channel function in *CNR C* (R. Ramanathan, R. D. Hinrichsen, and C. Kung, personal communication).

2.6. Characterization of Ciliary Membrane Proteins

Paramecium can be reared in large quantities with relative ease, providing a readily available source of pure ciliary membrane. At first glance, it would appear to be a relatively simple task to solubilize the protein constituents of this membrane, separate them electrophoretically, and then assign individual proteins specific functions according to their presence or absence in mutants known to be defective in ciliary membrane excitability. This aspiration has spurred a number of elaborate and occasionally conflicting surveys. First attempts at characterizing the ciliary membrane proteins of *P. tetraurelia* identified 12–15 peptides by SDS-PAGE with molecular masses of 25–150 kDa (Hansma and Kung, 1975). The major entity (75%), as in all subsequent studies, is a 250-kDa protein identified as the "*i*-antigen" (Fig. 5). This protein forms a 20-nm-thick fuzzy layer over the entire surface of the cell (Wyroba and Przelecka, 1973), so that antibodies raised against whole paramecia cause adjacent cilia to cross-link and clump, hence the term *i*mmobilization antigen. Similar coats are found covering other protists (see Bloodgood, this volume; Williams, this volume). Although the functions of these surface coats have yet to be defined, it is possible that they buffer the cell against minor chemical and ionic changes in the environment to provide a relatively stable milieu for

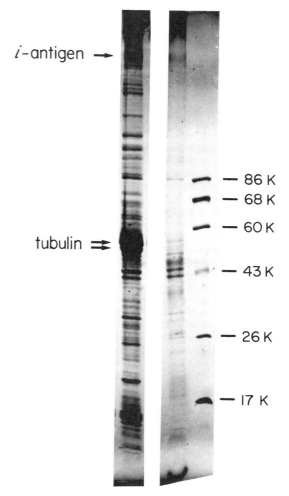

Figure 5. SDS gel of ciliary proteins. Whole cilia (left lane) and purified ciliary membrane vesicles (middle lane) from *Paramecium* were solubilized in 2% SDS. Samples were electrophoresed on a "long" SDS polyacrylamide gel system (6–20%), using methods described by Adoutte *et al.* (1980). Highly purified ciliary membrane vesicles were obtained by sucrose gradient centrifugation. Arrows indicate the presence of tubulins in whole cilia, and *i*-antigen in both whole cilia and ciliary membrane vesicles. Right lane contains molecular weight standards. Figure kindly provided by Dr. D. L. Nelson.

optimal activity of intramembranous components (discussed by Preer, 1986). In addition to the 250-kDa entity, the ciliary membrane of *Paramecium* contains a set of at least four *i*-antigen-related proteins with molecular masses in the 40-kDa range. The exact relationship between the 250- and 40-kDa polypeptides is uncertain, but it is possible that the smaller proteins are degradation products of the 250-kDa entity, or perhaps serve as membrane anchors for this largely external protein. The possibility of the *i*-antigen being involved in cell excitability was first suggested by the studies of Merkel *et al.* (1981), who compared ciliary membrane proteins from wild-type and mutant *P. tetraurelia* (*pwA, B, C,* and *PaA*) in one-dimensional SDS-PAGE. The only obvious differences in protein staining patterns was that a 43-kDa band from *pawn B* cilia reproducibly stained with greater intensity than the equivalent protein from the wild type. This finding was later suggested by Adoutte *et al.* (1983) to result from natural variations in the relative amounts of the various *i*-antigen proteins, but there is evidence that the antigens may be important for normal Ca^{2+} channel function. Antisera raised to partially purified i-antigen and to ciliary membranes containing the 42- to 45-kDa proteins specifically reduce the amplitude of the peak Ca^{2+} current by 75–85%, without direct effects on any other conductances (Eisenbach et al., 1983; Ramanathan et al., 1983). Cell immobilization is not required for this effect, since monovalent Fab fragments produce a similar inhibition of the current. Additionally, reducing the amount of surface-associated i-antigen by mild protease treatment (Ramanathan et al., 1981) also reduces the Ca^{2+} current. The possibility that the i-antigen is a functional component of the Ca^{2+} channels is unlikely in view of the relative amounts and distributions of i-antigen and Ca^{2+} channels: these intriguing observations have so far defied plausible explanation.

Since Hansma and Kung's (1975) initial study, technical improvements and decreased contamination by trichocyst and soluble ciliary proteins enabled Brugerolle et al. (1980) to resolve 53 polypeptides with molecular masses of 10–350 kDa in SDS-PAGE separations of the ciliary membrane of P. tetraurelia, while Adoutte et al. (1980), using two-dimensional SDS-PAGE and isoelectric focusing (IEF), identified 70 polypeptides in the range of 15–250 kDa. The latter also noted that a 31-kDa polypeptide appears in the ciliary membrane with the onset of cell starvation (Fig. 5), and may be related to the acquisition of mating reactivity in Paramecium (see Watanabe, this volume). A later survey by Adoutte et al. (1983) compared ciliary membrane proteins from the wild type and 33 mutant strains of P. tetraurelia, but only PaA and fast-2 (fna) demonstrate mutation-specific aberrations in protein staining patterns. IEF gels of cilia from PaA mutants show an extra band at pH 4.7, while fast-2 membranes yield a band at pH 4.4–4.5 that is not present in gels of wild-type cilia. These observations have not been pursued further, but it is interesting to note that although the authors were unaware of it at the time of publication, these two mutants are both defective in Ca^{2+}-dependent Na^+ permeability (Saimi, 1986). Paranoiac A has a larger Na^+ current than the wild type under voltage-clamp, while this current is missing in fast-2.

Clearly, the ciliary membrane is more complex than anticipated, and the pawns have failed to reveal obvious differences in protein staining pattern that could have identified the Ca^{2+} channel. This type of survey should not be readily dismissed, however. The molecular nature of the defects caused by pw mutations is unknown, so there remains the unfortunate possibility that pawn Ca^{2+} channels, though nonfunctional, are inserted into the ciliary membrane nevertheless.

3. Enzymatic Activity Associated with the Ciliary Membrane

The previous section reviewed what is known of the mechanism of Ca^{2+} entry into the cilium in response to excitation. Once inside the cilium, Ca^{2+} triggers a series of interacting events that culminate in a reorientation of the ciliary power stroke. Only recently have the complexities of this system been fully appreciated. As in other organisms (reviewed by Rasmussen and Barrett, 1984), Ca^{2+} functions as a second messenger of membrane excitation, with many of its actions being mediated by Ca^{2+}-sensitive enzymes. Paramecium cilia contain a number of enzymes that are Ca^{2+}-sensitive. Several of these are stimulated by micromolar levels of Ca^{2+}, but as the Ca^{2+} concentration is raised, these same enzymes are inhibited. This complexity may reflect the

Table I. Enzymes Associated with Cilia and Their Known Properties

Enzyme	Apparent molecular mass (kDa)	Nucleotide specificity (K_m, μM)	Substrate	Ca²⁺ (μM) effects	Localization	References
Soluble Ca-ATPase	68	ATP > GTP > UTP		Stimulated: Ca > Sr > Ba	Ciliary cytosol	Riddle et al. (1982), Travis and Nelson (1986)
Membranous Ca-ATPase	60–70	ATP > GTP > UTP (CaATP = 9)		Stimulated: half-max. = 10 Ca > Sr > Ba	Ciliary membrane	Doughty (1978), Andrivon et al. (1983), Doughty and Kaneshiro (1985), Travis and Nelson (1986)
Adenylate cyclase		(ATP = 67)		Stimulated: half-max. = 0.9 Inhibited: > 5	Ciliary membrane	Schultz and Klumpp (1983), Klumpp et al. (1984), Gustin and Nelson (1987), Schultz et al. (1987)
Guanylate cyclase		GTP >> ATP (GTP = 108)		Stimulated: half-max. = 8 Inhibited: 300–500	Ciliary membrane	Schultz and Klumpp (1980), Klumpp and Schultz (1982)
cAMP-dependent protein kinase I	70		Histone, protamine, casein	Inhibited: half-max. = 2	Ciliary membrane and axoneme	Lewis and Nelson (1980), Schultz and Jantzen (1980), Mason (1987)
cAMP-dependent protein kinase II	220	(ATP = 10)		Inhibited: half-max. = 2	Ciliary membrane and axoneme	Lewis and Nelson (1980), Schultz and Jantzen (1980), Mason (1987)
cGMP-dependent protein kinase	77–88	GTP = ATP (ATP = 10, GTP = 34)	Histone	Inhibited: K_i = 10		Schultz and Jantzen (1980), Miglietta and Nelson (1988)
Ca-dependent protein kinase	51, 55	(ATP = 17–35)	Casein	Stimulated: half-max. = 1		Gundersen and Nelson (1987)
cAMP-phosphodiesterase[a]		(cAMP = 5)		None		Kudo et al. (1986)
cGMP-phosphodiesterase[a]		(cGMP = 2.5)		None		Kudo et al. (1986)

[a]The enzyme activity was characterized in *Tetrahymena*.

fact that ciliary beat frequency, as well as orientation, is under a fine enzymatic control that may also have a Ca^{2+} sensitivity. In the present section, the enzymatic components of the ciliary motility system are individually examined (summarized in Table I), and their possible actions and interactions are discussed.

3.1. Calmodulin and Calmodulin-Binding Proteins

The ubiquitous Ca^{2+}-binding protein calmodulin (CaM) is believed to be the Ca^{2+} sensor of many Ca^{2+}-stimulated molecules (see review by Stoclet *et al.*, 1987), so its association with cilia is to be expected. *Paramecium* CaM has been purified to homogeneity and its structure defined (Rauh and Nelson, 1981; Walter and Schultz, 1981; Schaefer *et al.*, 1987). The protein has three amino acid substitutions that are unique to *Paramecium*, and a novel posttranslational modification, N^ϵ N^ϵ-dimethyl lysine at position 13. Immunocytochemistry identifies CaM as being widespread throughout the cell body and the length of the cilium (Maihle *et al.*, 1981; Klumpp *et al.*, 1983b; Momayezi *et al.*, 1986). Parallel studies on *Tetrahymena* have identified both CaM (Ohnishi *et al.*, 1982) and a new Ca^{2+}-binding protein, TCBP-10 (Ohnishi and Watanabe, 1983), that may also be involved in Ca^{2+}-sensing by this ciliate.

$[^{125}I]$-CaM labels at least 36 polypeptides in nitrocellulose blots of *Tetrahymena* ciliary membranes (Hirano and Watanabe, 1985), perhaps pointing to multiple involvement of this regulatory protein in ciliary function. A similar study of *Paramecium* cilia (Evans and Nelson, 1989) identified eight or nine polypeptides that are reproducibly labeled by CaM in the presence of Ca^{2+}. Two of these (63 and 126 kDa) are axonemal, and one (36 kDa) is soluble (or easily dislodged from ciliary structures). The latter is also identified in small amounts in purified ciliary membrane vesicles, along with 63-, 70-, and 120-kDa proteins. This study is particularly intriguing in that the binding of $[^{125}I]$-CaM to two ciliary proteins with molecular masses of 95 and 105 kDa is *inhibited* by > 2 μM Ca^{2+}. Ca^{2+} inhibition of CaM binding is rare, and suggests a further level of complexity in the regulation of ciliary function by Ca^{2+}.

3.2. Ca-ATPases

Triton-PAGE of whole cells separates 15 proteins with ATPase activity (Andrivon *et al.*, 1983; Doughty and Kaneshiro, 1983), of which at least 5 are associated with cilia. Two are preferentially activated by Mg^{2+} over Ca^{2+} and represent the axonemal dyneins that power ciliary beating (see Witman, this volume). A third is a soluble Ca-ATPase that is stimulated by Ba^{2+} and Sr^{2+}, but not by Mg^{2+} (Riddle *et al.*, 1983; Doughty and Kaneshiro, 1985; Travis and Nelson, 1986). ATPases associated with the ciliary membrane have been described by several authors (Doughty, 1978; Andrivon *et al.*, 1983; Doughty and Kaneshiro, 1983; Travis and Nelson, 1986): the major entity is a 60- to 70-kDa Ca-ATPase. This enzyme has a similar divalent cation sensitivity and nucleotide specificity as the soluble ATPase, but Travis and Nelson (1986) provided convincing evidence that the soluble and membrane-associated activities are immunologically distinct. Travis and Nelson (1986) further suggested that there may be a second membrane-associated ATPase activity that is preferentially stimulated by Mg^{2+}. Ca-ATPases are often associated with Ca^{2+} pumping in higher organisms, so it is reasonable to assume that one or more of the ciliary ATPase activities is involved in Ca^{2+} homeostasis; at least two studies have shown that $^{45}Ca^{2+}$ is actively extruded from *Paramecium* (Browning and Nelson, 1976; Martinac and Hildebrand, 1981). Travis and Nelson (1986) noted, however, that the Ca^{2+} concentration dependence and drug resistance of the known ciliary membrane Ca-ATPase activities are inconsistent with the established properties of Ca^{2+} pumps, and pointed out that the activity of a Ca^{2+} extrusion mechanism in *Paramecium* may be below the current limits of resolution. In this regard, the recent isolation of a new group of mutants called *K-shy* may help identify a Ca^{2+} pump (Evans *et al.*, 1987). *K-shy* swims backward for abnormally long periods when depolarized, suggesting either a defective repolarization mechanism and/or an excessive buildup of Ca^{2+} within the cilium. Voltage-clamp analyses favor the latter, since the delayed rectifying K^+ current activates normally in the mutant, but the Ca^{2+}-activated currents are increased in amplitude and their residual currents upon repolarization are prolonged compared with the wild type. These observations are consistent with *K-shy* being a mutant with a defective Ca^{2+} extrusion mechanism.

3.3. Adenylate Cyclase

In multicellular organisms, cyclases often function to transduce sensory signals, so the occurrence of these enzymes in cilia of *Paramecium* is of particular interest. Adenylate cyclase (AC) is frequently reported to be associated with hormone receptors, where its activity is regulated by interaction with GTP-binding proteins that either stimulate or inhibit the synthesis of cAMP. The association of AC with the ciliary membrane of *Paramecium* was demonstrated by Schultz and Klumpp (1983). Its activity is stimulated by Ca^{2+} at submicromolar concentrations, and inhibited by Ca^{2+} concentrations of 5 μM and above (Gustin and Nelson, 1987). A similar AC activity that is also inhibited by high Ca^{2+} is found in *Tetrahymena* cilia (Kudo *et al.*, 1985). Gustin and Nelson (1987) and Schultz *et al.* (1987) reported the AC activity from *Paramecium* to be insensitive to NaF, forskolin, or guanine nucleotides, which may indicate that the isolated enzyme acts independently of GTP-binding proteins, and that it may be more closely related to the AC found in sperm (Kopf *et al.*, 1986) than that associated with hormone receptors. Surprisingly, Schultz's group (Klumpp *et al.*, 1984; Schultz *et al.*, 1987) found the enzyme to be stimulated *in vitro* by K^+, with a half-maximal effect at 3 mM.

3.4. Guanylate Cyclase

A guanylate cyclase (GC) activity is found in association with the ciliary membranes of both *P. tetraurelia* (Schultz and Klumpp, 1980), and *T. pyriformis* (Schultz *et al.*, 1983). In an earlier section of this chapter, studies localizing Ca^{2+} permeability to a low-density population of ciliary membrane vesicles from *Paramecium* were described. The same report (Thiele *et al.*, 1982) identified a high-density, multilamellar vesicle population as being relatively impermeable to Ca^{2+} but greatly enriched in GC, suggesting that Ca^{2+} channels and GC activity are spatially segregated within the ciliary membrane. In its native state, the ciliary GC is tightly bound to CaM (Klumpp *et al.*, 1983a). This association is resistant to osmotic shock, mechanical disruption, and to lowering of Ca^{2+} concentration. Addition of purified CaM fails to increase GC activity, suggesting that the majority of the enzyme exists in the associated state (Klumpp *et al.*, 1983a). As expected from this association, GC activity is strongly Ca^{2+}-dependent (Table I) and, like AC, GC is inhibited by high Ca^{2+} concentrations (300–500 μM). Adding La^{3+} to ciliary membranes causes CaM to dissociate from the catalytic subunit of GC, and the enzyme is inactivated (Schultz and Klumpp, 1982). Klumpp *et al.* (1983a; Schultz and Klumpp, 1984) tested CaM from a variety of sources for its ability to restore activity to dissociated GC, but while pig brain and soybean CaM could reactivate the enzyme, only CaM from *Paramecium* and *Tetrahymena* is able to restore activity fully. However, even when purified *Paramecium* CaM is used for reactivation, the kinetic properties and ion specificity of the reconstituted enzyme are changed, suggesting that once the regulatory subunit has been stripped from the catalytic subunit, it is difficult to reconstitute the former into the holoenzyme.

Cilia contain both cAMP- and cGMP-phosphodiesterases (PDE) (Schultz *et al.*, 1985; Gustin and Nelson, 1987). Gustin and Nelson (1987) reported two PDE activities with K_m values of ca. 1 and 5 μM in cilia from *P. tetraurelia*. Both are inhibited by isobutylmethylxanthine (IBMX) and papaverine at millimolar concentrations. A study of PDE activity in *T. pyriformis* (Kudo *et al.*, 1986) identified a soluble cGMP-PDE (K_m 2.5 μM), and a weakly membrane-associated cAMP-PDE (K_m 5 μM), both of which are partially inhibited by IBMX and theophylline.

3.5. Protein Kinases

Protein kinases (PKs) are frequently the effectors of Ca^{2+}- or cyclic nucleotide-stimulated modifications of cell activity; they are the workhorses of change. This has been well demonstrated for neurons, where increasingly there are reports of ion channels being modulated in both the short and long term by phosphorylation or dephosphorylation (see Rossie and Catterall, 1987; Schwartz and Greenberg, 1987). The discovery of PK activity in cilia from *Paramecium* is thus of particular interest, since the activity of this excitable membrane may be subject to similar regulation.

Paramecium contains a variety of PK activities, including three stimulated by cyclic nucleotides (Lewis and Nelson, 1980; Schultz and Jantzen, 1980; Mason, 1987; Miglietta and Nelson, 1988), two Ca^{2+}-dependent PKs (Gundersen and Nelson, 1987), and a casein kinase. The latter acts independently of cyclic nucleotides, Ca^{2+}, or Ca^{2+}/CaM, but has not been investigated in further detail (R. Ramanathan, R. D. Hinrichsen, and C. Kung, personal communication).

Paramecium cilia are enriched in a cGMP-dependent PK (cGPK) (Schultz and Jantzen, 1980; Miglietta and Nelson, 1988). This enzyme has a molecular mass of 77 kDa, is inhibited by micromolar concentrations of Ca^{2+}, and autophosphorylates (Miglietta and Nelson, 1988). The cGPK is unusual in that GTP and ATP serve equally well as substrates for its activity, and that it exists as a monomer: cGPKs from higher organisms are generally dimeric. Cilia contain two cAMP-dependent PKs (cAPK) (Schultz and Jantzen, 1980; Mason, 1987) that appear similar to the type I and II cAPKs from higher organisms. The activities of both are inhibited by high Ca^{2+}. cAPK II has been observed in 71- and 220-kDa forms, which may indicate that the native enzyme exists as a tetramer of two regulatory and two catalytic subunits (Mason, 1987). Similarly, cyclic nucleotide-dependent protein kinases are found in *Tetrahymena* cilia; Murofushi (1973, 1974) reported a single cAPK, and three kinases activated to varying extents by cGMP.

Cyclic nucleotide-stimulated PKs are common to many cell types and are the best understood, but generally there are two additional classes of Ca^{2+}-dependent PKs. The first class is activated by a Ca^{2+}/CaM complex, while the second functions independently of CaM but is dependent on phospholipid interaction (protein kinase C). *Paramecium* is unusual in that it does not appear to contain either class of Ca^{2+}-stimulated PK (D. L. Nelson, personal communication) but instead possesses a novel PK activity that is Ca^{2+}-dependent without being stimulated by CaM or diacylglycerol (Gundersen and Nelson, 1987).

Lewis and Nelson (1981) and Eistetter *et al.* (1983) examined endogenous substrates for PK activity in cilia from *Paramecium*. Lewis and Nelson (1981) identified 15 proteins that are reproducibly phosphorylated in whole cilia, four of which are membrane proteins. The rest, including tubulin, are axonemal. Of the four membrane proteins, the phosphorylation of three (320, 45, and 31 kDA) is specifically stimulated by cGMP. Lewis and Nelson (1981) also surveyed several mutants of *P. tetraurelia* for abnormal patterns of protein phosphorylation: cGMP-stimulated cGPK activity was found to be decreased in *PaA*. Eistetter *et al.* (1983) noted cyclic nucleotide-stimulated phosphorylation of eight proteins in cilia and ciliary fractions, with 43-, 41-, and 39-kDa proteins being located in a membrane vesicle population. The 43- and 41-kDa proteins are tentatively identified as autophosphorylating regulatory subunits of cAPK I and II, while an 85-kDa protein (also reported by Lewis and Nelson, 1981) may be the cGPK, since photoaffinity labeling of the protein with 8-azido-[³²P]-cAMP is specifically inhibited by cGMP.

Cilia from *P. tetraurelia* probably contain one or more phosphoprotein phosphatases. Lewis and Nelson (1980, 1981) reported an activity that is partially inhibited by NaF, while Klumpp *et al.* (1983b) showed that an antibody to calcineurin binds to somal membranes and cilia. Momayezi *et al.* (1986) also implicated a calcineurinlike protein in dephosphorylation of a 65-kDa polypeptide in the cell body immediately prior to trichocyst exocytosis.

4. Regulation of Cell Motility: Calcium Sensitivity

Although the enzymes described in the previous section are predominantly ciliary membrane-associated, the primary physiological target of these enzymes is axonemal function. Electron micrographs of detergent-extracted models of *Paramecium* show that while the integrity of the axoneme is preserved, these cells lack all surface membranes (Lieberman *et al.*, 1988). Yet these membrane- and soluble ciliary matrix-free axonemes can be reactivated to beat, and in the presence of >1 μM Ca^{2+}, they reverse their beating direction, suggesting that the basic mechanisms for ciliary motility are intrinsic to the axonemal structure. The membrane-associated and soluble enzyme activities may therefore be involved in the dynamic regulation of ciliary beat frequency and direction. Efforts directed toward understanding how this regulation is achieved, and how the enzyme activities involved in this regulation are themselves regulated by membrane potential changes, have focused on both intact cells and, increasingly, on detergent-extracted models. However, in interpreting these data, it should be remembered that there is in all likelihood a compartmentalization of ion fluxes and biochemical events within the intact cell. When these compartments are broken down during detergent treatment, the axoneme becomes accessible to factors that it may not otherwise encounter. A detailed discussion of the intricacies of regulation of ciliary motility is beyond the scope of this review, and the interested reader is referred to Bonini et al. (1989).

Two parameters of ciliary activity are under membrane control. Early studies by Machemer (1974) established that membrane hyperpolarization causes both an increase in ciliary beat frequency, and a shift in the direction of the effective power stroke toward the posterior end of the cell. This drives the cell forward at an increased speed. Membrane depolarization, however, causes an initial slowing of ciliary beat, a reorientation of the effective power stroke toward the anterior of the cell, and then increased, reversed ciliary beating frequency. Thus, changes in membrane potential cause changes in both ciliary beat frequency and beat direction. As discussed below, these events may be biochemically distinct.

4.1. Ciliary Responses to Hyperpolarization

It has long been known that hyperpolarization of *Paramecium* results from a K^+ efflux (Naitoh and Eckert, 1973), but the means by which the electrical response is transduced to effect increased ciliary beat frequency has only recently been investigated.

The membrane potential of *Paramecium* is sensitive to changes in external ion concentration, so it is possible to hyperpolarize a population of cells by transferring them from a medium of high cation concentration to a medium of low cation concentration. This transfer can elicit up to a threefold increase in intracellular cAMP concentration, along with an increase in forward swimming speed (Bonini et al., 1986). When cAMP is applied to permeabilized models (lacking either membrane or membrane potential), the cells respond with faster forward swimming (Nakaoka and Ooi, 1985; Bonini et al., 1986; Bonini and Nelson, 1988), suggesting that cAMP augments ciliary activity by direct interaction with the axoneme. These effects can be reproduced in live cells by injecting cAMP (Hennessey et al., 1985), or by incubating live paramecia with membrane-permeant derivatives of cAMP, such as N^6-monobutyryl cAMP and N^6-benzoyl cAMP, or with the PDE inhibitor IBMX (Bonini et al., 1986). All three drugs increase swimming speed after varying time lags, while cGMP and other cAMP analogues are without effect (Bonini et al., 1986). Taken together, these observations suggest that membrane hyperpolarization stimulates AC activity to raise intraciliary cAMP concentration, which in turn causes augmented ciliary beat frequency. This conclusion is apparently at odds with the findings of Schultz et al. (1984b) that cAMP levels are increased upon Ca^{2+}-induced *depolarization* of *P. tetraurelia*. The observations of Bonini et al. (1986) and Schultz et al. (1984b) may be reconciled by the fact that divalent cations are very effective in screening negative charges on the outer surface of the membrane, Ca^{2+} particularly so. Thus, upon addition of Ca^{2+} to the medium, one may expect a net *hyperpolarization* of the true potential difference that exists across the immediate inner and outer surfaces of the lipid bilayer (this potential difference cannot be measured using conventional microelectrode techniques; for a more detailed explanation, see Eckert and Brehm, 1979; Bonini et al., 1986; Machemer, 1988). This potential change is thought to trigger increased AC activity (Bonini *et al.*, 1986) and faster forward swimming (Nakaoka *et al.*, 1983). While the above points toward a role for cAMP as a second messenger of membrane hyperpolarization in *Paramecium*, it is possible the nucleotide also stimulates ion fluxes directly. Hennessey *et al.* (1985) noted that injection of cAMP into live paramecia elicits a 5- to 7-mV hyperpolarization: similar cAMP-dependent ion conductances have been demonstrated in a variety of organisms (see Hockberger and Swandulla, 1987).

What role, if any, does Ca^{2+} have in the ciliary response to hyperpolarization? Brehm and Eckert (1978a) showed that increased ciliary beating upon hyperpolarization occurs independently of a rise in intraciliary Ca^{2+} levels, consistent with the idea that AC activity in the ciliary membrane is stimulated by hyperpolarization directly, or by some other Ca^{2+}-independent mechanism.

Finally, it should be noted that increases in forward swimming speed can result from increased ciliary beat frequency and/or from a shift in the effective ciliary power stroke toward the posterior end of the cell. In the absence of a detailed analysis of ciliary movement, it is not possible to determine which of these mechanisms is responsible for changes in forward swimming speed. This may explain the conflicting observations that cAMP causes faster forward swimming of models (Nakaoka and Ooi, 1985; Bonini *et al.*, 1986), but that injecting cAMP into voltage-clamped cells is without effect on ciliary beat frequency (Hennessey *et al.*, 1985); cAMP may cause ciliary reorientation without a frequency change.

The axonemal target of cAMP has not been identified with certainty. Since one result of hyperpolarization is increased ciliary beat frequency, dyneins must ultimately be involved, and indeed, cAMP stimulates vanadate-sensitive ATPase activity in models (Bonini, 1987), which is consistent with cAMP stimulation of dynein activity. It is likely that the effects of cAMP are mediated by PKs that phosphorylate dyneins or dynein-associated regulatory proteins: cAMP stimulates the phosphorylation by endogenous kinases of proteins that sediment with dyneins from permeabilized cells (Bonini, 1987) and isolated cilia (Travis and Nelson, 1988).

Evidence suggests that extreme posteriorly directed ciliary orientation during hyperpolarization may also be effected by phosphorylation (Nakaoka and Ooi, 1985; Majima *et al.*, 1986; Bonini, 1987). Nakaoka and Ooi (1985) found that cAMP antagonizes Ca^{2+}-induced ciliary reversal in models and thereby renormalizes ciliary beat orientation. Addition of a cAPK inhibitor to the reactivation medium prevents cAMP from overriding the reversal response, suggesting that cAMP-dependent protein phosphorylation is required for posteriorly directed ciliary beating.

4.2. Ciliary Responses to Depolarization

The ciliary response to depolarization is more complex. When the Ca^{2+} channels open, there is a rapid increase in intraciliary Ca^{2+} concentration from about 0.2 μM to 40 μM or above (Brehm and Eckert, 1978b). The Ca^{2+}-dependence of AC suggests that its activity would be inhibited under these conditions (Table I), and indeed, there is no detectable increase in cAMP during ciliary reversal (Bonini *et al.*, 1986). Depolarization of intact cells is accompanied by small increases in cGMP concentration (Majima *et al.*, 1986; Schultz *et al.*, 1986), but only after a lag of up to 10 sec (Majima *et al.*, 1986). Since reversal of the ciliary power stroke occurs within 4 msec of peak membrane depolarization (Machemer and Eckert, 1973), it is unlikely that GC is involved in this response. It is possible, therefore, that Ca^{2+} acts on the reversal mechanism directly. CaM antagonists such as trifluoperazine block backward swimming in living cells (Otter *et al.*, 1984) and models (Otter *et al.*, 1984; Izumi and Nakaoka, 1987), suggesting that CaM may be the Ca^{2+} sensor for ciliary reversal.

Since phosphorylation reactions have been implicated in biasing ciliary orientation during hyperpolarization, it is possible that reversal of the ciliary power stroke may also involve activation of protein kinases. Presumably such activity would be Ca^{2+}- or Ca^{2+}/CaM-dependent, since there is no increase in cAMP concentration during reversal, and cGMP levels rise only after an appreciable time lag. Evidence in support of kinase involvement in ciliary reversal was provided by Nakaoka and Ooi (1985), who observed that models reactivated with ATPγS respond to Ca^{2+} with ciliary reversal, but the cilia are found to be locked in this position upon Ca^{2+} removal. ATPγS is used in a phosphatase-resistant thiophosphorylation of proteins, suggesting that dephosphorylation is required for renormalization of ciliary orientation. The precise function of phosphorylation in ciliary reversal is not clear, however. Bonini and Nelson (1988) showed four major proteins to be phosphorylated during Ca^{2+}-induced ciliary reversal in models, but not during Ba^{2+}- or Sr^{2+}-induced reversal. This observation is consistent with the known insensitivity of the characterized CaPK to Ba^{2+} and Sr^{2+} (Gundersen and Nelson, 1987). Thus, either these four proteins are not involved in reversal, or the reversal mechanism operates independently of PK activity (Bonini, 1987).

What is the role of cGMP? The facts that (1) the ciliary GC requires comparatively high concentrations of Ca^{2+} for activation (Table I), (2) elevation of intraciliary cGMP levels occurs with a latency of some seconds during prolonged depolarization (Majima *et al.*, 1986), and (3) cGMP antagonizes Ca^{2+}-induced ciliary reversal (Majima *et al.*, 1986), suggest that the cyclic nucleotide may be involved in the renormalization of ciliary orientation during lengthy bouts of backward swimming.

5. Perspectives

It is clear from this review that the ciliary membrane is a complex assembly, mediating a variety of interdependent electrical and biochemical events (summarized in Fig. 6). These events ultimately govern a cell's response to a stimulus, either fast forward swimming, or backing away from an obstacle in its swimming path.

With the discovery of at least five kinase activities in *Paramecium* cilia, it seems unlikely that the axoneme is the sole target of their actions. Increasingly, there are reports of ion channels in higher organisms that can be opened by interaction with GTP binding proteins (reviewed by Neer and Clapham, 1988) or cyclic nucleotides directly (reviewed by Hockberger and Swandulla, 1987), or ion channels whose activity is modified by Ca^{2+}- and cyclic nucleotide-stimulated phosphorylation (see Chad et al., 1987; Levitan, 1988). It is possible, therefore, that the activity of the ion channels within the ciliary membrane of *Paramecium* may also be regulated by covalent modification.

The ability to generate behavioral mutants has greatly facilitated dissection of the complexities of membrane excitation and motility regulation in *Paramecium*. Genetic manipulation is a powerful tool in investigations of this nature. The ability to delete

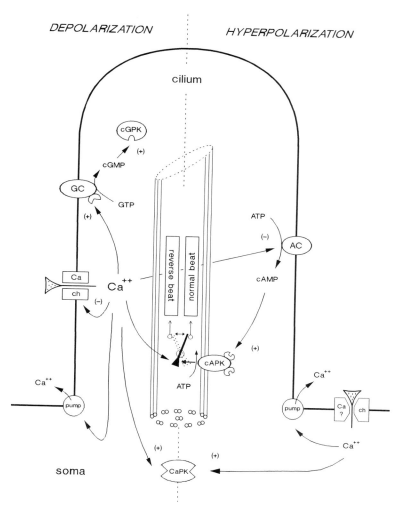

Figure 6. Actions and interactions of Ca²⁺ and enzyme activities in cilia. This diagram represents a cilium and its responses to membrane depolarization (left side) and membrane hyperpolarization (right side). *Upon depolarization,* voltage-sensitive Ca²⁺ channels open to admit Ca²⁺ into the soluble ciliary matrix. Ca²⁺ activates a ciliary reversal mechanism, whose nature is unknown, and the cell swims backward. The Ca²⁺ channels inactivate in response to Ca²⁺, which is presumed then to be sequestered and eventually extruded from the cell. This pathway is represented in the figure by a pump, but the entity responsible and its location within the cell have not been determined. Ca²⁺ stimulates the activity of a Ca²⁺-dependent protein kinase (CaPK) and guanylate cyclase (GC) directly; the resulting increase in intraciliary cGMP concentration activates a cGMP-dependent protein kinase (cGPK). The molecular targets of the CaPK and cGPK are unknown. *Membrane hyperpolarization* is accompanied by the activation of a membrane-associated adenylate cyclase (AC) with a resulting increase in intraciliary cAMP concentration. At present, it is not clear as to whether this increase causes or results from the membrane potential change. cAMP activates a cAMP-dependent protein kinase (cAPK) to cause increased ciliary beat frequency, possibly through phosphorylation of dynein. cAMP, or perhaps phosphorylation, also favors posteriorly directed ciliary beat, antagonizing the reversal mechanism. *Paramecium* contains a Ca²⁺ channel that is activated by hyperpolarization, but the channel and the associated Ca²⁺ flux may be restricted to the cell soma.

specific components of a sensory transduction or regulatory pathway is limited only by the imagination in devising screening methods for a specific mutant; it is likely that mutants will continue to feature prominently in future studies of membrane excitation and its modulation in *Paramecium*.

ACKNOWLEDGMENTS. We thank our colleagues, Drs. N. M. Bonini, T. C. Evans, M. Gustin, C. Kung, and D. L. Nelson, for their help in critically reviewing the manuscript. We are particularly grateful to Dr. David Nelson and members of his laboratory for many enlightening discussions and for providing us with copies of manuscripts in advance of their publication. Supported by NIH Grants GM 22714 and GM 36386.

References

Adoutte, A., Ramanathan, R., Lewis, R. M., Dute, R. R., Ling, K.-Y, Kung, C., and Nelson, D. L., 1980, Biochemical studies of the excitable membrane of *Paramecium tetraurelia*. III. Proteins of cilia and ciliary membranes, *J. Cell Biol.* **84**: 717–738.

Adoutte, A., Ling, K.-Y, Chang, S-Y., Huang, F., and Kung, C., 1983, Physiological and mutational protein variations in the ciliary membrane of *Paramecium, Exp. Cell Res.* **148**:387–404.

Andrivon, C., Brugerolle, G., and Delachambre, D., 1983, A specific Ca^{2+}-ATPase in the ciliary membrane of *Paramecium tetraurelia, Biol. Cell.* **47**:351–364.

Bardele, C. F., 1981, Functional and phylogenetic aspects of the ciliary membrane: A comparative freeze-fracture study, *BioSystems* **14**:403–421.

Berger, J. D., 1976, Gene expression and phenotypic change in *Paramecium tetraurelia* exconjugants, *Genet. Res.* **27**:123–134.

Boheim, G., Hanke, W., Methfessel, C., Eibl, H., Kaupp, U. B., Maelicke, A., and Schultz, J. E., 1982, Membrane reconstitution below lipid phase transition temperature, in: *Transport in Biomembranes: Model Systems and Reconstitution* (R. Antolini, A. Gliozzi, and A. Gorio, eds.), Raven Press, New York, pp. 87–97.

Bonini, N. M., 1987, Regulation of ciliary motility in *Paramecium* by cyclic AMP and cyclic GMP: Behavioral and biochemical studies, Ph. D. dissertation, University of Wisconsin, Madison.

Bonini, N. M., and Nelson, D. L., 1988, Differential regulation of *Paramecium* ciliary motility by cAMP and cGMP, *J. Cell Biol.* **106**:1615–1623.

Bonini, N. M., Gustin, M. C., and Nelson, D. L., 1986, Regulation of ciliary motility in *Paramecium*: A role for cyclic AMP, *Cell Motil. Cytoskel.* **6**:256–272.

Bonini, N. M., Evans, T. C., Miglietta, L. A. P., and Nelson, D. L., 1989, The regulation of ciliary motility in *Paramecium* by Ca^{2+} and cyclic nucleotides, *Adv. Second Messenger Phosphoprotein Res.* (in press).

Brehm, P., and Eckert, R., 1978a, An electrophysiological study of the regulation of ciliary beating frequency in *Paramecium, J. Physiol. (London)* **283**:557–568.

Brehm, P., and Eckert, R., 1978b, Calcium entry leads to inactivation of calcium channel in *Paramecium, Science* **202**:1203–1206.

Brehm, P., Eckert, R., and Tillotson, D., 1980, Calcium-mediated inactivation of calcium current in *Paramecium, J. Physiol. (London)* **306**:193–203.

Browning, J. L., and Nelson, D. L., 1976, Biochemical studies of the excitable membrane of *Paramecium aurelia*. I. $^{45}Ca^{2+}$ fluxes across the resting and excited membrane. *Biochim. Biophys. Acta* **448**:338–351.

Brugerolle, G., Andrivon, C., and Bohatier, J., 1980, Isolation, protein pattern and enzymatic characterization of the ciliary membrane of *Paramecium tetraurelia, Biol. Cell.* **37**:251–260.

Byrne, B. J., and Byrne, B. C., 1978a, An ultrastructural correlate of the membrane mutant 'paranoiac' in *Paramecium, Science* **199**:1091–1093.

Byrne, B. J., and Byrne, B. C., 1978b, Behavior and the excitable membrane in *Paramecium, Crit. Rev. Microbiol.* **6**:53–108.

Byrne, B. J., Tanner, A. P., and Dietz, P. M., 1988, Phenotypic characterization of paranoiac and related mutants in *Paramecium tetraurelia, Genetics* **118**:619–626.

Chad, J. E., Eckert, R., and Ewald, D., 1984, Kinetics of Ca-dependent inactivation of calcium current in neurones of *Aplysia californica, J. Physiol. (London)* **347**:279–300.

Chad, J., Kalman, D., and Armstrong, D., 1987, The role of cyclic AMP-dependent phosphorylation in the maintenance and modulation of voltage-activated calcium channels, in: *Cell Calcium and the Control of Membrane Transport* (L. J. Mandel and D. C. Eaton, eds.), Rockefeller University Press, New York, pp. 153–186.

Connolly, J. G., and Kerkut, G. A., 1981, The membrane potentials of *Tetrahymena vorax, Comp. Biochem. Physiol.* **69C**:265–273.

Connolly, J. G., and Kerkut, G. A., 1983, Ion regulation and membrane potential in *Tetrahymena* and *Paramecium, Comp. Biochem. Physiol.* **76A**:1–16.

Connolly, J. G., and Kerkut, G. A., 1984, An electrogenic component of the membrane potential of *Tetrahymena, Comp. Biochem. Physiol.* **77A**:335–344.

Doughty, M. J., 1978, Ciliary Ca^{2+}-ATPase from the excitable membrane of *Paramecium*. Some properties and purification by affinity chromatography, *Comp. Biochem. Physiol.* **60B**:339–345.

Doughty, M. J., and Kaneshiro, E. S., 1983, Divalent cation-dependent ATPase associated with cilia and other subcellular fractions of *Paramecium*. An electrophoretic characterization on Triton-polyacrylamide gels, *J. Protozool.* **30**:565–573.

Doughty, M. J., and Kaneshiro, E. S., 1985, Divalent cation-dependent ATPase activities in ciliary membranes and other surface structures in *Paramecium tetraurelia:* Comparative in vitro studies, *Arch. Biochem. Biophys.* **238**:118–128.

Dunlap, K., 1977, Localization of calcium channels in *Paramecium caudatum, J. Physiol. (London)* **271**:119–133.

Dute, R., and Kung, C., 1978, Ultrastructure of the proximal region of somatic cilia in *Paramecium tetraurelia, J. Cell Biol.* **78**:451–464.

Eckert, R., and Brehm, P., 1979, Ionic mechanisms of excitation in *Paramecium, Annu. Rev. Biophys. Bioeng.* **8**:353–383.

Eckert, R., and Chad, J. E., 1984, Inactivation of Ca channels, *Prog. Biophys. Mol. Biol.* **44**:215–267.

Eckert, R., Tillotson, D. L., and Brehm, P., 1981, Calcium-mediated control of Ca and K currents, *Fed. Proc.* **40**:2226–2232.

Ehrlich, B. E., Finkelstein, A., Forte, M., and Kung, C., 1984, Voltage-dependent calcium channels from *Paramecium* cilia incorporated into planar lipid bilayers, *Science* **225**:427–428.

Ehrlich, B. E., Jacobson, A. R., Hinrichsen, R. D., Sayre, L. M., and Forte, M. A., 1988, *Paramecium* calcium channels are blocked by a family of calmodulin antagonists, *Proc. Natl. Acad. Sci. USA* **85**:5718–5722.

Eisenbach, L., Ramanathan, R., and Nelson, D. L., 1983, Biochemical studies of the excitable membrane of *Paramecium tetraurelia*. IX. Antibodies against ciliary membrane proteins, *J. Cell Biol.* **97**:1412–1420.

Eistetter, H., Seckler, B., Bryniok, D., and Schultz, J. E., 1983, Phosphorylation of endogenous proteins of cilia from *Paramecium tetraurelia* in vitro, *Eur. J. Cell Biol.* **31**:220–226.

Evans, T. C., and Nelson, D. L., 1989, The cilia of *Paramecium* contain both Ca^{2+}-dependent and Ca^{2+}-inhibitable calmodulin binding proteins, *Biochem. J.* **259**:385–396.

Evans, T. C., Hennessey, T., and Nelson, D. L., 1987, Electrophysiological evidence suggests a defective Ca^{2+} control mechanism in a new *Paramecium* mutant, *J. Membr. Biol.* **98**:275–283.

Fox, A. P., 1981, Voltage-dependent inactivation of a calcium channel, *Proc. Natl. Acad. Sci. USA* **78**:953–956.

Gundersen, R. E., and Nelson, D. L., 1987, A novel Ca^{2+}-dependent protein kinase from *Paramecium tetraurelia, J. Biol. Chem.* **262**:4602–4609.

Gustin, M. C., and Nelson, D. L., 1987, Regulation of ciliary adenylate cyclase by Ca^{2+} in *Paramecium, Biochem. J.* **246**:337–345.

Haga, N., and Hiwatashi, K., 1982, A soluble gene product controlling membrane excitability in *Paramecium, Cell Biol. Int. Rep.* **6**:295–300.

Haga, N., Forte, M., Saimi, Y., and Kung, C., 1982, Microinjection as a test of complementation in *Paramecium, J. Cell Biol.* **82**:559–564.

Haga, N., Saimi, Y., Takahashi, M., and Kung, C., 1983, Intra- and interspecific complementation of membrane inexcitable mutants of *Paramecium, J. Cell Biol.* **97**:378–382.

Haga, N., Forte, M., Ramanathan, R., Hennessey, T., Takahashi, M., and Kung, C., 1984a, Characterization and purification of a soluble protein controlling Ca-channel activity in *Paramecium, Cell* **39:**71–78.

Haga, N., Forte, M., Ramanathan, R., Saimi, Y., Takahashi, M., and Kung, C., 1984b, Purification of a soluble protein controlling Ca^{2+} channel activity in *Paramecium, Biophys. J.* **45:**130–132.

Haga N., Forte, M., Saimi, Y. and Kung, C., 1984c, Characterization of cytoplasmic factors which complement Ca^{2+} channel mutations in *Paramecium tetraurelia, J., Neurogenet.* **1:**259–274.

Hamasaki, T., and Naitoh, Y., 1985, Localization of calcium-sensitive reversal mechanism in a cilium of *Paramecium, Proc. Jpn. Acad. Ser. B* **61:**140–143.

Hanke, W., Eibl, H., and Boheim, G., 1981, A new method for membrane reconstitution: Fusion of protein-containing lipids with planar lipid bilayers below lipid phase transition temperature, *Biophys. Struct. Mech.* **7:**131–137.

Hansma, H. G., and Kung, C., 1975, Studies of the cell surface of *Paramecium, Biochem. J.* **152:**523–528.

Hennessey, T. M., 1987, A novel calcium current is activated by hyperpolarization in *Paramecium tetraurelia, Soc. Neurosci. Abstr.* **13:**108.

Hennessey, T. M., and Kung, C., 1984, An anticalmodulin drug, W-7, inhibits the voltage-dependent calcium current in *Paramecium caudatum, J. Exp. Biol.* **110:**169–181.

Hennessey, T. M., and Kung, C., 1985, Slow inactivation of the calcium current of *Paramecium* is dependent on voltage and not internal calcium, *J. Physiol. (London)* **365:**165–179.

Hennessey, T. M., Machemer, H., and Nelson, D. L., 1985, Injected cyclic AMP increases ciliary beat frequency in conjunction with membrane hyperpolarization, *Eur. J. Cell Biol.* **36:**153–156.

Hinrichsen, R. D., and Kung, C., 1984, Genetic analysis of axonemal mutants in *Paramecium tetraurelia* defective in their response to calcium, *Genet. Res.* **43:**11–20.

Hinrichsen, R. D., and Saimi, Y., 1984, A mutation that alters properties of the calcium channel in *Paramecium tetraurelia, J. Physiol. (London)* **351:**397–410.

Hinrichsen, R. D., Saimi, Y., Hennessey, T., and Kung, C., 1984a, Mutants in *Paramecium tetraurelia* defective in their axonemal response to calcium, *Cell Motil.* **4:**283–295.

Hinrichsen, R. D., Saimi, Y., and Kung, C., 1984b, Mutants with altered Ca^{2+}-channel properties in *Paramecium tetraurelia:* Isolation, characterization and genetic analysis, *Genetics* **108:**545–558.

Hirano, J., and Watanabe, Y., 1985, Studies on calmodulin-binding proteins (CaMBPs) in the cilia of *Tetrahymena, Exp. Cell Res.* **157:**441–450.

Hiwatashi, K., Haga, N., and Takahashi, M., 1980, Restoration of membrane excitability in a behavioral mutant of *Paramecium caudatum* during conjugation and by microinjection of wild-type cytoplasm, *J. Cell Biol.* **84:**476–480.

Hockberger, P. E., and Swandulla, D., 1987, Direct ion channel gating: A new function for intracellular messengers, *Cell. Mol. Neurobiol.* **7:**229–236.

Izumi, A., and Nakaoka, Y., 1987, cAMP-mediated inhibitory effect of calmodulin antagonists on ciliary reversal of *Paramecium, Cell Motil. Cytoskel.* **7:**154–159.

Jennings, H. S., 1906, *Behavior of the Lower Organisms,* Columbia University Press, New York.

Kass, R. S., and Scheuer, T., 1982, Slow inactivation of calcium channels in the cardiac Purkinje fiber, *J. Mol. Cell. Cardiol.* **14:**615–618.

Klumpp, S., and Schultz, J. E., 1982, Characterization of a Ca^{2+}-dependent guanylate cyclase in the excitable ciliary membrane from *Paramecium, Eur. J. Biochem.* **124:**317–324.

Klumpp, S., Kleefeld, G., and Schultz, J. E., 1983a, Calcium/calmodulin-regulated guanylate cyclase of the excitable ciliary membrane from *Paramecium.* Dissociation of calmodulin by La^{3+}: Calmodulin specificity and properties of the reconstituted guanylate cyclase, *J. Biol. Chem.* **248:**12455–12459.

Klumpp, S., Steiner, A. L., and Schultz, J. E., 1983b, Immunocytochemical localization of cyclic GMP, cGMP-dependent protein kinase, calmodulin and calcineurin in *Paramecium tetraurelia, Eur. J. Cell Biol.* **32:**164–170.

Klumpp, S., Gierlich, D., and Schultz, J. E., 1984, Adenylate cyclase and guanylate cyclase in the excitable ciliary membrane from *Paramecium:* Separation and regulation, *FEBS Lett.* **171:**95–99.

Kopf, G. S., Woolkaiis, M. J., and Gerton, G. L., 1986, Evidence for a guanine nucleotide-binding regulatory protein in invertebrate and mammalian sperm, *J. Biol. Chem.* **261:**7327–7331.

Kudo, S., Muto, Y., and Nozawa, Y., 1985, Regulation by calcium of hormone-insensitive adenylate cyclase and calmodulin-dependent guanylate cyclase in *Tetrahymena* plasma membrane, *Comp. Biochem. Physiol.* **80B:**813–816.

Kudo, S., Nagao, S., Muto, Y., Takahashi, M., and Nozawa, Y., 1986, Characterization of cyclic AMP and cyclic GMP phosphodiesterases in *Tetrahymena* cilia, *Comp. Biochem. Physiol.* **83B**:99–102.

Kung, C., 1971a, Genic mutants with altered system of excitation in *Paramecium aurelia*. I. Phenotypes of the behavioral mutants, *Z. Vgl. Physiol.* **71**:142–164.

Kung, C., 1971b, Genic mutants with altered system of excitation in *Paramecium aurelia*. II. Mutagenesis, screening and genetic analysis of the mutants, *Genetics* **69**:29–45.

Kung, C., 1979, Biology and genetics of *Paramecium* behavior, in: *Neurogenetics: Genetic Approaches to the Nervous System* (X. O. Breakefield, ed.), Elsevier, Amsterdam, pp. 1–26.

Kung, C., and Saimi, Y., 1982, The physiological basis of taxes in *Paramecium*, *Annu. Rev. Physiol.* **44**:519–534.

Kung, C., Chang, S.-Y., Satow, Y., Van Houten, J., and Hansma, H., 1975, Genetic dissection of behavior in *Paramecium*, *Science* **188**:898–904.

Levitan, I. B., 1988, Modulation of ion channels in neurons and other cells, *Annu. Rev. Neurosci.* **11**:119–136.

Lewis, R. M., and Nelson, D. L., 1980, Biochemical studies of the excitable membrane of *Paramecium*. IV. Protein kinase activities of cilia and ciliary membrane, *Biochim. Biophys. Acta* **615**:341–353.

Lewis, R. M., and Nelson, D. L., 1981, Biochemical studies of the excitable membrane of *Paramecium tetraurelia*. VI. Endogenous protein substrates for in vitro and in vivo phosphorylation in cilia and ciliary membranes, *J. Cell Biol.* **91**:167–174.

Lieberman, S. J., Hamasaki, T., and Satir, P., 1988, Ultrastructure and motion analysis of permeabilized *Paramecium* capable of motility and regulation of motility, *Cell Motil. Cytoskel.* **9**:73–84.

Machemer, H., 1974, Frequency and directional responses of cilia to membrane potential changes in *Paramecium*, *J. Comp. Physiol.* **92**:293–316.

Machemer, H., 1988, Electrophysiology, in: *Paramecium* (H.-D. Görtz, ed.), Springer-Verlag, Berlin, pp. 185–215.

Machemer, H., and Deitmer, J. W., 1987, From structure to behaviour: *Stylonychia* as a model system for cellular physiology, *Prog. Protistol.* **2**:213–330.

Machemer, H., and Eckert, R., 1973, Electrophysiological control of reversed ciliary beating in *Paramecium*, *J. Gen. Physiol.* **61**:572–587.

Machemer, H., and Ogura, A., 1979, Ionic conductances of membranes in ciliated and deciliated *Paramecium*, *J. Physiol. (London)* **296**:49–60.

Maihle, N. J., Dedman, J. R., Means, A. R., Chafouleas, J. G., and Satir, B. H., 1981, Presence and indirect immunofluorescent localization of calmodulin in *Paramecium tetraurelia*, *J. Cell Biol.* **89**:695–699.

Majima, T., Hamasaki, T., and Arai, T., 1986, Increase in cellular cyclic GMP level by potassium stimulation and its relation to ciliary orientation in *Paramecium*, *Experientia* **42**:62–64.

Martinac, B., and Hildebrand, E., 1981, Electrically induced Ca^{2+} transport across the membrane of *Paramecium caudatum* measured by means of flow-through technique, *Biochim. Biophys. Acta* **649**:244–252.

Martinac, B., Saimi, Y., Gustin, M. C., and Kung, C., 1986, Single-channel recording in *Paramecium*, *Biophys. J.* **49**:167a.

Martinac, B., Saimi, Y., Gustin, M. C., and Kung, C., 1988, Ion channels of three microbes: *Paramecium*, yeast and *Escherichia coli*, in: *Calcium and Ion Channel Modulation* (A. D. Grinnell, D. Armstrong, and M. B. Jackson, eds.), Plenum Press, New York, pp. 415–430.

Mason, P. A., 1987, Purification and properties of the cyclic AMP-dependent protein kinases of *Paramecium tetraurelia*, Ph.D. dissertation, University of Wisconsin, Madison.

Merkel, S. J., Kaneshiro, E. S., and Gruenstein, E. I., 1981, Characterization of the cilia and ciliary membrane proteins of wild-type *Paramecium tetraurelia* and a pawn mutant, *J. Cell Biol.* **89**:206–215.

Miglietta, L. A. P., and Nelson, D. L., 1988, A novel cGMP-dependent protein kinase from *Paramecium*, *J. Biol. Chem.* **263**:16096–16105.

Momayezi, M., Kersken, H., Gras, U., Vilmart-Seuwen, J., and Plattner, H., 1986, Calmodulin in *Paramecium tetraurelia*: Localization from the in vivo to the ultrastructural level, *J. Histochem. Cytochem.* **34**:1621–1638.

Moss, A. G., and Tamm, S. L., 1987, A calcium regenerative potential controlling ciliary reversal is propagated along the length of ctenophore comb plates, *Proc. Natl. Acad. Sci. USA* **84**:6476–6480.

Murofushi, H., 1973, Purification and characterization of a protein kinase in *Tetrahymena* cilia, *Biochim. Biophys. Acta* **327**:354–364.

Murofushi, H., 1974, Protein kinases in *Tetrahymena* cilia. II. Partial purification and characterization of adenosine 3',5'-monophosphate-dependent and guanosine 3',5'-monophosphate-dependent protein kinases, *Biochim. Biophys. Acta* **370:**130–139.

Naitoh, Y., 1974, Bioelectric basis of behavior in protozoa, *Am. Zool.* **14:**883–893.

Naitoh, Y., and Eckert, R., 1973, Sensory mechanisms in *Paramecium.* II. Ionic basis of the hyperpolarizing mechanoreceptor potential, *J. Exp. Biol.* **59:**53–65.

Naitoh, Y., and Kaneko, H., 1972, ATP-Mg^{2+} reactivated Triton-extracted models of *Paramecium:* Modification of ciliary movement by calcium ions, *Science* **176:**523–524.

Naitoh, Y., and Kaneko, H., 1973, Control of ciliary activities by adenosine triphosphate and divalent cations in Triton-extracted models of *Paramecium caudatum, J. Exp. Biol.* **58:**657–676.

Nakaoka, Y., and Ooi, H., 1985, Regulation of ciliary reversal in Triton-extracted *Paramecium* by calcium and cyclic adenosine monophosphate, *J. Cell Sci.* **77:**185–195.

Nakaoka, Y., Oka, T., Serizawa, K., Toyotama, H., and Oosawa, F., 1983, Acceleration of *Paramecium* swimming velocity is effected by various cations, *Cell Struct. Funct.* **8:**77–84.

Neer, E. J., and Clapham, D. E., 1988, Roles of G protein subunits in transmembrane signalling, *Nature* **333:** 129–134.

Nock, A. H., Kretschmar, M., Lipps, H.-J., and Schultz, J. E., 1982, Restoration of membrane excitability by microinjection of cytoplasmic wild-type RNA into *Paramecium tetraurelia pawn C* mutants, *FEMS Microbiol. Lett.* **13:**275–277.

Oertel, D., Schein, S. J., and Kung, C., 1977, Separation of membrane currents using a *Paramecium* mutant, *Nature* **268:**120–124.

Oertel, D., Schein, S. J., and Kung, C., 1978, A potassium conductance activated by hyperpolarization in *Paramecium, J. Membr. Biol.* **43:**169–185.

Ogura, A., and Takahashi, K., 1976, Artificial deciliation causes loss of calcium dependent responses in *Paramecium, Nature* **264:**170–172.

Ohnishi, K., and Watanabe, Y., 1983, Purification and some properties of a new Ca^{2+} binding protein (TCBP-10) present in *Tetrahymena* cilium, *J. Biol. Chem.* **258:**13978–13985.

Ohnishi, K., Suzuki, Y., and Watanabe, Y., 1982, Studies on calmodulin isolated from *Tetrahymena* cilia and its localization within the cilium, *Exp. Cell Res.* **137:**217–227.

Onimaru, H., Ohki, K., Nozawa, Y., and Naitoh, Y., 1980, Electrical properties of *Tetrahymena,* a suitable tool for studies of membrane excitation, *Proc. Jpn. Acad. Ser. B* **56:**538–543.

Oosawa, Y., and Sokabe, M., 1985, Cation channels from *Tetrahymena* cilia incorporated into planar lipid bilayers, *Am. J. Physiol.* **249:**C177–C179.

Oosawa, Y., Sokabe, M., and Kasai, M., 1988, A cation channel for K$^+$ and Ca^{2+} from *Tetrahymena* cilia in planar lipid bilayers, *Cell Struct. Funct.* **13:**51–60.

Otter, T., Satir, B. H., and Satir, P., 1984, Trifluoperazine-induced changes in swimming behavior of *Paramecium:* Evidence for two sites of drug action, *Cell Motil.* **4:**249–267.

Plattner, H., 1975, Ciliary granule plaques: Membrane-intercalated particle aggregates associated with Ca^{2+}-binding-sites in *Paramecium, J. Cell Sci.* **18:**257–269.

Preer, J. R., Jr., 1986, Surface antigens of *Paramecium,* in: *The Molecular Biology of Ciliated Protozoa* (J. G. Gall, ed.), Academic Press, New York, pp. 301–339.

Preston, R. R., and Newman, T. M., 1986, Rectilinear particle arrays in freeze-fracture replicas of the surface membrane of *Paramecium tetraurelia, J. Cell Sci.* **83:**269–291.

Preston, R. R., and Usherwood, P. N. R., 1988, Characterization of a specific L-[^{3}H]glutamic acid binding site on cilia isolated from *Paramecium tetraurelia, J. Comp. Physiol. B* **158:**345–352.

Ramanathan, R., Adoutte, A., and Dute, R. R., 1981, Biochemical studies of the excitable membrane of *Paramecium tetraurelia.* V. Effects of proteases on the ciliary membrane, *Biochim. Biophys. Acta* **641:** 349–365.

Ramanathan, R., Saimi, Y., Peterson, J. B., Nelson, D. L., and Kung, C., 1983, Antibodies to the ciliary membrane of *Paramecium tetraurelia* alter membrane excitability, *J. Cell Biol.* **97:**1421–1428.

Rasmussen, H., and Barrett, P. Q., 1984, Calcium messenger system: An integrated view, *Physiol. Rev.* **64:** 938–983.

Rauh, J. J., and Nelson, D. L., 1981, Calmodulin is a major component of extruded trichocysts from *Paramecium tetraurelia, J. Cell Biol.* **91:**860–865.

Richard, E. A., Saimi, Y., and Kung, C., 1986, A mutation that increases a novel calcium-activated potassium conductance of *Paramecium tetraurelia, J. Membr. Biol.* **91:**173–181.

Riddle, L. M., Rauh, J. J., and Nelson, D. L., 1982, A Ca^{2+}-activated ATPase specifically released by Ca^{2+} shock from *Paramecium tetraurelia, Biochim. Biophys. Acta* **688:**525–540.

Rossie, S., and Catterall, W. A., 1987, Regulation of ionic channels, in: *The Enzymes,* 3rd ed., Volume 18B (R. D. Boyer and E. G. Krebs, eds.), Academic Press, New York, pp. 335–358.

Saimi, Y., 1986, Calcium-dependent sodium currents in *Paramecium:* Mutational manipulations and effects of hyper- and depolarization, *J. Membr. Biol.* **92:**227–236.

Saimi, Y., and Kung, C., 1982, Are ions involved in the gating of calcium channels? *Science* **218:**153–156.

Saimi, Y., and Kung, C., 1987, Behavioral genetics of *Paramecium, Annu. Rev. Genet.* **21:**47–65.

Saimi, Y., and Martinac, B., 1989, A calcium-activated potassium channel in *Paramecium* studied under patch clamp, *J. Membr. Biol.* (in press).

Satow, Y., and Kung, C., 1980, Membrane currents of pawn mutants of the *pwA* group in *Paramecium tetraurelia, J. Exp. Biol.* **84:**57–72.

Satow, Y., Chang, S.-Y., and Kung, C., 1974, Membrane excitability: Made temperature dependent by mutations, *Proc. Natl. Acad. Sci. USA* **71:**2703–2706.

Schaefer, W. H., Lukas, T. J., Blair, I. A., Schultz, J. E., and Watterson, D. M., 1987, Amino acid sequence of a novel calmodulin from *Paramecium tetraurelia* that contains dimethyllysine in the first domain, *J. Biol. Chem.* **262:**1025–1029.

Schein, S. J., 1976a, Calcium channel stability measured by gradual loss of excitability in pawn mutants of *Paramecium aurelia, J. Exp. Biol.* **65:**725–736.

Schein, S. J., 1976b, Non-behavioral selection for pawns, mutants of *Paramecium aurelia* with decreased excitability, *Genetics* **84:**453–468.

Schultz, J. E., and Jantzen, H. M., 1980, Cyclic nucleotide-dependent protein kinases from cilia of *Paramecium tetraurelia, FEBS Lett.* **116:**75–78.

Schultz, J. E., and Klumpp, S., 1980, Guanylate cyclase in the excitable ciliary membrane of *Paramecium, FEBS Lett.* **122:**64–66.

Schultz, J. E., and Klumpp, S., 1982, Lanthanum dissociates calmodulin from the guanylate cyclase of the excitable ciliary membrane from *Paramecium, FEMS Microbiol. Lett.* **13:**303–306.

Schultz, J. E., and Klumpp, S., 1983, Adenylate cyclase in cilia from *Paramecium.* Localization and partial characterization, *FEBS Lett.* **154:**347–350.

Schultz, J. E., and Klumpp, S., 1984, Calcium/calmodulin-regulated guanylate cyclases in the ciliary membranes from *Paramecium* and *Tetrahymena, Adv. Cyclic Nucleotide Protein Phosphorylation Res.* **17:**275–283.

Schultz, J. E., Schönefeld, U., and Klumpp, S., 1983, Calcium/calmodulin-regulated guanylate cyclase and calcium-permeability in the ciliary membrane from *Tetrahymena, Eur. J. Biochem.* **137:**89–94.

Schultz, J. E., Boheim, G., Gierlich, D., Hanke, W., Von Hirschhausen, R., Kleefeld, G., Klumpp, S., Otto, M. K., and Schönefeld, U., 1984a, Cyclic nucleotides and calcium in *Paramecium:* A neurobiological model organism, *Horm. Cell Regul.* **8:**99–114.

Schultz, J. E., Grunemund, R., Von Hirschhausen, R., and Schönefeld, U., 1984b, Ionic regulation of cyclic AMP levels in *Paramecium tetraurelia* in vivo, *FEBS Lett.* **167:**113–116.

Schultz, J. E., Klumpp, S., and Gierlich, D., 1985, Involvement of cyclic nucleotides in sensing and response in *Paramecium tetraurelia,* in: *Sensing and Response in Microorganisms* (M. Eisenbach and M. Balaban, eds.), Elsevier, Amsterdam, pp. 159–173.

Schultz, J. E., Pohl, T., and Klumpp, S., 1986, Voltage-gated Ca^{2+} entry into *Paramecium* linked to intraciliary increase in cyclic GMP, *Nature* **322:**271–273.

Schultz, J. E., Uhl, D. G., and Klumpp, S., 1987, Ionic regulation of adenylate cyclase from the cilia of *Paramecium tetraurelia, Biochem. J.* **246:**187–192.

Schwartz, J. H., and Greenberg, S. M., 1987, Molecular mechanisms for memory. Second messenger modifications of protein kinases in nerve cells, *Annu. Rev. Neurosci.* **10:**459–476.

Shusterman, C. L., 1981, Potassium resistant mutants and adaptation in *Paramecium tetraurelia,* Ph.D. dissertation, University of Wisconsin, Madison.

Stoclet, J.-C., Gerard, D., Kilhoffer, M.-C., Lugnier, C., Miller, R., and Schaeffer, P., 1987, Calmodulin and its role in intracellular calcium regulation, *Prog. Neurobiol.* **29:**321–364.

Takahashi, M., 1979, Behavioral mutants in *Paramecium caudatum, Genetics* **91**:393–408.

Takahashi, M., 1988, Behavioral genetics in *P. caudatum,* in: *Paramecium* (H.-D. Görtz, ed.), Springer-Verlag, Berlin, pp. 271–281.

Takahashi, M., and Naitoh, Y., 1978, Behavioral mutants of *Paramecium caudatum* with defective membrane electrogenesis, *Nature* **271**:656–659.

Takahashi, M., Haga, N., Hennessey, T., Hinrichsen, R. D., and Hara, R., 1985, A gamma ray-induced non-excitable membrane mutant in *Paramecium caudatum:* A behavioral and genetic analysis, *Genet. Res.* **46:** 1–10.

Thiele, J., and Schultz, J. E., 1981, Ciliary membrane vesicles of *Paramecium* contain the voltage-sensitive calcium channel, *Proc. Natl. Acad. Sci. USA* **78**:3688–3691.

Thiele, J., Klumpp, S., Schultz, J. E., and Bardele, C. F., 1982, Differential distribution of voltage-dependent calcium channels and guanylate cyclase in the excitable membrane of *Paramecium tetraurelia, Eur. J. Cell Biol.* **28**:3–11.

Thiele, J., Otto, M. K., Deitmer, J. W., and Schultz, J. E., 1983, Calcium channels of the excitable ciliary membrane from *Paramecium:* An initial biochemical characterization, *J. Membr. Biol.* **76**:253–260.

Travis, S. M., and Nelson, D. L., 1986, Characterization of Ca^{2+}- or Mg^{2+}-ATPase of the excitable ciliary membrane from *Paramecium tetraurelia:* A comparison with a soluble Ca^{2+}-dependent ATPase, *Biochim. Biophys. Acta* **862**:39–48.

Travis, S. M., and Nelson, D. L., 1988, Purification and properties of dyneins from *Paramecium* cilia, *Biochim. Biophys. Acta* **966**:73–83.

Van Houten, J., and Van Houten, J., 1982, Computer simulation of *Paramecium* chemokinesis behavior, *J. Theor. Biol.* **98**:453–468.

Walter, M. F., and Schultz, J. E., 1981, Calcium receptor protein calmodulin isolated from cilia and cells of *Paramecium tetraurelia, Eur. J. Cell Biol.* **24**:97–100.

Wyroba, E., and Przelecka, A., 1973, Studies on the surface coat of *Paramecium aurelia.* I. Ruthenium red staining and enzyme treatment, *Z. Zellforsch. Mikrosk. Anat.* **143**:343–353.

Structure, Turnover, and Assembly of Ciliary Membranes in *Tetrahymena*

Norman E. Williams

1. Introduction

Tetrahymena cells (Fig. 1) have upwards of 600 cilia depending upon the species and the stage in the cell cycle (Nanney, 1971). Of these, about 20% are in the oral apparatus (Williams and Bakowska, 1982). The cilia are easily detached (less so the oral cilia) by the application of mild shearing forces after various chemical treatments (reviewed by Nozawa, 1975). This, combined with the fact that some species will approach 10^6 cells/ml before entering stationary phase, has made *Tetrahymena* cilia a favored material for biochemical investigation. Studies of the surface membrane of *Tetrahymena* cilia will be reviewed here with an emphasis on biogenesis.

Reviews containing information on the ciliary membranes of *Tetrahymena* have been published by Thompson (1972), Holz and Conner (1973), Thompson and Nozawa (1977), Allen (1978), and Aufderheide *et al.* (1980). It may be helpful to readers of included citations to point out that the species name ''*Tetrahymena pyriformis*'' applied to a sibling species swarm prior to 1976. This was recognized following the application of breeding tests and modern molecular procedures. In 1976, Nanney and McCoy named all species known at the time to be valid by the newer criteria. One consequence of this is that many, though not all, of the membrane studies reported prior to 1976 involved what is today regarded as *T. thermophila,* instead of *T. pyriformis* as stated in the original reports.

2. Lipid Composition

The lipid composition of *T. thermophila* ciliary membranes is compared with the lipid composition of whole cells, microsomes, and pellicles from the same species in

Norman E. Williams • Department of Biology, University of Iowa, Iowa City, Iowa 52242.

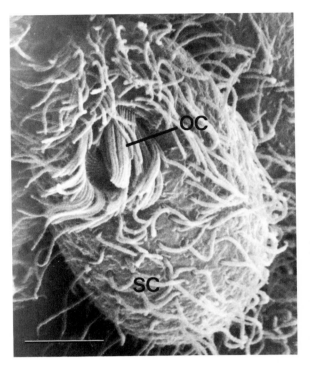

Figure 1. Scanning electron micrograph of a *Tetrahymena pyriformis* cell showing the arrangement of cilia at the cell surface. The somatic cilia (SC) cover the body and are organized into longitudinal rows. The oral cilia (OC) are arranged into four tightly packed clusters associated with the feeding organelle system. Bar = 10 μm. From Williams (1975) with permission from Academic Press.

Table I. *Tetrahymena* is unusual in having high levels of 2-aminoethylphosphonolipid (AEPL). This compound, characterized by possession of a C–P bond instead of the more usual P–O–C bond, was early recognized as being highly concentrated in ciliary membranes (Kennedy and Thompson, 1970; Smith *et al.*, 1970; Jonah and Erwin, 1971). Phosphonolipid is resistant to attack by endogenous lipolytic enzymes, and it has been suggested that this may be of importance in conferring stability to the surface membranes (Kennedy and Thompson, 1970).

The principal neutral lipid in *Tetrahymena* is a pentacyclic triterpenoid, "tetrahymanol" (Mallory *et al.*, 1963). This compound serves the function fulfilled in mammalian cells by cholesterol. The data in Table I show that the molar ratio of tetrahymanol to phospholipid in cilia is ∼ fivefold higher than in microsomal membranes. There is good evidence to suggest that the high levels of tetrahymanol in ciliary membranes serve to maintain constant physical properties of the membrane in the face of changing environmental influences. Using electron spin resonance techniques, Nozawa *et al.* (1974) found the ciliary membranes to be less fluid than those of microsomal membranes from the same cell. In addition, they showed by direct measurement that tetrahymanol damps the effect of temperature on the fluidity of phospholipids. Added to this, Kitajima and Thompson (1977) showed that intramembranous particle aggregation, a response of internal membranes to cold, does not occur in *Tetrahymena* ciliary membranes under the same conditions.

**Table I. The Lipid Composition of *T. thermophila* Strain NT-1,
Whole Cells and Selected Organelles**[a,b]

Lipid component	Whole cells	Cilia	Pellicles	Microsomes
Tetrahymanol phospholipid (molar ratio)	0.078	0.260	0.090	0.055
Individual phospholipids				
Cardiolipin	5.6–0.4	0.4–0.5	0.3–0.4	4.8–1.1
2-Aminoethylphosphonolipid	25.4–2.0	37.0–5.9	32.9–1.5	20.8–0.7
Ethanolamine glycerophosphatides	34.4–1.6	21.4–2.6	36.2–1.6	35.5–1.1
Lysophosphatidylethanolamine, lyso-2-aminoethylphosphonolipid, and ceramide-aminoethylphosphonate	5.4–0.9	14.0–3.0	4.6–1.6	3.7–0.5
Choline glycerophosphatides	26.2–0.5	18.1–1.8	21.4–2.0	27.6–1.3
Lysophosphatidylcholine	2.6–1.8	1.5–1.0	2.7–1.6	3.8–1.2

[a]From Thompson and Nozawa (1977).
[b]Phospholipid values are reported as a percentage of total lipid phosphorous.

Other characteristics of ciliary membranes relative to other cellular membranes, particularly internal membranes, are small amounts of phosphatidylcholine, small amounts of phosphatidylethanolamine, and practically no cardiolipin (Smith *et al.*, 1970; Jonah and Erwin, 1971; Nozawa and Thompson, 1971a). As in the case of the polar head groups, the fatty acid compositions of different membranes in the cells are also distinctive. Jonah and Erwin (1971) presented a comparison of the fatty acids present in cilia with those in other membrane-containing fractions of *Tetrahymena*. These authors also provided evidence that the fatty acid composition of a given phospholipid may vary as a function of the organelle from which the latter is obtained.

We may conclude from the above that the lipid composition of *Tetrahymena* ciliary membranes is substantially different from that of many of the other membranes within the cell. As all lipids are presumed synthesized by microsomal enzymes, the mechanism that results in the individuality of the ciliary membrane represents a major unsolved problem in lipid metabolism.

In this connection, one unresolved question of *Tetrahymena* is how different the ciliary membranes may be from the surface membrane over the nonciliated regions of the cell. This is because no one has yet been able to isolate nonciliary plasma membrane in the absence of the tightly associated alveolar membranes that underlie it in the cell. The alveolar sac membranes are labeled (AS) in Fig. 2, though they are not well resolved here. The nature of these membranes and their relationship to the cell surface have been described in numerous publications, perhaps most completely in recent publications by Satir and Wissig (1982) and Williams (1983). The isolated pellicle fractions used in the lipid studies described above are "contaminated" with these alveolar membranes. It is not likely that major similarities in lipid composition found between cilia and pellicles, such as high levels of AEPL and tetrahymanol, reflect alveolar membrane properties alone. Nevertheless, the degree to which ciliary membranes resemble the nonciliary plasmalemma in *Tetrahymena* cannot be determined with certainty until pure plasma membrane is isolated from this cell type.

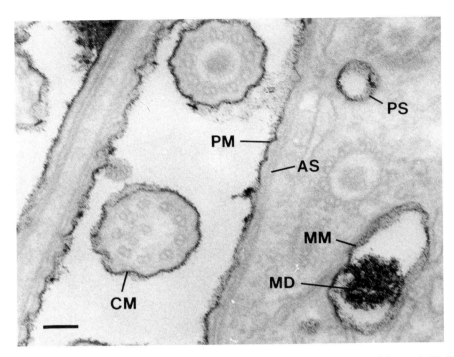

Figure 2. Section of ruthenium red-stained *Tetrahymena* cell showing surface coat material around cilia (CM), and over the somatic plasma membrane (PM). Depressions in the surface called parasomal sacs are coated (PS). Mucocysts in the process of discharging at the time of staining also react with the dye. Both the mucocyst discharge (MD) and a coating on the internal surface of the mucocyst membrane (MM) can be seen. The subsurface alveolar sac membranes (AS) are coated internally, but do not stain in these preparations unless the surface is ruptured. Bar = 0.1 μm. Unpublished micrograph by Ruth Jaeckel-Williams.

3. Protein Components

A direct approach to the identification of ciliary membrane proteins, unlike the identification of ciliary membrane lipids, requires preparation of a membrane fraction from cilia. This is usually done by applying shearing forces to isolated cilia in appropriate buffers. Subbaiah and Thompson (1974) were the first to use this approach in conjunction with SDS-PAGE in an attempt to identify ciliary membrane proteins in *Tetrahymena*. They reported major component proteins of 63, 53, 44, and 29 kDa. Adoutte *et al.* (1980) using a similar procedure but without detergent recovered major proteins of 64 and 20 kDa. They suggested that bands seen by others (Subbaiah and Thompson, 1974; Butzel and Decaprio, 1978; Dentler, 1978, 1980a) in gels at the molecular weight positions of tubulin and/or dynein might represent entrapment of axonemal proteins within ciliary membrane vesicles made in the presence of detergents. Recent work indicates that the bulk of the tubulinlike protein in *Tetrahymena* ciliary membranes reported by Dentler (1980a) is not tubulin, although there is some suggestion that a small amount of tubulin may be present (W. L. Dentler, personal communication; Dentler, this volume; Stephens, this volume).

Ciliary membrane proteins have also been identified in *Tetrahymena* by lactoperoxidase iodination in conjunction with cell fractionation and protein separation techniques. After radioiodination of living cells, only externally exposed proteins are labeled. *Tetrahymena* cells were treated this way, then cilia and pellicles were prepared separately and compared by autoradiography after one- and two-dimensional electrophoresis (Williams *et al.*, 1980). The same array of labeled proteins was found in the two preparations. The major labeled components were a pair of 47- to 48-kDa proteins with different isoelectric points (Williams, 1983). Over 20 less-heavily labeled proteins were found in both fractions.

Specific proteins known as "immobilization antigens" are found on the ciliary and other surface membranes of *T. thermophila*. The individual protein expressed by a given cell is a function of both the genotype and environmental influences. Ordinarily, a single antigen is present on the surface at a time. The *SerH3* gene product, expressed at 28°C in *T. thermophila*, has been purified and shown to be an acidic protein with an apparent molecular mass of 52 kDa (Williams *et al.*, 1985; Doerder and Berkowitz, 1986). A slightly higher value was reported by Bannon *et al.* (1986). Doerder and Berkowitz (1986) have also identified and partially purified the H1 (52 kDa), H2 (44 kDa), and H4 (49 kDa) antigens. These antigens, all expressed at 28°C, and those expressed at high (Bannon *et al.*, 1986) and low temperatures, have been studied to a lesser extent. There is, however, a wealth of genetic information based upon cells expressing these phenotypes (Doerder, 1979).

4. Ultrastructure

A coat on the outer surface of *Tetrahymena* ciliary membranes can be demonstrated with ruthenium red (Fig. 2). This indicates that these membranes are coated with polyanions of high charge density, believed to reflect the presence of sugar polymers attached to proteins in the membrane (Luft, 1976). Such a coat is also present on the nonciliated cell surface, on the inner surfaces of the alveolar sac membranes (Williams *et al.*, 1980; Williams, 1983), and on the inner surfaces of the secretory organelle (mucocyst) membranes (Fig. 2). All three have been suggested as possible sources of ciliary membrane (discussed below). The surface coat on the plasma membrane and around the cilia of *Tetrahymena* has also been shown by alcian blue staining (Nilsson and Behnke, 1971).

The glycoproteins responsible for the surface staining reaction have not been identified. Con A inhibits conjugation in *T. thermophila* (Frisch and Loyter, 1977), and Pagliaro and Wolfe (1987) have identified a 23-kDa glycoprotein as a candidate for the mating-type receptor. However, neither Con A binding nor the presence of this glycoprotein has yet been associated with ciliary membranes. Interestingly, Smith and Lepak (1982) isolated a phosphonic acid-containing glycoprotein from *Tetrahymena* with a molecular mass of 22.5 kDa. They reported its presence in surface membranes, but evidence was not presented. Richmond (1976) reported the presence of a glycoprotein in the surface membranes of *Tetrahymena*, but it was not characterized. Some proteins associated with the surface membrane that may be weakly glycosylated are dynein (Dentler, 1980a) and the H3 immobilization antigen (Ron and Williams, unpublished).

Ultrastructural studies have also revealed the presence of very fine elements linking the ciliary membrane of *Tetrahymena* to the axoneme (Sattler and Staehelin, 1974; Dentler *et al.*, 1980) and to the microtubule capping structures (Dentler, 1980b). Some of these linkers and the consequence of their presence are illustrated in Fig. 3.

A striking feature of the ciliary membranes of *Tetrahymena* is the presence of regular arrays of intramembranous particles that can be revealed by freeze-etching procedures (Fig. 4). The first of these to be reported was the ciliary necklace (Satir *et al.*, 1972; Wunderlich and Speth, 1972). This differentiation, consisting of two rows of closely spaced particles near the base of the cilium (Fig. 4A), had already been described in other cell types. It appears that all *Tetrahymena* strains and species have ciliary necklaces, and they occur on both oral and somatic cilia (Sattler and Staehelin, 1974). Wunderlich and Speth (1972) also described small groups of particles, termed plaques, immediately distal to the necklace (Fig. 4A). There are nine plaques, one connected to each of the ciliary doublet microtubules in the axoneme by linking structures. The necklace particles are also connected to the axoneme by linkers. Unlike the necklace, ciliary plaques are absent from some strains or species of *Tetrahymena*. There is some evidence from studies of *Paramecium* and *Tetrahymena* to suggest that the proximal intramembranous particles of cilia may be involved in cation movement between the cilium and the surrounding medium (Baugh *et al.*, 1976; Fisher *et al.*, 1976; Dute and Kung, 1978).

Three patterns of intramembranous particles have been described distal to the necklace

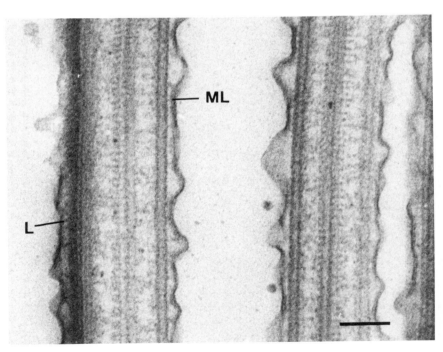

Figure 3. Longitudinal section through oral cilia showing stretches of membrane linked by thin connecting strands (ML) to ciliary doublet microtubules. The cilium on the left is one having an accessory body of unknown function located between the doublet microtubules and the membrane on one side. Linkers (L) also appear to connect the membrane with the accessory body in these cilia. Bar = 0.1 μm.

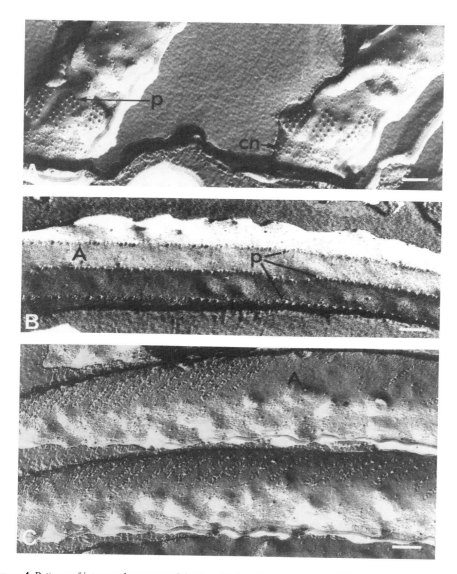

Figure 4. Patterns of intramembranous particles found in the ciliary membranes of *Tetrahymena*. (A) Patches of particles called plaques (p), three particles wide and nine particles long, are located on the cytoplasmic face of the ciliary membrane immediately distal to the ciliary necklace (cn). Intramembranous particles distal to the plaques are randomly distributed in somatic and some oral cilia. (B) One type of oral cilium shows three parallel longitudinal rows of particles (p) on the cleaved face of the cytoplasmic membrane leaflet (A). The spacing between the rows suggests that the particles could be associated with elements of the underlying microtubule doublets. (C) Another type of oral cilium shows short parallel rows of particles on the external face of the ciliary membrane arranged at an angle of about 48° with the main ciliary axis. These particles attach to short bristles which protrude from the ciliary membrane surface into the surrounding medium. The pattern of these intramembranous particles is reflected in the distribution of holes in the cytoplasmic face of the ciliary membrane in this micrograph. Bar = 0.1 μm. From Sattler and Staehelin (1974) with permission from the Rockefeller University Press.

and plaques, depending upon the location of the cilia on the cell surface (Sattler and Staehelin, 1974). Randomly distributed particles occur in quantity in somatic cilia, and sparsely in certain oral cilia. Some of the oral cilia have parallel longitudinal rows of particles on one side of the cilium (Fig. 4B), and oral cilia located in the outside row of the first membranelle (only) have patches of particles arranged in short rows oriented obliquely to the main axis of the cilium (Fig. 4C). The latter particles appear to connect toward the outside to a curious bristle differentiation postulated to play some role in the feeding process (Williams and Luft, 1968; Sattler and Staehelin, 1974). They appear to connect to the inside to an accessory body that underlies the bristles (Fig. 3). Other intramembranous particles connect with the axoneme, and some of these are postulated to affect the form of the ciliary beat (Sattler and Staehelin, 1974).

5. Turnover

It is a remarkable fact that the building of a membrane never stops. This is shown for *Tetrahymena* ciliary membrane by the studies summarized here, and it has been shown for mammalian cell membranes in the classic studies of Omura *et al.* (1967) and Warren and Glick (1969). The basic observation is that membranes continue to show the incorporation of labeled precursors of both lipids and proteins at significant rates when neither growing nor differentiating.

The incorporation of [^{14}C]palmitate into *Tetrahymena* membranes in nongrowing cells has been studied extensively by Nozawa and Thompson (1972). They found the rates of incorporation into both whole cell and ciliary lipids to be nearly identical to those found in growing cells. In contrast, it has been found that the rate of incorporation of ^{23}P into whole cell and ciliary phospholipid is two orders of magnitude slower than that found in growing cells (Table II; Skriver and Williams, 1980). A greater retention of ^{32}P-labeled lipids in nongrowing cells relative to growing cells was also found by Nozawa and Thompson (1972). Overall, the results appear to indicate that phospholipids turn over in

Table II. Phospholipid Synthesis in Nongrowing *Tetrahymena*[a,b]

Experiment	Time in starvation medium (hr)	$10^{-7} \times$ number of cells harvested	^{32}P added (μCi)	$10^{-4} \times$ radioactivity in lipid (cpm)	Lipid P (μg)	Specific activity (cpm/μg lipid P)
Logarithmic cells	—	0.96	35	340	66.0	5200
Nongrowing control	5.5	1.0	35	0.248	56.0	44.3
Nongrowing deciliated	5.5	0.92	35	0.157	41.0	38.3
Nongrowing control	6.5	1.8	150	1.62	116.5	139.0
Nongrowing deciliated	6.5	1.6	150	1.24	112.5	110.2

[a]From Skriver and Williams (1980).
[b]The incorporation of ^{32}P into whole cell phospholipids was measured in exponentially growing, starved, and starved deciliated cells. Two 20-ml aliquots from each of the cultures were labeled with carrier-free ^{32}P for 30 min after which the incorporation of label was stopped by pouring the cells into chloroform/methanol (6:1, v/v). The lipids were extracted by the method of Bligh and Dyer (1959), washed twice according to Folch *et al.* (1957), and the lipid specific activity determined as cpm/μg lipid P.

the membrane at low rates and fatty acids exchange within existing phospholipids at high rates.

The turnover of ciliary membrane proteins has also been demonstrated in *Tetrahymena* (Williams *et al.*, 1980). Nongrowing cells were labeled by growing in the presence of [^{14}C]leucine, then starved and labeled again with [^{3}H]leucine. The ^{3}H/^{14}C ratios determined in four identified ciliary membrane proteins established that turnover was occurring, and that the turnover rates were not identical within the group of four proteins. It was concluded that the ciliary membrane proteins are not likely to be coordinately assembled into the membrane. As in the case of other systems, there is also evidence to suggest that lipids and proteins are not coordinately assembled into the membrane (Nozawa and Thompson, 1972).

Why do membrane components undergo turnover? One reason seems to be that it provides cells with a way of altering the molecular composition of membranes in the absence of net synthesis. This makes it possible for cells to maintain critical membrane characteristics, such as fluidity, in the face of environmental change, and to undergo differentiation, e.g., develop mating competence, in the absence of growth. These adaptive responses will be considered in the next section.

6. Modulation

This selection deals with modulation, or differentiation, of the ciliary membrane, by which is meant any significant change in the molecular composition of the membrane. This occurs when membranes of an alternate type are assembled in growing cells, but it can also be accomplished by turnover in nongrowing cells.

In *Tetrahymena* membranes generally, a low-temperature shock causes a lipid phase separation that affects function and is reflected in intramembranous particle aggregation. The cell responds to restore fluidity in these membranes primarily by fatty acid desaturation, and the particles again randomize (Erwin and Bloch, 1963; Wunderlich *et al.*, 1973; Fukushima *et al.*, 1976; Kitajima and Thompson, 1977). The ciliary membranes are less fluid than other membranes, presumably because of their high tetrahymanol content, and do not respond to low temperature by aggregating their intramembranous particles. Nevertheless, Fukushima *et al.* (1976) have shown that low temperature induces ciliary membrane modulation, notably an increase in fatty acid unsaturation and an increase in lysophospholipids. Both changes are believed to result in an increase in membrane fluidity.

The phospholipid alterations appear primarily related to maintaining physical properties constant that are necessary for proper membrane function. Membrane protein changes may confer new structural, enzymatic, or interactive properties. Like phospholipid changes, these can occur in both growing and nongrowing cells.

One of the most significant changes in surface membrane proteins that occurs requires starvation; this is the development of mating competence. Specific gene expression during conjugation has been documented (Martindale and Bruns, 1983), but none of the gene products has yet been identified as a surface membrane protein. A direct biochemical approach to the problem (Van Bell and Williams, 1984) has revealed that mating-competent *T. thermophila* cells have lost three surface membrane proteins that are present in nonmating cells. Studies with *Paramecium* also support a notion of mating as a disinhibited state (Haga and Hiwatashi, 1981).

Serotype transitions provide the best documented examples of ciliary membrane protein modulation. It is disappointing that no known function exists for the antigens involved, yet they may still provide a model for mechanisms. Like trypanosomes, *Paramecium* and *Tetrahymena* cells express a variety of surface antigens, one at a time. Unlike trypanosomes, expression-linked gene duplications and rearrangements are not involved (Forney *et al.*, 1983). Moreover, expression is under the control of the investigator in the ciliates. The H antigens in *T. thermophila* are specified by multiple alleles at the *SerH* locus, and are expressed from 20 to 36°C. Above 36°C, antigens specified by the *SerT* locus are expressed. The antigens coat the entire surface of the cell, including all cilia. Antisera directed against these antigens will immobilize living cells, hence the name "immobilization antigens."

The transition from expressing T to expressing H3, effected by a temperature shift from 40°C to 28–30°C, has been studied in *T. thermophila* using both nongrowing (Williams *et al.*, 1985) and growing cells (Bannon *et al.*, 1986.) Synthesis of the H3 antigen was detected within 2 hr after the temperature shift in growing cells, but within 30 min in nongrowing cells. The difference is probably due to differences in procedure. No transition occurred in the presence of either cycloheximide or actinomycin D.

The appearance of H3 antigen at the cell surface was detected by immunofluorescence in nongrowing cells as early as 30 min after the temperature shift, and the intensity of the fluorescent reaction increased over the next 90 min. Electron microscopic observations following indirect immunoferritin staining of temperature-shifted cells showed that the H3 antigen appeared first over nonciliated areas of the cell, then gradually spread distally over the cilia (Williams *et al.*, 1985). A cell midway in this process is shown in Fig. 5. Antigen deployment at the surface was not sensitive to either colchicine or cytochalasin B. The details of the distal migration in cilia remain to be disclosed. Kinetic studies on the movement of induced antigen over the ciliary surface would seem to offer an excellent opportunity to probe underlying mechanisms of deployment. It will be of especial interest to see how the movement of antigens over the ciliary membranes of *Tetrahymena* compares with movement of glycoproteins in other ciliary/flagellar membranes (Bloodgood, 1987, this volume).

Other important questions concern the internal sites of origin and passage for new antigen, the precise site of delivery to the cell surface, and the fate of the old antigen during transition. In growing cells, the old antigen may simply be diluted out. In nongrowing cells, it may be somehow internalized, or possibly shed into the medium. Methods are now available for getting answers to at least some of these questions. A recent study of the H-to-T conversion has shown that mRNA stability plays a role in regulating serotype transitions (Love *et al.*, 1988).

Another favorable opportunity for the study of ciliary membrane protein dynamics in *Tetrahymena* is offered by the discovery by Doerder (1981) that each partner in a conjugating pair of *T. thermophila* cells gains the surface antigen of its partner while coupled. This process is cycloheximide insensitive. Doerder has considered that the transfer is most likely via mixing of the internal cytoplasms of the conjugants, but has recognized that a lateral migration of antigen from the surface of one cell to the surface of its partner is also

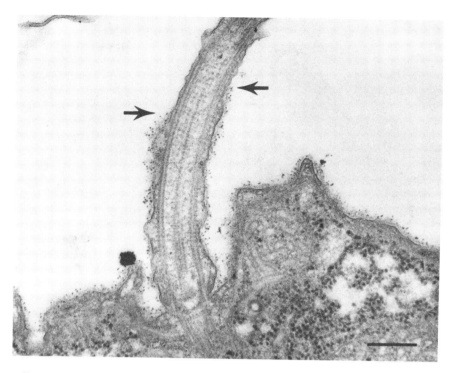

Figure 5. *Tetrahymena* cell initiating expression of the H3 antigen. The cell was stained with anti-H3 serum followed by ferritin-conjugated anti-rabbit immunoglobulin. New antigens first appear throughout the nonciliated surface membrane, then spread over the ciliary membranes, base to tip. The transforming cell shown here has H3 antigen on the somatic plasma membrane and over the proximal portion of the cilium shown, but none distal to the point marked by the arrows. Bar = 0.25 μm.

a possibility. Mutants are available that transfer in one direction only. Further studies of this interesting process may contribute to our understanding of the mechanisms involved in regulating the insertion and/or migration of ciliary membrane proteins in *Tetrahymena*.

7. Assembly

Cilia are formed anew in growing *Tetrahymena* cells. Morphological studies have revealed that the first sign of ciliogenesis in growing cells is the appearance of a ring of intramembranous particles in the surface membrane lying just above the new basal body (Satir *et al.*, 1972; Hufnagel, 1983). This is believed to be a precursor of the ciliary necklace. Other intramembranous particle arrays form later within the ciliary membrane.

The synthesis of phospholipids has been studied in growing *Tetrahymena* cells (Thompson, 1967; Nozawa and Thompson, 1971b). The incorporation pattern into all membranes, including ciliary membranes, was essentially the same whether labeled with side chain ($[^{32}P]$palmitate) or polar head group ($[^{3}H]$chimyl alcohol; ^{32}Pi) precursors. This suggests that, in growing cells, phospholipid molecules may move intact from their biosynthetic site on the endoplasmic reticulum to the outlying membranes (Thompson, 1972). Characteristically, the ciliary membranes acquired labeled phospholipids more slowly than the other membranes studied. This presents a problem, for ciliary membranes are not assembled at rates lower than other membranes in growing cells. This was initially interpreted as evidence for a relatively isolated precursor pool of ciliary phospholipids somewhere within the cell (Nozawa and Thompson, 1971b), and later it was suggested to result from the greater accessibility of internal membranes to exchange reactions (Thompson, 1972). Skriver (1979) has shown the migration of somatic plasma membrane phospholipids labeled with trinitrobenzene sulfonic acid into the membranes of regenerating cilia, which suggests that the postulated pool of ciliary phospholipids may, in fact, be the pellicle membranes (discussed below). Both explanations may apply, however.

The incorporation of [^{3}H]leucine into ciliary membrane proteins has also been studied and compared with phospholipid data obtained from growing cells. This has led to the formulation of a ciliary matrix model of ciliary membrane assembly (Subbaiah and Thompson, 1974). It was found that bands in electrophoretic gels corresponding to putative ciliary membrane proteins were more heavily labeled at early times following radioactive pulses in the ciliary matrix fraction than they were in the ciliary membrane fraction. A preliminary analysis of phospholipids extracted from cilia following a 5-min labeling with [^{14}C]palmitate indicated matrix lipids were also two to three times more radioactive than membrane lipids. Subbaiah and Thompson (1974) concluded it is likely that new lipids as well as new proteins are transported into ciliary membranes via the ciliary matrix. Lipids and proteins exhibited kinetic differences suggesting they are not coassembled into the membrane.

An alternative view considers that ciliary membrane lipids and proteins are first inserted into nonciliated regions of the cell surface, then migrate laterally to form the ciliary membranes. The i-antigen modulation studies, discussed above, support this view. Further evidence comes from the study of ciliary regeneration.

When *Tetrahymena* cilia are detached from the cell, they break off near the base at a specific point between the ciliary necklace (which stays with the cell) and the intramembranous particle plaques (Satir *et al.,* 1976). The residual membranes seal over the ciliary stubs. New membrane next grows out above the necklace, and the plaque particles appear much later. Surface-labeling studies have suggested the lateral migration of both phospholipids and proteins from the somatic plasma membrane into the newly forming membranes of regenerating cilia. Skriver (1979) used trinitrobenzene sulfonic acid to label externally exposed phospholipids, and Williams (1983) used lactoperoxidase iodination to label externally exposed proteins. In both experiments, the cells were deciliated, allowed to regenerate cilia, then examined for the presence of the probe in newly formed cilia. Significant quantities of label were found in the regenerated cilia in both cases.

The results of studies on induced synthesis during ciliary regeneration support the notion of an independent regulation of ciliary membrane phospholipids and proteins. It was found that nongrowing cells regenerate a full complement of cilia in the absence of the induced synthesis of phospholipids (Table II; Skriver and Williams, 1980), whereas a significant amount of *de novo* synthesis of identified ciliary membrane proteins occurred under the same conditions (Williams, 1983). The extent of labeling in these experiments was also different in different membrane proteins, suggesting the absence of coordinate regulation of individual membrane proteins.

The regeneration of considerable quantities of ciliary membrane in nongrowing cells without a concomitant synthesis of significant amounts of phospholipid implies the existence of an intracellular reservoir of ciliary phospholipids somewhere within the cell. Careful measurements showed that the observed effects cannot be explained solely by a redistribution of the surface membrane remaining after deciliation (Skriver and Williams, 1980). Free phospholipids, bound to carrier proteins, are one possibility. However, cytological studies have suggested the possible involvement of one or the other of two separate preformed membrane stores that exist within the cell. One is the alveolar membrane system (Williams, 1983), and the other is the mucocyst membrane (Satir *et al.*, 1976). The alveoli have ruthenium red-positive coats to the inside, as has been seen in ruptured alveoli. The mucocysts also have internally coated membranes, as can be seen by staining during discharge (Fig. 2, MM). The internal coats in both cases would be on the outside after fusing with the cell surface membrane. The study of cells regenerating cilia by electron microscopy (Williams, 1983) has produced images suggestive of the origin of alveolar membranes and their fusion with the cell surface membranes. Satir *et al.* (1976) reported the discharge of large numbers of mucocysts concomitant with dibucaine-induced deciliation of one strain of *Tetrahymena*. This was not seen, however, when other strains were deciliated by other means (Williams, 1983).

One problem in developing an integrated picture of the regeneration of ciliary membranes in *Tetrahymena* based on the above considerations is the apparent contradiction between the flow of precursor membrane to the surface and evidence supporting the independent regulation of the molecular components of the ciliary membrane. It has been suggested as one possibility for resolution that the precursor membranes may not be fully differentiated, but might instead become so as they are moved into place (Williams, 1983). This is consistent with a previous suggestion that exchange reactions in membranes may make wholly independent modes of transport for different molecules more apparent than real (Subbaiah and Thompson, 1974). Still, the routes of passage, putative storage depots, and sites of joining for the molecular materials found in the ciliary membrane elude us. Though some progress has been made, further studies will be required before we have an adequate understanding of the assembly of ciliary membranes in *Tetrahymena*.

8. Concluding Remarks

The major objective in many of the studies reviewed here has been to contribute to the understanding of ciliary membrane biogenesis and modulation in eukaryotic cells generally. These studies have raised a wide range of fundamental questions for the future and have shown the potential that studies with *Tetrahymena* have for contributing to their resolution. Among these are (1) how ciliary membranes gain and maintain their unique lipid compositions, (2) the activation of desaturase activity by low temperature, (3) the sites of phospholipid stores and the mechanisms controlling their deployment, (4) the nature of the information targeting proteins to the ciliary membranes, (5) the signals for specific gene expression during modulation and regeneration, (6) the topographic pathways to the surface and the combining sites within the cell for membrane components, (7) the mechanisms of the lateral migration of phospholipids and proteins into ciliary membranes, and (8) the fate of the molecules replaced during turnover and modulation.

It is important to note additionally that the initiation of ciliary membrane assembly is controlled both temporally and spatially within the cell. For example, it has been observed that cilia grow out asynchronously in deciliated *Tetrahymena* cells (Hadley and Williams, 1981), and it has been shown that unciliated basal bodies in certain regions of the *Tetrahymena* cell sprout cilia in a restricted portion of the cell cycle (Frankel *et al.*, 1981). It will therefore also be important in the future to understand the temporal and spatial controls that regulate ciliogenesis in *Tetrahymena* cells.

References

Adoutte, A., Ramanathan, R., Lewis, R. M., Dute, R. R., Ling, K.-Y., Kung, C., and Nelson, D. L., 1980, Biochemical studies of the excitable membrane of *Paramecium tetraurelia*. III. Proteins of cilia and ciliary membranes, *J. Cell Biol.* **84:**717–738.

Allen, R. D., 1978, Membranes in ciliates: Ultrastructure and fusion, in: *Membrane Fusion: Cell Surface Reviews*, Volume 5 (G. Poste and G. L. Nicolson, eds.), Elsevier/North-Holland, pp. 657–763.

Aufderheide, K., Frankel, J., and Williams, N. E., 1980. Formation and positioning of surface-related structures in protozoa, *Microbiol. Rev.* **44:**252–302.

Bannon, G. A., Perkins-Dameron, R., and Allen-Nash, A., 1986, Structure and expression of two temperature-specific surface proteins in the ciliated protozoan *Tetrahymena thermophila, Mol. Cell Biol.* **6:**3240–3245.

Baugh, L. C., Satir, P., and Satir, B., 1976, A ciliary membrane protein Ca++ ATPase: A correlation between structure and function, *J. Cell Biol.* **70:**66a.

Bligh, E. G., and Dyer, W. J., 1959, A rapid method of total lipid extraction and purification, *Can. J. Biochem. Physiol.* **37:**911–917.

Bloodgood, R. A., 1987, Glycoprotein dynamics in the *Chlamydomonas* flagellar membrane, *Adv. Cell Biol.* **1:** 97–130.

Butzel, H. M., and Decaprio, A., 1978, Ciliary membrane proteins of *Tetrahymena thermophila, J. Protozool.* **25:**267–269.

Dentler, W. L., 1978, Isolation and characterization of *Tetrahymena* ciliary membranes, *J. Cell Biol.* **79:**284a.

Dentler, W. L., 1980a, Microtubule–membrane interactions in cilia. I. Isolation and characterization of ciliary membranes from *Tetrahymena pyriformis, J. Cell Biol.* **84:**364–380.

Dentler, W. L., 1980b, Structures linking the tips of ciliary and flagellar microtubules to the membrane, *J. Cell Sci.* **42:**207–220.

Dentler, W. L., Pratt, M. M., and Stephens, R. E., 1980, Microtubule–membrane interactions in cilia. II. Photochemical cross-linking of bridge structures and the identification of a membrane-associated dynein-like ATPase, *J. Cell Biol.* **84:**381–403.

Doerder, F. P., 1979, Differential expression of immobilization antigen genes in *Tetrahymena thermophila*. I. Genetic and epistatic relationships among recessive mutations which alter normal exprssion of i-antigens. *Immunogenetics* **9:**551–562.

Doerder, F. P., 1981, Differential expression of immobilization antigen genes in *Tetrahymena thermophila*. II. Reciprocal and non-reciprocal transfer of immobilization antigen during conjugation and expression of immobilization antigen genes during macronuclear development, *Cell Differ.* **10:**299–307.

Doerder, F. P., and Berkowitz, M. S., 1986, Purification and partial characterization of the H immobilization antigens of *Tetrahymena thermophila, J. Protozool.* **33:**204–208.

Dute, R., and Kung, C., 1978, Ultrastructure of the proximal region of somatic cilia in *Paramecium tetraurelia, J. Cell Biol.* **78:**451–464.

Erwin, J., and Bloch, K., 1963, Lipid metabolism in ciliated protozoa, J. Biol. Chem. **238:**1618–1624.

Fisher, G., Kaneshiro, E. S., and Peters, P. D., 1976, Divalent cation affinity sites in *Paramecium aurelia, J. Cell Biol.* **69:**429–442.

Folch, J., Lees, M., and Sloane-Stanley, G. H., 1957, A simple method for the isolation and purification of total lipids from animal tissues, *J. Biol. Chem.* **226:**497–509.

Forney, J. D., Epstein, L. M., Preer, L. B., Rudman, B. M., Widmayer, D. J., Klein, W. H., and Preer, J. R., Jr., 1983, Structure and expression of genes for surface proteins in *Paramecium, Mol. Cell Biol.* **3:**466–474.

Frankel, J., Nelsen, E. M., and Martel, E., 1981, Development of the ciliature of *Tetrahymena thermophila*. II. Spatial subdivision prior to cytokinesis, *Dev. Biol.* **88**:39–54.

Frisch, A., and Loyter, A., 1977, Inhibition of conjugation in *Tetrahymena pyriformis* by concanavalin A. Localization of concanavalin A-binding sites, *Exp. Cell Res.* **110**:337–346.

Fukushima, H., Martin, C. E., Iida, H., Kitajima, Y., Thompson, G. A., Jr., and Nozawa, Y., 1976, Changes in membrane lipid composition during temperature adaptation by a thermotolerant strain of *Tetrahymena pyriformis*, *Biochim. Biophys. Acta* **431**:165–179.

Hadley, G. A., and Williams, N. E., 1981, Control of the initiation and elongation of cilia during ciliary regeneration of *Tetrahymena*, *Mol. Cell Biol.* **1**:865–870.

Haga, N., and Hiwatashi, K., 1981, A protein called immaturin controlling sexual maturity in *Paramecium*, *Nature* **289**:177–179.

Holz, G. G., Jr., and Conner, R. L., 1973, The composition, metabolism, and roles of lipids in *Tetrahymena*, in: *The Biology of Tetrahymena* (A. M. Elliott, ed.), Dowden, Hutchinson, & Ross, Stroudsburg, Pa., pp. 99–122.

Hufnagel, L. A., 1983, Freeze-fracture analysis of membrane events during early neogenesis of cilia in *Tetrahymena*: Changes in fairy-ring morphology and membrane topography, *J. Cell Sci.* **60**:137–156.

Jonah, J., and Erwin, J. A., 1971, The lipids of membranous cell organelles isolated from the ciliate, *Tetrahymena pyriformis*, *Biochim. Biophys. Acta* **231**:80–92.

Kennedy, K. E., and Thompson, G. A., Jr., 1970, Phosphonolipids: Localization in surface membranes of *Tetrahymena*, *Science* **168**:989–991.

Kitajima, Y., and Thompson, G. A., Jr., 1977, *Tetrahymena* strives to maintain the fluidity interrelationships of all its membranes constant. Electron microscopical evidence, *J. Cell Biol.* **72**:744–755.

Love, H. D., Jr., Allen-Nash, A., Zhao, O., and Bannon, G. A., 1988, mRNA stability plays a major role in regulating the temperature-specific expression of a *Tetrahymena thermophila* surface protein, *Mol. Cell Biol.* **8**:427–432.

Luft, J. H., 1976, The structure and properties of the cell surface coat, *Int. Rev. Cytol.* **45**:291–382.

Mallory, F. B., Gordon, J. T., and Conner, R. L., 1963, The isolation of a pentacyclic triterpenoid alcohol from a protozoan, *J. Am. Chem. Soc.* **85**:1362–1363.

Martindale, D. W., and Bruns, P. J., 1983, Cloning of abundant mRNA species present during conjugation of *Tetrahymena thermophila*: Identification of mRNA species present exclusively during meiosis, *Mol. Cell Biol.* **3**:1857–1865.

Nanney, D. L., 1971, The pattern of replication of cortical units in *Tetrahymena*, *Dev. Biol.* **26**:296–305.

Nanney, D. L., and McCoy, J. W., 1976, Characterization of the *Tetrahymena pyriformis* complex, *Trans. Am. Microsc. Soc.* **95**:664–682.

Nilsson, J. R., and Behnke, O., 1971, Studies on a surface coat of *Tetrahymena*, *J. Ultrastruct. Res.* **36**:542–544.

Nozawa, Y., 1975, Isolation of subcellular membrane components from *Tetrahymena*, in: *Methods in Cell Physiology*, Volume X (D. Prescott, ed.), Academic Press, New York, pp. 105–133.

Nozawa, Y., and Thompson, G. A., Jr., 1971a, Studies of membrane formation in *Tetrahymena pyriformis*. II. Isolation and lipid analysis of cell fractions, *J. Cell Biol.* **49**:712–721.

Nozawa, Y., and Thompson, G. A., Jr., 1971b, Studies of membrane formation in *Tetrahymena pyriformis*. III. Lipid incorporation into various cellular membranes of logarithmic phase cultures, *J. Cell Biol.* **49**:722–730.

Nozawa, Y., and Thompson, G. A., Jr., 1972, Studies of membrane formation in *Tetrahymena pyriformis*. V. Lipid incorporation into various cellular membranes of stationary phase cells, starving cells, and cells treated with metabolic inhibitors, *Biochim. Biophys. Acta* **282**:93–104.

Nozawa, Y., Iida, H., Fukushima, H., Ohki, K., and Ohnishi, S., 1974, Studies on *Tetrahymena* membranes: Temperature-induced alterations in fatty acid compositions of various membrane fractions in *Tetrahymena pyriformis* and its effect on membrane fluidity as inferred by spin-label study, *Biochim. Biophys. Acta* **367**:134–147.

Omura, T., Siekevitz, P., and Palade, G. E., 1967, Turnover of constituents of the endoplasmic reticulum membranes of rat hepatocytes, *J. Biol. Chem.* **242**:2389–2396.

Pagliaro, L., and Wolfe, J., 1987, Concanavalin A binding induces association of possible mating-type receptors with the cytoskeleton in *Tetrahymena*, *Exp. Cell Res.* **168**:138–152.

Richmond, J. E., 1976, Biosynthesis of membrane and membrane glycoprotein in *Tetrahymena pyriformis*, *Comp. Biochem. Physiol.* **55**:61–63.

Satir, B. H., and Wissig, S. L., 1982, Alveolar sacs of *Tetrahymena:* Ultrastructural characteristics and similarities of subsurface cisterns of muscle and nerve, *J. Cell Sci.* **55**:13–33.

Satir, B., Schooley, C., and Satir, P., 1972, The ciliary necklace in *Tetrahymena, Acta Protozool.* **11**:291–293.

Satir, B., Sale, W. S., and Satir, P., 1976, Membrane renewal after dibucaine deciliation of *Tetrahymena, Exp. Cell Res.* **97**:83–91.

Sattler, C. A., and Staehelin, L. A., 1974, Ciliary membrane differentiation in *Tetrahymena pyriformis, J. Cell Biol.* **62**:473–490.

Skriver, L., 1979, Membrane renewal during ciliary regeneration, *Proc. Int. Congr. Biochem.* **11**:403.

Skriver, L., and Williams, N. E., 1980, Regeneration of cilia in starved *Tetrahymena thermophila* involves induced synthesis of ciliary proteins but not synthesis of membrane lipids, *Biochem. J.* **188**:695–704.

Smith, J. D., and Lepak, N. M., 1982, Purification and characterization of a phosphonic acid-containing glycoprotein from the cell membranes of *Tetrahymena, Arch. Biochem. Biophys.* **213**:565–572.

Smith, J. D., Snyder, W. R., and Law, J. H., 1970, Phosphonolipids in *Tetrahymena* cilia, *Biochem. Biophys. Res. Commun.* **39**:1163–1169.

Subbaiah, P. V., and Thompson, G. A., Jr., 1974, Studies of membrane formation in *Tetrahymena pyriformis*. The biosynthesis of proteins and their assembly into membranes of growing cells, *J. Biol. Chem.* **249**: 1302–1310.

Thompson, G. A., Jr., 1967, Studies of membrane formation in *Tetrahymena pyriformis*. I. Rates of phospholipid biosynthesis, *Biochemistry* **6**:2015–2022.

Thompson, G. A., Jr., 1972, *Tetrahymena pyriformis* as a model system for membrane studies, *J. Protozool.* **19**:231–236.

Thompson, G. A., Jr., and Nozawa, Y., 1977, *Tetrahymena:* A system for studying dynamic membrane alterations within the eucaryotic cell, *Biochim. Biophys. Acta* **472**:55–92.

Van Bell, C. T., and Williams, N. E., 1984, Membrane protein differences correlated with the development of mating competence in *Tetrahymena thermophila, J. Protozool.* **31**:112–116.

Warren, L., and Glick, M. C., 1969, Membranes of animal cells. II. The metabolism and turnover of the surface membrane, *J. Cell Biol.* **37**:729–746.

Williams, N. E., 1975, Regulation of microtubules in *Tetrahymena, Int. Rev. Cytol.* **41**:59–86.

Williams, N. E., 1983, Surface membrane regeneration in deciliated *Tetrahymena, J. Cell Sci.* **62**:407–417.

Williams, N. E., and Bakowska, J., 1982, Scanning electron microscopy of cytoskeletal elements in the oral apparatus of *Tetrahymena, J. Protozool.* **29**:382–389.

Williams, N. E., and Luft, J. H., 1968, Use of nitrogen mustard derivative in fixation for electron microscopy and observations on the ultrastructure of *Tetrahymena, J. Ultrastruct. Res.* **25**:271–292.

Williams, N. E., Subbaiah, P. V., and Thompson, G. A., Jr., 1980, Studies of membrane formation in *Tetrahymena*. The identification of membrane proteins and turnover rates in non-growing cells, *J. Biol. Chem.* **255**:296–303.

Williams, N. E., Doerder, F. P., and Ron, A., 1985, Expression of a cell surface immobilization antigen during serotype transformation in *Tetrahymena thermophila, Mol. Cell Biol.* **5**:1925–1932.

Wunderlich, F., and Speth, V., 1972, Membranes in *Tetrahymena*. I. The cortical pattern, *J. Ultrastruct. Res.* **41**:258–269.

Wunderlich, F., Speth, V., Butz, W., and Kleinig, H., 1973, Membranes of *Tetrahymena*. III. The effect of temperature on membrane core structure and fatty acid composition, *Biochim. Biophys. Acta* **298**:39–49.

Ciliary Membrane Tubulin

R. E. Stephens

1. Introduction

Reports of membrane-associated tubulin, isolated from membrane fractions of nervous tissue, date back nearly to the initial discovery of tubulin itself while tubulin associated with ciliary membranes was reported a decade later. The literature dealing with neuronal membrane tubulin is comparatively large, sometimes contradictory, and somewhat controversial. The history of the membrane-associated tubulin question, with emphasis on membrane tubulin from a diversity of chiefly neuronal sources, has been reviewed in detail recently (Stephens, 1986a). This chapter will deal primarily with ciliary membrane systems, emphasizing tubulin from molluscan gill cilia and including some recent comparative work with sea urchin embryonic cilia as well.

2. Ciliary versus Flagellar Membranes

The cilia from the scallop (*Aequipecten irradians*) gill can be isolated in essentially quantitative yield by brief hypertonic salt treatment of the excised tissue, while sperm flagella can be isolated by simple mechanical shearing of the tails from live sperm (Stephens and Linck, 1969; Linck, 1973a). Early work by Linck (1973b) demonstrated very striking biochemical differences in the fractionation of these morphologically indistinguishable ciliary and flagellar axonemes, prepared by nonionic detergent solubilization of the surrounding membrane. Specifically, low-ionic-strength dialysis, a procedure devised by Gibbons (1965) to prepare dynein from *Tetrahymena* cilia, releases all of the dynein and solubilizes both central pair members from sperm flagellar axonemes, leaving insoluble the nine outer doublet microtubules (as expected from the *Tetrahymena* fractionation). In contrast, low-ionic-strength dialysis of cilia releases one-half of the dynein but solubilizes only one central pair member and all of the B-subfiber microtubules, leaving a ring of nine (singlet) A tubules.

Later, an analysis of the detergent-solubilized membrane from the two organelles also demonstrated some remarkable differences (Stephens, 1977a). Although the mem-

R. E. Stephens • Marine Biological Laboratory, Woods Hole, Massachusetts 02543.

brane fraction in both cases represents about one-fifth of the total protein of the respective organelle, the proteins that comprise this fraction are totally different. In sperm flagella, the membrane is rapidly dispersed by Triton X-100 at concentrations only slightly in excess of the critical micelle concentration. Analyzed by SDS-PAGE, the dominant membrane constituent is a glycoprotein of about 250 kDa and, in spite of the relative lability of both central pair members and the potential presence of unpolymerized tubulin in the periaxonemal space, comparatively little tubulin is detectable in the flagellar membrane extract (Fig. 1a,b). In contrast, multiple extractions with much higher Triton X-100 concentrations are required to totally remove the ciliary membrane. The dominant membrane constituents are a pair of equimolar polypeptides with an average molecular mass of 55 kDa, electrophoretically comigrating with authentic tubulin of the axoneme (Fig. 1c,d). Similar results are obtained from the gill cilia and sperm flagella of the deep-sea scallop *Placopecten magellanicus* (Stephens, unpublished) and also from two separable classes of gill cilia from the blue mussel *Mytilus edulis* (Stommel, 1984b) and its sperm flagella (Stephens, unpublished), although in none of these cases has any biochemical study been done on the presumptive tubulin subunits.

For *Aequipecten,* the identity of gill ciliary membrane tubulin as such is based on a number of criteria in addition to electrophoretic properties (Stephens, 1977a). Isolated by excision of protein bands from polyacrylamide gels, the presumptive α and β chains from the membrane fraction yield cyanogen bromide fragments in identical number and of identical size to those cleaved from bona fide tubulin subunits derived from the axoneme,

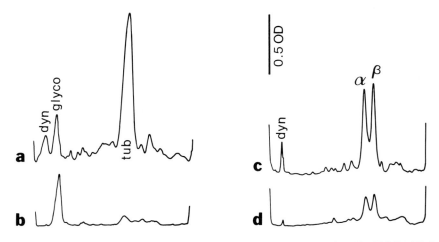

Figure 1. Scallop sperm flagellar and ciliary membrane and axoneme fractions analyzed by SDS-PAGE. Sperm tails and gill cilia were isolated by methods originated by Linck (1973a), the membranes were solubilized with 1% Triton X-100, and the respective fractions were run on either continuous Na-phosphate (a, b) or Tris-glycine-based (c, d) polyacrylamide-SDS gels and stained with Fast Green. (a) Whole sperm flagella; tubulin (tub), dynein (dyn), and a 250-kDa glycoprotein (glyco) dominate the pattern. (b) Sperm flagellar membrane fraction; the 250-kDa glycoprotein is the major constituent, with only a trace of tubulin present. (c) Ciliary axoneme; α- and β-tubulin chains, resolved by differential SDS binding on a Tris-glycine system, and dynein are the major components. (d) Ciliary membrane fraction loaded at 2.5× the stoichiometric ratio; bands comigrating with tubulin chains are the principal components and no 250-kDa protein is present. Adapted from Stephens (1977a).

indicating that at least the methionine residues are positioned identically in the respective chain pairs. High-resolution, two-dimensional tryptic peptide mapping shows near but not total identity, suggesting that, although the lysine and arginine residue positions are highly conserved, local nonpolar differences probably exist since certain peptides are reciprocally shifted. An apparent major difference is that ciliary membrane tubulin, subjected to electrophoresis designed for carbohydrate staining, is mildly PAS (periodic acid–Schiff)-positive while tubulin from the axoneme is not. The carbohydrate moiety is labile, however, and it is readily removed by acetone precipitation or by alkaline reduction and alkylation (Stephens, 1977a). The individual subunits, when resolved by discontinuous SDS-PAGE, are not PAS-positive (Stephens, unpublished), an important point discussed further in Section 5.

A more detailed comparative analysis of ciliary membrane and axonemal tubulin subunits confirms that nonpolar differences distinguish the respective tubulins (Stephens, 1981). Since electrophoretic purification using one-dimensional gels runs the risk of coisolating comigrating contaminants, tubulin subunits may be reduced and either carboxymethylated or carboxamido-methylated with iodoacetate or iodoacetamide, respectively, before isolation. Although these two derivatizations yield tubulin chain pairs that each migrate quite differently with respect to the unmodified tubulin chains, not only do the membrane and axonemal tubulin subunits comigrate but their respective tryptic peptide maps (Fig. 2) continue to show the same types of characteristic nonpolar peptide differences reported previously (Stephens, 1977a). Consistent with this, comparative

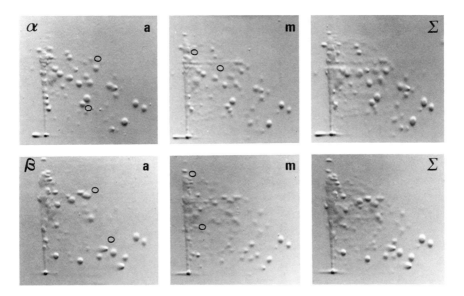

Figure 2. Comparative two-dimensional tryptic peptide mapping of α- and β-tubulin chains derived from scallop gill ciliary axoneme and membrane. Top row: α chains from the axoneme (a), membrane (m), and a mixture of the two (Σ). Circles denote peptides absent from one chain type that are characteristic of the homologous chain. Bottom row: the same for β chains. Coincidence of peptides indicates > 95% homology between membrane and axonemal tubulins in both chain types. Bas-relief-enhanced; based on data from Stephens (1981).

amino acid analysis also detects small but statistically significant differences in both polypeptide chains, most notably as a much higher isoleucine content in the membrane tubulin β chain. Finally, the differential binding of the detergents CTAB (cetyl trimethylammonium bromide) and DOC (deoxycholate) in the presence of excess Triton X-100 indicates that membrane tubulin is an amphiphilic protein whereas dialysis-solubilized 6 S B-subfiber tubulin is not, using the method and criteria established by Helenius and Simons (1977). However, later results indicating that membrane tubulin is complexed with lipids and with other proteins, even in the presence of moderate detergent concentrations (see Sections 5 and 6), now render this result somewhat equivocal since tightly bound lipid or mass differences could also contribute to the observed electrophoretic shifts.

The nonpolar differences that are evident in both of the membrane tubulin chains would suggest that they might be distinguishable from their axonemal counterparts on the basis of reverse-phase HPLC (high-performance liquid chromatography). This method can very efficiently separate TFA (trifluoroacetic acid)-solubilized axonemal tubulin α and β chains on a conventional C_{18} column, using linear TFA–water–acetonitrile gradients (Stephens, 1984). However, when ciliary membrane tubulin is delipidated by Triton X-114 detergent condensation and then solubilized with TFA–guanidine-HCl, it fails to chromatograph, in contrast to its well-behaved axonemal counterpart, identically solubilized. Retained on the column by self-aggregation, the membrane tubulin may be released with a pulse of DMSO (dimethylsulfoxide), after which both membrane tubulin chains chromatograph with retention times indistinguishable from those of axonemal tubulin (Stephens, 1986b). Such a result indicates that the respective membrane and axonemal tubulin chains must share common surface domains that interact identically with the alkyl groups of the column packing, at least in their final, eluted conformations, but it also suggests that the membrane tubulins may have internal nonpolar domains involved in either self-association or lipid binding that prevent normal chromatographic elution. An alternative interpretation is that the membrane tubulin chains retain tightly bound lipid to the extent that one cycle of HPLC and/or pure DMSO are required to remove the lipid and free the tubulin subunits.

The stark contrast in protein fractionation behavior between molluscan gill cilia and sperm flagella noted above is characteristic of sea urchin embryonic cilia and sperm flagella as well (Stephens, 1977b, 1978, 1986c), as are differences in the composition of the respective membranes. The major membrane protein of sea urchin sperm flagella is guanylate cyclase, present as a 160/150-kDa doublet of phosphorylated and dephosphorylated forms (Ward et al., 1986). The major detergent-solubilized membrane proteins of embryonic cilia, isolated by hypertonic shock from both a tropic and an arctic sea urchin species, are an equimolar pair of polypeptides of about 55 kDa, comigrating with the subunits of axonemal tubulin and cross-reacting with antibodies to sea urchin egg and vertebrate brain tubulin (Stephens, 1987). The fraction of tubulin found in the ciliary membrane of sea urchin embryos is similar to that of scallop gill and remains relatively constant from swimming blastula to at least the pluteus stage in both urchin species. A more detailed biochemical characterization of this sea urchin ciliary membrane-associated tubulin, with emphasis on protein synthetic differences between membrane and axonemal tubulin, is currently in progress but some significant basic results on metabolism, posttranslational modification, and reconstitution are discussed below (Sections 5 and 6).

3. Protozoan Cilia and Flagella

The cilia of *Tetrahymena* were the first to be analyzed biochemically (Gibbons, 1965) and remain today the most widely studied. In addition, because of extensive electrophysiological work, the cilia of *Paramecium* have also received much attention. Comparative studies on ciliary membranes from both organisms have been carried out by Adoutte *et al.* (1980) and Schultz *et al.* (1983), while Brugerolle *et al.* (1980) and Merkel *et al.* (1981) have investigated the membranes of mutant and wild-type *Paramecium* cilia. In all of these studies, in spite of the variety of methods used to isolate and purify the ciliary membranes, tubulin, if found at all, was only a minor membrane protein constituent. In contrast, Dentler (1980) reported a major *Tetrahymena* ciliary membrane protein that had electrophoretic properties similar to tubulin and was classified as tubulinlike on the basis of two-dimensional tryptic peptide mapping of its unresolved subunits. However, current work based on detergent phase-partitioning and antibody cross-reactivity now shows that the bulk of this apparent tubulinlike protein is not tubulin, although a small amount of tubulin is present in the membrane plus periaxonemal matrix fraction (Dentler, 1988, this volume).

The paired flagella of *Chlamydomonas* have long served as a model for flagellar assembly. The fact that an unequally sheared pair of flagella can coordinately resorb and regrow (Coyne and Rosenbaum, 1970) would indicate an exquisitely controlled equilibrium with unpolymerized axonemal components. In spite of this potential equilibrium, tubulin has never been reported as more than a trace constituent of the detergent-solubilized membrane plus periaxonemal matrix fraction (e.g., Witman *et al.*, 1972). The one claim for a bona fide membrane tubulin (Adair and Goodenough, 1978) was not later substantiated (Monk *et al.*, 1983). The α chain of *Chlamydomonas* flagellar tubulin is posttranslationally modified prior to assembly by the acetylation of an unmodified cytoplasmic form (L'Hernault and Rosenbaum, 1983, 1985). This modification must take place just prior to or upon assembly at the growing tip of the flagellum since the minor amount of tubulin found in the detergent-solubilized fraction, presumably derived from either the membrane or the periaxonemal matrix, is not acetylated (L'Hernault and Rosenbaum, 1983; Piperno and Fuller, 1985).

4. Definitions and Origin

The identification of a "membrane" tubulin (or for that matter, any purported tubulin variant with unique properties or derivation) presents a paradox, for if sufficient differences are found to rigorously prove its uniqueness, then one can argue that the protein is not tubulin. The ultimate criterion would be the ability to form bona fide microtubules but, as of this writing, this has not been accomplished and it is not clear (depending upon one's view on the function or state of membrane tubulin) whether this, in fact, need be a property of a membrane tubulin per se. In the case of scallop ciliary membrane tubulin, electrophoretic comigration of *both* chains under all conditions tested and CNBr peptide identity of *both* chains by peptide size and number would point to strict identity. In addition, isoelectric focusing indicates charge identity for *both* chain types (Stephens, 1981). Three widely popular tubulin antibodies cross-react equally well with *either or both* chain types from scallop (and sea urchin) ciliary membrane and axoneme fractions (Fig. 3 and Stephens *et al.*, 1987). By these four independent criteria, *both* of the scallop gill ciliary membrane protein chains can be defined as tubulins.

The two membrane tubulin subunits were described originally as being essentially equimolar (Stephens, 1977a; Fig. 1d), but more recent work using discontinuous SDS-PAGE would appear to indicate otherwise, with the α subunit predominating (see Fig. 5d). This is primarily an artifact of the discontinuous electrophoretic system, where the migration and proportionality of membrane tubulin chains are dependent on system geometry (Stephens, 1981) and on competing salt and detergent concentrations (Stephens, 1983a), although the axonemal tubulins behave more conventionally. A similar phenomenon has been reported by Bibring and Baxandall (1974) for a discontinuous, urea-containing gel system, whereupon the β chain of axonemal tubulin is progressively underestimated as sample load decreases. In addition, the α chain appears to be preferentially associated with the membrane and some true loss of the β chain may take place as ciliary membranes are reconstituted through several cycles (Stephens, 1985a; also Section 8).

Even though the identity of ciliary membrane tubulin may be convincing, its origin presents a different set of problems. As with membrane tubulin isolated from brain tissue, where contamination by tubulin derived from neurotubules has always clouded the issue of reality (see Stephens, 1986a), the unavoidable presence of axonemal microtubules and the potential presence of unpolymerized periaxonemal tubulin has made the origin of

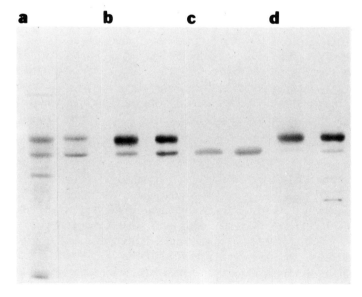

Figure 3. Antibody recognition of the two major scallop ciliary membrane proteins as tubulins. Western blots of equivalent loadings of membrane (left lane) and axonemal (right lane) tubulin, stained for protein (a), and with a polyclonal antibody against sea urchin egg tubulin (b; Polysciences #17870–1 : 5000 dilution), a monoclonal antibody against rat brain β-tubulin (c; Miles/ICN #63-781—1 : 500 dilution), and a polyclonal antibody against chick brain tubulin (d; Miles/ICN #65-095—1 : 2500 dilution). Protein loading in (b) is one-half, (c) is twice, and (d) is equal to that in (a). Homologous membrane and axonemal tubulin chains react equivalently with all three antibodies. Original data in (d); based on Stephens *et al.* (1987).

ciliary membrane tubulin unclear. The fact that little or no tubulin is detectable in the detergent-solubilized membrane fraction from either mollusk or sea urchin sperm *flagella,* wherein the central pair microtubules are quite labile, is not a totally satisfactory argument against *ciliary* membrane tubulin arising from central pair solubilization. Quantitatively, comparatively little breakdown of ciliary central pair microtubules is seen in isolated scallop cilia and both central pair members would have to be fully solubilized to account for even the minimal estimate of "membrane" tubulin actually observed. This argument also holds for sea urchin embryonic cilia, although the somewhat greater lability of their central pair and B-subfiber microtubules could lead to greater tubulin contamination of the membrane fraction.

Since axonemal tubulin can be solubilized by high salt, it should be pointed out that the hypertonic salt treatment used to isolate sea urchin embryo and scallop gill cilia cannot be the cause for the high tubulin content of the detergent-solubilized membrane plus periaxonemal matrix fraction. When cilia are isolated from either scallop gill or sea urchin embryos by the rapid, low-pH, isotonic, calcium–ethanol method of Auclair and Siegel (1966), essentially the same proportion of tubulin (and other proteins) occurs in the detergent-solubilized membrane plus periaxonemal matrix fraction (Stephens, unpublished).

However, the periaxonemal space in cilia (in contrast to that of flagella?) could contain the observed tubulin in unpolymerized form, possibly in equilibrium with that of the axoneme. This tubulin would be released upon detergent extraction and erroneously considered a membrane component. Such an argument would have two interesting but unlikely consequences. First, both the stable cilia from terminally differentiated scallop gill cells and the constantly elongating or resorbing cilia from sea urchin blastomeres would each have to have similar amounts of unpolymerized periaxonemal tubulin since the proportion of "membrane" to axonemal tubulin is similar in both cilia types. Second, in both kinds of organism, although the easily solubilized sperm flagellar membranes are very rich in protein, their poorly soluble ciliary membrane counterparts would have to be composed of nearly pure lipid (which actually should be more readily solubilized by detergent).

5. Metabolic Relationship between Membrane and Axonemal Tubulin

The ability to pulse–chase label proteins during growth or regeneration of sea urchin blastula cilia allows investigation of any potential precursor–product relationship between membrane and axonemal tubulin (Stephens, 1987). Parallel cultures of developing sea urchin embryos may be pulse–chase labeled with [^{3}H]leucine at various times of development, prior to isolation and regeneration of cilia. The isolated virgin and regenerate cilia may be fractionated into detergent-soluble membrane and detergent-insoluble axoneme fractions, resolved by SDS-PAGE, and the relative specific activity of the membrane and axonemal tubulin subunits determined by densitometry and fluorography. Regardless of the developmental stage of labeling, membrane tubulin has a considerably higher specific activity than axonemal tubulin in virgin cilia but the two proteins attain nearly equal specific activities upon ciliary regeneration. The former result would argue against any direct precursor–product relationship, since the membrane-associated tubulin contains considerably more label than the tubulin of the assembling axoneme, but the fact that both have nearly the same degree of label after an additional regeneration might suggest a precursor pool, compartmentalization, and complex processing.

The question of acylation of ciliary membrane tubulin after synthesis also can be approached with developing sea urchin embryos (Stephens, 1987). Parallel cultures may be set up such that one is regenerating cilia while the other undergoes steady-state maintenance and growth. Each are labeled with either [³H]palmitate, myristate, or leucine, the cilia are isolated, and the embryos are then allowed to regenerate a second population of cilia. During either steady-state or regeneration, only a high-molecular-weight membrane protein is acylated with palmitate (but not with myristate) and this protein's specific activity is the same in both steady-state and regeneration, as is also true of leucine-labeled tubulin. Two conclusions can be drawn from these results. First, sea urchin ciliary membrane tubulin is clearly not an acylated protein. Second, the fact that leucine-labeled membrane tubulin and the acylated membrane protein each have the same specific activity in membranes of both steady-state and regenerating cilia would indicate a very rapid physical turnover of membrane proteins relative to membrane growth during ciliary elongation.

The apparent presence of carbohydrate bound to ciliary membrane tubulin reported in earlier work was noted above and a discussion of covalently bound carbohydrate on neuronal membrane-associated tubulins has been presented in detail elsewhere (Stephens, 1986a). It should be pointed out that tubulin has at least the potential for being glycosylated. A single site within the ∝ chain of rat brain tubulin can undergo glycosylation when the membrane anchor domain of asialoglycoprotein is engineered onto the COOH-terminal portion of rat brain ∝-tubulin and the construct is translated *in vitro* in the presence of microsomes (Spiess adn Lodish, 1986).

No PAS-reactivity is observed when either scallop gill or sea urchin embryo ciliary membrane tubulin is subjected to discontinuous SDS-PAGE for the purpose of resolving the two tubulin subunits (Stephens, unpublished). This is still consistent with the original interpretation that sugars may be associated with tubulin through alkali-labile O-glycosidic linkages (Stephens, 1977a), since the running pH of such gels is quite alkaline. However, an alternative interpretation now seems more likely. In scallop membranes, the lipids PIP and PIP$_2$ (phosphatidylinositol mono- and bisphosphates) are present in rather significant amounts, migrating on discontinuous SDS-PAGE between the leading dodecylsulfate and the trailing glycinate anion fronts, and they are retained within the gel after fixation and staining (see Stommel and Stephens, 1985). On the continuous SDS-PAGE systems used previously, wherein sharp voltage, pH and SDS gradients are absent, these lipids can remain bound to proteins. Upon periodate oxidation, the inositol ring is cleaved to a dialdehyde (Grado and Ballou, 1961), giving rise to PAS-positive material which, in reality, is protein-bound lipid rather than carbohydrate. Whether bound PIP or PIP$_2$ is a natural characteristic of membrane-associated tubulin or merely an artifactual consequence of detergent solubilization of a membrane that is high in phosphoinositides has yet to be proven. However, one intriguing report notes that colchicine inhibits Con A-induced PI turnover in lymphocytes (Schellenberg and Gillespie, 1977), the authors favoring the hypothesis that this inhibition is somehow mediated by a membrane-associated tubulin. Conversely, phosphatidylinositol can modulate the assembly of microtubules through interaction with MAP-2 (Yamauchi and Purich, 1987).

Two posttranslational modifications of the α chain of tubulin distinguish cytoplasmic tubulin from tubulin of the axoneme. Paralleling the case of *Chlamydomonas* flagella noted above, the tubulin of sea urchin embryonic ciliary axonemes is acetylated while twice-recycled egg cytoplasmic tubulin is not (Fig. 4, Ac), as demonstrated here with a monoclonal antibody specific for acetylated α-tubulin (Piperno and Fuller, 1985). Antibodies prepared against synthetic peptides having the same sequence as the tyrosinated and nontyrosinated COOH terminus of α-tubulin can be used to quantitate tyrosinated and detyrosinated tubulin (Gunderson and Bulinski, 1986). Evaluated with these antibodies, sea urchin cytoplasmic tubulin exists almost exclusively with tyrosine as its α-chain

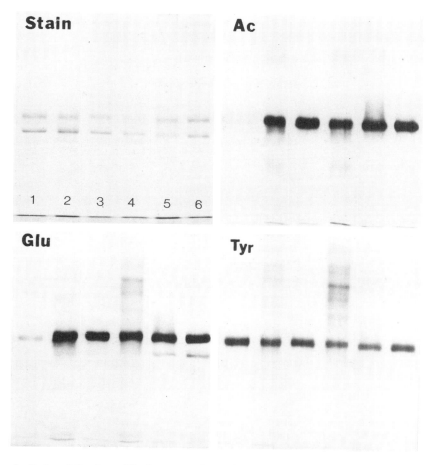

Figure 4. Posttranslational modification state of membrane and axonemal tubulin α chains. *Stain:* Approximately equal loading of twice-repolymerized sea urchin egg tubulin (1), sea urchin ciliary membrane (2) and axonemal tubulins (3), sucrose gradient-purified reconstituted sea urchin ciliary membrane (4), and scallop gill ciliary membrane (5) and axonemal (6) tubulins. *Ac:* Western blot of same, using a monoclonal antibody against acetylated α-tubulin (Ab 6-11B-1 of Piperno and Fuller, 1985; 1:100 dilution). *Glu* and *Tyr:* Western blots of same, using peptide antibodies against tyrosinated (Tyr) and detyrosinated (Glu) α-tubulin (Gunderson and Bulinski, 1986; each at 1:5000 dilution). Cytoplasmic α-tubulin is nonacetylated and contains approximately 3% detyrosinated tubulin. The remaining α- tubulins are all acetylated to approximately the same extent and contain approximately the same amounts of tyrosinated and detyrosinated tubulin.

COOH-terminal amino acid whereas ciliary axonemal tubulin contains approximately equal amounts of Tyr- and Glu-tubulin (Fig. 4, Glu/Tyr), as determined by a quantitative dilution series (not shown). The same holds true for scallop axonemal tubulin; it is both acetylated and tyrosinated to essentially the same degree as its sea urchin axonemal counterpart, although the state of the cytoplasmic tubulin in this case is not known. In both organisms, however, the ciliary membrane-associated tubulin cannot be distinguished unequivocally from that of the axoneme by these three highly specific antibodies for acetylated, tyrosinated, or nontyrosinated tubulin.

These observations are surprising for two reasons. First, Piperno and Fuller (1985) observed that tubulin in the detergent-solubilized *Chlamydomonas* membrane plus periaxonemal matrix fraction, like cytoplasmic tubulin, is nonacetylated. In fact, this is the primary evidence for the idea that tubulin synthesized in the cell body is acetylated before incorporation into the axoneme. One might then expect that if membrane-associated tubulin is destined for the axoneme, it might still be nonacetylated. Clearly, the sea urchin ciliary membrane tubulin must already be fully processed. Furthermore, the tubulin cannot be simply an unpolymerized cytoplasmic contaminant since cytoplasmic tubulin is principally nonacetylated. Second, Nath and Flavin (1978) noted that neuronal membrane-associated tubulin appeared to be nontyrosinated but could accept a COOH-terminal tyrosine when solubilized from the membrane, suggesting that this was the primary structural distinction between membrane and cytoplasmic tubulin. However, in both scallop gill and sea urchin embryo ciliary membrane (and axoneme), the tubulins exists in both forms, in roughly equal amounts. This suggests no clear distinction of ciliary membrane tubulin as a nontyrosinated entity, although both it and axonemal tubulin are readily distinguishable from the almost fully tyrosinated cytoplasmic tubulin in sea urchin embryos. Later, Nath and Flavin (1980) modified their view somewhat, concluding that there is, in fact, a tyrosinated tubulin fraction in the membrane and that the COOH-terminal tyrosines of both membrane and cytoplasmic tubulin turn over at the same rate, but that the membrane and cytoplasmic tubulins differ somehow in their ability to accept tyrosine in *in vitro* assays.

6. Reconstitution of Ciliary Membranes

Solubilization of scallop gill ciliary membranes with either Triton X-100 or Nonidet P-40 results in the production of a heterodisperse 4–6 S species that is clearly distinct from that of pure detergent micelles (Stephens, 1977a). When the detergent is removed from the solubilized membrane plus periaxonemal matrix components by adsorption to SM-2 BioBeads, a solution of presumptive protein–lipid micelles results. When such a solution is frozen and thawed, or else concentrated by ultrafiltration, uniform membrane vesicles are formed (Fig. 5). Such reconstituted vesicles have the same lipid and protein composition, and consequently the same buoyant density, as the native ciliary membranes from which they were derived. The membranes are stable to 1 M KI or KSCN extraction, with no change in either protein composition or buoyant density (Stephens, 1983a). The only major protein of the initial membrane plus periaxonemal matrix extract that does not reconstitute to form membrane vesicles is calmodulin, thought to serve primarily as a soluble calcium buffer and potentially as a calcium control protein (Stommel *et al.*, 1982). A minor subfraction of tubulin also remains soluble. Since this varies with the prepara-

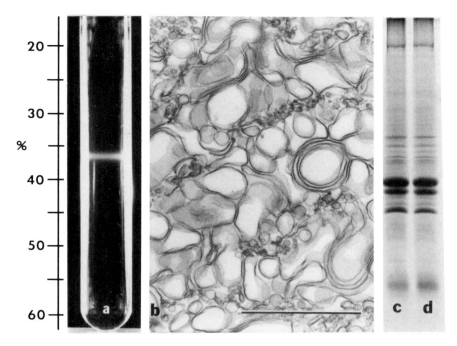

Figure 5. Reconstituted scallop gill ciliary membrane tubulin vesicles. A monodisperse population of vesicles is formed when Triton X-100 is removed from a detergent-solubilized ciliary membrane fraction and the preparation is frozen and thawed. Isopycnic centrifugation (a) and electron microscopy (b) of the vesicles demonstrate uniformity of the product; bar = 1 μm. SDS-PAGE reveals no appreciable difference in protein composition before (c) or after (d) sucrose gradient purification. Adapted from Stephens (1983a).

tion, it may represent inefficient incorporation of membrane tubulin or perhaps residual central pair or periaxonemal tubulin. Interestingly, attempts to polymerize microtubules from the detergent-solubilized membrane using the microtubule-inducing drug Taxol result only in the formation of protein-depleted membrane vesicles, in spite of the continued presence of detergent (Stephens, 1983b).

Membrane tubulin derived from sea urchin embryonic cilia will also undergo membrane reconstitution when detergent is removed and the preparation is frozen and thawed or concentrated. When analyzed by isopycnic sucrose gradient centrifugation and electron microscopy, a very uniform membrane population results, having essentially the same buoyant density as vesicles reconstituted from solubilized scallop gill ciliary membrane, although in this case open membrane leaflets rather than closed vesicles are the primary product. The membrane reconstitution is virtually quantitative; no detectable proteins reman soluble at the top of the gradient nor is any insoluble protein sedimented through the gradient (Stephens, in preparation). Such gradient-purified reconstituted sea urchin ciliary membrane vesicles are used as a test sample in Fig. 4.

Two lines of evidence argue against the artifactual incorporation of soluble tubulin during ciliary membrane reconstitution. First, if the freeze–thaw reconstitution is carried out in the presence of added brain tubulin, no brain tubulin is incorporated into the membrane vesicles (Stephens, 1985a). Conversely, if brain tubulin is polymerized in the presence of the detergent-free membrane extract, although the microtubules which form contain some ciliary membrane tubulin, the vesicles which also form upon warming contain no brain tubulin (Fig. 6). These experiments were carried out with an elasmobranch brain tubulin whose subunits could be distinguished from those of molluscan cilia by isoelectric focusing. The results suggest organelle specificity of tubulin function but one could still argue that species differences might play some role in this selective incorporation.

Second, if the freeze–thaw reconstitution is carried out using detergent-solubilized ciliary membrane lipids (produced by chloroform–methanol extraction of cilia) and soluble 6 S B-subfiber/central pair ciliary axonemal tubulin (produced by low-ionic-strength dialysis), the tubulin is only weakly adsorbed to the resulting pure lipid vesicles since it is removed by a simple low-ionic-strength wash (Fig. 7). This is the primary evidence against the possibility that ciliary membrane lipids, solubilized by Triton X-100, incorporate soluble axonemal tubulin when either the detergent is removed or the solution is frozen and thawed, potentially denaturing the tubulin (Stephens, 1985a).

The detergent-mediated solubilization/freeze–thaw reconstitution can be cycled a number of times with no qualitative change in protein composition (Stephens, 1985a).

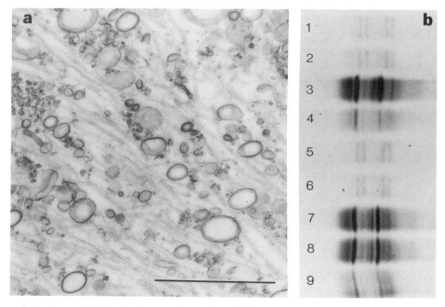

Figure 6. Sorting out of tubulin during coformation of ciliary membrane vesicles and brain microtubules. Electron micrograph (a) of detergent-solubilized ciliary membrane tubulin and skate brain tubulin warmed together in repolymerization buffer shows that micro- (and macro-) tubules and vesicles form; bar = 1 μm. Isoelectric focusing of the starting materials and separated final products (b) demonstrates organelle specificity. Lanes 1 and 2 are the pellet and supernatant of the solubilized membrane preparation warmed alone; lanes 3 and 4 are the same for brain tubulin preparation. Cold depolymerization of the copolymerization product solubilizes only the microtubules. The resulting (washed) membrane pellet (lanes 5 and 6) contains only the scallop (membrane) tubulin isoforms; the cold supernatant (lanes 7 and 8) contains mainly skate (brain microtubule) isoforms. The final lane (9) is the supernatant from the initial copolymerization and contains both sets of tubulin isoforms. Based on Stephens (1983a).

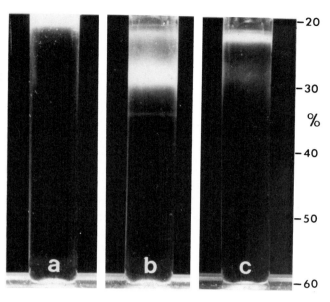

Figure 7. Isopycnic sucrose gradient analysis of vesicle reconstitution in the presence of 6 S ciliary axonemal tubulin. When ciliary lipids are dissolved in Triton X-100, followed by detergent removal and freeze–thaw, liposomes of low density result (a). When the same is carried out in the presence of dialysis-solubilized B-subfiber plus central pair tubulin, liposomes of moderate density form (b). Following a low-ionic-strength wash, recentrifugation of such vesicles to equilibrium results again in a population whose density is characteristic of nearly pure lipid (c). Based on Stephens (1985a).

There is some progressive but indiscriminant loss of protein, the inevitable result of denaturation, but the vesicle population density distribution remains uniform (though decreasing) at each cycle. Conversely, increasing detergent concentrations (whether Triton X-100, NP-40, or octyl glucoside) solubilize all ciliary membrane proteins uniformly, suggesting that membrane tubulin and the various minor proteins of the ciliary membrane form some sort of a complex with each other and with lipids. Supporting this concept is the additional evidence that only a very-high-molecular-weight complex, and not free tubulin dimer, is detectable by sucrose density gradient centrifugation, gel exclusion chromatography, and native polyacrylamide gel electrophoresis, all in the presence of excess solubilizing detergent (Stephens, 1985a).

7. Micellarization with Detergents and Interaction with Lipids

Members of the Triton series of detergents come out of solution as a micellar phase when warmed above a certain critical temperature, the "cloud point." Bordier (1981) has taken advantage of this phenomenon as a means for judging whether proteins are integral membrane proteins based on their ability to partition into the micellar or condensed phase of the detergent Triton X-114, which has a cloud point near room temperature. Regula *et al.* (1986) demonstrated the ability of a Triton X-100-soluble, Taxol-precipitated subfraction of bovine brain tubulin to partition into micellar Triton X-114. In contrast to the behavior of this neuronal membrane tubulin, Triton X-114-solubilized tubulin from gill ciliary membranes (plus most of the associated minor membrane components) remains with the aqueous phase when subjected to the Bordier procedure (Stephens, 1985b). The exception is a single protein of about 20 kDa. The micellar phase contains most, if not all, of the membrane lipids while the still-soluble but now-delipidated membrane tubulin and minor membrane components evidently are complexed with the detergent in some non-condensable form. A similar situation exists with the acetylcholine receptor which, although unquestionably an integral membrane protein complex, does not partition into the micellar phase of Triton X-114 (Maher and Singer, 1985).

One possible explanation for the inability of either membrane tubulin or the acetylcholine receptor proteins to partition into a detergent micellar phase is that the protein complex is too large to be encompassed within or condensed as a detergent–protein mixed micelle. Ciliary membrane tubulin is not impressively soluble in Triton X-100 (see Stephens, 1985a) but it is even less soluble in Triton X-114, with the latter serving as a very convenient means for delipidating the protein (Stephens, 1985b). This leads to a second possibility, not mutually exclusive from the first, namely that Triton X-114 may more fully displace protein-bound lipid than does Triton X-100, possibly explaining the extreme ease of membrane vesicle reconstitution by Triton X-100 removal. Such an argument would suggest that condensation from Triton X-100 might be a more effective means for probing the possible integral nature of ciliary membrane tubulin.

When Triton X-100 is substituted for Triton X-114 in the Bordier (1981) procedure, temperatures in excess of 60°C are required to bring about detergent condensation. Although this treatment would appear to be extreme, there is little evidence of denaturation when the procedure is applied to dialysis- or high-salt-solubilized ciliary axonemal tubulin since almost no tubulin appears in the condensate, either through incorporation or coagulation (Fig. 8a). In contrast, when the procedure is applied to Triton X-100-solubilized ciliary membrane, most of the tubulin partitions into the condensate (Fig. 8b). Condensation of tubulin with the detergent phase at higher temperatures does not occur with either Triton X-114 or Triton N-101 (not shown). When salt is added to lower the cloud point to the same temperature range used for Triton X-114 condensation, nearly all of the membrane tubulin still partitions into the detergent-rich condensate phase while many minor proteins remain in the soluble phase, producing a result nearly identical to the simple high-temperature condensation (Fig. 8c). Complicating such a seemingly ideal outcome is the fact that soluble axonemal tubulin and minor proteins totally partition into the detergent condensate under such high-salt conditions (Fig. 8d). Simplistically, one might conclude from this that soluble axonemal tubulin is even more hydrophobic than is membrane tubulin, but it is far more likely that dimeric axonemal tubulin, because of its size and lack of bound lipid, is salted out as a condensable detergent complex, in clear contrast to its opposite behavior in the absence of high salt at higher temperature. Perhaps significantly, membrane tubulin can partition into the detergent-rich phase under either condition.

It should be pointed out that dimeric cytoplasmic tubulin is quite capable of binding a surprisingly high amount of nonionic detergent. Andreu and Munoz (1986) demonstrated that calf brain tubulin will bind a minimum of 60 octyl glucoside or 95 DOC molecules. Hydrodynamic determinations are consistent with a detergent-induced tubulin self-association in octyl glucoside and a simple tubulin dimer–detergent complex in DOC. Earlier work by Andreu (1982) showed a comparable interaction with Triton X-100, although the stoichiometry was not ascertained due to this detergent's optical properties.

Several lines of evidence, based on liposome interaction and detergent competition, indicate a natural association of tubulin with lipids. Daleo et al. (1977) showed that phospholipids could inhibit in vitro microtubule polymerization and that the effect was reversible by detergents whose action correlated with their ability to remove phos-

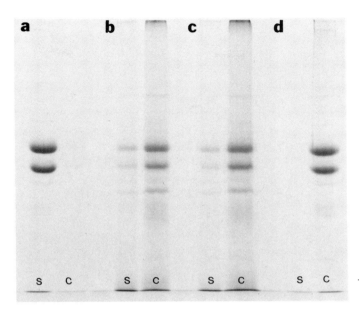

Figure 8. Integral membrane protein cocondensation with detergent using Triton X-100. (a) Aqueous supernatant (S) and detergent-rich condensate (C) phases obtained by warming dialysis-solubilized axonemal tubulin in 1% Triton X-100 warmed to 70°C. Nearly all protein remains in the aqueous phase. (b) Ciliary membrane proteins treated in the same manner. Most protein condenses with the detergent. (c) Ciliary membrane proteins warmed to 30°C in the presence of 2.5 M NaCl. Essentially the same partitioning takes place (in this case the condensate floats due to the density of the salt solution). (d) Dialysis-solubilized axonemal tubulin warmed to 30°C in the presence of 2.5 M salt. In contrast to (a), all protein condenses with the detergent phase.

pholipids. Reaven and Azhar (1981) demonstrated inhibition of microtubule polymerization by various hepatic membrane fractions and the lipids extracted from them, suggesting that tubulin might be involved in phospholipid transport. Mesland and Spiele (1984) found that extensive microtubule formation took place in hepatocytes treated with low concentrations of certain detergents. The detergent Triton N-101, which does not reverse the phospholipid-mediated inhibition of microtubule polymerization, had no effect, implying that hepatocytes may contain tubulin–phospholipid complexes from which tubulin could be released by removal of lipid. More recently, Hargreaves and McLean (1988) demonstrated that highly purified microtubule protein preparations contain a unique subset of phospholipids, distinct from whole-brain homogenate, and that tubulin and MAPs also differ in their associated lipids. Collectively, these observations based on cytoplasmic tubulin would suggest that the interaction of tubulin or tubulin isoforms with specific lipids might be an important determinant in lipid or protein localization, a point dealt with further in Section 9.

8. Tubulin as an Integral Membrane Structural Element

As pointed out above, concentrations of Triton X-100 at many times the critical micelle concentration are needed to fully solubilize ciliary membrane tubulin and there is no selective release of any particular proteins during the solubilization process. This is true for both native membranes and reconstituted vesicles. Evidently, just above the critical micelle concentration, lipid is selectively removed from the membrane, leaving membrane-associated proteins as a sedimentable complex, as judged from SDS-PAGE analysis of the solubilization process (see Stephens, 1985a).

Direct observation of Triton X-100 action confirms these conclusions and reveals a "membrane skeleton" containing essentially all of the membrane tubulin and associated minor proteins (Stephens *et al.*, 1987). When applied to isolated cilia, this is perhaps the strongest single argument against membrane tubulin being a soluble periaxonemal contaminant. In the case of both reconstituted membrane vesicles (Fig. 9a–c) and intact cilia still attached to the gill (Fig. 9d), a "skeleton" having a structure similar to the original membrane remains after detergent extraction of lipids but before solubilization of most

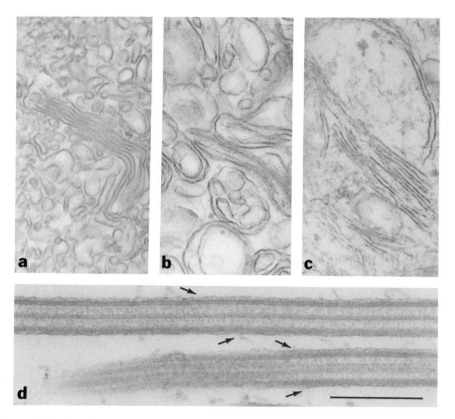

Figure 9. Membrane skeleton formation by Triton X-100 extraction without protein solubilization. Reconstituted ciliary membrane vesicles before extraction (a), and after extraction with (b) 0.02% detergent (> critical micelle concentration), and (c) 0.06% detergent (four times the critical micelle concentration). The latter detergent concentration extracts little protein but leaves a granuloreticular remnant of the original membrane vesicles. Similar results are obtained when gill cilia are gently extracted with 0.06% detergent *in situ* (d), where a granular membrane "sleeve" (arrows) remains around the axoneme. Bar = 0.5 μm. Adapted from Stephens *et al.* (1987).

proteins. The fact that a stable skeleton is obtained under the same conditions for both intact gill cilia and reconstituted vesicles would argue that the insoluble membrane skeleton seen in reconstituted vesicles is not a consequence of denaturation during the freeze–thaw process. Particularly in the case of the membrane vesicles, this skeleton appears to have a definite granularity, and at the stage of solubilization which represents near-complete delipidation, the granular elements can be shown by antitubulin/immunogold labeling to contain tubulin (Stephens *et al.*, 1987).

Although the final stages of detergent solubilization of ciliary membranes reveal the presence of a granuloreticular structure and similar material can be observed after detergent removal (Stephens, 1985b) or during membrane reconstitution (Stephens *et al.*, 1987), and even though a very-high-molecular-weight complex exists after complete detergent solubilization, there is no evidence that a native particle of membrane tubulin and associated proteins exists within the ciliary membrane. Freeze-fracture analysis reveals that, with the exception of the particles in the basal region which comprise the necklace complex and doublet–membrane linkages, the ciliary membrane proper is comparatively smooth. The membrane "particles" observed at late stages of solubilization or early stages of reconstitution most likely represent large, discrete lipid–protein–detergent micelles (Stephens, 1985a; Stephens *et al.*, 1987).

Preliminary vectorial labeling studies with both isolated and *in situ* scallop gill cilia indicated that each of the membrane tubulin subunits could be tagged by hydrophobic probes, that the α subunit was preferentially labeled, and that the β subunit could be preferentially cleaved by trypsin (Stephens, 1977c). Similar evidence for the preferential association of α-tubulin with presynaptic membranes was later presented by Gozes and Littauer (1979) while the hydrophobic cross-linking studies of Dentler *et al.* (1980) gave further evidence for tubulin–tubulin interaction within or just beneath the lipid bilayer. Parallel work with reconstituted ciliary membrane vesicles has proven equivocal since the vesicles are not fully sealed nor do the proteins appear to be uniformly inserted (Stephens, unpublished). Further details of ciliary membrane tubulin topology will most likely depend upon the use of more specific probes, applied to fully intact and unperturbed ciliary membranes *in situ*.

9. Summary and Discussion

All evidence points to the existence of bona fide tubulin α and β chains as the two primary protein constituents of the detergent-solubilized membrane plus periaxonemal fraction in cilia of both molluscan gill and sea urchin embryos. This contrasts with their sperm flagellar counterparts, and also with protozoan cilia or flagella, none of which contain appreciable amounts of tubulin in the detergent-solubilized membrane fraction. Scallop gill ciliary membrane tubulin is distinguishable from that of the axoneme by local, nonpolar amino acid substitutions, but the methionine positions within homologous chains are identical, the lysine and arginine positions are nearly so, and both proteins appear to share common surface domains. A variety of antibodies are unable to distinguish scallop gill or sea urchin embryo ciliary membrane tubulin chains from their axonemal tubulin counterpart. Detergent-solubilized scallop gill or sea urchin embryonic ciliary membranes may be reconstituted by detergent removal and freeze–thaw to a uniform vesicle population of similar density and protein composition. The tubulin of scallop gill ciliary membranes may be reconstituted into vesicles to the exclusion of polymerization-competent brain tubulin or soluble ciliary axonemal tubulin. Once reconstituted into membranes, the tubulin is resistant to extraction by various chaotropic salts. Metabolically, sea urchin ciliary membrane tubulin behaves differently than that of the axoneme but there is currently no evidence for any simple posttranslational modification that would render it so. Rather than existing as integral monomeric subunits or dimers in the membrane, tubulin appears to form tight associations with itself, with lipids, and with various minor proteins of the scallop gill ciliary membrane, giving rise to a comparatively rigid membrane skeleton.

The function of membrane tubulin is purely speculative at this point but the fact that it is found in the cilia and not the flagella of only certain organisms may provide some useful clues. Molluscan gill lateral cilia have the need to directly drive rapid currents of water, in contrast to the cilia of ciliates, which propel the cell bodies, or typical epithelial cilia, which propel a layer of mucus. The scallop ciliary membrane is tightly associated with the axoneme and, in fact, photochemical cross-linking studies indicate that it is tethered to the axoneme, possibly through a linkage involving a dyneinlike protein (Dentler et al., 1980; Dentler, this volume). The same need for a tethered ciliary membrane may hold for sea urchin embryos which must swim rapidly while, at the same time, elongating their existing cilia and replacing lost ones. Sperm flagellar membranes normally would not be subjected to the same degree of shear as cilia and, in fact, one can generally strip much of the membrane from flagella (but not cilia of the same species) by mechanical shear. Clearly, the non-tubulin-containing flagellar membranes are less tightly associated with the axoneme.

The second feature that tubulin-containing cilia have in common is the fact that they are mechanically sensitive. Scallop and mussel gill lateral cilia arrest as a consequence of a calcium-based electrical response and calcium influx (Stommel et al., 1980; Murakami and Machemer, 1982) whereas mussel gill abfrontal cilia are activated as a consequence of similar calcium-based phenomena (Stommel, 1984a,b). Also, sea urchin embryonic cilia are sensitive to touch and are even capable of reversal, at least in the pluteus stage (e.g., Mackie et al., 1968). One could argue that a relatively rigid, mechanically resistant ciliary membrane could function in the transduction process, either directly as a stretch receptor or indirectly as a means to transmit force to the basal or somatic membrane region. In fact, it has been demonstrated that rendering molluscan statocyst ciliary membranes rigid through photochemical cross-linkage amplifies the organ's electrical response to mechanical force transmitted along the ciliary membrane to the plasma membrane of the hair cell (Stommel et al., 1980). It should be added that the bioelectric control of Paramecium cilia, which have no significant membrane tubulin and an easily removed membrane, appears to be a much more complex phenomenon, with specific directional mechanoreceptor channels residing on the cell's somatic membrane but voltage-sensitive calcium channels along the ciliary membrane (Machemer and Ogura, 1979; Preston and Saimi, this volume).

Retinal rods, being ciliary derivatives, may eventually provide some insight into membrane tubulin function. Cook and co-workers (1987) identified a 63-kDa protein as the cGMP-dependent cation channel responsible for hyperpolarization in response to light. Matesic and Liebman (1987), however, claim a 39-kDa protein serves this function. In an intriguing preliminary report, Matesic and Liebman (1989) prepared the Cook group's protein and demonstrated that it could be resolved into two equimolar polypeptides, identified as tubulin chains on the basis of antibody cross-reactivity and one-dimensional peptide maps. Furthermore, this protein is tightly associated with rod outer segment plasma membranes and can be reconstituted into liposomes, thus classifying it as a membrane-associated tubulin. Although the question of which protein is actually the cGMP-dependent cation channel has yet to be resolved, these observations suggest a possible role for membrane tubulin in channel function or integrity (see Stephens, 1986).

Further circumstantial support for some sensory-related role for ciliary membrane tubulin comes from the work of Chen and Lancet (1984), comparing frog olfactory cilia with respiratory cilia. Detergent-solubilized membranes of the former contain at least five times more protein than does the latter. One of the two dominant proteins derived from olfactory ciliary membrane is a nonglycosylated tubulin, identified by monoclonal antibody cross-reactivity, which is absent in the respiratory ciliary membrane. Although these workers note that the protein may serve some sensory role, they also point out that membrane tubulin could mediate microtubule–membrane interactions important for the structural integrity of these very long olfactory cilia or that the tubulin may be involved in confinement of specific proteins to this specialized sensory membrane. Recent work from this group favors a unique transmembrane glycoprotein as the actual olfactory receptor (Chen *et al.*, 1986).

An interesting phenomenon that also needs to be considered is that of ciliary membrane-mediated movement. This was first described as a rapid shuttling of externally attached beads along the surface of flagella from a *Chlamydomonas* paralyzed mutant (Bloodgood, 1977). The phenomenon also occurs along the immotile apical tuft cilia of sea urchin embryos (Bloodgood, 1980). The interaction of a dyneinlike ATPase bridging the membrane and axoneme with tubulin attached to the membrane (Dentler *et al.*, 1980) could provide a mechanism for such movement although the minimal tubulin in the *Chlamydomonas* membrane fraction would argue against such a model. Whatever the cause of this movement (see Bloodgood, this volume), both this intriguing phenomenon and the metabolic results discussed above indicate that the ciliary membrane is highly dynamic.

In addition to such physical and physiological arguments, one can also argue that ciliary membrane tubulin is simply a precursor to tubulin in the axoneme, adding to the distally growing outer doublets (see Dentler and Rosenbaum, 1977; Dentler, 1981). The fact that ciliary membrane tubulin cannot be distinguished by specific antibodies from tubulin of the axoneme in terms of acetylation or tyrosination (in contrast to cytoplasmic tubulin which lacks acetylation and is almost full tyrosinated) is certainly consistent with such a role. However, the fact that membrane tubulin labeled during ciliary regeneration has a much higher specific activity than does that of the growing axoneme would suggest that it is not an immediate precursor. One could argue that the ciliary membrane tubulin fraction is dominated by newly synthesized protein but that the axoneme can only be assembled from previously synthesized and posttranslationally modified tubulin. The fact that axonemal tubulin synthesized during one regeneration is not fully utilized until the next is consistent with such a stored precursor model (Stephens, 1977b). Even if not a direct or even an indirect precursor, ciliary membrane tubulin could be involved in mediating the transport of axonemal proteins, including a distinct axonemal tubulin, to the growing ciliary tip. However, still remaining is the question of why the cilia from terminally differentiated molluscan gill cells should have a similar proportion of membrane tubulin as cilia from actively differentiating sea urchin embryos. Perhaps the answer is that gill cilia turn over much more rapidly than one would suspect (see Stephens, 1988) and, conversely, that those cilia or flagella showing little detergent-soluble tubulin do not.

As noted above, tubulin appears to be complexed with lipids in hepatic cells, leading to the suggestion that such a complex is involved in either the delivery of lipids or the deployment of microtubules. A related proposal concerning actin and the origin of the acrosomal process in certain sperm was made over a decade ago by Sardet and Tilney (1977). In this case, based on staining with hydrophobic probes, it was proposed that lipids were complexed with profilamentous actin in the periacrosomal region and were released upon activation. The resultant free actin would then assemble to form the filamentous process while the nascent lipids would contribute to the extensive acrosomal membrane. Such an attractive packaging model could be applied to the ciliary membrane tubulin–lipid complex as well. This concept is made even more intriguing by the possibility that PIP or PIP_2 may be associated with membrane tubulin (see Section 5). There is now considerable evidence for the functional, reversible interaction of normally soluble cytoplasmic proteins, including many cytoskeletal proteins, with these or other specific lipids. Such proteins are referred to as amphitropic (Burn, 1988) and their reversible association with lipids may have important implications for cytoskeletal–membrane interactions and transmembrane signaling. The potential modulation of tubulin–lipid or tubulin–membrane association by specific events in the PI cycle opens up a considerable number of functional possibilities.

Finally, there is the simple question of fundamental differences in cell type. Is plasma membrane-associated tubulin a characteristic of only the ciliary membrane proper or of the whole cell? Sea urchin egg and blastomere plasma membranes have been purified and studied extensively, but detailed analyses (e.g., Kinsey et al., 1980; Dasgupta and Garbers, 1983) have shown these membranes to be quite different in protein composition from the ciliary membrane described here. As for scallop gill lateral cell plasma membranes, one can easily isolate such cells after deciliation in order to compare ciliary and cytoplasmic proteins (Stommel et al., 1982). When plasma membranes are purified from deciliated lateral cells and compared with reconstituted ciliary membranes, the differences are quite clear (Fig. 10). The plasma membrane polypeptide composition is very complex but nevertheless similar to that of many other cell types, for example liver cells prepared by the same method (Hertzberg, 1984). The somatic plasma membrane contains only a trace of tubulinlike polypeptides, based upon antibody cross-reactivity, whereas tubulin comprises the bulk of the ciliary membrane protein.

Such ciliary membrane protein differences with respect to the somatic plasma membrane are not restricted to sea urchin or molluscan cells. Noting that major compositional differences exist between flagellar and cell body membranes, and utilizing several independent arguments concerning the localization of antigens specifically for flagellar and cell body membranes, Musgrave et al. (1986) present strong evidence for a functional membrane barrier in the transition zone between the flagellum and cell body of Chlamydomonas gametes. The question of how such striking topographic differentiation in any of these systems is initially created and continually maintained is not at all obvious.

It has been suggested often that the ciliary necklace may prevent the intermingling of ciliary and somatic membrane components (see Gilula and Satir, 1972; Dentler, 1981) but there is no direct evidence to support this attractive though primarily teleological argument. Specific topological targeting or localization of membrane proteins represents a

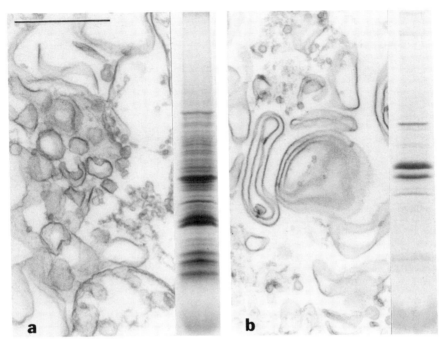

Figure 10. Comparison of scallop gill epithelial cell and ciliary membranes. (a) Electron micrograph and SDS-PAGE analysis of a gradient-purified deciliated epithelial cell plasma membrane fraction, prepared by the method of Hertzberg (1984). Bar = 0.5 μm. (b) Same of a gradient-purified reconstituted ciliary membrane fraction, prepared by method of Stephens (1983a).

current major area of interest (see Kaprelyants, 1988) and it is argued that membrane proteins are directed to specific locations through membrane–cytoskeletal interactions. In this particular case, trafficking of membrane proteins specifically to a region beyond the ciliary necklace would present some major steric difficulties, but the fact remains that the ciliary and somatic membranes frequently contain quite different lipid and protein components (see Dentler, 1981; Bouck *et al.*, this volume; Williams, this volume; Kaneshiro, this volume).

From a number of independent viewpoints, the membranes of cilia and flagella appear to encompass very dynamic systems in terms of motility, morphogenesis, and metabolism. Associated with certain ciliary membranes, the presence of a bona fide tubulin, generically considered a contractile or motile protein, implies that membrane tubulin could serve some function in this dynamism. However, no concrete evidence for any such role currently exists, still leaving the function of ciliary membrane tubulin as a major unanswered question. The fact that membrane tubulin is not a universal ciliary constituent would indicate that its role may be quite specialized, but its presence in both molluscan gill and echinoderm embryonic cilia would suggest that it is more than a simple evolutionary anomaly.

ACKNOWLEDGMENTS. The author's work reported herein was supported by USPHS Grants GM 20644 and 29503 from the National Institute of General Medical Sciences. I thank Drs. Gianni Piperno and Jeannette Chloe Bulinski for their generous contributions of antibodies to modified tubulins, Dr. Susan Oleszko-Szuts for excellent technical support in obtaining new data for this review, and Mr. Michael J. Good for the preparation, purification, and analysis of deciliated epithelial cell plasma membranes.

References

Adair, W. S., and Goodenough, U. W., 1978, Identification of a membrane tubulin in *Chlamydomonas* flagella, *J. Cell Biol.* **79:**54a.

Adoutte, A., Ramanathan, R., Lewis, R. M., Dute, R. R., Ling, K.-Y., Kung, C., and Nelson, D., 1980, Biochemical studies of the excitable membrane of *Paramecium tetraurelia*. III. Proteins of cilia and ciliary membranes, *J. Cell Biol.* **84:**717–738.

Andreu, J. M., 1982, Interaction of tubulin with non-denaturing amphiphiles, *EMBO J.* **1:**1105–1110.

Andreu, J. M., and Munoz, J. A., 1986, Interaction of tubulin with octyl glucoside and deoxycholate. 1. Binding and hydrodynamic studies, *Biochemistry* **25:**5220–5230.

Auclair, W., and Siegel, B. W., 1966, Cilia regeneration in the sea urchin embryo: Evidence for a pool of ciliary proteins, *Science* **154:**913–915.

Bibring, T., and Baxandall, J., 1974, Tubulins 1 and 2. Failure of quantitation in polyacrylamide gel electrophoresis may influence their identification, *Exp. Cell Res.* **86:**120–126.

Bloodgood, R. A., 1977, Motility occurring in association with the surface of the *Chlamydomonas* flagellum, *J. Cell Biol.* **75:**983–989.

Bloodgood, R. A., 1980, Direct visualization of dynamic membrane events in cilia, *J. Exp. Zool.* **213:**293–295.

Bordier, C., 1981, Phase separation of integral membrane proteins in Triton X-114 solution, *J. Biol. Chem.* **256:**1604–1607.

Brugerolle, G., Andrivon, C., and Bohatier, J., 1980, Isolation, protein pattern, and enzymatic characterization of the ciliary membrane of *Paramecium tetraurelia*, *Biol. Cell.* **37:**251–260.

Burn, P., 1988, Amphitropic proteins: A new class of membrane proteins, *Trends Biochem. Sci.* **13:**79–83.

Chen, Z., and Lancet, D., 1984, Membrane proteins unique to vertebrate olfactory cilia: Candidates for sensory receptor molecules, *Proc. Natl. Acad. Sci. USA* **81:**1859–1863.

Chen, Z., Pace, U., Ronen, D., and Lancet, D., 1986, Polypeptide gp95. A unique glycoprotein of olfactory cilia with transmembrane receptor properties, *J. Biol. Chem.* **261:**1299–1305.

Cook, N. J., Hanke, W., and Kaupp, U. B., 1987, Identification, purification, and functional reconstitution of the cyclic GMP-dependent channel from rod photoreceptors, *Proc. Natl. Acad. Sci. USA* **84:**585–589.

Coyne, B., and Rosenbaum, J. L., 1970, Flagellar elongation and shortening in *Chlamydomonas*. II. Reutilization of flagellar proteins, *J. Cell Biol.* **47:**777–781.

Daleo, G. R., Piras, M. M., and Piras, R., 1977, The effect of phospholipids on the *in vitro* polymerization of rat brain tubulin, *Arch. Biochem. Biophys.* **180:**288–297.

Dasgupta, J. D., and Garbers, D. L., 1983, Tyrosine protein kinase activity during embryogenesis, *J. Biol. Chem.* **258:**6174–6178.

Dentler, W. L., 1980, Microtubule–membrane interactions in cilia. I. Isolation and characterization of ciliary membranes from *Tetrahymena pyriformis*, *J. Cell Biol.* **84:**364–380.

Dentler, W. L., 1981, Microtubule–membrane interactions in cilia and flagella, *Int. Rev. Cytol.* **72:**1–47.

Dentler, W. L., 1988, Fractionation of *Tetrahymena* ciliary membranes with Triton X-114 and the identification of a ciliary membrane ATPase, *J. Cell Biol.* **107:**2679–2688.

Dentler, W. L., and Rosenbaum, J. L., 1977, Flagellar elongation and shortening in *Chlamydomonas*. III. Structures attached to the tips of flagellar microtubules and their relationship to the directionality of flagellar assembly, *J. Cell Biol.* **74:**747–759.

Dentler, W. L., Pratt, M. M., and Stephens, R. E., 1980, Microtubule–membrane interactions in cilia. II. Photochemical cross-linkage of bridge structures and the identification of a membrane-associated dynein-like ATPase, *J. Cell Biol.* **84:**381–403.

Gibbons, I. R., 1965, Chemical dissection of cilia, *Arch. Biol.* **76:**317–352.

Gilula, N. B., and Satir, P., 1972, The ciliary necklace. A membrane specialization, *J. Cell Biol.* **53:**494–509.

Gozes, I., and Littauer, U. Z., 1979, The α-subunit of tubulin is preferentially associated with presynaptic membranes, *FEBS Lett.* **95**:169–172.

Grado, C., and Ballou, C. E., 1961. Myo-inositol phosphates obtained by alkaline hydrolysis of beef brain phosphoinositides, *J. Biol. Chem.* **236**:54–60.

Gunderson, G. G., and Bulinski, J. C., 1986, Distribution of tyrosinated and nontyrosinated α-tubulin during mitosis, *J. Cell Biol.* **102**:1118–1126.

Hargreaves, A. J., and McLean, W. G., 1988, The characterization of phospholipids associated with microtubules, purified tubulin, and microtubule associated proteins *in vitro*, *Int. J. Biochem.* **20**:1133–1138.

Helenius, A., and Simons, K., 1977, Charge-shift electrophoresis: Simple method for distinguishing between amphiphilic and hydrophilic proteins, *Proc. Natl. Acad. Sci. USA* **74**:529–532.

Hertzberg, E. L., 1984, A detergent-independent procedure for the isolation of gap junctions from rat liver, *J. Biol. Chem.* **259**:9936–9943.

Kaprelyants, A. S., 1988, Dynamic spatial distribution of proteins in the cell, *Trends Biochem. Sci.* **13**:43–46.

Kinsey, W. H., Decker, G. L., and Lennarz, W. H., 1980, Isolation and partial characterization of the plasma membrane of the sea urchin egg, *J. Cell Biol.* **87**:248–254.

L'Hernault, S. W., and Rosenbaum, J. L., 1983, *Chlamydomonas* α-tubulin is post-translationally modified in the flagella during flagellar assembly, *J. Cell Biol.* **97**:258–263.

L'Hernault, S. W., and Rosenbaum, J. L., 1985, *Chlamydomonas* α-tubulin is post-translationally modified by acetylation on the ε-amino group of a lysine, *Biochemistry* **24**:473–478.

Linck, R. W., 1973a, Comparative isolation of cilia and flagella from the lamellibranch mollusc, *Aequipecten irradians*, *J. Cell Sci.* **12**:345–367.

Linck, R. W., 1973b, Chemical and structural differences between cilia and flagella from the lamellibranch mollusc, *Aequipecten irradians*, *J. Cell Sci.* **12**:951–981.

Machemer, H., and Ogura, A., 1979, Ionic conductances of membranes in ciliated and deciliated *Paramecium*, *J. Physiol. (London)* **296**:49–60.

Mackie, G. O., Spencer, A. N., and Strathmann, R., 1968, Electrical activity associated with ciliary reversal in an echinoderm larva, *Nature* **223**:1384–1385.

Maher, P. A., and Singer, S. J., 1985, Anomalous interaction of the acetylcholine receptor protein with the nonionic detergent Triton X-114, *Proc. Natl. Acad. Sci. USA* **82**:958–962.

Matesic, D. F., and Liebman, P. A., 1987, cGMP-dependent cation channel of retinal rod outer segments, *Nature* **326**:600–603.

Matesic, D. F., and Liebman, P. A., 1989, Identification and chemical characterization of a putative cation channel protein in ROS plasma membranes, *Biophys. J.* **55**:455a.

Merkel, S. J., Kaneshiro, E. S., and Grunstein, E. I., 1981, Characterization of cilia and ciliary membrane proteins of wild-type *Paramecium tetraurelia* and a prawn mutant, *J. Cell Biol.* **89**:206–215.

Mesland, D. A. M., and Spiele, H., 1984, Brief extraction with detergent induces the appearance of many plasma membrane-associated microtubules in hepatocytic cells, *J. Cell Sci.* **68**:113–137.

Monk, B. C., Adair, W. S., Cohen, R. A., and Goodenough, U. W., 1983, Topography of *Chlamydomonas*: Fine structure and polypeptide components of the gametic flagellar membrane surface and the cell wall, *Planta* **158**:517–533.

Murakami, A., and Machemer, H., 1982, Mechanoreception in the lateral cells of *Mytilus, J. Comp. Physiol.* **145**:351–362.

Musgrave, A., de Wildt, P., van Etten, I., Pijst, H., Scholma, C., Kooyman, R., Homan, W., and van den Ende, H., 1986, Evidence for a functional membrane barrier in the transition zone between the flagellum and cell body of *Chlamydomonas eugametos* gametes, *Planta* **167**:544–553.

Nath, J., and Flavin, M., 1978, A structural difference between cytoplasmic and membrane-bound tubulin of brain, *FEBS Lett.* **95**:335–338.

Nath, J., and Flavin, M., 1980, An apparent paradox in the occurrence, and the *in vivo* turnover, of C-terminal tyrosine in membrane-bound tubulin of brain, *J. Neurochem.* **35**:693–706.

Piperno, G., and Fuller, M. T., 1985, Monoclonal antibodies specific for an acetylated form of α-tubulin recognize the antigen in cilia and flagella from a variety of organisms, *J. Cell Biol.* **101**:2085–2094.

Reaven, E., and Azhar, S., 1981, Effect of various hepatic membrane fractions on microtubule assembly—with special emphasis on the role of membrane phospholipids, *J. Cell Biol.* **89**:300–308.

Regula, C. S., Sager, P. R., and Berlin, R. D., 1986, Membrane tubulin, *Ann. N. Y. Acad. Sci.* **466**:832–842.

Sardet, C., and Tilney, L. G., 1977, Origin of the membrane for the acrosomal process: Is actin complexed with membrane precursors? *Cell Biol. Int. Rep.* **1**:193–200.

Schellenberg, R. R., and Gillespie, E., 1977, Colchicine inhibits phosphatidylinositol turnover induced by concanavalin A, *Science* **265**:741–742.

Schultz, J. E., Schonefeld, V., and Klumpp, S., 1983, Calcium/calmodulin-regulated guanylate cyclase and calcium-permeability in the ciliary membrane from *Tetrahymena, Eur. J. Biochem.* **137**:89–94.

Spiess, M., and Lodish, H. F., 1986, An internal signal sequence: The asialoglycoprotein receptor membrane anchor, *Cell* **44**:177–185.

Stephens, R. E., 1977a, Major membrane protein differences in cilia and flagella: Evidence for a membrane-associated tubulin, *Biochemistry* **16**:2047–2058.

Stephens, R. E., 1977b, Differential protein synthesis and utilization during cilia formation in sea urchin embryos, *Dev. Biol.* **61**:311–329.

Stephens, R. E., 1977c, Vectorial labeling of ciliary membrane tubulin, *J. Cell Biol.* **75**:233a.

Stephens, R. E., 1978, Primary structural differences among tubulin subunits from flagella, cilia, and the cytoplasm, *Biochemistry* **17**:2882–2891.

Stephens, R. E., 1981, Chemical differences distinguish ciliary membrane and axonemal tubulins, *Biochemistry* **20**:4716–4723.

Stephens, R. E., 1983a, Reconstitution of ciliary membranes containing tubulin, *J. Cell Biol.* **96**:68–75.

Stephens, R. E., 1983b, Ciliary membrane reconstitution: Membrane tubulin–lipid interaction, *Biophys. J.* **41**:20a.

Stephens, R. E., 1984, Separation and analysis of organelle-specific tubulin subunits by reverse-phase HPLC, *J. Cell. Biol.* **99**:37a.

Stephens, R. E., 1985a, Evidence for a tubulin-containing lipid–protein structural complex in ciliary membranes, *J. Cell Biol.* **100**:1082–1090.

Stephens, R. E., 1985b, Ciliary membrane tubulin and associated proteins: A complex stable to Triton X-114 dissociation, *Biochim. Biophys. Acta* **821**:413–419.

Stephens, R. E., 1986a, Membrane tubulin, *Biol. Cell* **57**:95–110.

Stephens, R. E., 1986b, Analysis of tubulin-containing ciliary membranes by reverse-phase HPLC, *Biophys. J.* **49**:418a.

Stephens, R. E., 1986c, Isolation of embryonic cilia and sperm flagella, *Methods Cell Biol.* **27**:217–227.

Stephens, R. E., 1987, Membrane tubulin in sea urchin blastula cilia is derived from a pool distinct from axonemal tubulin, *J. Cell Biol.* **105**:34a.

Stephens, R. E., 1988, Rapid incorporation of architectural proteins into terminally differentiated molluscan gill cilia, *J. Cell Biol.* **107**:20a.

Stephens, R. E., and Linck, R. W., 1969, A comparison of muscle actin and ciliary microtubule protein in the mollusk *Pecten irradians, J. Mol. Biol.* **40**:497–501.

Stephens, R. E., Oleszko-Szuts, S., and Good, M. J., 1987, Evidence that tubulin forms an integral membrane skeleton in molluscan gill cilia, *J. Cell Sci.* **88**:527–535.

Stommel, E. W., 1984a, Calcium regenerative potentials in *Mytilus edulis* gill abfrontal ciliated epithelial cells, *J. Comp. Physiol.* **155A**:445–456.

Stommel, E. W., 1984b, Calcium activation of mussel gill abfrontal cilia, *J. Comp. Physiol.* **155A**:457–469.

Stommel, E. W., and Stephens, R. E., 1985, Calcium-dependent phosphatidylinositol phosphorylation in lamellibranch gill lateral cilia, *J. Comp. Physiol.* **157A**:441–449.

Stommel, E. W., Stephens, R. E., and Alkon, D. L., 1980, Motile statocyst cilia transmit rather than directly transduce mechanical stimuli, *J. Cell Biol.* **87**:652–662.

Stommel, E. W., Stephens, R. E., Masure, H. R., and Head, J. F., 1982, Specific localization of scallop gill epithelial calmodulin in cilia, *J. Cell Biol.* **92**:622–628.

Ward, G. E., Moy, G. W., and Vacquier, V. D., 1986, Phosphorylation of membrane-bound guanylate cyclase of sea urchin spermatozoa, *J. Cell Biol.* **103**:95–101.

Witman, G. B., Carlson, K., Berliner, J., and Rosenbaum, J. L., 1972, *Chlamydomonas* flagella. I. Isolation and electrophoretic analysis of microtubules, matrix, membranes, and mastigonemes. *J. Cell Biol.* **54**:507–539.

Yamauchi, P. S., and Purich, D. L., 1987, Modulation of microtubule assembly and stability by phosphatidylinositol action on microtubule-associated protein-2, *J. Biol. Chem.* **262**:3369–3375.

Lipids of Ciliary and Flagellar Membranes

Edna S. Kaneshiro

1. Preparations of Cilia and Flagella and Their Membranes

The lipids of isolated cilia and flagella have been analyzed in only a few organisms. There is a substantial number of reports on the ciliary lipids of *Tetrahymena*. Vast differences in proteins and lipids exist between strains formerly grouped together as *T. pyriformis*. Thompson's and Nozawa's laboratories did their earlier work mainly with *T. pyriformis* amicronucleate strains W, E, or GL or the sexual strain, formerly known as WH-14 or mating type II, variety 1 and now designated as *T. thermophila*. These investigators have used the extremely thermotolerant strain, NT-1, in most of their recent studies. Most studies done by Conner and his colleagues have been on the amicronucleate strain, *T. pyriformis* W. These major contributors to our current understanding of *Tetrahymena* ciliary membrane lipids have done some of their analyses on isolated membrane preparations; all other workers have analyzed the entire organelle. Smith *et al.* (1970) and Kennedy and Thompson (1970) were the first to report on lipids of isolated cilia using *T. thermophila* and *T. pyriformis* E, respectively. These initial studies were followed by a report on *T. thermophila* by Jonah and Erwin (1971) who provided more detailed analyses of ciliary lipids, including fatty acid data. Thus, in addition to differences in analytical techniques, species and strain differences must be taken into consideration when apparent discrepancies occur in the *Tetrahymena* lipid literature.

Analyses of the ciliary lipids of *Paramecium tetraurelia* began later than the *Tetrahymena* studies and some detailed compositional studies have been completed although some components are still uncharacterized. Aside from *Tetrahymena* and *Paramecium*, only single reports exist in the literature on lipids in cilia from the gills of the scallop, *Aequipecten irradians* (Stephens, 1983); ciliary fragments snipped off comb plates of the ctenophores, *Leucothoe harmata* and *Beroë ovata* (Morris and Bone, 1985); and ciliary fragments snipped off the endostyle of the tunicate, *Ciona intestinalis* (Morris and Bone, 1983). Similarly, only single studies have been done on the lipids of flagella isolated from

Edna S. Kaneshiro • Department of Biological Sciences, University of Cincinnati, Cincinnati, Ohio 45221-0006.

Chlamydomonas reinhardtii (Gealt *et al.,* 1981), *Ochromonas danica* (Chen *et al.,* 1976), and *Euglena gracilis* (Chen and Bouck, 1984).

The manner in which cilia and flagella are isolated influences the purity of the preparation. Chemical methods such as the use of local anesthetics (Thompson *et al.,* 1974), Ca^{2+}, or pH shock can achieve deciliation or deflagellation without lysis of the cell, preventing contamination of lipids from cytoplasmic organelles. Mechanical methods such as vortex agitation can cause lysis of some cells, which increases the probability of lipids from other cell compartments contaminating the preparations. Methods in which cells are lysed, then organelles separated (Nozawa and Thompson, 1971a), generally result in cross contamination. The use of organic solvents, e.g., ethanol (Child, 1959), or surfactants, e.g., Triton X-100 (Witman *et al.,* 1972), during these procedures may cause the removal of some lipids from the preparations. One estimate of the purity of ciliary or flagellar preparations is the absence of cardiolipin, which is a mitochondrial marker lipid (Jonah and Erwin, 1971; Kaneshiro *et al.,* 1979).

When ciliary or flagellar membrane fractions have been isolated, the procedures typically involve the use of hypotonic conditions and chelating agents such as EDTA in conjunction with vortex agitation (Chen *et al.,* 1976; Brugerolle *et al.,* 1980; Adoutte *et al.,* 1980). In some studies, membranes have been extracted from cilia using detergents (Stephens, 1983) or solubilization of axonemal proteins by high salt (Gibbons, 1963). Isotonic conditions for the subfractionation of *Paramecium* cilia have been accomplished by hydrostatic pressure disruption employing a French pressure cell after which membrane vesicles are isolated by Percoll gradient centrifugation (Thiele *et al.,* 1982, 1983). The subfractionation procedures for physically separating components of these organelles result in several subfractions. For example, Brugerolle *et al.* (1980) isolated axonemes and two membrane fractions from sucrose gradients. Adoutte *et al.* (1980) designate two membrane-containing fractions isolated from sucrose gradients as membranes and incompletely demembranated cilia. The isotonic, French press, Percoll gradient method for subfractionation of *Paramecium* cilia produces a heavier and a lighter membrane fraction which are further distinguished not only by their ultrastructure but also by their biochemical composition (Thiele *et al.,* 1982, 1983). The heavier membrane fraction contains multilamellar structures and guanylate cyclase activity and the lighter membrane fraction contains single vesicles, voltage-dependent Ca^{2+} channels, and lacks guanylate cyclase activity (Thiele *et al.,* 1982, 1983). Hence, it is possible that lipids reported for ciliary or flagellar membrane subfractions actually represent those of microdomains of the membranes. Nonetheless, on the level of detection used for lipid analyses thus far, isolated membrane lipids are generally similar to those extracted from the whole organelle.

Reports on the lipids of ciliary or flagellar membrane fractions, separated from axonemal and other components of these organelles, are sparse. Thus, for the most part, this chapter will consider the lipids of the whole organelle. The assumption that the lipids in the organelle are restricted to its membrane, however, is unwarranted. For example, Subbaiah and Thompson (1974) provide evidence that lipids exist in compartments other than the membrane. Following a 5-min exposure of *Tetrahymena* to [^{14}C]palmitic acid, cilia were isolated and subciliary fractions prepared. The matrix fraction contained phospholipids that were two to three times more radioactive than phospholipids in the membrane fraction, which cannot be explained simply by contamination of the matrix fraction with membrane lipids. Triglycerides and quinones were detected in ciliary preparations from *Paramecium* that did not contain cardiolipin but were undetectable in ciliary membrane fractions (Kaneshiro *et al.*, 1983). The triglycerides extracted from *Paramecium* cilia had a fatty acid composition distinct from that of whole cell triglycerides, suggesting that these lipids are found within the cilia. These lipids are presumably located in the ciliary matrix and may include lipids in transit from their sites of synthesis to the ciliary membrane and/or intermediates involved in the tailoring of membrane lipids.

2. Lipid Composition

Ciliary and flagellar membranes generally contain about equal amounts of lipid and protein. Chen *et al.* (1976) report a lipid/protein ratio of 1.3 for *Ochromonas* flagellar membranes physically separated from the rest of the flagellar components. Stephens (1983) isolated cilia from the gill of the scallop and then extracted them with detergent. He designated the detergent extract as the ciliary membrane fraction although it contains both membrane and matrix components. This fraction had a lipid/protein ratio of 1.2.

In all cases in which lipids of cilia or flagella have been compared with the rest of the cell, the lipid composition of the organelle has been found to be distinct. There are usually quantitative differences, as are found in *Tetrahymena* and *Paramecium*. However, besides the understandable absence of cardiolipin in cilia and flagella as discussed above, a striking difference between cell and flagellar lipids was reported in *Ochromonas* (Chen *et al.*, 1976). Phospholipids were not detected in the flagella and flagellar membrane preparations isolated from this organism whereas phospholipids were identified in the lipids extracted from the cell.

Generally, free fatty acids and lyso compounds are present in only low concentrations within biomembranes and therefore cilia and flagella and membranes isolated from them are expected to contain small amounts of these lipids. Those that are present probably are intermediates, precursors, or products for the *in situ* tailoring of the membrane. When present in significant amounts, they are often suspected to be the result of contamination from other cell components, products of degradation occurring during the preparation of the cellular or suborganellar fraction, or due to degradation during the preparation of the lipid samples. For example, free fatty acids were not detected in preparations of cilia from *Paramecium* whereas they were present in ciliary membrane preparations (Kaneshiro *et al.*, 1983), illustrating the degradative processes that can occur when relatively lengthy procedures are required to separate the components of the organelle. A possible exception is the report on *Ochromonas* flagella and flagellar membrane lipids in which free fatty acids comprised 54.2% of the neutral lipids (Chen *et al.*, 1976). High levels of free fatty acids may actually occur in the surface membrane of this organism, which cannot grow at pH values greater than pH 5, at which free fatty acids are not charged (Chen *et al.*, 1976).

Table I compares the relative amounts of all lipid classes presently identified in cilia of late-log-phase *Paramecium* with those in whole cell extracts. (See Williams, this volume, for the lipid composition of *Tetrahymena* whole cells, cilia, and other fractions.)

2.1. Sterols

The types of sterol found in ciliary and flagellar membranes are those synthesized by the organism or taken up as dietary sterols, which often can be further metabolized to

Table I. Composition of *Paramecium tetraurelia* Whole Cell and Ciliary Lipids[a]

	Weight%[b]			
	Cells		Cilia	
Monoglycerides, diglycerides, and fatty alcohols	0.9		0.9	
Sterols	14.8		18.7	
Cholesterol		3.6		5.2
7-Dehydrocholesterol		0.5		1.0
Stigmasterol		9.7		10.3
7-Dehydrostigmasterol		1.0		2.2
Free fatty acids	2.3		0.0	
Quinones	2.1		0.4	
Triglycerides	12.8		3.6	
Steryl esters	8.0		5.3	
Cholesterol		0.9		2.0
7-Dehydrocholesterol		0.2		0.1
Stigmasterol		5.3		2.0
7-Dehydrostigmasterol		1.7		1.3
Neutral sphingolipids (sum of two ceramides and four glycosphingolipids)	7.2		4.1	
Phosphatidylcholine (PC)	14.2		7.1	
Diacyl PC		7.7		4.9
Alkacyl PC		4.7		2.1
Phosphatidylethanolamine (PsE)	15.5		7.5	
Diacyl PsE		12.0		5.9
Alkacyl PsE		1.7		1.6
Ethanolamine phosphonolipid (PnE)	11.8		26.4	
Diacyl PnE		1.4		11.1
Alkacyl PnE		10.3		15.3
Phosphatidylserine (PS)	1.2		4.2	
Diacyl PS		0.9		2.8
Alkacyl PS		0.3		1.4
Phosphatidylinositol (PI)	1.9		2.0	
Diacyl PI		1.8		1.5
Alkacyl PI		0.2		0.5
Phosphatidylinositol monophosphate (PIP)	0.2		ND	
Phosphatidylinositol bisphosphate (PIP$_2$)	0.3		ND	
Choline sphingolipids	2.1		2.5	
Ethanolamine sphingolipids	3.8		15.5	
Dihydrosphingosine, phosphoryl (DPsE)		0.2		0.9
Sphingosine, phosphoryl (SPsE)		0.6		1.6
Phytosphingosine, posphoryl (PPsE)		0.5		1.8
Dihydrosphingosine, phosphonyl (DPnE)		0.2		1.5
Sphingosine, phosphonyl (SPnE)		1.1		4.4
Phytosphingosine, phosphonyl (PPnE)		1.3		5.2
Cardiolipin	0.8		0.0	
Lyso PC, PsE, PnE	0.4		2.2	
Phosphatidic acid	0.1		ND	
CDP-diacylglycerol	0.1		ND	
Total neutral lipids	48.1		32.9	
Total phospholipids	51.9		67.2	

[a]From Kaneshiro (1987).

[b]Analyses were done on cells grown axenically in the crude medium for 5 days at 25°C.

other sterols. *Tetrahymena* synthesizes tetrahymanol, which is not a sterol but a pentacyclic triterpenoid solid alcohol (Mallory *et al.*, 1963), and this compound is found in its ciliary membrane (Jonah and Erwin, 1971; Conner *et al.*, 1971; Thompson *et al.*, 1971). *Paramecium* cannot synthesize sterols but incorporates exogenous sterols into its lipids. Stigmasterol and poriferasterol best suit this nutritional requirement (Conner and Van Wagtendonk, 1955; Van Wagtendonk, 1974) but others such as cholesterol can be metabolized and incorporated into its membranes (Conner *et al.*, 1971; Kaneshiro *et al.*, 1983; Hennessey *et al.*, 1983). *Chlamydomonas* flagellar sterols are comprised of 55–63% ergosterol and 37–45% 7-dehydroporiferasterol (Gealt *et al.*, 1981). *Ciona* endostyle cilia contain 64–75% cholesterol and 25–36% cholestanol (Morris and Bone, 1983). The gill cilia of *Aequipecten* are reported to also contain cholesterol (Stephens, 1983) but apparently only colorimetric procedures were employed and hence the unambiguous identification of sterols in the cilia of this organism remains to be done. *Ochromonas* cells contain ergosterol, brassicasterol, 22-dihydrobrassicasterol, clionasterol, poriferasterol, and 7-dehydroporiferasterol (Gershengorn *et al.*, 1968). These are reported to occur in its flagellar membranes in the same proportions as they occur in the whole cell (Chen *et al.*, 1976). Employing gas chromatography (GC) and mass spectrometry (MS), Morris and Bone (1985) identified 19 sterols in lipids from *Beroë* cilia, some present in only trace amounts. The two major components were C_{29} ($\Delta 5$) (44.8%) and C_{27} ($\Delta 5$) (23.9%) sterols.

Sterols are often found in the free form in ciliary and flagellar membranes; however, high concentrations of steryl esters were identified in *Paramecium* cilia and ciliary membranes (Kaneshiro *et al.*, 1983). Also, *Euglena* flagellar lipids contain high concentrations of steryl glucosides, presumably containing ergosterol, the major sterol of the organism (Chen and Bouck, 1984; see Bouck *et al.*, this volume).

High concentrations of sterols generally occur in flagella and cilia; this is indicated by the low phospholipid/sterol ratio of lipids extracted from the organelle and/or membrane as compared to the ratio of lipids from the rest of the cell, which can be six times greater (Conner *et al.*, 1971). The phospholipid/tetrahymanol ratio is 2.4 and 2.1 in *T. pyriformis* W cilia and ciliary membranes, respectively (Conner *et al.*, 1971). It is 4.23, 3.8, and 2.56 in ciliary lipids of *T. pyriformis* NT-1 grown at 39, 24, and 15°C, respectively (Thompson and Nozawa, 1977; Ramesha and Thompson, 1982). It is 3.1 and 2.1 in *T. pyriformis* E for cilia from log-phase cells and 93-hr-starved cells, respectively (Thompson *et al.*, 1972). The phospholipid/sterol ratio in *P. tetraurelia* cilia from late-log-phase cells is 3.8 (Kaneshiro, 1987) but can vary dramatically with culture age from 5.1 to 13.0 (Hennessey *et al.*, 1983). Calculations from values available in two reports indicate that the phospholipid/sterol ratio is 2.8–3.8 in *Ciona* cilia (Morris and Bone, 1983) and 2.4–2.7 in *Aequipecten* cilia (Stephens, 1983). A chlorosulfolipid/sterol ratio of 5.8 can be calculated from data on *Ochromonas* flagellar lipids which are reported to lack phospholipids (Chen *et al.*, 1976).

2.2. Fatty Acid Composition

With the exception of *Tetrahymena* (Table II) and *Paramecium* (Table III), the fatty acid compositions of individual lipid classes are not known. The nature of fatty acid data reported for ciliary or flagellar lipids of other organisms is varied and/or the source of the

Table II. Fatty Acid Compositions of Individual Lipids from Cilia Isolated from *Tetrahymena pyriformis* NT-1 Cells Grown at Either 39 or 15°C

Fatty acid	PC[a] 39°C	PC[a] 15°C	PsE[a] 39°C	PsE[a] 15°C	PnE[a] 39°C	PnE[a] 15°C	SP$_1$[b] 39°C	SP$_2$[b] 39°C	SP$_2$[b] 15°C
12:0	1.7	0.9	c						
14:0	18.9	26.8	25.7	20.5	3.2	2.8	1.8	18.4	0.4
iso 15:0	6.4	2.8	8.9	2.6					
15:0	3.8		4.2		0.9		0.5	2.5	
16:0	19.4	15.8	24.0	23.9	4.1	2.3	5.8	44.3	57.0
16:1	10.1	9.0	15.5	18.2	4.7	2.5			
iso 17:0							1.2	14.2	39.6
17:0							0.9	3.8	2.1
16:2 + 17:0 + 17:1	3.6	2.8	4.6	4.5	1.8				
iso 18:0									0.2
18:0	1.3	1.2	2.1	1.2			0.6	5.5	0.5
18:1	5.2	4.4	4.9	12.9	2.5	2.4			
18:2 (Δ6,11)		3.2			4.4	16.0			
18:2 (Δ9,12)	5.1	3.9	7.0	9.8	4.4	3.5			
18:3	6.5	9.7	2.1	5.5	23.9	23.5			
iso 19:0							0.8	5.7	
OH-16:0							71.6	5.7	0.2
OH-iso 17:0							7.5		
OH-17:0							5.4		
OH-18:0							3.9		

[a]Ester-linked fatty acid values are expressed as mole%. Data are from Ramesha and Thompson (1982).
[b]SP$_1$ and SP$_2$ are both ethanolamine sphingophosphonolipids. The amide-linked fatty acid values are expressed as weight%. Data are from Kaya *et al.* (1984a).
[c]Blank spaces indicate that fatty acids were not detected or were less than 0.5%.

Table III. Fatty Acid Compositions of *Paramecium* Ciliary Lipids[a]

Fatty acid	Percent by weight[b] Ester-linked Total	TG	SE	PC	PsE	PnE	PS	PI	Amide-linked ESL
14:0	0.8	1.3	4.1	0.5	1.6	c	1.1	3.8	1.1
15:0	0.5	0.7	1.0		0.9				
16:0	12.3	29.0	25.9	14.9	42.8	2.8	33.8	23.5	65.8
16:1	0.5	0.7	3.4	0.6	1.3	0.1	0.5	1.5	2.4
17:0	0.2	0.8	0.7						6.5
18:0	2.6	26.3	9.9	3.5	5.8	1.7	6.8	5.5	19.6
18:1	6.7	16.6	36.2	8.7	11.9	0.4	7.7	11.7	4.7
18:2	8.6	5.9	24.0	25.6	0.3	6.6	23.9		
18:3	3.2	3.8	1.8	11.2	7.9	0.2	3.9	20.5	
20:1	1.0	1.2	1.4				8.7		
20:3	1.6	0.9	0.8	3.8		0.4	0.8	0.9	
20:4	53.3	5.6	2.0	25.2	2.1	82.6	22.8	2.5	
20:5	7.5	0.9	2.4			10.3	2.3	0.1	
22:6	0.1	1.6	0.8						

[a]From Kaneshiro (1987).
[b]Cilia were isolated from *Paramecium tetraurelia* 51s grown in an enriched crude medium for 5 days at 25°C. TG, triglycerides; SE, steryl esters; PC, phosphatidylcholine; PsE, phosphatidylethanolamine; PnE, ethanolamine glycerophosphonolipid; PS, phosphatidylserine; PI, phosphatidylinositol; ESL, total ethanolamine sphingolipids.
[c]Blank spaces indicate that fatty acids were not detected or were less than 0.5% (by weight).

Table IV. Fatty Acid Compositions of Total Lipids from *Chlamydomonas* Cells and Flagella, and Phospholipids of Cilia Isolated from *Ciona*, *Beroë*, and *Leucothoe*

Fatty acid	Chlamydomonas[a]		Ciona[b]	Beroë[c]	Leucothoe[c]
	Cells	Flagella			
12:0	d				7.9
14:0	0.7		12.7	18.9	17.5
iso 15:0					0.9
anteiso 15:0					0.9
15:0				8.0	9.5
16:0	7.0	12.8	41.1	38.7	34.9
16:1	1.5	20.0	12.9	1.4	2.2
16:2	2.0				
16:3	33.0				
17:0					2.5
18:0	2.0	13.8	13.2	13.2	14.0
18:1	5.8	13.3	13.5	19.8	5.9
18:2	1.5	17.1	13.5		5.9
18:3	46.4	22.9	1.5		
20:0					0.3
20:1					0.6
20:5					0.8
22:0					0.4
22:1					0.4
22:6					0.4

[a]Total lipid fatty acids. Data are from Gealt *et al.* (1981).
[b]Phospholipid fatty acids. Data are from Morris and Bone (1983).
[c]Phospholipid fatty acids. Data are from Morris and Bone (1985).
[d]Blank spaces indicate that fatty acids were not detected or were present in trace amounts.

fatty acids in the analyses unclear. Studies on *Paramecium* and *Tetrahymena* demonstrate that ciliary membrane lipids are at least quantitatively different from those of membranes of other organelles and substantial concentrations of hydrophobic "tails" such as amide-linked (alkaline-stable) fatty acids in sphingolipids and ether-linked fatty alcohols in glycerolipids are present, in addition to fatty acids linked by ester bonds (alkaline-labile) in glycerolipids.

The reported *Chlamydomonas* flagellar fatty acid composition represents total lipid, alkaline-labile, saponifiable fatty acids that are ester linked, plus any free fatty acids potentially present, since whole flagella were directly saponified (Gealt *et al.*, 1981) (Table IV). Reports containing information on the total phospholipid fatty acid compositions for cilia from *Ciona* (Morris and Bone, 1983), *Leucothoe,* and *Beroë* (Morris and Bone, 1985) do not provide sufficient information to determine whether these represent only the saponifiable fatty acids or total fatty acids (Table IV). No fatty acid composition data are currently available for flagella of *Euglena* and *Ochromonas* and cilia of *Aequipecten.*

2.3. Glycerolipids

With the exception of *Ochromonas* flagella and flagellar membranes, all other ciliary and flagellar preparations thus far examined contain typical glycerophospholipids: phosphatidylethanolamine (PE), phosphatidylcholine (PC), phosphatidylserine (PS), and phosphatidylinositol (PI). The latter two acidic phospholipids are often not detected or are present in trace amounts. PS can exist only as a short-lived intermediate in the formation of PE (Dennis and Kennedy, 1970). Although plant and fungal cells can have substantial amounts of inositol lipids (Lester et al., 1978), PI is present in low amounts in many cell types examined. The polyphosphoinositides, phosphatidylinositol monophosphate (PIP) and phosphatidylinositol bisphosphate (PIP$_2$) have been detected in some samples of Paramecium ciliary lipids (Suchard et al., 1989) but probably go undetected in many analyses because they are there in only trace amounts, require special procedures for complete extraction, and are known to turn over rapidly in various cell types and are highly susceptible to phospholipase attack. Thus, during the procedures used to isolate cilia and flagella and their membranes, it is very likely that most polyphosphoinositides present would become degraded. It is unlikely that any quantitation of isolated organellar or membrane fraction polyphosphoinositides represents the control values occurring in situ in the intact cell.

The cilia and ciliary membranes of Tetrahymena and Paramecium have several similar features in their glycerophospholipid compositions, but some differences occur. In cilia from both genera, PC and ethanolamine glycerophosphatides are present in significant concentrations (Kaneshiro, 1987; Nozawa and Thompson, 1979; see Williams, this volume). An analogue of PE that contains a phosphonyl, or direct, P–C bond (herein abbreviated as PnE) is present in much higher concentrations than that with a phosphoryl, or P–O–C, bond (herein abbreviated as PsE). Furthermore, in cilia from both species, PnE is highly enriched in ether-linked fatty alcohols at the C-1 position of the glycerol backbone. Both PS and PI have been identified and characterized in lipids from Paramecium (Tables I and III) but have not been detected in Tetrahymena cilia.

There is no phospholipid class information for cilia from Ciona, Beroë, Leucothoe, or Chlamydomonas. Stephens (1983) tentatively identified PC, PE, PS, and lyso PC in Aequipecten ciliary lipids by one-dimensional thin-layer chromatographic (TLC) and I$_2$ staining techniques. The major glycerophospholipids in Euglena flagella are PE and PS (Bouck, personal communication; see Bouck et al., this volume).

2.3.1. Glycerophosphonolipids. The first identification of the presence of phosphono bonds in Tetrahymena lipids was by Kandatsu and Horiguchi (1962) who gave the trivial name ciliatine to 2-aminoethylphosphonic acid (AEP). Later, Smith et al. (1970) and Kennedy and Thompson (1970) found PnE in high concentrations in ciliary lipids. This was also found to be true of the ciliary lipids of Paramecium (Rhoads and Kaneshiro, 1979; Andrews and Nelson, 1979).

The fatty acid compositions of these glycerolipids are distinct, and because PnE dominates the polar lipid fraction of cilia from Tetrahymena and Paramecium the overall saponifiable fatty acid composition of the membrane is greatly affected by that of PnE (Tables I and II). The major end products of two separate fatty acid synthetic pathways in Tetrahymena are 18:3 (Δ6,9,12), γ-linolenate, and 18:2 (Δ6,11), cilienic acid (Koroly and Conner, 1976). The major end product of fatty acid synthesis in Paramecium is 20:4 (Δ5,8,11,14), arachidonate (Kaneshiro et al., 1979; Rhoads and Kaneshiro, 1984). These fatty acids occur in high concentrations in PnE of the respective ciliates, specifically at the C-2 position (Pieringer and Conner, 1979; Kaneshiro, 1980). Another feature of PnE in these ciliates is that the C-1 position is generally occupied by an ether-linked fatty alcohol (see below).

2.3.2. Glyceryl Ethers.

Takemoto (1961) was the first to report the presence of ether lipids in *T. pyriformis* W cells. Later, Berger *et al.* (1972) found that PC and PnE of this ciliate contained high concentrations of ethers and they identified the glyceryl ether as 1-hexadecyl glycerol, chimyl alcohol. The glyceryl ether concentration is higher in ciliary phospholipids compared to phospholipids from whole cells and intracellular organelles (Nozawa and Thompson, 1979).

Paramecium also contains high concentrations of ethers in PnE (Kaneshiro *et al.*, 1987). All other glycerolipids examined in the cilia of this ciliate, although dominated by diacyl species, contain alkyenyl species (Table I). In addition to chimyl alcohol, *Paramecium* lipids also contain 1-*O*-octadec-*cis*-11-enyl glycerol, paramecyl alcohol (Rhoads *et al.*, 1981; Kaneshiro *et al.*, 1987), and both glyceryl ethers have been detected in all glycerolipid classes analyzed from ciliary lipids.

2.4. Sphingophospholipids and Sphingophosphonolipids

Carter and Gaver (1967) were first to identify sphingolipids in *T. pyriformis* W and characterized the presence of AEP as a head group in these lipids. They also identified 15-methyl-C_{16} sphingosine and 17-methyl-C_{18} sphingosine in the long-chain base (LCB) moiety of these lipids and characterized the amide-linked fatty acids present. Ferguson *et al.* (1972) analyzed the sphingolipid fatty acids in the same ciliate species that was grown in a crude medium and found 16:0, iso 17:0, and 18:0 as well as the corresponding α-hydroxy fatty acids. These workers failed to detect hydroxy-fatty acid-containing ceramide AEP in the lipids of cells grown in a chemically defined medium. Ceramide AEP was not detected in *T. thermophila* grown in a chemically defined medium (Jonah and Erwin, 1971) or a crude medium (Nozawa and Thompson, 1979), but Sugita *et al.* (1979) identified ethanolamine sphingophosphonolipids in these cells grown in an enriched crude medium. In an attempt to determine the role of strain, temperature, and nutrition on the accumulation of hydroxy fatty acids by these cells, a broad range of *Tetrahymena* strains, including *T. thermophila* and *T. pyriformis* W, were grown in different crude and chemically defined media at different temperatures. However, in all cases examined, ethanolamine sphingophosphonolipids as well as their normal and α-hydroxy fatty acid components were detected in the cellular lipids (Kaneshiro and Witte, unpublished observations).

Sphingolipids in *T. pyriformis* NT-1 have been found concentrated in ciliary lipids (Kaya *et al.*, 1984b) comprising 30–40 mole% of the phospholipids (Kaya *et al.*, 1984a). Two AEP-containing sphingolipids in *T. pyriformis* NT-1 can be separated by thin-layer chromatography (TLC) due to different fatty acid compositions (Kaya *et al.*, 1984a). The fatty acids in both are all saturated and include iso and hydroxy acids; however, one is rich in hydroxy fatty acids (OH-16:0, OH-17:0, iso OH-17:0, and OH-18:0) and the other contains mainly 16:0, iso 17:0, and 14:0 and less than 6% OH-16:0. The LCB of both the "hydroxy" and "non-hydroxy" fatty acid-containing sphingolipids are associated with both normal and iso C_{16}, C_{18}, and C_{19} sphingosines (Kaya *et al.*, 1984a).

Ciliary lipids of *Paramecium* are also enriched in sphingolipids (Rhoads and Kaneshiro, 1979; Andrews and Nelson, 1979; Kaneshiro *et al.*, 1984) (Table I). Six ethanolamine sphingolipids from *Paramecium* have been resolved by two-dimensional TLC. They have similar fatty acid compositions mainly consisting of the saturated acids, 16:0 and 18:0 (Table III). The six separate by TLC because three contain phosphoryl bonds and the other three contain phosphonyl bonds in the head group, and two contain the LCB, C_{18} dihydrosphingosine, two contain C_{18} sphingosine, and two contain C_{18} phytosphingosine (Kaneshiro *et al.*, 1984). Choline sphingophospholipids (sphingomyelin), which may also be enriched in the ciliary fraction, have also been identified. Furthermore, a mixture of neutral sphingolipids that include several glycosphingolipids have been detected in *Paramecium* cells but detailed characterization and their distribution within cilia have yet to be reported (Table I). The choline sphingolipid and neutral sphingolipid fractions isolated from whole cells are unlike those of the ethanolamine sphingophospholipids and sphingophosphonolipids. These contain long-chain fatty acids, some of which are hydroxy fatty acids (Kaneshiro, 1987).

Sphingomyelin has been tentatively identified in the ciliary lipids of *Aequipecten* by one-dimensional TLC and I_2 staining (Stephens, 1983). It comprises about 5% of the phospholipids of the cilia (Stephens, 1983).

2.5. Chlorosulfolipids

Although the high levels of free fatty acids that were found in flagella and flagellar membrane preparations of *Ochromonas* could be attributed to degradation of complex lipids (see above), the presence of chlorosulfolipids in the lipids extracted cannot be artifactual (Chen *et al.*, 1976). These unusual lipids include 1,14-docosanediol-1,14-disulfide and 1,15-tetradocosanediol-1,15-disulfate that contain from zero up to six H on the hydrocarbon chain that are substituted by Cl. These highly polar compounds actually partition into the aqueous, rather than the organic phase during the isolation and purification of lipids (Haines, 1973). Thus, these compounds would ordinarily go undetected during routine lipid extraction and analysis procedures. It has been estimated that these polychlorosulfolipids make up 71 mole% of the total lipids and 91 mole% of the polar lipids of the flagellar membrane. The presence of two ionic groups at the ends of the hydrocarbon straight chains as well as halogen atoms along the chain, poses a challenge for the explanation of the presumably bilayer organization of lipids in this membrane (Chen *et al.*, 1976; Haines, 1973), if indeed these compounds are part of the membrane lipid bilayer. It is conceivable that chlorosulfolipids are not located totally in the membrane bilayer but exist as part of the glycocalyx of the flagellar membrane.

2.6. Lipoconjugates

Lipids in the membranes of cilia and flagella are often covalently bound to carbohydrates and/or proteins. In addition to the glycosphingolipids in *Paramecium* described above, glycolipids have been reported to be present in *Aequipecten* cilia constituting 7% of total lipids (Stephens, 1983). *Ochromonas* flagella and flagellar membranes contain six uncharacterized lipids comprising 9% of the polar lipids, five of which are glycosylated (Chen *et al.*, 1976). Several glycolipids occur in *Euglena* flagellar lipids. Chen and Bouck

(1984), employing isolated flagellar preparations, have demonstrated that lipids are glycosylated *in situ* (see Bouck *et al.,* this volume).

Detergent extracts of cilia from *Aequipecten* spontaneously re-form membranes after removal of the detergent followed by several cycles of freezing and thawing (Stephens, 1985). It was argued that proteins and lipids are held together in tight complexes within this membrane since they reassociated with or retained most of the original components throughout these procedures. The nature of the interactions between lipids and proteins, however, was not determined.

The immobilization antigens (i-Ag) of *Paramecium* (Capceville *et al.,* 1986) and *T. thermophila* (Chen and Conner, personal communication) ciliary membranes appear to be anchored into the membrane by phosphoinositides. The inositide–glycan–protein structure is similar to that reported for *Leishmania* (Bordier *et al.,* 1986), *Trypanosoma* (Ferguson *et al.,* 1985), as well as other cell types and organisms (Low and Saltiel, 1988). PI-specific phospholipase C cleaves the i-Ag, and antibodies directed against the cross-reacting determinant of the trypanosome variant surface glycoprotein react with these major antigenic components of the cilia of these organisms. Hansma (1975) provided evidence that the i-Ag of *Paramecium* had endogenous proteolytic activity. Similar activities have since been shown for several of these phosphoinositide-anchored glycoproteins including that of *Leishmania* (Etges *et al.,* 1986).

3. Enzymes and Lipid Metabolism

Although many enzymes that are involved with the *de novo* synthesis of lipids are generally restricted to the microsomal fraction of cells (Fukushima *et al.,* 1977; Thompson and Nozawa, 1984), several enzymes that act on lipidic substrates have been identified in flagella and cilia allowing for the refinement, tailoring/retailoring, and uniqueness of ciliary and flagellar lipid compositions. As mentioned above, isolated *Euglena* flagella contain glycosyltransferases that glycosylate lipids in its membrane (Chen and Bouck, 1984; see Bouck *et al.,* this volume). Also, *Paramecium* and *Tetrahymena* i-Ag molecules cleave themselves apparently leaving the phosphoinositide anchor within the membrane and releasing the glycoprotein into the medium (Hansma, 1975; Chen and Conner, personal communication).

PI and PIP kinase activities were detected in cilia isolated from gills of *Aequipecten* and from gills of the mussel *Mytilus* (Stommel and Stephens, 1985). Whole cilia were treated with the detergent NP-40 and then incubated with [γ-^{32}P]-ATP; this resulted in the incorporation of radiolabel into the products, PIP and PIP$_2$. At 10 mM MgCl$_2$, the phosphorylation reactions exhibited micromolar Ca^{2+}-dependence.

Recently, PI and PIP kinase activities in *Paramecium* cilia were characterized using *in vitro* incubations with exogenous soybean PI, [γ-^{32}P]-ATP, and isolated cilia (Suchard *et al.,* 1989). The enzyme(s) were found to be restricted to the subfractions of the cilia that contained membranes; the soluble and axonemal fractions were without activity. The phosphorylation reactions were maximal at neutral pH values, were stimulated by millimolar MgCl$_2$ but not millimolar CaCl$_2$, and were not affected by the presence of EGTA. Decreases in radioactive PIP and PIP$_2$ that were observed with incubation time using whole cilia suggested that phosphatase activity might also be present. These results demonstrate that phosphorylation, and perhaps dephosphorylation, of phosphoinositides occur *in situ* in the *Paramecium* ciliary membrane. Furthermore, PIP and PIP_2 levels in cilia are probably higher than indicated by direct analyses of cilia or their membranes due to degradative enzymes within the ciliary membrane. The presence of polyphosphoinositides in ciliary membranes suggests that transmembrane signal transduction involving these lipids may occur in the cilia. Thus, the ciliary membrane appears to have the capacity to phosphorylate and dephosphorylate phosphoinositides (Suchard *et al.*, 1989) as well as to cleave the head group off phosphoinositides by phospholipase C activity associated with the i-Ag (Capceville *et al.*, 1986).

Thompson *et al.* (1971) incubated isolated *T. pyriformis* E cilia with complex lipids labeled with [^{14}C]palmitate. Since 12% of the radioactivity that had been incorporated was released as free fatty acids, they concluded that the cilia contain lipolytic enzymes.

Isolated cilia from *T. pyriformis* NT-1 cells grown at 39°C were incubated at 15°C and found to exhibit fatty acid compositional changes in their phospholipids similar to those seen *in vivo* (Ramesha and Thompson, 1983). The occurrence of *in vitro* acylation of lipids in cilia was demonstrated by incubation of cilia with ATP, coenzyme A (CoA), and [^{14}C]palmitic acid (Ramesha and Thompson, 1983). Less than 1% of the radioactivity from palmitic acid was incorporated but without ATP and CoA in the *in vitro* assay, no incorporation occurred and therefore the authors concluded that the enzymes for acylation and deacylation were present in cilia.

Although enzymes for desaturation of fatty acids were reported to be restricted to microsomal fractions of *Tetrahymena* (Fukushima *et al.*, 1977; Thompson and Nozawa, 1984), Ramesha and Thompson (1983) observed radioactivity from [^{14}C]16:0 not only in 16:0 but also in 16:1, 18:1, and 18:2 after the *in vitro* incubations. More recently, these workers reported that the ciliary membrane contained calcium-dependent phospholipase A activity (Ramesha and Thompson, 1984). It was not determined whether this activity was due to phospholipase A_1 or A_2 activity. Also, the presence of acyl-CoA synthetase activity was indicated by the *in vitro* formation of radioactive fatty acid-CoA during incubations with radiolabeled palmitate and linoleate (Ramesha and Thompson, 1984). Acyltransferase activity was identified by the incorporation of radioactivity from [^{14}C]16:0-CoA or [^{14}C]18:2-CoA into isolated ciliary phospholipids in the presence of supplementary ATP and CoA.

Kapoulas *et al.* (1969a,b) characterized the *in vivo* and *in vitro* cleavage and degradation of [^{3}H]chimyl alcohol by *T. pyriformis* W cells. The major products were fatty acids but fatty aldehyde and fatty alcohol intermediates were also detected. Their *in vitro* studies did not include ciliary fractions and these organelles were probably present in their crude mitochondrial fraction. Of interest is their observation that the *Tetrahymena* "etherase" activity was present in all subcellular fractions, leaving the possibility that the enzymes for ether lipid degradation occur in cilia where these lipids are present in great abundance. Direct assays of cilia for the ability to cleave and degrade ethers have not been done.

Ethanolamine sphingolipids are highly concentrated in the ciliary membrane of *Paramecium* (Rhoads and Kaneshiro, 1979; Andrews and Nelson, 1979) as well as that of *T. pyriformis* NT-1 (Ramesha and Thompson, 1982). Recently, the neutral and choline sphingolipid fractions of *Paramecium* cells were analyzed and found to lack any phytosphingosine LCB (Kaneshiro, 1987). These observations suggest the possibility that modification of the LCB moiety of ethanolamine sphingolipids could occur *in situ* in the cilia. *In vitro* assays testing this hypothesis have not been reported.

4. Alterations in Lipid Composition

The lipids of *Tetrahymena* and *Paramecium* cilia undergo changes under different conditions of growth of the organisms. In addition, lipid compositions have been manipulated by supplementation of the organisms' culture media as well as by the use of drugs. Furthermore, mutants of *Paramecium* with altered lipid compositions have been described. In all studies comparing various cellular organelles and/or membranes of *Tetrahymena*, the cilia exhibited the lowest rate of incorporation of radiolabeled precursors, the lowest turnover rates of membrane components, and the lowest degree of alterations in lipid composition as a result of experimental manipulations, as well as being the least fluid (Fig. 1) (Nozawa and Thompson, 1971b, 1972, 1979; Thompson *et al.*, 1972; Nozawa *et al.*, 1974, 1980; Martin *et al.*, 1976; Kitajima and Thompson, 1977; Martin and Thompson, 1978; Iida *et al.*, 1978; Shimonaka *et al.*, 1978; Nandini-Kishore *et al.*, 1979).

4.1. Culture Age

When inoculated into fresh medium at 1:10 dilutions, axenic cultures of *Paramecium* exhibit lag phase for 1 day. Days 2–4 are typically log phase; day 5 is the transition from

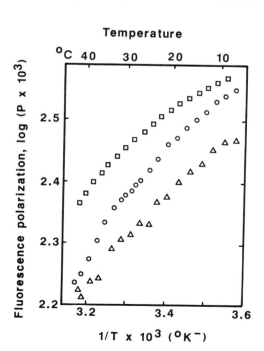

Figure 1. Temperature dependence of the degree of fluorescence polarization of 1,6-diphenyl 1,3,5-hexatriene presented as log P versus $1/T$ of various membrane fractions from *Tetrahymena pyriformis* NT-1. □, cilia; ○, pellicles; △, microsomes. From Shimonaka *et al.* (1978) with permission.

late log into early stationary. Stationary phase is followed by death phase at day 8 (Kaneshiro *et al.*, 1979). Almost all lipid classes examined in *Paramecium* cilia undergo quantitative changes during culture aging (Kaneshiro, 1987). These changes result in relative increases of phosphonolipids (Rhoads and Kaneshiro, 1979), ether lipids (Kaneshiro *et al.*, 1987), and ethanolamine sphingolipids (Kaneshiro *et al.*, 1984) within the cilia. Among the ethanolamine sphingolipids, those containing the LCB dihydrosphingosine and sphingosine, and those containing phosphono head groups, accumulate to greater percentages during culture aging (Kaneshiro *et al.*, 1984).

Among the glycerophospholipids, the PnE percentage increases as PsE, PC, and PS decrease as cultures age (Rhoads and Kaneshiro, 1979). Only small changes are seen in the fatty acid compositions of these lipids but since PnE contains high concentrations of polyunsaturated fatty acids (PUFA), the relative percentage of saponifiable PUFA increases in ciliary lipids (Kaneshiro *et al.*, 1979). The unsaturation index of saponifiable fatty acids from ciliary phospholipids increases from 143 at day 3 to 205 at day 7. However, when the substantial amounts of ether-linked fatty alcohols, ethanolamine sphingolipid LCB, and amide-linked fatty acids present in this membrane are also considered, the unsaturation index difference is insignificant: day-3, 104 and day-7, 114 (Kaneshiro, 1987).

Studies on *Tetrahymena* indicate that only slight changes in lipid composition occur in the ciliary membrane as a result of the nutritional state of the cultures. There is a slight increase in the relative concentration of 18:3 while other fatty acids remain the same or decrease slightly; the original values are restored upon refeeding (Nozawa *et al.*, 1980). For *T. pyriformis* E in death phase and those starved in a nongrowth solution for 4 days, the phospholipid/tetrahymanol ratio of ciliary lipids decreases but such changes are apparent only under these extreme conditions. This ratio in log cells is 3.1 and in cilia of cells starved for 93 hr, it is 2.1 (Thompson *et al.*, 1972).

4.2. Dietary Supplementation

Supplementation of *T. pyriformis* W cultures with high levels of ergosterol inhibits the synthesis of tetrahymanol by these cells and ergosterol is incorporated into cilia and ciliary membranes (Conner *et al.*, 1971) resulting in a significantly lower phospholipid/sterol ratio from controls. With this qualitative alteration in lipid composition, these sterol-containing cells were killed by the polyene antibiotics filipin and nystatin, whereas control cells containing tetrahymanol were resistant to these drugs. These observations were confirmed in *T. thermophila* by freeze-fracture electron microscopy showing that ergosterol-containing ciliary membranes treated with filipin contained 25 to 30-nm clumps or aggregates that were absent in tetrahymanol membranes (Fig. 2) (Nozawa *et al.*, 1975). These workers did not detect significant changes in phospholipid/sterol ratios of this strain and concluded that there was a one-to-one replacement involved. Ergosterolsupplemented *T. pyriformis* NT-1 cells had less fluid pellicle and microsomal membranes than controls as determined by fluorescence polarization measurements (Shimonaka *et al.*, 1978). Although values for control ciliary membranes were presented, the effect of ergosterol supplementation on the fluidity of these membranes was not reported.

Ergosterol supplementation is accompanied by changes in cellular fatty acid compositional and phospholipid head group changes (Ferguson *et al.*, 1971, 1975; Nozawa *et*

Figure 2. Freeze-etch electron microscopy of the ergosterol-replaced cilia of *Tetrahymena thermophila* untreated (A) and treated (B) with 76.4 µM filipin. Bars = 0.1 µm. From Nozawa *et al.* (1975) with permission.

al., 1975). There is a relative increase in 18:2 (Δ6,11) as well as fatty acids in the palmitoleic acid pathway, 16:0 → 18:2 (Δ6,11) (Koroly and Conner, 1976; see above) with a concomitant decrease in 18:2 (Δ9,12) and fatty acids in the stearic acid pathway, 16:0 → 18:3 (Δ6,9,12) (Ferguson *et al.*, 1975). Sphingolipid fatty acids changed dramatically with relative increases in C_{16} and C_{17} acids and a decrease in C_{18} acids and an increase in the total hydroxy fatty acid. These workers concluded that ergosterol replacement of tetrahymanol results in fatty acid compositions differing in desaturation and chain length, but it is unlikely that ergosterol directly affects desaturases or the enzymes of chain elongation. They hypothesized that ergosterol replacement affects fatty acid synthesis pathways, favoring the palmitoleic pathway.

Ciliary lipids of *T. pyriformis* NT-1 cells grown with 1.6% ethanol show less fatty acid compositional and phospholipid class changes than do other organelles analyzed (Nandini-Kishore *et al.*, 1979). There was an apparent relative decrease in PnE concentrations in ciliary lipids. The fatty acid composition of ethanol-treated ciliary lipids contained high concentrations of 16:0 and relatively less 16:1, 16:2, and 18:3 as compared to controls. Thus, since ethanol has a fluidizing effect on membranes, the ciliary membrane fatty acid composition, unlike that of other organelles, apparently underwent compensatory changes to make it less fluid by increasing saturated fatty acids and decreasing unsaturated fatty acids.

Supplementation of *Paramecium* (Kaneshiro *et al.*, 1987) and *Tetrahymena* (Fukushima *et al.*, 1976; Shimonaka *et al.*, 1978) cultures with chimyl alcohol increased the glyceryl ether concentrations in cellular lipids. Its effects on the lipids of several organelles, excluding cilia, of *T. pyriformis* NT-1 have been reported; ciliary membrane lipid alterations remain unreported for both genera of ciliates. In *T. pyriformis* NT-1, chimyl alcohol supplementation increases the glyceryl ether content of PC to the greatest extent (Fukushima *et al.*, 1976). Pellicle preparations from chimyl alcohol-supplemented cells were found to be more fluid than controls (Shimonaka *et al.*, 1978). Supplementation of *T. pyriformis* NT-1 cells with the choline analogue methylethanolamine caused the synthesis and accumulation of phosphatidylmonomethylethanolamine which could be as high as 34% of the total cellular phospholipids (Nozawa and Thompson, 1979). The cells compensated for the presence of high concentrations of this lipid by a relative decrease in PsE. Ciliary lipids, however, were apparently not analyzed. Supplementation of *T. thermophila* cultures with 2-aminoethylphosphonate increased the phosphonolipid content up to 37% of cellular phospholipids (Smith and O'Malley, 1978) but ciliary lipids were not evaluated.

T. pyriformis NT-1 cultures supplemented with 18:2 increased the concentration of this fatty acid in its ciliary phospholipids (Kasai *et al.*, 1976). Increases in iso acids with odd numbers of carbons were observed in cellular lipids of *T. pyriformis* W cultures supplemented with isovalerate (Conner and Reilly, 1975). Fukushima and his colleagues (Nozawa and Thompson, 1979) analyzed a strain of *Tetrahymena* that normally contains up to 27% iso fatty acids. Upon supplementation with isovalerate, 74% of its fatty acids were iso acids, half of which were saturated. Thus, it was concluded that branched-chain fatty acids could serve as a fluidizing influence (Nozawa and Thompson, 1979). The effect of iso fatty acid supplementation on ciliary lipids in any organism has yet to be reported.

4.3. Drugs and Inhibitor Compounds

Triparanol inhibits cholesterol synthesis in mammalian cells and inhibits desaturation of fatty acids in *Tetrahymena* (Holz *et al.*, 1962; Pollard *et al.*, 1964) and *Paramecium* (Rhoads and Kaneshiro, 1984). The main effect of triparanol on *Paramecium* was the inhibition of desaturation of dietary oleate, which is a nutritional requirement of this organism. Thus, triparanol treatment reduced the concentration of PUFA in the organism's ciliary lipids which was interpreted as the basis for the alteration of the ciliate's response to ionic stimulation (Rhoads and Kaneshiro, 1984).

Potassium tetralylphenoxyisobutyric acid inhibits acetyl-CoA carboxylase and the incorporation of radiolabeled acetate in *T. pyriformis* E lipids (Nozawa and Thompson, 1972). Inhibited cells, however, could incorporate radiolabeled palmitate into all subcellular fractions with the same pattern as control cells (Nozawa and Thompson, 1972). Treatment of cells with this inhibitor apparently had no effect on the ciliary lipid composition (Nozawa and Thompson, 1979). Cerulenin also effectively inhibits *de novo* fatty acid synthesis in *Paramecium* (Rhoads *et al.*, 1987) and *Tetrahymena* (Nozawa and Thompson, 1979) but analysis of its effect on the ciliary lipid composition of these organisms remains to be done.

4.4. Temperature Shifts

Ciliary lipids of *Paramecium* grown at 35 or 25°C did not have different unsaturation indexes or total unsaturations taking into account all aliphatic moieties of lipids (Hennessey and Nelson, 1983). Thus, the ciliary membrane of this organism appears to be maintained constant despite changes in temperature and culture age (see above). Similarly, Conner and Stewart (1976) examined the effects on chilling of *T. thermophila* cellular lipids. Although they observed no change in the unsaturation index of fatty acids from cells grown at 35 or 15°C, important fatty acid compositional changes were noted. All fatty acids in the palmitoleic acid pathway increased in all temperature downshift experiments and all compounds in the stearic acid pathway decreased. These workers concluded that, in this organism, temperature downshifts mainly affected the activity of different fatty acid synthesis pathways and that chain elongase activity converting 16:0 to 18:0 was depressed and/or desaturase activity converting 16:0 to 16:1 (Δ9) was stimulated.

When *Tetrahymena* cells are cooled to a point where cellular membranes exhibit phase separation as indicated by the aggregation of intramembranous particles (Wunderlich *et al.*, 1973), the ciliary membrane shows no such alterations (Kitajima and Thompson, 1977). Furthermore, decrease in the growth temperature results in an increase in the unsaturated/saturated fatty acid ratio in all fractions except the cilia of *T. thermophila* (Nozawa *et al.*, 1974). Cilia from cells grown at 34°C have less fluid membranes than those from cells grown at 15°C as indicated by microviscosity measurements using 5-nitroxystearate as a probe (Nozawa *et al.*, 1974). The resistance of the ciliary membrane to freezing was explained by the dual role that sterols play in membrane fluidity and the high concentrations of tetrahymanol in this membrane. These workers compared the change in fluidity of whole cell total lipids with the phospholipid fraction that lacked tetrahymanol employing the 5-nitroxysterate probe. Below the transition temperature, the total lipid extract was more fluid than the total phospholipid fraction; above the transition temperature, the total lipid extract was less fluid than the total phospholipid fraction (Nozawa *et al.*, 1974). Ferguson *et al.* (1982) examined another aspect of phase behavior of *Tetrahymena* ciliary membrane lipids and subjected PsE and PnE to ^{31}P-nuclear magnetic resonance (NMR), X-ray diffraction, and freeze-fracture electron microscopy. They observed that PsE assumed the hexagonal phase above 10°C but PnE, which dominates ciliary membrane lipids, retained the lamellar phase up to 20°C and assumed the hexagonal phase at 4°C. Furthermore, a small amount of PC added to the PnE preparation had a dramatic stabilizing effect on the lamellar phase. Analysis of cilia at 30°C by ^{31}P-NMR suggested that PnE existed in the lamellar phase and these workers concluded that the presence of PC in the natural membrane kept PnE in the lamellar phase. The observation that the ciliary membrane is resistant to freezing and phase separations is also consistent with the presence of intimate lipid–protein interactions (Stephens, 1985) and/or rigid structural organization at the molecular level within ciliary membranes. Ramesha and Thompson (1982) demonstrated by fluorescence polarization procedures that the ciliary membrane of *T. pyriformis* NT-1 is indeed less fluid than the total lipids extracted from cilia (Fig. 3).

More recent studies on ciliary lipids employing *T. pyriformis* NT-1 indicate that this strain exhibits different responses to temperature shifts. Unlike studies with other *Tetrahymena* strains, this thermotolerant strain exhibited larger changes in the lipids of the cilia compared to those in microsomes (Ramesha and Thompson, 1982). The study by Fukushima *et al.* (1976) describing lipid class and fatty acid compositional differences in cells grown at different temperatures was later refined by Ramesha and Thompson (1982). These workers identified 18:2 (Δ6,11, cilienic acid) and sphingolipid hydroxy fatty acids in these studies on the NT-1 strain, which was not the case in their earlier studies on this strain or their studies on other strains. The relative amounts of PC, PsE, and ethanolamine sphingophospholipids [subsequently identified as hydroxy fatty acid-containing ethanolamine sphingophosphonolipids by Kaya *et al.* (1984a)] were lower in cilia of 15°C-grown cells and ethanolamine sphingophosphonolipids [subsequently identified as non-hydroxy fatty acid-containing ethanolamine sphingophosphonolipids by Kaya *et al.* (1984a)] were tenfold higher in these cilia. There were relatively more sphingolipids in the cilia of cells grown at 15°C and the phospholipid/tetrahymanol ratio was higher in these cells (Ramesha and Thompson, 1982). Phospholipid fatty acid composition showed relative increases in 18:2 (Δ6,11) and 18:3 (Δ6,9,12) and an overall increase in the

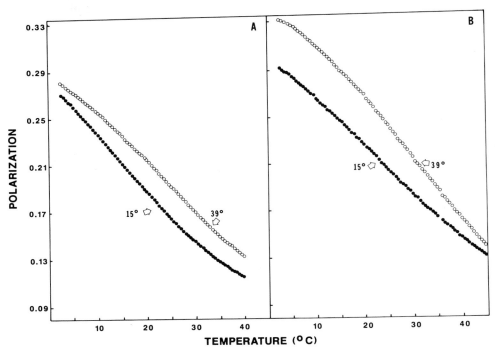

Figure 3. Influence of temperature on 1,6-diphenyl 1,3,5-hexatriene polarization in multibilayer vesicles of total lipids extracted from cilia (A) and in ciliary membrane vesicles (B) prepared from *Tetrahymena pyriformis* NT-1 cells. Cilia were isolated from cells grown at either 39°C (○) or at 15°C (●). Redrawn from Ramesha and Thompson (1982) with permission.

percent of unsaturated fatty acids. Even before changes in overall unsaturation and chain lengths of fatty acids are detected, alterations occur in individual molecular species in the phospholipids of cells exposed to chilling (Ramesha *et al.*, 1982; Maruyama *et al.*, 1982; Ramesha and Thompson, 1983).

The mechanism by which the cell compensates for the decreased fluidity of its ciliary membrane imposed by chilling is believed to involve activation of a desaturase in the microsomes and deacylation and acylation *in situ* in the cilia resulting in insertion of more unsaturated fatty acids into ciliary membrane lipids (Nozawa and Thompson, 1979; Thompson and Nozawa, 1984). These workers hypothesize that physical changes caused by freezing activate or expose palmitoyl-CoA and stearyl-CoA desaturase activities as well as the synthesis of these enzymes in the microsomes. However, stimulation of the palmitoleic acid versus stearic acid pathways also appears to be involved in temperature acclimation and adaptation (Conner and Stewart, 1976; see above). Ramesha and Thompson (1982) observed only one consistent pattern of fatty acid compositional changes in *T. pyriformis* NT-1 cells subjected to chilling. This pattern was the first change detected, which is seen at 8–12 hr after the shift. The concentration of 16:1 increased, reached a maximum, and then decreased, which is consistent with the hypothesis that chilling stimulates the palmitoleic acid pathway. In cilia of these cells, the end product of this pathway, 18:2 (Δ6,11), increased the most in comparison to cilia from cells grown at 39°C. It was further observed that palmitoyl-CoA desaturase activity in the microsomes was most active 2 hr after the downshift (Nozawa and Kasai, 1978).

Deacylation and reacylation of resident phospholipids in the cilia are thought to occur resulting in the retailoring of the membrane lipids (Ramesha and Thompson, 1984). Relative activities of the phospholipiase A, acyl-CoA synthetase, and acyltransferase enzymes in cilia (see above) of cells grown at the two temperature extremes of this organism vary with the temperature of the assays (Table V) (Ramesha and Thompson, 1984).

The relative amounts of two sphingophosphonolipids in *T. pyriformis* NT-1, which differ in their non-hydroxy and hydroxy fatty acid compositions, change at the two temperature extremes of this organism (Ramesha and Thompson, 1982; Kaya et al., 1984a,b). The percent of hydroxy fatty acids decreases at the low temperature and increases at the high temperature suggesting that *in situ* hydroxylation occurs in the ciliary membrane (Kaya et al., 1984a). However, *in vitro* hydroxylation of fatty acids using isolated cilia did not occur (Kaya et al., 1984b) and hence the reactions occurring *in situ* at this membrane remain an open question. Furthermore, the presence of bulky hydroxyl groups in the bilayer would tend to fluidize membranes, but increased hydroxy fatty acid concentrations were found during temperature upshifts; thus, the role of fatty acid hydroxylation in response to temperature shifts remains unclear.

Upon chilling, there appears to be a fast response in which *T. pyriformis* GL cells were observed to alter the unsaturation in their fatty acids (Wunderlich et al., 1973). After 16 hr these cells become acclimated and exhibit normal swimming behavior and intramembrane particle patterns. Analysis of cilia from *T. thermophila* cells grown at 15°C indicated that their membranes were more fluid than those from cells grown at 34°C (Nozawa et al., 1974). However, the number of double bonds in fatty acids of cells kept at high and low temperatures for prolonged periods of time do not differ (Wunderlich et al., 1973). Thus, an adaptation phase following the initial acclimation phase probably occurs during which time another fluidizing influence presumably enters the acclimated mem-

Table V. Activities of Enzymes Involved in Lipid Metabolism in Cilia Isolated from *Tetrahymena pyriformis* NT-1 Cells Grown at Either 39° or 15°C[a]

	Phospholipase A activity			Acyltransferase activity			
				Without lysophospholipid		With 1-acyllyso-PsE	
	Incubation temperature			Incubation temperature			
Source of cilia	37°C	15°C	Acyl-CoA added	37°C	15°C	37°C	15°C
39°C-grown cells	15.7	1.3	[¹⁴C]16:0	10.0	1.3	15.5	10.0
			[¹⁴C]18:2	5.9	0.1	27.4	6.1
15°C-grown cells	49.2	27.0	[¹⁴C]16:0	16.4	4.7	35.7	15.9
			[¹⁴C]18:2	18.5	6.6	ND[b]	21.4

[a]Modified from Ramesha and Thompson (1984).
[b]ND, not determined.

branes allowing fatty acid compositions to return to those observed in cells kept at normal temperatures. Our understanding of the mechanisms involved in short-term acclimation and long-term adaptation may improve when precise biochemical events have been examined over short and long periods of time after temperature shifts.

4.5. Mutations

Mutants of *P. tetraurelia* with altered lipid compositions have been described. The *baA* or barium-shy mutant is reported to have altered ethanolamine sphingolipid and glycerolipid compositions in its ciliary membrane (Forte *et al.*, 1981). The phenotype of this mutant is characterized by prolonged backward swimming in response to Ba^{2+}, decreased voltage-sensitive Ca^{2+} and K^+ membrane conductances (Forte *et al.*, 1981), and a lack of thermal avoidance behavior (Hennessey and Nelson, 1983).

Polyene antibiotic-resistant mutants of *Paramecium* also exhibit prolonged backward swimming in Ba^{2+}-containing solutions but are genetically distinct from the barium-shy mutants (Forte *et al.*, 1986). Resistance of these mutants to polyene antibiotics implies alterations in membrane sterols. Changes in sterols have not yet been demonstrated by direct analysis of the ciliary membrane lipids of this mutant.

5. Conclusions

The membranes surrounding cilia and flagella are clearly domains of the cell surface that are distinct with respect to lipid composition. It has been suggested that in *Tetrahymena*, the unique lipid composition of the cell-surface ciliary membrane serves a protective function (Kennedy and Thompson, 1970; Florin-Christensen *et al.*, 1986). The ciliary membrane of *Tetrahymena* and *Paramecium* is enriched in lipids with resistant bonds. The direct P–C bond in phosphonolipids is stronger than phosphoryl (P–O–C) bonds. Ether bonds linking fatty alcohols to glycerol backbones of ether lipids are stronger than corresponding ester bonds of fatty acids. Amide-linked fatty acids in sphingolipids are more resistant to hydrolysis than are ester-linked fatty acids of glycerolipids. It has been suggested that specific lipids are directed by cellular membrane trafficking mechanisms to their specific membranes and then enzymes secreted into the environment (Florin-Christensen *et al.*, 1986) and/or enzymes *in situ* in the ciliary membrane adjust the lipid composition of this cell-surface membrane (Nozawa and Thompson, 1979) for increased resistance to hydrolytic events and lysis of the cell. It is, however, reasonable to assume that the lipid compositions of these specialized membranes do not only have the property of resistance to physical and enzymatic attack but are also chemically unique to serve the specialized functions of these motile organelles. Their roles in ciliary and flagellar functions are not known but it is already evident from the available data that ciliary membranes and their lipids are unlike those of other membranes of the cell and have chemical and physical properties quite unlike any other membrane system of the cell. Specific lipid species such as PIP_2 are known to be directly involved in transmembrane signaling events that lead to the mobilization of Ca^{2+} in several cell types examined. Since the concentration of Ca^{2+} is of central importance in ciliary and flagellar activity, it would not be surprising if PIP_2 metabolism has a role in the control of these organelles. The information currently available on lipid mutants indicates that lipids play important roles in the proper functioning of ionic channels of *Paramecium* surface membranes (Forte *et al.*, 1981, 1986). With the availability of such mutants, it should be feasible to correlate the mutational defects with lipid changes and hence the specific role of lipids in ciliary membranes. A comparative approach to finding common denominators that could correlate lipid membrane composition with ciliary and flagellar functions is, however, unfeasible until more membranes of these organelles are examined and more is known about their lipid compositions and the enzymes that act on them. This review demonstrates that there is currently a severe lack of information on the lipids of membranes surrounding cilia and flagella of cells other than *P. tetraurelia* and about three strains of *Tetrahymena*.

ACKNOWLEDGMENT. This work was supported in part by NIH grant GM32425 and the University of Cincinnati Research Council.

References

Adoutte, A., Ramanathan, R., Lewis, R. M., Dute, R. R., Ling, K.-Y., Kung, C., and Nelson, D. L., 1980, Biochemical studies of the excitable membrane of *Paramecium tetraurelia*. III. Proteins of cilia and ciliary membranes, *J. Cell Biol.* **84:**717–738.

Andrews, D., and Nelson, D. L., 1979, Biochemical studies of the excitable membrane of *Paramecium tetraurelia*. II. Phospholipids of ciliary and other membranes, *Biochim. Biophys. Acta* **550:**174–187.

Berger, H., Jones, P., and Hanahan, D. J., 1972, Structural studies on lipids of *Tetrahymena pyriformis* W, *Biochim. Biophys. Acta* **260:**617–629.

Bordier, C. R., Etges, J., Ward, J., Turner, M. J., and Cardoso de Almeida, M. L., 1986, *Leishmania* and *Trypanosoma* surface glycoproteins have a common glycophospholipid membrane anchor, *Proc. Natl. Acad. Sci. USA* **83:**5988–5991.

Brugerolle, G., Andrivon, C., and Bohatier, J., 1980, Isolation, protein pattern and enzymatic characterization of the ciliary membrane of *Paramecium tetraurelia*, *Biol. Cell.* **37:**251–260.

Capceville, Y., Baltz, T., Deregnaucourt, C., and Keller, A.-M., 1986, Immunological evidence of a common structure between *Paramecium* antigens and *Trypanosoma* variant surface glycoproteins, *Exp. Cell Res.* **167:**75–86.

Carter, H. E., and Gaver, R. C., 1967, Branched-chain sphingosines from *Tetrahymena pyriformis*, *Biochem. Biophys. Res. Commun.* **29:**886–891.

Chen, L. L., Pousada, M., and Haines, T. H., 1976, The flagellar membrane of *Ochromonas danica*. Lipid composition, *J. Biol. Chem.* **251:**1835–1842.

Chen, S.-J., and Bouck, G. B., 1984, Endogenous glycosyltransferases glucosylate lipids in flagella of *Euglena*, *J. Cell Biol.* **98:**1825–1835.

Child, F. M., 1959, The characterization of the cilia of *Tetrahymena pyriformis*, *Exp. Cell Res.* **28:**258–267.

Conner, R. L., and Reilly, A. E., 1975, The effects of isovalerate supplementation on growth and fatty acid composition of *Tetrahymena pyriformis* W, *Biochim. Biophys. Acta* **398:**209–216.

Conner, R. L., and Stewart, B. Y., 1976, The effect of temperature on the fatty acid composition of *Tetrahymena pyriformis* WH-14, *J. Protozool.* **23:**193–196.

Conner, R. L., and Van Wagtendonk, W. J., 1955, Steroid requirements of *Paramecium aurelia*, *J. Gen. Microbiol.* **12:**31–36.

Conner, R. L., Mallory, F. B., Landrey, J. R., Ferguson, K. A., Kaneshiro, E. S., and Ray, E., 1971, Ergosterol replacement of tetrahymanol in *Tetrahymena* membranes, *Biochem. Biophys. Res. Commun.* **44:**995–1000.

Dennis, E. A., and Kennedy, E. P., 1970, Enzymatic synthesis and decarboxylation of phosphatidylserine in *Tetrahymena pyriformis*, *J. Lipid Res.* **11:**391–403.

Etges, R., Bouvier, J., and Bordier, C., 1986, The major surface protein of *Leishmania* promastigotes is a protease, *J. Biol. Chem.* **261:**9098–9101.

Ferguson, K. A., Conner, R. L., and Mallory, F. B., 1971, The effect of ergosterol on fatty acid composition of *Tetrahymena pyriformis*, *Arch. Biochem. Biophys.* **144:**448–450.

Ferguson, K. A., Conner, R. L., Mallory, F. B., and Mallory, C. W., 1972, α-Hydroxy fatty acids in sphingolipids of *Tetrahymena, Biochim. Biophys. Acta* **270**:111–116.

Ferguson, K. A., Davis, F. M., Conner, R. L., Landrey, J. R., and Mallory, F. B., 1975, Effect of sterol replacement *in vivo* on the fatty acid composition of *Tetrahymena, J. Biol. Chem.* **250**:6998–7005.

Ferguson, K. A., Hui, S. W., Stewart, T. P., and Yeagle, P. L., 1982, Phase behavior of the major lipids of *Tetrahymena* ciliary membranes, *Biochim. Biophys. Acta* **684**:179–186.

Ferguson, M. A. J., Low, M. G., and Cross, G. A. M., 1985, Glycosyl-*sn*-1,2-dimyristylphosphatidyl inositol is covalently linked to *Trypanosoma brucei* variant surface glycoprotein, *J. Biol. Chem.* **260**:14547–14555.

Florin-Christensen, J., Florin-Christensen, M., Knudsen, J., and Rasmussen, L., 1986, Phospholipases and phosphonolipids in a ciliate: An attack and defence system? *Trends Biochem. Sci.* **11**:354–355.

Forte, M., Satow, Y., Nelson, D., and Kung, C., 1981, Mutational alteration of membrane phospholipid composition and voltage-sensitve ion channel function in *Paramecium, Proc. Natl. Acad. Sci. USA* **78**: 7195–7199.

Forte, M., Hennessey, T., and Kung, C., 1986, Mutations resulting in resistance of polyene antibiotics decrease voltage-sensitive calcium channel activity in *Paramecium tetraurelia, J. Neurogenet.* **3**:75–86.

Fukushima, H., Martin, C. E., Iida, H., Kitajima, Y., Thompson, G. A., Jr., and Nozawa, Y., 1976, Changes in membrane lipid composition during temperature adaptation by a thermotolerant strain of *Tetrahymena pyriformis, Biochim. Biophys. Acta* **431**:165–179.

Fukushima, H., Nagao, S., Okano, Y., and Nozawa, Y., 1977, Studies on *Tetrahymena* membranes. Palmitoyl-coenzyme A desaturase, a possible key enzyme for temperature adaptation in *Tetrahymena* microsomes, *Biochim. Biophys. Acta* **488**:442–453.

Gealt, M. A., Adler, J. H., and Nes, W. R., 1981, The sterols and fatty acids from purified flagella of *Chlamydomonas reinhardi, Lipids* **16**:133–136.

Gershengorn, M. C., Smith, A. R. H., Goulston, G., Goad, L. J., and Goodwin, T. W., 1968, The sterols of *Ochromonas danica* and *Ochromonas malhamensis, Biochemistry* **7**:1698–1706.

Gibbons, I. R., 1963, Studies on the protein components of cilia from *Tetrahymena pyriformis, Proc. Natl. Acad. Sci. USA* **50**:1002–1010.

Haines, T. H., 1973, Halogen- and sulfur-containing lipids of *Ochromonas, Annu. Rev. Microbiol.* **27**:403–411.

Hansma, H. G., 1975, The immobilization antigen of *Paramecium aurelia* is a single polypeptide chain, *J. Protozool.* **22**:257–259.

Hennessey, T. M., and Nelson, D. L., 1983, Biochemical studies of the excitable membrane of *Paramecium tetraurelia.* VIII. Temperature-induced changes in lipid composition and in thermal avoidance behavior, *Biochem. Biophys. Acta* **728**:145–158.

Hennessey, T. M., Andrews, D., and Nelson, D. L., 1983, Biochemical studies of the excitable membrane of *Paramecium tetraurelia.* VII. Sterols and other neutral lipids of cells and cilia, *J. Lipid Res.* **24**:575–587.

Holz, G. G., Jr., Erwin, J., Rosenbaum, N., and Aaronson, S., 1962, Triparanol inhibition of *Tetrahymena,* and its prevention by lipids, *Arch. Biochem. Biophys.* **98**:312–322.

Iida, H., Maeda, T., Ohki, K., and Nozawa, Y., 1978, Transfer of phosphatidylcholine between different membranes in *Tetrahymena* as studied by spin labeling, *Biochim. Biophys. Acta* **508**:55–64.

Jonah, M., and Erwin, J. A., 1971, The lipids of membraneous cell organelles isolated from the ciliate, *Tetrahymena pyriformis, Biochim. Biophys. Acta* **231**:80–92.

Kandatsu, M., and Horiguchi, M., 1962, Occurrence of ciliatine (2-aminoethylphosphonic acid) in *Tetrahymena, Agric. Biol. Chem.* **26**:721–722.

Kaneshiro, E. S., 1980, Positional distribution of fatty acids in the major glycerophospholipids of *Paramecium tetraurelia, J. Lipid Res.* **21**:559–570.

Kaneshiro, E. S., 1987, Lipids of *Paramecium, J. Lipid Res.* **28**:1241–1258.

Kaneshiro, E. S., Beischel, L. S., Merkel, S. J., and Rhoads, D. E., 1979, The fatty acid composition of *Paramecium aurelia* cells and cilia: Changes with culture age, *J. Protozool.* **26**:147–158.

Kaneshiro, E. S., Meyer, K. B., and Reese, M. L., 1983, The neutral lipids of *Paramecium tetraurelia:* Changes with culture age and the detection of steryl esters in ciliary membranes, *J. Protozool.* **30**:392–396.

Kaneshiro, E. S., Matesic, D. F., and Jayasimhulu, K., 1984, Characterizations of six ethanolamine sphingophospholipids from *Paramecium* cells and cilia, *J. Lipid Res.* **25**:369–377.

Kaneshiro, E. S., Meyer, K. B., and Rhoads, D. E., 1987, The glyceryl ethers of *Paramecium* phospholipids and phosphonolipids, *J. Protozool.* **34**:357–361.

Kapoulas, V. M., Thompson, G. A., Jr., and Hanahan, D. J., 1969a, Metabolism of α-glyceryl ethers by *Tetrahymena pyriformis*. I. Characterization of the *in vivo* degradation system, *Biochim. Biophys. Acta* **176:**237–249.

Kapoulas, V. M., Thompson, G. A., Jr., and Hanahan, D. J., 1969b, Metabolism of α-glyceryl ethers by *Tetrahymena pyriformis*. II. Properties of a cleavage system *in vitro*, *Biochim. Biophys. Acta* **176:**250–264.

Kasai, R., Kitajima, Y., Martin, C. E., Nozawa, Y., Skriver, L., and Thompson, G. A., Jr., 1976, Molecular control of membrane properties during temperature acclimation. Membrane fluidity regulation of fatty acid desaturase action? *Biochemistry* **15:**5228–5233.

Kaya, K., Ramesha, C. S., and Thompson, G. A., Jr., 1984a, Temperature-induced changes in the hydroxy and non-hydroxy fatty acid-containing sphingolipids abundant in the surface membrane of *Tetrahymena pyriformis* NT-1, *J. Lipid Res.* **25:**68–74.

Kaya, K., Ramesha, C. S., and Thompson, G. A., Jr., 1984b, On the formation of α-hydroxy fatty acids. Evidence for a direct hydroxylation of nonhydroxy fatty acid-containing sphingolipids, *J. Biol. Chem.* **259:**3548–3553.

Kennedy, K. E., and Thompson, G. A., Jr., 1970, Phosphonolipids: Localization in surface membranes of *Tetrahymena, Science* **168:**989–991.

Kitajima, Y., and Thompson, G. A., Jr., 1977, *Tetrahymena* strives to maintain the fluidity interrelationships of all its membranes constant. Electron microscope evidence, *J. Cell Biol.* **72:**744–755.

Koroly, M. J., and Conner, R. L., 1976, Unsaturated fatty acid biosynthesis in *Tetrahymena*. Evidence for two pathways, *J. Biol. Chem.* **251:**7588–7592.

Lester, R. L., Becker, G. W., and Kaul, K., 1978, Phosphoinositides of fungi and plants, in: *Cyclitols and Phosphoinositides* (W. W. Wells and F. Eisenberg, Jr., eds.), Academic Press, New York, pp. 83–99.

Low, M. G., and Saltiel, A. R., 1988, Structural and functional roles of glycosylphosphatidylinositol in membranes, *Science* **239:**268–275.

Mallory, F. B., Gordon, J. T., and Conner, R. L., 1963, The isolation of a pentacyclic triterpenoid alcohol from a protozoan, *J. Am. Chem. Soc.* **85:**1362–1363.

Martin, C. E., and Thompson, G. A., Jr., 1978, Use of fluorescence polarization to monitor intracellular membrane changes during temperature acclimation. Correlation with lipid compositional and ultrastructural changes, *Biochemistry* **17:**3581–3586.

Martin, C. E., Hiramitsu, K., Kitajima, Y., Nozawa, Y., Skriver, L., and Thompson, G. A., Jr., 1976, Molecular control of membrane properties during temperature acclimation. Fatty acid desaturase regulation of membrane fluidity in acclimating *Tetrahymena* cells, *Biochemistry* **24:**5218–5227.

Maruyama, H., Banno, Y., Watanabe, T., and Nozawa, Y., 1982, Studies on thermal adaptation in *Tetrahymena* membrane lipids: Modification of positional distribution of phospholipid acyl chains in plasma membranes, mitochondria, and microsomes, *Biochim. Biophys. Acta* **711:**229–244.

Morris, R. J., and Bone, Q., 1983, Metazoan lipids: an unusual association of saturated sterols with relatively saturated fatty acids in the cilia of *Ciona intestinalis, Lipids* **18:**900–901.

Morris, R. J., and Bone, Q., 1985, Highly saturated lipid composition of ctenophore cilia: Possible indication of low membrane permeability, *Lipids* **20:**933–935.

Nandini-Kishore, S. G., Mattox, S. M., Martin, C. E., and Thompson, G. A., Jr., 1979, Membrane changes during growth of *Tetrahymena* in the presence of ethanol, *Biochim. Biophys. Acta* **551:**315–327.

Nozawa, Y., and Kasai, R., 1978, Mechanism of thermal adaptation of membrane lipids in *Tetrahymena pyriformis* NT-1. Possible evidence for temperature-mediated induction of palmitoyl-CoA desaturase, *Biochim. Biophys. Acta* **529:**54–66.

Nozawa, Y., and Thompson, G. A., Jr., 1971a, Studies of membrane formation in *Tetrahymena pyriformis*. II. Isolation and lipid analysis of cell fractions, *J. Cell Biol.* **49:**712–721.

Nozawa, Y., and Thompson, G. A., Jr., 1971b, Studies of membrane formation in *Tetrahymena pyriformis*. III. Lipid incorporation into various cellular membranes of logarithmic phase cultures, *J. Cell Biol.* **49:**722–729.

Nozawa, Y., and Thompson, G. A., Jr., 1972, Studies of membrane formation in *Tetrahymena pyriformis*. V. Lipid incorporation into various cellular membranes of stationary phase cells, starving cells and cells treated with metabolic inhibitors, *Biochim. Biophys. Acta* **282:**93–104.

Nozawa, Y., and Thompson, G. A., Jr., 1979, Lipids and membrane organization in *Tetrahymena*, in: *Biochemistry and Physiology of Protozoa*, 2nd ed., Volume 2 (M. Levandowsky and S. H. Hutner, eds.), Academic Press, New York, pp. 275–338.

Nozawa, Y., Iida, H., Fukushima, H., Ohki, K., and Ohnishi, S., 1974, Studies on *Tetrahymena* membranes: Temperature-induced alterations in fatty acid compositions of various membrane fractions in *Tetrahymena pyriformis* and its effect on membrane fluidity as inferred by spin-label study, *Biochim. Biophys. Acta* **367**: 134–147.

Nozawa, Y., Fukushima, H., and Iida, H., 1975, Studies on *Tetrahymena* membranes. Modification of surface membrane lipids by replacement of tetrahymanol by exogenous ergosterol in *Tetrahymena pyriformis, Biochim. Biophys. Acta* **406**:248–263.

Nozawa, Y., Kasai, R., and Sekiya, T., 1980, Modifications of membrane lipid composition following a nutritional shift-up of starved cells. A comparison with membrane biogenesis in *Tetrahymena, Biochim. Biophys. Acta* **603**:347–365.

Pieringer, J., and Conner, R. L., 1979, Positional distribution of fatty acids in the glycerolipids of *Tetrahymena pyriformis, J. Lipid Res.* **20**:363–370.

Pollard, W. O., Shorb, M. S., Lund, P. G., and Vasaitis, V., 1964, Effect of triparanol on synthesis of fatty acids by *Tetrahymena pyriformis, Proc. Soc. Exp. Biol. Med.* **116**:539–543.

Ramesha, C. S., and Thompson, G. A., Jr., 1982, Changes in lipid composition and physical properties of *Tetrahymena* ciliary membranes following low-temperature acclimation, *Biochemistry* **21**:3612–3617.

Ramesha, C. S., and Thompson, G. A., Jr., 1983, Cold stress induces in situ phospholipid molecular species changes in cell surface membranes, *Biochim. Biophys. Acta* **731**:251–260.

Ramesha, C. S., and Thompson, G. A., Jr., 1984, The mechanism of membrane response to chilling. Effect of temperature on phospholipid deacylation and reacylation reactions in the cell surface membrane, *J. Biol. Chem.* **259**:8706–8712.

Ramesha, C. S., Dickens, B. F., and Thompson, G. A., Jr., 1982, Phospholipid molecular species alterations in *Tetrahymena* ciliary membranes following low-temperature acclimation, *Biochemistry* **21**:3618–3622.

Rhoads, D. E., and Kaneshiro, E. S., 1979, Characterizations of phospholipids from *Paramecium tetraurelia* cells and cilia, *J. Protozool.* **26**:329–338.

Rhoads, D. E., and Kaneshiro, E. S., 1984, Fatty acid metabolism in *Paramecium:* Oleic acid metabolism and inhibition of polyunsaturated fatty acid synthesis by triparanol, *Biochim. Biophys. Acta* **795**:20–29.

Rhoads, D. E., Meyer, K. B., and Kaneshiro, E. S., 1981, Isolation and preliminary characterization of 1-O-octadec-*cis*-11-enyl glycerol from *Paramecium* phospholipids, *Biochem. Biophys. Res. Commun.* **98**:858–865.

Rhoads, D. E., Honer-Schmid, O., and Kaneshiro, E. S., 1987, Metabolism of saturated fatty acids by *Paramecium tetraurelia, J. Lipid Res.* **28**:1424–1433.

Shimonaka, H., Fukushima, H., Kawai, K., Nagao, S., Okano, Y., and Nozawa, Y., 1978, Altered microviscosity of *in vivo* lipid-manipulated membranes in *Tetrahymena pyriformis:* A fluorescence study, *Experientia* **34**:586–587.

Smith, J. D., and O'Malley, M. A., 1978, Control of phosphonic acid and phosphonolipid synthesis in *Tetrahymena, Biochim. Biophys. Acta* **528**:394–398.

Smith, J. D., Snyder, W. R., and Law, J. H., 1970, Phosphonolipids in *Tetrahymena* cilia, *Biochem. Biophys. Res. Commun.* **39**:1163–1169.

Stephens, R. E., 1983, Reconstitution of ciliary membranes containing tubulin, *J. Cell Biol.* **96**:68–75.

Stephens, R. E., 1985, Evidence for a tubulin-containing lipid–protein structural complex in ciliary membranes, *J. Cell Biol.* **100**:1082–1090.

Stommel, E. W., and Stephens, R. E., 1985, Calcium-dependent phosphatidylinositol phosphorylation in lamellibranch gill lateral cilia, *J. Comp. Physiol. A* **157**:441–449.

Subbaiah, P. V., and Thompson, G. A., Jr., 1974, Studies of membrane formation in *Tetrahymena pyriformis*. The biosynthesis of proteins and their assembly into membranes of growing cells, *J. Biol. Chem.* **249**: 1302–1310.

Suchard, S. J., Rhoads, D. E., and Kaneshiro, E. S., 1989, The inositol lipids of *Paramecium tetraurelia* and preliminary characterizations of phosphoinositide kinase activity in the ciliary membrane, *J. Protozool.* **36**: 185–190.

Sugita, M., Fukunaga, Y., Ohkawa, K., Nozawa, Y., and Hori, T., 1979, Structural components of sphingo-phosphonolipids from the ciliated protozoan, *Tetrahymena pyriformis* WH-14, *J. Biochem.* **86**:281–288.

Takemoto, T., 1961, Phospholipids in *Tetrahymena pyriformis* W, *Z. Allg. Mikrobiol.* **1**:331–340.

Thiele, J., Klumpp, S., and Schultz, J. E., 1982, Differential distribution of voltage-dependent calcium channels and guanylate cyclase in the excitable membrane from *Paramecium tetraurelia, Eur. J. Cell Biol.* **28**:3–11.

Thiele, J., Otto, M. K., Deitmer, J. W., and Schultz, J. E., 1983, Calcium channels of the excitable ciliary membrane from *Paramecium:* An initial biochemical characterization, *J. Membr. Biol.* **76:**253–260.

Thompson, G. A., Jr., and Nozawa, Y., 1977, *Tetrahymena:* A system for studying dynamic membrane alterations within the eukaryotic cell, *Biochim. Biophys. Acta* **472:**55–92.

Thompson, G. A., Jr., and Nozawa, Y., 1984, The regulation of membrane fluidity in *Tetrahymena,* in: *Membrane Fluidity* (M. Kates and G. L. Manson, eds.), Plenum Press, New York, pp. 397–432.

Thompson, G. A., Jr., Bambery, R. J., and Nozawa, Y., 1971, Further studies of the lipid composition and biochemical properties of *Tetrahymena pyriformis* membrane systems, *Biochemistry* **10:**4441–4447.

Thompson, G. A., Jr., Bambery, R. J., and Nozawa, Y., 1972, Environmentally produced alterations in the tetrahymanol:phospholipid ratio in *Tetrahymena pyriformis* membrane, *Biochim. Biophys. Acta* **260:**630–638.

Thompson, G. A., Jr., Baugh, L. C., and Walker, L. F., 1974, Nonlethal deciliation of *Tetrahymena* by a local anesthetic and its utility as a tool for studying cilia regeneration, *J. Cell Biol.* **61:**253–257.

Van Wagtendonk, W. J., 1974, Nutrition of *Paramecium,* in: *Paramecium: A Current Survey* (W. J. Van Wagtendonk, ed.), Elsevier, Amsterdam, pp. 339–376.

Witman, G. B., Carlson, K., Berliner, J., and Rosenbaum, J. L., 1972, *Chlamydomonas* flagella. I. Isolation and electrophoretic analyses of microtubules, matrix, membranes and mastigonemes, *J. Cell Biol.* **54:**507–555.

Wunderlich, F., Speth, V., Batz, W., and Kleining, H., 1973, Membranes of *Tetrahymena.* III. The effect of temperature on membrane core structures and fatty acid composition of *Tetrahymena* cells, *Biochim. Biophys. Acta* **298:**39–49.

Flagellar Surfaces of Parasitic Protozoa and Their Role in Attachment

Keith Vickerman and Laurence Tetley

Flagella are thought of primarily as propulsive structures but in the lives of certain parasitic protozoa they have another important role as organelles of attachment to host surfaces. There is now good evidence that such attachment is vital to the survival of the parasite, not simply by anchoring it in a preferred environment, but also by ensuring its transmission to another host. Most practical interest attaches to the pathogenic trypanosomes and leishmanias, as transmission of these important causative agents of major diseases in man and his domestic animals in the tropics and subtropics depends upon flagellar attachment in the insect vector. Prevention of such attachment would, in theory, be a means of controlling the spread of disease. Little is known of the mechanisms operating in flagellar attachment, however, or of the relationship of attachment to parasite morphogenesis. The purpose of this review is to draw attention to an important aspect of parasite cell biology which is now ripe for further investigation.

1. Developmental Cycles of Kinetoplastid Protozoa

The ability of the flagellum to function as a host-attachment organelle would appear to be unique to protozoa of the flagellate order Kinetoplastida Honigberg, 1963. In some kinetoplastids the flagellum is also attached to the body of the flagellate for much of its length, the surface of the cell being drawn up into crests by the beating flagellum, forming an ''undulating membrane.'' Such body attachment is not unique to kinetoplastids, however, and spectacular undulating membranes occur in *Trichomonas, Pyrsonympha,* and other higher zooflagellates. Nor is attachment to host surfaces the prerogative of kinetoplastids: a sophisticated sucking disk enables diplomonads of the genus *Giardia* to attach to host gut epithelium (Holberton, 1974; Erlandsen and Feely, 1984), and certain

Keith Vickerman and Laurence Tetley • Department of Zoology, University of Glasgow, Glasgow G12 8QQ, Scotland, United Kingdom.

species of *Pyrsonympha* (Oxymonadida) have a specialized anterior attachment organelle of unknown homology which anchors them to the cuticularized hindgut of termites of the genus *Reticulitermes* (Bloodgood, 1975).

Kinetoplastid flagellates are distinguished by the possession of a single mitochondrion in which one or more kinetoplasts, stainable masses of mitochondrial DNA, are evident, usually close to the base of the flagellar apparatus which emerges from a distinct pocket in the cell body (Fig. 1). The order has two families—the biflagellate Bodonidae, which includes free-living and parasitic forms, and the uniflagellate Trypanosomatidae, which are wholly parasitic. In both families, representatives with one host (homoxenous) and two host (heteroxenous) life cycles are found (reviewed in Vickerman, 1976, 1989). The latter, in particular, provide excellent examples of how a parasitic cell can adapt its structure and metabolism to drastic changes in its surroundings (Vickerman, 1985).

Although bodonid parasites are widespread among invertebrates and lower vertebrates, they are not of major economic importance and have been little studied. Some species of *Cryptobia* live in the spermatheca of their gastropod mollusk host and are transmitted during copulation. The parasites anchor themselves to the microvilli of the host epithelium by the anterior flagellum (Current, 1980); the posterior (recurrent) flagellum in this genus adheres to the flagellate's body along its length. Other homox-

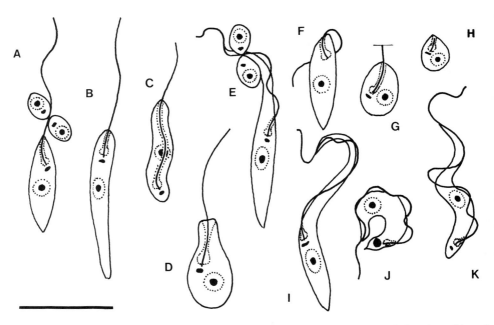

Figure 1. Different stages in the life cycles of trypanosomatid flagellates showing variations in position of kinetoplast, nucleus, flagellum, and flagellar pocket. (A) *Leptomonas oncopelti* promastigote with "straphanger" amastigote cysts. (B, C) *Herpetomonas muscarum* promastigote (B) and opisthomastigote (C). (D) *Crithidia fasciculata* choanomastigote. (E) *Blastocrithidia familiaris* epimastigote with "straphanger" amastigote cysts. (F–H) *Leishmania major* promastigote (F), paramastigote haptomonad (G), and amastigote (H). (I, J) *Trypanosoma cruzi* epimastigote (I) and trypomastigote blood form (J). (K) *T. brucei* trypomastigote slender blood form. After Vickerman (1989).

enous species of *Cryptobia* live in the foregut and on the gills of teleost fishes; here the posterior flagellum serves in attachment (Lom, 1980). Heteroxenous species of *Cryptobia* (often regarded as a separate genus, *Trypanoplasma*) live in the blood of fishes and amphibians and undergo a cycle of development in leeches; whether attachment plays a part in production of the fish-infective metacyclic stage in the leech has not been determined (Woo, 1987).

In the complex life cycles of the Trypanosomatidae, various developmental forms are recognized, according to the position of the kinetoplast and flagellar base in relation to the nucleus and posterior or anterior end of the elongate body, and according to whether the single flagellum (which corresponds to the anterior flagellum of bodonids) is free or attached to the body (Fig. 1). Thus, we can distinguish amastigote (no emergent flagellum), promastigote (anterior emergent flagellum, kinetoplast in front of the nucleus), paramastigote (anterior emergent flagellum, kinetoplast alongside the nucleus), opisthomastigote (anterior emergent flagellum, kinetoplast behind the nucleus), epimastigote (laterally emergent flagellum attached to the body, kinetoplast in front of the nucleus), and trypomastigote (laterally emergent flagellum attached to the body, kinetoplast behind the nucleus). The choanomastigote form is distinguished by its broad flagellar pocket, anterior emergent flagellum, and anterolateral kinetoplast. Attached stages are sometimes referred to as haptomonads in contrast to the free-swimming or nectomonad stages. Flagellar beating often continues in the haptomonad stages.

Different genera and species of Trypanosomatidae are distinguished by a characteristic spectrum of developmental forms in the life cycle. In the genus *Leishmania*, the parasite multiplies in man or other mammalian host as the amastigote form inside macrophages. On being taken up by the bloodsucking sandfly vector (genera *Phlebotomus*, *Lutzomyia*), the amastigotes at first multiply extracellularly in the gut lumen, but soon transform into elongate promastigotes within the peritrophic membrane (PTM) of the insect's abdominal midgut (Fig. 2). With the breakup of the PTM, the nectomonad promastigotes migrate to the fly's thoracic midgut, where they insert their flagella between the midgut microvilli. Some *Leishmania* spp. (described as peripylarian) then invade the terminal part of the midgut (pylorus) and the hindgut (ileum) and attach to the cuticular lining as paramastigotes; the majority of species (suprapylarian), however, produce attached nondividing pro- or paramastigotes on the stomodaeal valve at the entrance to the midgut (Fig. 2), in the esophagus, and occasionally in the crop. Anterior spread of attached parasites to the pharynx occurs. Development terminates with the appearance in the fly's proboscis of small metacyclic promastigotes which are infective to the mammalian host. For a more detailed account of development, see Molyneux and Killick-Kendrick (1987).

The pathogenic trypanosomes of mammals develop to the infective metacyclic stage either in the hindgut (stercorarian species) or in the mouthparts or salivary glands (salivarian species) of the vector. *Trypanosoma cruzi* (stercorarian), causative agent of Chagas' disease in man, multiplies as an intracellular amastigote inside muscle or mononuclear phagocytic cells, but moves from cell to cell via the blood as a motile trypomastigote which is the form ingested by the reduviid bug (Hemiptera) vector. In the bug's midgut (Fig. 2), multiplication occurs in the epimastigote form; invasion of the hindgut is followed by attachment of the epimastigotes by their flagella to its chitinous lining, by further division and eventually by production of nondividing metacyclic trypomastigotes;

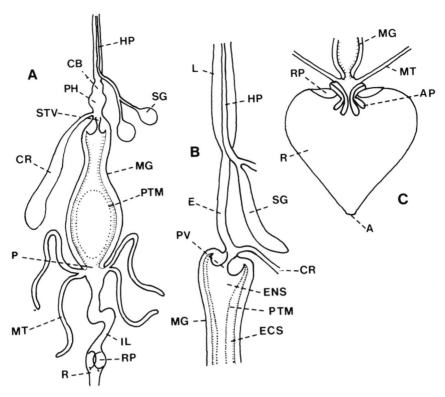

Figure 2. Diagrams of alimentary systems of insect vectors. (A) Sandfly, *Phlebotomus* sp. (Diptera), vector of *Leishmania* spp.; midgut region shown in section. The peritrophic membrane is a temporary structure surrounding the recently ingested bloodmeal. After various authors. (B) Tsetse fly, *Glossina* sp. (Diptera), vector of *Trypanosoma brucei, T. congolense,* and *T. vivax*; anterior part of gut (in section), proboscis, and one salivary gland shown. The peritrophic membrane is a continuous structure secreted as a tube from the proventriculus. After Hoare (1949). (C) Kissing bug, *Triatoma infestans* (Hemiptera), vector of *Trypanosoma cruzi*; hindgut region (in section) only, showing the rectal pads and processes of the ampulla cells. After Böker and Schaub (1984). A, anus; AP, ampullar cell processes; CB, cibarium (prepharynx); CR, crop; E, esophagus; ECS, ectotrophic space; ENS, endotrophic space; HP, hypopharynx; IL, ileum; L, labrum; MG, midgut; MT, Malpighian tubules; P, pylorus; PH, pharynx; PTM, peritrophic membrane; PV, proventriculus; R, rectum, RP, rectal pads; SG, salivary gland; STV, stomodaeal valve.

these lose their hold on the substratum and are released free into the lumen to be deposited by the bug in its feces when it feeds. Infection of the mammal is by contamination of the bite with fecal metacyclics (reviewed by Vickerman, 1985).

In contrast, the metacyclic forms of the African (salivarian) trypanosomes are discharged in the saliva of the vector, a species of tsetse fly (*Glossina* spp.). *Trypanosoma brucei* parasitizes ungulates largely, but some strains infect man causing sleeping sickness. The free-swimming trypomastigotes divide in the blood of man or other mammals and are ingested into the tsetse fly midgut (Fig. 2) where they transform into larger "procyclic" trypomastigotes. After multiplication in the midgut, procyclics become nondividing mesocyclic trypomastigotes which migrate via the esophagus and proboscis to the salivary glands; there they become attached to the gland epithelium microvilli and multiply as epimastigote forms. From the attached epimastigotes, free metacyclic trypomastigotes arise and lie free in the gland lumen (reviewed by Vickerman, 1985). In *T. congolense* (the major pathogenic trypanosome of cattle in Africa), migration from the midgut stops in the proboscis, where epimastigotes multiply while attached to the chitinous wall of the labrum. The metacyclic trypanosomes developing from them invade the hypopharynx (common salivary duct) so that they can be injected with the salivary flow into a new mammalian host (Thevenaz and Hecker, 1980). *T. vivax* (another important parasite of cattle) appears to omit the midgut multiplicative phase; after ingestion by the tsetse fly as a bloodstream trypomastigote, it quickly transforms to an attached epimastigote whose development to the metacyclic stage in the fly's proboscis resembles that of *T. congolense* (Vickerman, 1973).

The homoxenous trypanosomatids, species of *Leptomonas*, *Herpetomonas*, *Crithidia*, and *Blastocrithidia*, live as nectomonads and haptomonads in the gut of their invertebrate (commonly insect) host. Infections of terrestrial hosts are transmitted through the feces by small encysted forms which usually arise by a process of unequal binary fission, the amastigote daughter clinging to the flagellum of her still-flagellated sister (Fig. 1). Sometimes several cysts are attached to the same flagellum and are known as "straphangers." Alternatively, in aquatic hosts, nectomonad forms may swim from one host to another without the intervention of an encysted stage (reviewed by Vickerman, 1989).

Broadly speaking, nectomonad stages are easier to obtain in *in vitro* culture (Taylor and Baker, 1987). Thus, the procyclic stage of *Trypanosoma brucei* and the vector midgut promastigote form of *Leishmania* spp. are readily grown in culture; the later stages in the vector developmental cycle, including the attached stages, are more difficult to culture.

2. Flagellar Surfaces and Their Relation to Other Surface Domains

2.1. Flagellar Structure and Function in Kinetoplastids

The flagella of kinetoplastid protozoa have characteristic structural peculiarities as seen by transmission electron microscopy (Figs. 3, 4). The axoneme has the usual 9 doublet plus 2 singlet microtubule organization and its basal body has the conventional 9 triplet-microtubule structure; the two are separated by a transition zone (at the base of the extracellular flagellum and within the flagellar pocket) where 9 doublets lacking dynein arms are linked to the flagellar membrane by radial connectives. The transition zone is separated from the basal body by one incomplete plate and from the axoneme by another—the terminal plate; a fine cylindrical "collarette" (Fig. 5) encases the transitional zone. Beyond the terminal plate a latticelike cord, the paraxial or paraflagellar rod, runs alongside the axoneme terminating before the axonemal doublets toward the flagellar tip. In trypanosomatids the rod is often not present until the flagellum has emerged from the flagellar pocket; it is always subtended by axonemal doublets 4–7 and lies in the plane bisecting the axoneme through its 2 central microtubules. In the bodonids the rod bears a similar relationship to the axoneme in the anterior flagellum, but in the recurrent flagellum is subtended by doublets 2–5 or 3–6; the paraxial rods of the two emergent flagella face one another and an interflagellar membrane connective links their origins at the level of

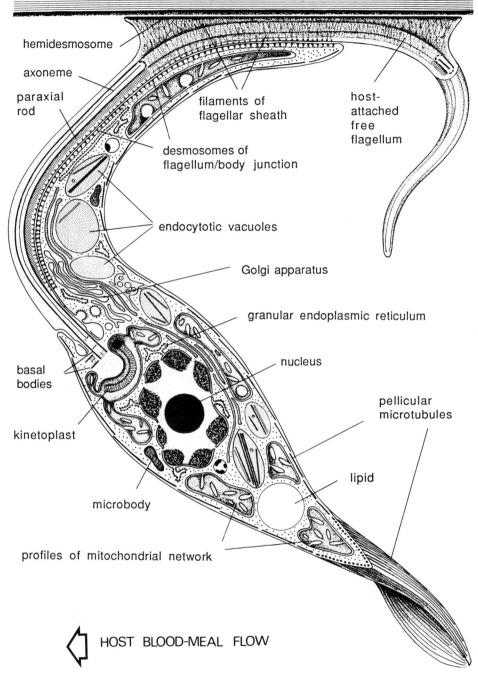

LABRUM

hemidesmosome

axoneme

paraxial
rod

filaments of
flagellar sheath

host-
attached
free
flagellum

desmosomes of
flagellum/body junction

endocytotic vacuoles

Golgi apparatus

granular endoplasmic reticulum

nucleus

basal
bodies

pellicular
microtubules

kinetoplast

lipid

microbody

profiles of mitochondrial network

HOST BLOOD-MEAL FLOW

Figure 3. *Trypanosoma vivax*. Schematic diagram showing fine structure of epimastigote stage attached directly to the wall of the proboscis (labrum) of the tsetse fly. The trypanosome is seen in sagittal section except for the posterior extremity of the body and the anterior part of the free flagellum. Based on Vickerman (1973).

the terminal plates. The recurrent flagellum of bodonids does not beat in synchrony with the anterior flagellum and usually serves as a steering device and skid on which the flagellate skates over the substratum. Mastigonemes (flagellar hairs) are occasionally found on the anterior flagellum and rarely on the proximal part of the posterior flagellum in bodonids. For a more detailed account of the kinetoplastid flagellar apparatus, see Vickerman and Preston (1976).

A remarkable feature of the flagella of trypanosomatids is their ability to propagate waveforms from tip to base of the flagellum as well as from base to tip. The basal body is not necessary for bend initiations; bends may be initiated at any point on the flagellum (Goldstein, 1974). One difference between the classical axoneme plan and that of trypanosomatids is the presence of a septum dividing the B tubule of each doublet. The paraxial rod is not present in all trypanosomatids, being absent for some unknown reason from those that possess bacterial symbionts and from their aposymbiotic derivatives (Freymuller and Camargo, 1981). Although its constituent proteins have now been characterized (Russell et al., 1983), the three-dimensional organization of the filaments that compose the rod is still being elucidated (Farina et al., 1986). Its role in flagellar movement is unclear as locomotion in Crithidia oncopelti, which lacks the rod, appears to be essentially similar to that in C. fasciculata, which has a paraxial rod.

The membrane which envelops the flagellar structure is separated from the axoneme–paraxial rod complex by a cortex (the flagellar sheath) of low electron density but occasionally traversed by loose filamentous material. This material is more organized in flagella which show obvious attachment, either to the body of the same organism, to other flagella or to host surfaces. Such attachments are discussed in greater detail in Sections 2 to 5.

During division of the flagellate, "daughter" flagella are formed alongside "parental" flagella within the flagellar pocket. Membrane for the growing flagellum or cell surface must be added from the flagellar pocket membrane; the pocket is now well characterized as a center of membrane addition and subtraction in the cell (Coppens et al., 1987; Duszenko et al., 1988; Olecnick et al., 1988). In flagellar resorption (such as occurs when nectomonad forms attach to become haptomonads, and promastigotes or trypomastigotes change to amastigote forms) it is possible that phagocytosis ("mastigophagy") of flagellar membrane plays a part (Fig. 5). An alternative way of shedding membrane from both flagellum and body is via the enigmatic streamers variously termed "filopodia" or plasmanemes—threadlike extensions of the surface which are discarded into the ambient by bloodstream trypanosomes when removed from their host (Ellis et al., 1976).

2.2. Functional Aspects of the Parasite Surface: Similarity of Flagellar and Body Membranes

Studies on the surface of trypanosomatids (there have been virtually none on bodonids) have been concerned very much with the host–parasite interface—the surface that the parasite presents to the host's defense systems, or in the case of intracellular forms the surface that the parasite presents to the host cells that it seeks to invade. With respect to these functions, flagellar and body membranes appear to have the same properties. Studies of lectin-binding surface carbohydrates, surface antigens, and membrane-associ-

Figures 4–6. Transmission electron micrographs of sections of attached African (salivarian) trypanosomes.

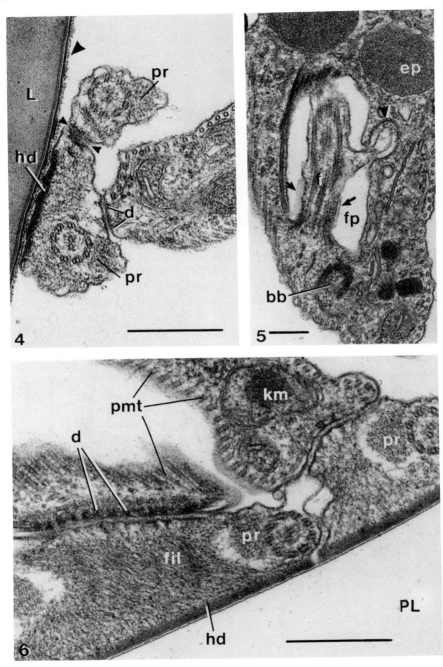

ated enzymes confirm this view (see De Souza, 1989, for review). There have been no attempts to characterize flagellar membranes as distinct from body membranes in terms of their composition, and what evidence we have for different membrane domains in these flagellates comes largely from electron microscopy (especially freeze-fracture studies) and to a lesser extent from cytochemical or antigen studies (see Section 2.3).

Undoubtedly the best studied surface of any parasitic protozoan is that of *Trypanosoma brucei;* the other salivarian trypanosomes are believed to be similar. *T. brucei* mammalian bloodstream forms have a protective surface coat which is composed of a monomolecular layer of glycoprotein enveloping both body and flagellum (Vickerman, 1969; Cross, 1975). The coat serves to protect the parasite against both nonspecific (phagocytosis, complement-mediated lysis) and specific (antibody) defenses of its host (reviewed by Vickerman and Barry, 1982). In the case of the latter, the coat forms a replaceable surface—a glycoprotein of one antigenic specificity being repeatedly replaced by another of differing specificity. The parasite has 100–1000 genes for different variant surface glycoproteins (VSGs) and only one gene is expressed at a time (reviewed by Pays, 1988); it avoids the host's immune response by antigenic variation. In the mammal, the VSG coat forms an impenetrable barrier to antibodies against other membrane components. When ingested by the tsetse fly, the VSG coat is lost and replaced with a major surface glycoprotein, "procyclin," which contains a dipeptide repeat (Richardson *et al.,* 1986; Roditi *et al.,* 1989). When the trypanosome transforms to the mammal-infecting metacyclic stage in the fly's salivary glands, procyclin is replaced with a VSG coat which once more protects the parasite in its encounter with mammalian host defenses (Tetley *et al.,* 1987; Vickerman *et al.,* 1988). Antibodies raised against a specific VSG or against procyclin stain both body and flagellum in immunofluorescence reactions. The oligosaccharide groups of VSG molecules are not exposed for lectin binding, but fluorescein-labeled Con A binds to the flagellar pocket of bloodstream forms and over the entire surface of procyclic forms; at 37°C the bound lectin redistributes to the flagellar pocket and flagellum, suggesting free movement of some glycoproteins over the flagellate's surface (Balber and Frommel, 1988).

In the life cycle of *Trypanosoma cruzi,* each stage would appear to have a variety of

---

Figure 4. *Trypanosoma vivax.* Transverse section of anterior end of epimastigote attached by a flagellar hemidesmosome (hd) to the flocculent deposit (large arrowhead) on the labrum (L) of *Glossina fuscipes.* Filamentous material in the flagellar sheath surrounds the paraxial rod (pr)–axoneme complex and connects the hemidesmosome on the one side and desmosome-like junctional complexes (d) between trypanosome body and flagellum, on the other. The free flagellum of a second trypanosome (above) forms an indistinct macular attachment (between small arrowheads) with the flagellum of the first trypanosome. Bar = 0.5 μm.

Figure 5. *T. vivax.* Vertical section of flagellar pocket (fp) region of epimastigote. A tongue of flagellar membrane is being captured by a coated vesicle (arrowhead) forming from the pocket wall; small arrows indicate collarette around flagellar transition region. bb, barren basal body; ep, endoeytosed protein vacuole; f, flagellum. Bar = 0.25 μm.

Figure 6. *T. congolense.* Parts of sections of two epimastigotes attached to Thermanox plastic coverslips (PL) by their flagella. A dense hemidesmosome (hd) is present wherever the flagellar membrane is in contact with the substratum. Filamentous material (fil) connecting the hemidesmosome and the flagellum-body desmosomes (d) is oriented in isotropic fashion in the flagellum on the right, but in a more anisotropic arrangement in the flagellum on the left. km, kinetoplast-mitochondrion; pmt, pellicular microtubules; pr, paraxial rod. Bar = 0.5 μm.

glycoproteins and glycolipids expressed on its surface. Certain glycoproteins may dominate the surface of a particular stage, however, and these may vary in their structure from stock to stock. Thus, a 90-kDa major glycoprotein of all mammalian stages appears to have an antiphagocytic function and glycosidase activity, the enzyme possibly serving to remove sugar residues on the nonspecific receptors responsible for parasite attachment and engulfment (Nogueira, 1983). In the vector stages (including metacyclics) the 90-kDa glycoprotein is replaced by a dominant 72-kDa glycoprotein which (unlike the 90 kDa) will activate complement and may be the receptor for a vector gut lectin which induces transformation to the metacyclic stage (Snary *et al.*, 1987). Cell-infective *T. cruzi* trypomastigotes develop receptors for the host protein fibronectin which acts as a bridge between parasite and host cell by virtue of the latter's possession of a similar receptor for this symmetrical molecule (Ouaissi, 1986). Again, these surface molecules do not appear to be confined to cell body or flagellum.

Leishmania promastigotes of several species have two major glycoconjugates on their surface, a 63-kDa glycoprotein (GP63) which is a protease and has projecting *N*-linked mannose-containing oligosaccharide groups (Etges *et al.*, 1987), and lipophosphoglycan (LPG) (Turco *et al.*, 1987). These molecules appear to participate in the uptake of the promastigote by a mammalian macrophage, so that the parasite may progress to the intracellular amastigote stage. The GP63 protease activates complement by the alternate pathways producing C3b and iC3b which are ligands for the CR1 and CR3 receptors respectively on the macrophage surface; the mannose-containing oligosaccharides bind to another macrophage receptor, the mannose fucose receptor, causing the parasite to adhere to the host cell (Russell and Wilhelm, 1986). The LPG can mediate attachment to an independent macrophage receptor (Handman and Goding, 1985). The distribution of GP63 and LPG on the flagellar and body membranes of *Leishmania* spp. is relevant to the controversial question of whether promastigote attachment is initiated by the flagellum only or whether attachment of any part of the flagellate's body to the macrophage host cell can induce engulfment. Zenian *et al.* (1979) suggested that promastigote motility is the deciding factor governing parasite orientation during uptake; as promastigotes swimming flagellum foremost exhibit a chemotactic response toward macrophages (Bray and Alexander, 1987), they are most likely to make contact with the flagellum. Quantitative *in vitro* studies of *L. donovani* promastigotes interacting with hamster peritoneal or human blood-derived macrophages, however, have shown that promastigotes become attached and internalized either flagellum or posterior end first, each with the same frequency (Chang, 1979; Pearson *et al.*, 1983). The lack of importance of promastigote orientation is in keeping with the distribution of the relevant receptor binding surface glycoconjugates on body or flagellar membranes.

2.3. Compositional Differences between Flagellar and Body Membranes

Electron microscopy of freeze-fracture replicas provides the best evidence that the flagellum and body membranes represent distinct domains. This evidence comes from counts of intramembranous particles (IMPs) and pits—believed to represent integral membrane proteins—and of lesions induced by the polyene antibiotic filipin—believed to represent a measure of the sterol content of membranes (Table I).

Flagellar membrane fracture faces generally have fewer IMPs than corresponding

Table I. Distribution of Intramembranous Particles and Filipin-Induced Complexes in Numbers per μm^2 of the Body and Flagellar Membranes of Trypanosomatids

Trypanosomatid	Body membrane		Flagellar membrane		Filipin-induced complexes		Reference
	PF	EF	PF	EF	Body	Flagellum	
Leishmania mexicana amazonensis							
Promastigote							
"Infective"[a]	1967	1090	608	224	nd	nd[b]	Pimenta and De Souza
"Noninfective"	1477	605					(1987)
Leishmania mexicana mexicana	3641	896	650	543	146	704	Tetley *et al.* (1986)
Promastigote							
Trypanosoma brucei							
Bloodstream slender	2353	600	nd	nd	5	36	Tetley (1986)
Bloodstream stumpy	2314	536	nd	nd	104	184	
Culture procyclic	3332	241	nd	nd	162	258	
Trypanosoma cruzi							
Trypomastigote	122	126	nd	nd	nd	nd	De Souza *et al.* (1978)
Epimastigote	1830	1450	nd	nd	150	nd[c]	Souto-Padron and De Souza (1983)
Leptomonas collosoma	1355	1476	116	1865	nd	nd	Linder and Staehelin
Promastigote							(1977)
Leptomonas samueli	1630	2350	427	1446	nd	nd	Souto-Padron *et al.*
Promastigote							(1980)

[a]The "infective" and "noninfective" promastigotes were defined as those of the same strain which could or could not, respectively, produce visible lesions when inoculated into the skin of hamsters.
[b]Unusually, the density of filipin-induced complexes was reported as higher in the cell body than in the flagellar membrane although no figures were given.
[c]Density of filipin-induced complexes reported as higher in the flagellar membrane than in the cell body membrane, although no figures were given.

body fracture faces (Figs. 8, 9), suggesting a reduced protein content in the flagellar domain. Exceptions are provided by certain homoxenous trypanosomatids. Thus, in *Leptomonas collosoma* in culture, the flagellar exoplasmic face (EF) has a greater IMP density than the plasma membrane (Linder and Staehelin, 1977), while in *L. samueli* and *Herpetomonas samuelpessoai* IMP densities in these two faces are similar (Souto-Padron *et al.*, 1980; De Souza *et al.*, 1979). It might be expected that some control of lateral mobility of proteins must be exercised to maintain two domains in a continuous membrane system and that the flagellar pocket region might contain a functional barrier. In *Leishmania mexicana amazonensis,* both promastigotes and amastigotes, the pocket membrane, as seen in freeze-fracture replicas, is divided into a high-density basal and a low-density apical region by a line of closely spaced IMPs (Pimenta and De Souza, 1987); in *T. brucei* bloodstream forms an assembly of IMPs has been noted at the neck of the pocket on the pocket membrane EF (L. Tetley, unpublished observation). These putative barriers cannot be observed in sectioned material.

Use of filipin in freeze-fracture cytochemistry to demonstrate the distribution of β-hydroxysterols in membranes has been reviewed by Severs and Robenek (1983). Sterol molecules, complexing with filipin as it diffuses laterally in the membrane, form aggregates which result in doughnut-shaped lesions (protuberances or pits) in freeze-fracture

Figures 7–9. *Leishmania mexicana mexicana.* Freeze-fracture replicas of surface membranes.

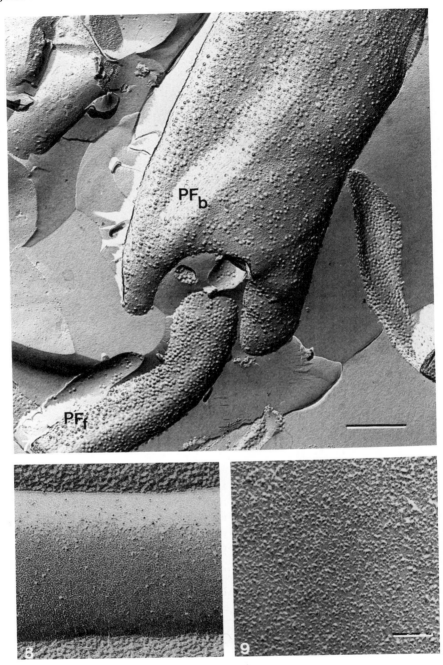

replicas. Studies on *T. brucei* have shown a higher density of filipin-induced lesions in the flagellar membrane than in the plasma membrane, for both bloodstream and procyclic developmental stages (Table I; Tetley, 1986). Similar results have been obtained with *L. mexicana mexicana* promastigotes (Fig. 7; Tetley *et al.*, 1986) and for *T. cruzi* epimastigotes (Souto-Padron and De Souza, 1983). These relative-density figures are the converse of those obtained for IMP distribution and suggest that in the flagellum the organism increases membrane sterol content while reducing integral protein content. The increased sterol content might be expected to make the flagellar membrane more rigid and hence less deformable during movement; the plasma membrane is supported by underlying cytoskeletal microtubules. It is possible that the reduced integral protein content of the flagellar membrane reflects the exclusion of transport sites for molecules not required for flagellar activity but necessary for bodily functions. The polarity of filipin-induced lesions in the flagellar membrane of *T. brucei* changes during transformation from bloodstream to procyclic form; in the former, lesions take the form of protuberances on the EF, in the latter of protuberances on the protoplasmic face (PF), suggesting a change in sterol distribution between the two halves of the membrane (Tetley, 1986). This change could be related to an adaptation in the membrane structure induced by temperature change; bloodstream forms live at 37°C and procyclic forms at around 26°C.

2.4. Flagellar Fractions and Flagellum-Specific Antigens

Surprisingly few attempts have been made to separate the flagellum of trypanosomatids or other kinetoplastids from the cell body as an isolated organelle. Whereas in phytoflagellates such as *Chlamydomonas, Ochromonas,* and *Euglena,* flagella can be isolated with high efficiency by rapidly lowering the pH of the suspension buffer (Witman *et al.,* 1972), by cold shock (Hyams, 1982; Chen and Haines, 1976), or by use of dibucaine (Witman *et al.,* 1978), in trypanosomatids these procedures have proved ineffective, presumably because the association of the flagellum with the cell body is stronger, the flagellar pocket and desmosome-like junctional complexes (see Section 3.1) buttressing this association. Even shearing forces—successful with the zooflagellate *Giardia lamblia* (Clark and Holberton, 1988)—were found by Pereira *et al.* (1977)* to be insufficient to deflagellate in the absence of detergent. Use of Lubrol PX in the presence of 3 mM Mg^{2+} by the same workers proved effective in deflagellating *Crithidia fasciculata, Herpetomonas samuelpessoai,* and *Leishmania tarentolae;* subsequent separation of flagella was achieved on sucrose density gradients. Using this protocol, however, most of the flagella had their membranes removed by the detergent. Pereira *et al.* (1978) used similar methods to obtain flagellar and membrane fractions of *T. cruzi* epimastigote

---

*According to Russell *et al.* (1983), shearing forces alone can detach flagella of *Crithidia fasciculata;* the duration and intensity of the treatment are critical, however.

Figure 7. PF fracture faces of promastigote body (PF$_b$) and flagellum (PF$_f$) membranes, showing relative distribution of lesions induced by filipin and suggesting a greater density of β-hydroxysterol molecules in the flagellar membrane. Bar = 0.5 μm.

Figures 8 and 9. PF fracture faces of flagellum (Fig. 8) and body (Fig. 9) membranes showing greater density of intramembranous particles in the latter. Bar = 0.1 μm.

forms. In the membrane fraction, two types of vesicle were recognized by freeze-fracture electron microscopy; the one with the fewer IMPs was tentatively identified as of flagellar origin. These authors found that the flagellar fraction obtained from the homoxenous trypanosomatids mentioned above conferred some protection on mice challenged with *T. cruzi* in terms of increased survival time.

Segura *et al.* (1976, 1977) obtained a flagellum-rich fraction of *T. cruzi* epimastigotes using a pressure–depressure method to disrupt cells followed by sucrose density fractionation. They found that this preparation conferred better protection of mice against *T. cruzi* than did a membrane fraction. Segura *et al.* (1986) also prepared monoclonal antibodies (MAbs) against the flagellar fraction of epimastigotes; one of these was active in complement-mediated lysis of bloodstream and metacyclic trypomastigotes, one was active only against metacyclics, and the third did not mediate any such activity. The first of these MAbs conferred 80% passive protection on mice against challenge with bloodstream trypomastigotes. There was no evidence, however, that the antigens recognized were strictly located on the flagella, as the MAbs bound to both surface and internal membranes of the trypanosome body in immunofluorescence reactions. The case for flagellar membrane antigens of *T. cruzi* being indispensibly involved in the parasite's entry into host cells is at present unconvincing.

Although MAbs against trypanosomatids have now been extensively developed with a view to distinguishing particular species or zymodemes (parasite populations which can be recognized by their isoenzyme spectrum), and many of them will distinguish one stage in the life cycle from another (Gibson and Miles, 1985), flagellum-specific MAbs have rarely been reported. Flint *et al.* (1984) illustrated the specificity of one such monoclonal which binds to flagellum and flagellar pocket of epimastigote and trypomastigote stages of *T. cruzi* in immunofluorescence, but no ultrastructural localization was attempted. Petry *et al.* (1986) described a MAb raised against the epimastigote stage of *Trypanosoma vespertilionis* (*T. cruzi*-like trypanosome of bats). Using indirect immunofluorescence they showed that this antibody reacted with the flagella of epimastigote and metacyclic trypomastigote stages of *T. cruzi*, *T. dionisii* (another bat parasite), and *T. vespertilionis* (all belonging to the subgenus *Schizotrypanum* Hoare) but not with *T. brucei*, *Leishmania* spp., or homoxenous trypanosomatid flagella (Petry *et al.*, 1987); the 19-kDa antigen was not shown to be located on the flagellar membrane, however, and it could be an internal component of the flagellum. The paraxial rod proteins of *T. brucei* were labeled by a MAb produced by Gallo and Schrevel (1985); this antibody also labeled rod proteins of *Euglena gracilis*. A monospecific polyclonal rabbit antibody to human erythrocyte spectrin also reacted with paraxial rod proteins of *T. brucei* (Schneider *et al.*, 1988), suggesting that a protein antigenically related to the major component of the erythrocyte membrane skeleton occupies an internal position in trypanosomes.

In *Leishmania major*, however, a 13.2-kDa protein recognized by three monoclonals has been localized in the flagellar membrane using MAbs and immunocytochemical techniques (Ismach *et al.*, 1989). This flagellum-specific protein is shared with *L. brasiliensis*, *L. mexicana amazonensis*, *T. cruzi*, and *Endotrypanum schaudinni*, and occurs in the flagellar pocket as well as along the flagellum shaft, suggesting that the pocket may store surface membrane for use in rapid flagellar growth and thus have a role in flagellum biogenesis; such storage has been shown unequivocally in *Euglena* (Rogalski and Bouck, 1982).

It is to be expected that further flagellum-specific macromolecules will be discovered in the near future. One might anticipate that such molecules would participate in (1) attachment of the flagellum to the body or to host surfaces (see Sections 3, 4, and 5), though few attempts have been made to identify specific molecules in attached trypanosomatids; (2) chemotaxis of motile flagellates, revised orientation of flagellar movement in response to flagellate encounters with chemotactic substances (Bray and Alexander, 1987); such responses might account for migratory movements within the vector or other host; (3) mating reactions; evidence for a sexual process in vector-transmitted trypanosomatids has only recently come to light (Jenni *et al.*, 1986); nothing is known of gamete stage interaction, but by analogy with mating in other flagellates (see van den Ende *et al.*, this volume) the intervention of flagellar surfaces seems probable, especially as mating will occur only between complementary stocks.

3. Flagellar Attachment to the Body or to Other Flagellates

3.1. Attachment to the Body

Flagellar attachment in kinetoplastids may or may not involve distinguishable junctional complexes between body and flagellum. Attachment (Fig. 10) may occur solely at the neck of the flagellar pocket (as in *Leptomonas, Crithidia, Herpetomonas,* or *Leishmania* spp.), or along the length of the flagellum (as in *Blastocrithidia, Endotrypanum,* or *Trypanosoma*). In the latter case, the beating flagellum may draw up the body surface into an "undulating membrane"; marked folds or ridges on the pellicle for flagellar attachment, such as are prominent in the undulating membranes of some trichomonad flagellates (Brugerolle, 1976), are not found in kinetoplastids. Attachment is usually related to cortical differentiation, however, in the form of a gap in the microtubule corset of the body and often to junctional complexes within the gap (Vickerman, 1969). These complexes are morphologically comparable with the desmosomes formed by epithelial cells of higher animals.

In the stercorarian *Trypanosoma lewisi* of rats, no elaboration of the contact zone is visible in transmission electron micrographs of sections, other than the gap, and dense material between the membranes of apposed body and flagellum (Brooks, 1978). In trypomastigotes of *T. brucei* and other salivarian trypanosomes, a series of macular desmosome-like junctions, each plaque 25 nm across and spaced at 95 nm, fills the gap. Filaments converge on the focal plaques on both cytoplasmic and flagellar sides. Those on the cytoplasmic side penetrate a cortical bed of fine filaments paralleling the surface of the trypanosome. Those on the flagellar side often show connections to the paraxial rod (Vickerman, 1969). In freeze-fracture images, clusters of 8-nm particles are seen at the sites of these junctional complexes only on the PF of the flagellum (Smith *et al.*, 1974; Hogan and Patton, 1976; Vickerman and Tetley, 1979). In the epimastigote stage of *T. vivax* and *T. congolense*, several parallel rows of these desmosome-like complexes may be present, the number of rows decreasing toward the tip of the flagellum (Figs. 4, 6; Vickerman, 1973). Junctional complexes in the attached *T. cruzi* flagellum are less clear. Macular densities are present but not prominent along the line of attachment; they lack the associated filaments and no dense material is present between apposed membranes. In

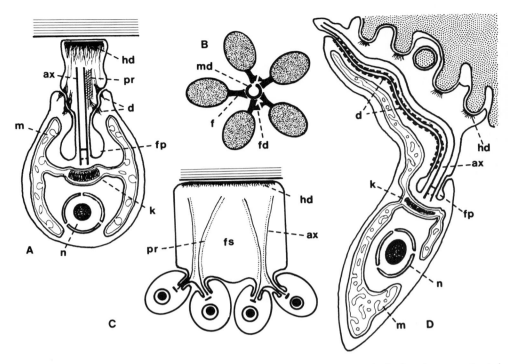

Figure 10. (A) *Crithidia fasciculata*. Schematic diagram of haptomonad in longitudinal section; reconstructed from transmission electron micrographs, showing hemidesmosome (hd) junction of flagellum with chitinous wall of mosquito hindgut, and desmosome (d) junctions at neck of flagellar pocket (fp).

(B) *C. fasciculata*. Diagram of rosette of flagellates from culture. Flagella (f) form junctional complexes with membranous debris (md) or, via flagellipodia (fd), with one another.

(C) *Trypanosoma lewisi*. Cluster of four epimastigotes with shared flagellar sheath (fs) which forms a hemidesmosome (hd) with the chitinous wall of the flea hindgut; the paraxial rods face the center of the flagellum. Based on Molyneux (1969a).

(D) *Trypanosoma brucei*. Schematic diagram of epimastigote attached to microvillar surface of tsetse fly salivary gland epithelium. Hemidesmosome complexes (hd) are formed where microvilli impinge on the branched flagellar outgrowths. Desmosomes (d) link flagellum and body. ax, axoneme; k, kinetoplast; m, mitochondrion; n, nucleus; pr, paraxial rods.

freeze-fracture replicas of epimastigote and trypomastigote stages, however, regularly spaced IMP clusters are present in both PF and EF of the flagellar membrane, and also in corresponding body membrane fracture faces in the trypomastigote (De Souza *et al.*, 1978).* De Souza (1984) has argued that these *T. cruzi* junctions are not comparable to desmosomes in higher animals, though in their freeze-fracture images they are closer than those of *T. brucei*.

Desmosome-like complexes are striking, however, at the neck of the flagellar pocket in *Crithidia* spp. (Fig. 10) and promastigote stages of investigated trypanosomatids. In *C.*

---

*Fixation has been shown to alter particle partitioning in some freeze-fracture preparations; it is possible that the extraordinary differences reported between *T. cruzi* and other trypanosomatids may be attributed to this.

fasciculata these complexes form a series of spiral linear arrays under the pocket membrane unsupported by a microtubule corset (Brooker, 1970). Freeze fracture of the related *C. guilhermei* (Soares *et al.*, 1986) reveals corresponding particle clusters on the flagellar PF only. Similar observations have been made for *Leptomonas collosoma* (Linder and Staehelin, 1977), *Herpetomonas samuelpessoai* (De Souza *et al.*, 1979), *Leptomonas samueli* (Souto-Padron *et al.*, 1980), and *Leishmania mexicana amazonensis* (Benchimol and De Souza, 1980). An unusual adhesion zone between flagellum and body has been described for the epimastigote of *Blastocrithidia culicis* (Soares and De Souza, 1987); this species lacks a flagellar paraxial rod. Although in transmission electron micrographs of sections there is little evidence of differentiated junctional complexes filling the flagellum-associated microtubule corset gap, in freeze-fracture replicas of the region five or six strands of IMPs are discernible in EF and PF of both body and flagellum; each strand consists of a double row of IMPs. It is possible that other body-attached flagella which show no junctional complexes in sections (e.g., those of the cryptobias; Brugerolle *et al.*, 1979) will reveal localized flagellum body specializations in freeze-fracture images.

How far the desmosome-like structures of the flagellum–body junction resemble the true desmosomes of vertebrate epithelia is uncertain. Desmosomes (*maculae adhaerentes*) are considered to play an important part in securing cell adhesion in the face of mechanical stress. As seen in transmission electron micrographs, they consist of electron-dense plaques, 10–25 nm thick, on either side of two apposed membranes. Tonofilaments (10-nm diameter) extend into the cytoplasm from each plaque, and amorphous material may be present in the intermembrane space as an electron-dense line. Characteristic proteins (desmoplakins, desmogleins, and desmocollins) have been found to be associated with desmosomes (reviewed by Skerrow, 1986). Within vertebrates, desmosomal proteins appear to be highly conserved (Cowin and Garrod, 1983). Desmosomes will form between embryonic mouse and chick cells (Overton, 1975) but there is as yet no evidence that they will form between kinetoplastid parasites and their host cells. The nature of the proteins forming trypanosomatid desmosomes is unknown.

Raised Ca^{2+} levels appear to promote the formation of desmosomes (Jones and Goldman, 1985). Trypanosomes dividing in citrated blood produce a completely free "daughter" flagellum while the "parent" flagellum remains attached, suggesting that Ca^{2+} (chelated by citrate) is necessary for initial adhesion of the developing flagellum, but not for the maintenance of attachment (Vickerman, 1969). It is at present impossible to draw further comparison between kinetoplastid and vertebrate desmosomes.

3.2. Attachment to Other Flagella

Where neighboring haptomonad flagellates are crowded together, striplike junctional complexes may be formed between flagella of different individuals. Brooker (1970) described such zonular complexes as type B desmosomes to distinguish them from the macular type A desmosomes formed between body and flagellum. He found them in *Crithidia fasciculata* (Fig. 10), situated on pedunculated processes arising from the main shaft of the flagellum; such junctions appear to maintain the integrity of rosettes of flagellates in the gut of the mosquito or in culture. Similar demosomes have been reported for *Herpetomonas mirabilis* (Brun, 1974), *Trypanosoma melophagium* (Molyneux, 1975), unidentified trypanosomatids in the hindgut of the flea *Peromyscopsylla* (Molyneux and Ashford, 1975), *Leptomonas* sp. in a similar location (Molyneux *et al.*, 1981), *Leptomonas lygaei* (Tieszen *et al.*, 1989), and *Crithidia flexonema* (Tieszen and Molyneux, 1989). Macular desmosomes are formed between flagella of *T. vivax* (Fig. 4; Vickerman, 1973). According to Brooker (1970), interflagellar desmosomes may be formed between *C. fasciculata* and *Herpetomonas muscarum* in culture, suggesting a relative lack of specificity of adhesion between trypanosomatid flagella. Brooker (1970) regarded type B desmosomes as temporary structures, essentially comparable to the hemidesmosomes formed with host surfaces (see Section 4.1) but in this case formed mutually between adjacent flagella.

3.3. Attachment to Cysts

Little can be said of the extraordinary unequal division of certain homoxenous flagellates that results in the production of straphanger cysts. Cysts are attached to lateral peduncular processes of the flagellum shaft in *Blastocrithidia triatomae* as seen by scanning electron microscopy (Schaub and Boker, 1986). Transmission electron micrographs of encysting flagellates show that interflagellar desmosomes are present initially at such attachments in *Leptomonas lygaei* (Tieszen *et al.*, 1989). Micrographs of sections of the cysts or of freeze-fracture replicas suggest a drastic reorganization of the cytoplasm associated with development of resistance to desiccation (Peng and Wallace, 1982; Reduth and Schaub, 1988). There is no true cyst wall; microtubules, flagella, and basal bodies disappear; and the surface membrane becomes thickened on its inner leaflet. In straphanger cysts of *B. familiaris* a surface coat is formed at encystation (Tieszen *et al.*, 1986). With such striking changes in the surface of the encysting organism, it would be interesting to know if the mature cyst is held in place with a conventional flagellar junctional complex.

4. Flagellar Attachment to Host Surfaces

4.1. Attachment to Chitin and Other Nonliving Surfaces

4.1.1. *In Vivo* Observations. The flagellum is most commonly developed as an attachment organelle in trypanosomatids which multiply while attached to cuticle-lined derivatives of the foregut or hindgut of insects. The flagellar sheath is expanded where its membrane contacts the substratum and special junctional complexes are characteristic of such cuticular attachments. These complexes are usually referred to as hemidesmosomes (Brooker, 1970) or hemidesmosome-like plaques (Molyneux *et al.*, 1987); as viewed in sections by transmission electron microscopy they resemble the junctional complexes formed between vertebrate epithelial cells and their supporting basal lamina (basement membrane).

Thus, in established infections with the homoxenous flagellate *Crithidia fasciculata* in the hindgut of the mosquito *Anopheles gambiae,* the flagellate occurs as haptomonad forms attached by their truncated flagella to the cuticular lining (Fig. 10), often in clumps arising from repeated binary fission. The tip of the flagellum is expanded up to six times its usual diameter and the axoneme and paraflagellar rod are replaced in the expanded portion by the junctional complex—a mass of 3-nm filaments converging on an electron-dense plate applied to the thickened inner leaflet of the flagellar membrane over the region of contact between flagellum and cuticle. The gap between flagellar membrane and cuticle is about 18–20 nm and occupied by a thin layer of fibrous material which lines the cuticle of the entire hindgut. In heavy infections, layer upon layer of closely packed flagellates may be attached to the gut, the length of the flagellum varying with the distance of the flagellate from the gut wall; longer flagella pursue a tortuous path between the bodies of other flagellates to achieve anchorage (Brooker, 1971b). Similar junctional complexes are found where flagella make contact with membranous debris in culture (Brooker, 1970) or in the gut lumen (Brooker, 1971b). Rosettes of flagellates in the gut lumen may arise centered around such supports.

Crithidia fasciculata does not form cysts and would appear to be transmitted by the swimming nectomonad form which arises from the above haptomonads. Experimentally, this transformation may be induced by the addition of distilled water to infected insect guts and is accompanied by flagellar detachment. After 5 min, reduction in the number of filaments converging on the attachment plaque precedes the formation of several deep invaginations of the apposed flagellar membrane and loss of the underlying dense band. The onset of flagellar motility causes the final release of the swimming nectomonad (Brooker, 1971b).

Similar attachments in the hindgut have been noted for other trypanosomatids (see Molyneux, 1983, for earlier observations). In Diptera (Fig. 2) a cuticularized ileum connects the pylorus (where the Malpighian tubules enter the gut) with the expanded rectum in which the rectal glands (rectal pads, rectal ampullae) form a favored site for trypanosomatid attachment. A cuticularized ileum is absent from Hemiptera, the Malpighian tubules discharging at the entrance to the rectum (Fig. 2). The glands contain cells specialized for water absorption (with extensively folded plasma membranes in close association with mitochondria) but whether it is this function, or the fact that less expansion and contraction occurs in the cuticle over the glands than elsewhere in the hindgut, that attracts the flagellates is unknown.

Preferential attachment to the rectal glands has been noted for several homoxenous trypanosomatids: *Crithidia fasciculata* in *Anopheles* (Brooker, 1971b); *C. flexonema* in *Gerris odontogaster* (Hemiptera) (Tieszen and Molyneux, 1989); *Leptomonas oncopelti* in *Oncopeltus fasciatus* (Hemiptera) (Lauge and Nishioka, 1977); *Leptomonas* sp. in Siphonaptera (Molyneux *et al.*, 1981); *Blastocrithidia familiaris* in *Lygaeus pandurus* (Hemiptera) (Tieszen *et al.*, 1986); *B. triatomae* in *Triatoma infestans* (Hemiptera) (Schaub and Böker, 1986). In addition, heteroxenous trypanosomatids which are transmitted via the posterior station appear to colonize the rectal glands first: *Trypanosoma cruzi* in its hemipteran vectors (*Dipetalogaster maximus, Triatoma dimidiata*, Zeledon *et al.*, 1977, 1984; *Triatoma infestans*, Böker and Schaub, 1984); *Trypanosoma leonidasdeanei* of bats in the sandfly (Diptera) *Lutzomyia beltrani* (Williams, 1976); and *Trypanosoma corvi* of rooks in the hippoboscid fly *Ornithomyia avicularia* (Mungomba *et al.*, 1989; Fig. 11). In all cases the flagellum is expanded in the region of attachment and a zonular hemidesmosome-like attachment plaque is present. In flagellates which have a body-attached flagellum (i.e, epimastigote or trypomastigote stages) the hemidesmosome-like plaque may be formed at the tip of the free flagellum, but is more usually lateral in the zone of flagellum body attachment, alongside the paraxial rod. In *T. lewisi* in the flea hindgut, the expanded flagellar membrane and sheath enclose the axoneme–paraxial rod complexes of several flagellates as a result of bizarre incomplete division (Fig. 10;

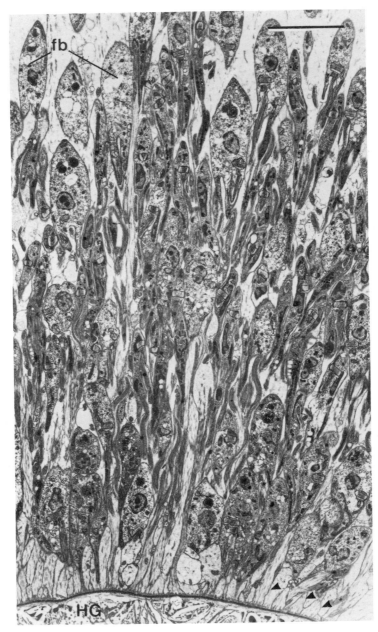

Figure 11. *Trypanosoma corvi.* Transmission electron micrograph of section of hindgut (ileum; HG) of the fly *Ornithomyia avicularia,* showing trypanosomes (mainly epimastigotes) attached to the cuticularized lining. The flagellate bodies (fb) lie in several layers; those farthest away from the gut wall have the longest flagella; the swollen flagellar tips (arrowheads) occupy all available attachment space. Micrograph courtesy of Dr. K. Wallbanks. Bar = 5.0 μm.

Molyneux, 1969a). More information on lateral flagellar attachments is available from trypanosomes which undergo metacyclogenesis in the anterior station, specifically the salivarian trypanosomes *Trypanosoma congolense* and *T. vivax* (Vickerman, 1973). These multiply as epimastigotes attached to the chitinous wall of the proboscis (labrum) of the tsetse fly (Figs. 2–4, 6) and migrate into the hypopharynx for final maturation to the metacyclic stage. The shearing force of the passing bloodmeal represents a threat to the dividing epimastigotes and in many of these the flagella appear to be aligned along the axis of the proboscis, their tips pointing toward its tip in order to minimize resistance to bloodmeal flow, though the large clumps which result from trypanosome multiplication in some cases almost block the proboscis lumen.

The zonular hemidesmosome of *T. vivax* consists of a 35-nm-thick dense plaque running beside axonemal doublets 6–9 and 1 along the shaft of the flagellum where the expanded flagellar membrane contacts the labrum wall (Figs. 3, 4). A 12-nm gap separates the flagellar membrane from a 15-nm-thick granular encrusting layer which lines the labrum. The plaque-associated filaments penetrate the mass of apparently isotropic filaments of the flagellar sheath and occasionally appear continuous with the filaments emerging from the macular desmosomes of the flagellum–body junction. These filaments are not well preserved in material processed for conventional transmission electron microscopy of sections. Although both Brooker (1971b) and Vickerman (1973) estimated the diameter of the finest filaments as 3 nm, most observers have refrained from commenting on their dimensions and their published micrographs are of insufficient resolution to permit measurements to be made. The junctional attachment of *T. congolense* to the tsetse proboscis (Fig. 6) appears to be essentially similar (Evans *et al.*, 1979; Thevenaz and Hecker, 1980). Transformation of attached epimastigotes to uncoated dividing premetacyclic trypomastigotes in the salivarian trypanosomes involves the acquisition of the variable antigen (VSG) coat (Vickerman, 1974) and proceeds via transformation of dividing uncoated epimastigotes to dividing uncoated premetacyclic trypomastigotes. When these cease to divide, they acquire the coat while still attached and become nascent metacyclics. Detachment of the fully coated trypanosome marks its acquisition of metacyclic status. This sequence of events, first described for *T. brucei* (Tetley and Vickerman, 1985), would appear to be similar in *T. vivax* and *T. congolense,* except that detachment of the epimastigote or premetacyclic must occur to allow migration to the hypopharynx where reattachment occurs and metacyclogenesis is completed. It seems reasonable to suppose that the attachment mechanism is weakened by the interpolation of the 12- to 15-nm-thick surface coat between flagellum and substratum so that the metacyclic trypanosome is easily released from its moorings by the host's salivary flow, but the actual stimulus for release is unknown.

The sequence of development of the leishmanias in the anterior station of their sandfly vectors is less certain. Infective metacyclic promastigotes with long flagella are found free-swimming, not attached, in the proboscis of the sandfly. These are supposedly derived from the heavy populations of promastigotes or paramastigotes* attached by

---

*The significance of the paramastigote morphology in the attached phase has been questioned by Lawyer *et al.* (1987), who found only unattached paramastigotes during the development of *L. mexicana mexicana* in *Lutzomyia diabolica* and *Lu. shannoni,* and these only in small numbers, the attached populations being promastigotes. Killick-Kendrick *et al.* (1988) have demonstrated the presence of abundant attached paramastigotes in *L. major* infections of a natural vector, *Phlebotomus papatasi.*

flagellar tip hemidesmosomes to the cuticularized stomodaeal valve, pharynx, and cibarium (Fig. 2; Killick-Kendrick *et al.*, 1974; Killick-Kendrick, 1979).

4.1.2. *In Vitro* Observations: The Mechanism of Attachment. Attachment of trypanosomatids to chitinous surfaces by hemidesmosome-like plaques can be mimicked *in vitro* by cultivating flagellates over suitable substrates. Thus, Brooker (1971a) found that Millipore filters formed such a substratum for *Crithidia fasciculata* (replacing the mosquito hindgut) and Hommel and Robertson (1976) utilized the polystyrene walls of plastic Falcon flasks for *Trypanosoma blanchardi* and certain other rodent trypanosomes. *T. congolense* vector proboscis stages (Figs. 6, 12) will also grow on a plastic substratum (Gray *et al.*, 1981). *T. vivax* proboscis forms will develop on a substratum of Green A dye immobilized on agarose beads (see Section 5.1.1).

Scratching the wall surface of plastic flasks encourages attachment of many species which are otherwise reluctant to adhere (see Section 5.1), especially *Leishmania* spp. (Maraghi *et al.*, 1987). The increased negative charge on the substratum induced by scratching (Molyneux *et al.*, 1987) has been taken to indicate the role of surface charge in attachment.

The properties of the attachment between trypanosomatid flagella and an artificial substratum, such as a glass coverslip, are amenable to study using the technique of interference reflection microscopy (IRM; Verschueren, 1985). The intensity of monochromatic light or the color of white light interference pattern reflected at the cell's underface gives information about the closeness of contact between cell (in this case attached flagellum) and the substratum. For incident white light, interference colors occur in strict sequence directly related to the optical path difference (Δ) between reflections from the glass–medium and medium–cell interfaces. At a value of 20 nm or less, a black image is obtained; as Δ increases, the image changes through gray to white; for values of Δ greater than 300 nm, chromatic images of the standard interference color series are encountered. Using short-duration (1–4 day) cultures of *T. congolense* vector proboscis forms at low inoculum density, Hendry (1987) was able to test the effect of various reagents on flagellar attachment to glass coverslips in this species. IRM images of attached flagella showed closest contact (gap of 20 nm or less) along the length of the body-attached flagellum, but excluding its tip. The area of contact was wider than the flagellum shaft, presumably due to expansion of the flagellar membrane, and extended into lateral branches (the "flagellipodia"). In Minimum Essential Medium without fetal calf serum, close attachments were still formed but their area of contact was reduced compared with controls, so serum components are not necessary for attachment. Trypsin treatment (with or without serum) widened the gap (as it does with attached tissue culture cells), suggesting a role for protein components in attachment. Varying K^+ concentration (10 nM to 100 nM) had no effect on attachment, suggesting that ionic charge does not play any part in the attachment process. Growth in Ca^{2+}/Mg^{2+}-free medium, or removal of divalent cations with 1–10 mM EDTA, caused a darkening of the IRM image (over the pH range 6–9). These findings contrast sharply with those for tissue culture cell attachment, where focal adhesions and desmosomes are disrupted by Ca^{2+} chelators (Borysenko and Revel, 1973). Hemidesmosome formation is dependent upon the presence of Ca^{2+} (Trinkhaus-Randall and Gipson, 1984). In the presence of chelating anticoagulants of tsetse saliva, Ca^{2+} deprivation might be expected to be a frequent condition of attached *T. congolense* [though this would create problems for the dividing trypanosome in forming flagellum–

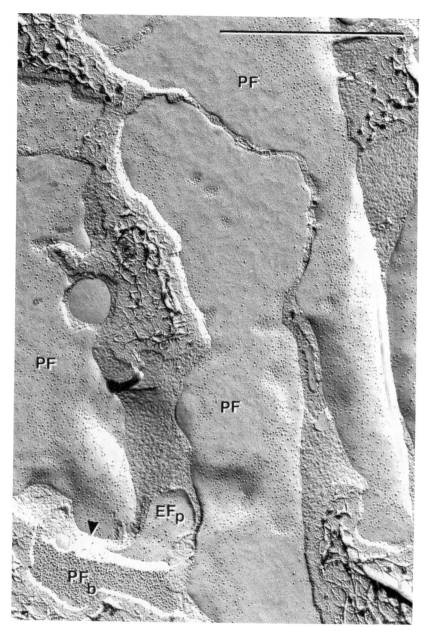

Figure 12. *Trypanosoma congolense.* Freeze-fracture replica of flagellar membranes of epimastigotes attached to a Thermanox coverslip. The expanded flagellar contacts with the substratum have been cleaved to expose the PFs of the flagellar membranes. Note the uneven distribution of intramembranous particles (IMPs) in the PFs. Substantial areas are free from particles (or pits) while in other areas, particles are clustered. This distribution suggests that the IMPs do not play an essential part in the hemidesmosome-like attachment (see Fig. 6). A flagellum cleaved at the point where it emerges from the flagellar pocket (arrowhead), on the left, allows comparison of IMP density in the flagellum PF and body PF (PF$_b$). EF$_p$, EF of pocket membrane. Bar = 1.00 μm.

body junctions (see Section 3.1)]. Molyneux *et al.* (1987), however, also found that chelating agents could not dislodge attached *T. grosi* or *Leishmania major* from plastic surfaces and so resistance to divalent cation deprivation may be a widespread feature of trypanosomatid attachments.

The possible role of lectins and their oligosaccharide ligands in specificity of attachment to the vector and in parasite morphogenesis has been extensively discussed by Molyneux and Killick-Kendrick (1987). The evidence and arguments supporting this hypothesis are (1) that lectins interacting with saccharides of the intestinal villus membrane form the basis of many bacterial attachments in the mammalian gut (Farthing, 1985); (2) the report by Pereira *et al.* (1981) that lectins in the gut of reduviid bugs trigger the transformation of epimastigotes to metacyclic trypomastigotes; differentiation of the metacyclic can be blocked by a MAb to a 72-kDa surface glycoprotein (Sher and Snary, 1982) and this glycoprotein may be a receptor controlling transformation; and (3) the distinct ability of different stages in the life cycle of *Leishmania* spp. (Sacks *et al.*, 1985) and *T. cruzi* (Pereira *et al.*, 1981) to be agglutinated by specific lectins. A weakness of this hypothesis is that, despite the voluminous literature on the binding of lectins to trypanosomatid surfaces [e.g., see Dwyer (1977) on *Leishmania*, Jackson *et al.* (1978) on *T. congolense*, De Souza (1984) on *T. cruzi*, and the review by De Souza (1989)], no specific localization of lectin binding sites to the flagellum has been noted; binding is over the entire surface of body and flagellum, whereas attachment to vector surfaces is via the flagellum only. The lectins could be on the flagellar membrane, however, in which case one might expect common oligosaccharide groups on the substrata to which the flagella attach.

Overton (1982) demonstrated that desmosome formation could be inhibited by lectin binding to surface glycoproteins, suggesting both a recogniton and an adhesive function for cell-surface carbohydrates in attachment. Hendry (1987) found that Con A, lentil lectin, and wheat germ agglutinin, which bind over the entire surface of *T. congolense* epimastigote forms, will weaken flagellar attachment as assessed by IRM. She also observed that tunicamycin (at a concentration 500 times greater than that affecting mammalian cell attachment) does not affect *T. congolense* flagellar attachment, suggesting that no turnover of *N*-linked oligosaccharide-containing glycoproteins occurs at the site.

The nature of the cytoskeletal filaments associated with the hemidesmosome-like flagellar attachment plaques is unknown. Brooker (1971b) and Vickerman (1973) found the finest filaments to be about 3 nm in diameter in transmission electron micrographs of sections of *Crithidia fasciculata* or *T. vivax*, while Hendry (1987) estimated the flagellar filaments of *T. congolense* (Fig. 6) epimastigotes to be 5 nm in diameter and so fall within the actin range. Although actin genes are present in trypanosomes (Ben Amar *et al.*, 1988), indirect immunofluorescence studies of attached *T. congolense* epimastigotes using actin-specific MAbs and NBD-phallacidin show only a diffuse staining of the trypanosome body and flagellum with no increase in brightness at flagellar attachment sites. Analogy of the stable flagellar junctional complexes of trypanosomatids with the more fleeting actin-based adhesion plaques of locomoting fibroblasts (Heath and Dunn, 1978) can therefore be discounted. Analogy with the hemidesmosomes of vertebrate cells, however, would lead us to expect filaments in the intermediate range (about 10 nm). Yet MAb which binds to intermediate filaments of all classes (Pruss *et al.*, 1981) and MAb against desmoplakins I and II (components of mammalian desmosomes) do not stain any part of the attached trypanosome (Hendry, 1987).

Silver-stained two-dimensional gels of *T. congolense* epimastigotes and corresponding extracted cytoskeletal preparations show a similar pattern of protein distribution for attached and unattached trypanosomes except in the vicinity of the paraflagellar rod proteins. The paraflagellar rod group of proteins (M_r 70,000) shows as a large spot of protein situated alongside at a more acidic pI (Hendry, 1987; Vickerman *et al.*, 1988) and this probably corresponds to the additional cytoskeletal proteins present in attached trypanosomes. Anti-intermediate filament MAb exhibits no binding to nitrocellulose blots of two-dimensional gels of attached cytoskeletons, however (Hendry, 1987).

Despite the preliminary nature of available evidence, it does appear that the attachments of trypanosomes to inert substrates and the junctional complexes associated with them are substantially different from the morphologically similar attachments of tissue culture cells to their substratum.

4.2. Attachment to Living Host Cells

4.2.1. *In Vivo* Observations: Invertebrate Microvillar Surfaces.
The insect midgut is lined with columnar cells with a brush border or microvillar surface to which many trypanosomatids attach, interdigitating their flagella between the microvilli. In certain species the developmental cycle leads to invasion of the Malpighian tubules or salivary glands where similar attachment occurs. Unlike flagellates attaching to cuticular surfaces, those attaching to microvillar surfaces may inflict overt damage on the host cell and attachment may be a prelude to cell invasion.

The simplest association of trypanosomatids with host microvilli involves interweaving of the unmodified flagellum with these structures in the absence of junctional complexes or even of close membrane association; indeed it is doubtful whether true attachment occurs in many such situations. Examples are provided by *Leishmania* promastigotes in the sandfly midgut after dissolution of the host's PTM (Killick-Kendrick, 1979), and by the sheep trypanosome, *T. melophagium,* in the midgut of *Melophagus ovinus* (Molyneux, 1975). A more definite attachment can be seen where the flagellar sheath is expanded to engulf a microvillus as in the homoxenous *Herpetomonas ampelophilae* in *Drosophila melanogaster* midgut (Rowton *et al.*, 1981), or the epimastigote stages of *Trypanosoma cobitis* in the crop of its leech vector, *Hemiclepsis marginata* (Lewis and Ball, 1979); no definite junctional complexes are present. An advance on this condition is shown by the expanded flagellar sheath flowing around microvilli to produce arborescent outgrowths allowing extensive interdigitation and close application of the flagellar membrane to that of the microvilli. This is the situation found in *Blastocrithidia familiaris* epimastigotes attaching to midgut microvilli of *Lygaeus pandurus* (Tieszen *et al.*, 1986); the entire luminal surface of the hemipteran midgut and ileum may be covered with attached flagellates. Attachment does not occur to microvilli which possess the so-called "double plasma membrane" (the outer membrane being a stage in the formation of the PTM), only directly to microvilli lacking this structure. Arborescent outgrowths or "flagellipodia" are a striking feature of the anterior flagellum of the bodonid flagellate *Cryptobia* sp. found in the aquatic snail *Triadopsis multilineata* (Current, 1980), so they are not found only in trypanosomatids attached to insect microvillar surfaces.

Hemidesmosome-like junctional complexes have been described, however, between flagellar and microvillar surfaces in the salivary gland stages of the life cycle of *Trypanosoma brucei* (Steiger, 1973; Tetley and Vickerman, 1985). Here the multiplicative epimastigotes develop ramifying flagellar outgrowths which interdigitate with microvilli of the host epithelium and where the tips of the microvilli impinge on the flagellar membrane, punctate hemidesmosome-like plaques 100–130 nm in diameter are visible in transmission electron micrographs of sections (Fig. 13). A 20-nm gap separates the confronting host and parasite membranes in the plaque region; between plaques the gap may be wider. The flagellar outgrowths become reduced, but the plaques persist, as the uncoated epimastigotes transform first to uncoated premetacyclic trypomastigotes and then to coated nascent metacyclics, still anchored to the salivary epithelium. The gland epithelial cells appear to become locally bereft of microvilli in regions of heavy trypanosome infestation, and flagella may then attach to the flattened cell membrane or become inserted into the cortex of the host cell. Such insertions of the flagellar tip are particularly common in nascent metacyclics, and punctate hemidesmosomes are still recognizable along the apposed membranes (Fig. 14). Freeze-fracture studies reveal no obvious modification of the flagellar membrane at junctional complex sites but the host membrane is strikingly altered in that its IMPs are clustered in tightly regimented arrays (Tetley and Vickerman, 1985).

A suspicion that flagellar attachment may lead to erosion of host microvilli also arises from studies on epimastigotes of the frog trypanosome *T. rotatorium* attached to the gastric epithelium microvilli of the leech *Batracobdella picta* by hemidesmosomal junctions (Desser, 1976), and on *Blastocrithidia gerridis* in the midgut of the water strider *Gerris odontogaster* (Tieszen *et al.*, 1983). In the latter association, junctional complexes are evident only where microvillar erosion has taken place.

Occasional flagellar insertion into the insect host midgut cells with the formation of junctional complexes occurs also in *Trypanosoma melophagium* in *Melophagus ovinus* (Molyneux, 1983), *Herpetomonas muscarum* in the fly *Hippelates pusio* (Bailey and Brooks, 1972), *Leishmania mexicana amazonensis* in *Lutzomyia longipalpis* (Molyneux *et al.*, 1975) and *Crithidia flexonema* in *Gerris odontogaster* (Tieszen and Molyneux, 1989); there is no evidence that such insertion causes overt damage to the host. When *Blastocrithidia triatomae* invades the Malpighian tubules of the Chagas' disease vectors *Triatoma infestans* or *Rhodnius prolixus,* however, similar flagellar insertions with focal plaque formation may be correlated with disturbed fluid secretion and larval mortality (Schaub and Schnitker, 1989); this flagellate which is transmitted by cysts in the bug's feces is therefore under study as a possible biological control agent for bloodsucking Hemiptera.

4.2.2. *In Vitro* Observations. An *in vitro* culture system with insect cells for the production of metacyclic *T. brucei* from procyclic (tsetse midgut) forms has recently been described by Kaminsky *et al.* (1987) but details of attached stages are not given. Similarly, an axenic system utilizing *N*-acetylchitosan and glycol chitosan as attachment substrates has been utilized by Wallbanks *et al.* (1989) to produce flagellates with the light microscopic appearance of epimastigote and metacyclic forms. Evidence that these *in vitro* forms correspond to those found in the tsetse salivary gland is lacking to date but, should it be forthcoming, the implication would be that a living microvillar surface is unnecessary for differentiation of the metacyclic stage.

Species of *Blastocrithidia* are difficult to cultivate in axenic media but satisfactory growth can be obtained with associated insect cells. When *B. triatomae* was cultured with a cell line of its host *Triatoma infestans,* good growth was obtained and the flagellate

Figures 13 and 14. *Trypanosoma brucei rhodesiense* (causative agent of East African sleeping sickness). Transmission electron micrographs of sections of salivary gland from an infected tsetse fly, *Glossina morsitans.*

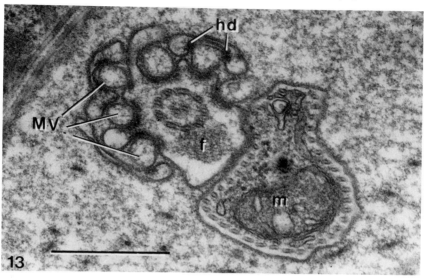

Figure 13. Anterior end of epimastigote form (cut in transverse section) embedded in dense host saliva. The flagellum (f) of the parasite has an expanded sheath with flagellar outgrowths embracing microvilli (MV) of the host's salivary epithelium. Hemidesmosomes (hd) are present where the microvilli impinge on the flagellar membrane. m, trypanosome mitochondrion. Bar = 0.25 μm.

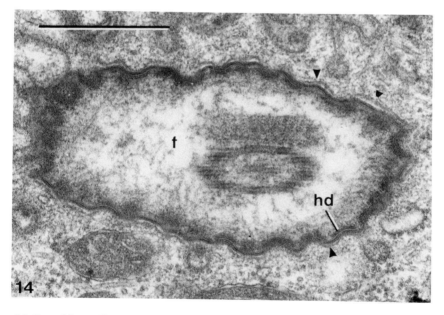

Figure 14. Part of host salivary epithelial cell with inserted tip of flagellum (f) of nascent metacyclic trypanosome. Host cell membrane (at arrowheads) is seen to be distant from the trypanosome surface coat between the numerous punctate hemidesmosomes (hd). Bar = 0.25 μm.

destroyed host cells by flagellar insertion (Reduth *et al.*, 1989) in the manner already described for *in vivo* infections of midgut and Malpighian tubules. No experimental studies on the nature of flagellum–host cell attachment have been conducted as yet.

4.2.3. Cell Penetration. *Trypanosoma rangeli*, a nonpathogenic parasite of man in South America, utilizes as vector reduviid bugs similar to those that transmit *T. cruzi*. *T. rangeli*, however, is transmitted via the anterior station and the metacyclic stage is generated in the salivary glands of the vector. *T. rangeli* reaches the glands by a direct route penetrating through the midgut wall, crossing the hemocoel, and invading the glands from outside (Ellis *et al.*, 1980). A similar route is followed in the vector by *Phytomonas* spp., trypanosomatids of the latex and phloem of plants, and transmitted by phytophagous Hemiptera, metacyclics arising in the salivary glands (Dollet, 1984). Several workers have suggested that *T. brucei* might follow a similar route to the salivary glands of the tsetse fly, but the evidence is unconvincing and consists largely of electron micrographs of trypanosomes which have invaded fly midgut cells (reviewed by Evans and Ellis, 1983). *T. congolense* (which does not invade the salivary glands) has also been observed to penetrate midgut cells. The significance of nonphagocytic cell penetration by these and other trypanosomatids [*Crithidia flexonema* (Tieszen and Molyneux, 1989), *Herpetomonas muscarum* (Bailey and Brooks, 1972), *Leishmania mexicana amazonensis* (Molyneux, 1975), *Trypanosoma lewisi* (Molyneux, 1969b)] is questionable; that the flagellum is active in achieving penetration seems likely though induced phagocytosis as a result of flagellar contact cannot be ruled out and whether the cell dies as a result of penetration or is already dead when entered is uncertain.

4.2.4. Attachment in the Mammalian Host. Although trypanosomatid attachment is probably a feature of all infections in the invertebrate host, attachment is only an occasional feature of trypanosome infections in the vertebrate host. Certain trypanosome species localize in specific regions of the vascular system; e.g., *T. musculi* in the vasa recta of the mouse kidney after host production of trypanocidal antibody (Targett and Viens, 1975). The only example which has been studied in any detail, however, is *T. congolense*, the most important pathogenic trypanosome of cattle in Africa, whose bloodstream trypomastigotes multiply while adhering to the endothelial lining of the mammalian host's microvasculature. Blood trypomastigotes will also bind to the red blood cells of certain host species; adhesion is in all cases by the anterior tip of the flagellate, occurs only with viable organisms, is dependent on temperatures above 4°C and on the presence of trypsin or chymotrypsin-sensitive sites on the trypanosome. Treatment of the host cell surface with neuraminidase, poly-L-lysine, and sodium periodate also inhibits binding, suggesting that sialic acid residues on the host cell serve as receptors (Banks, 1978, 1979). There is no evidence, however, that these receptors bind only to the flagellar membrane. *T. congolense* has no free flagellum and the electron micrographs of Büngener and Müller (1976) suggest that the anterior tips of both body and flagellum interact with the host cell membrane; junctional complexes or other signs of morphological differentiation are not evident at the attachment site. As both body and flagellum are covered with the VSG coat which projects to the exterior a different N-terminal region with each switch of variable antigen type, it seems probable that common membrane components projecting beyond the VSG molecules are responsible for adhesion.

5. Role of Host Attachment in Parasite Development and Transmission

5.1. Specificity of Attachment

The behavior of parasites in their hosts, and in particular the specificity of vector transmission and the location of parasite attachment in the invertebrate host, suggests that parasite–host attachments might depend upon the specific interactions of ligands and receptors on parasite flagellum and host surface. On the other hand, *in vitro* studies have revealed the ability of at least some trypanosomatids to adhere to a range of substrates (e.g., host chitin, Millipore filters, and membranous debris in *Crithidia fasciculata*). Simply scratching the plastic wall of culture flasks can increase attachment markedly, allowing trypanosome parasites of rodents (*T. grosi, T. microti, T. evotomys, T. lewisi, T. musculi*) and leishmania parasites (*L. major, L. aethiopica, L. donovani, L. hertigi*) which usually grow as rosettes in the supernatant to attach to the substratum (Maraghi *et al.*, 1987; Molyneux *et al.*, 1987).

5.1.1. Experimental Studies. The only detailed study of the specificity of attachment is that by Fish *et al.* (1987) on *Trypanosoma vivax* flagellar binding to immobilized organic dye surfaces on agarose beads. The trypanosome bead interaction mimics the *in vivo* attachment of this species to the tsetse fly proboscis wall with formation of hemidesmosomes. It was found possible to reduce or abolish flagellar attachment to Amicon Matrex™ Gel Green A dye beads by bloodstream *T. vivax* by removal of part or all of the Green A dye molecule; certain dyes structurally related to Green A also permitted binding to different degrees. The conclusion was that chemically the 1-amino, 2-sulfonic acid 9–10 anthroquinone structure most closely approximated the tsetse fly proboscis component responsible for binding; dye packing density appeared to be important to the recognition event as witnessed by the preferential binding to bead pockets with greater curvature and hence denser packing of dye molecules. Only certain bloodstream forms of *T. vivax* (late parasitemia, nondividing forms with subterminal kinetoplast) are able to infect the vector or bind to beads; it is reasonable to suppose that in these forms the flagellum has developed receptor molecules capable of interacting with the substratum ligand array.

5.1.2. Attachment to Mechanoreceptors and Effect on Host Feeding. *Leishmania*-infected sandflies appear to experience difficulty in obtaining a blood meal and probe (attempt to feed) more frequently than uninfected sandflies. Killick-Kendrick *et al.* (1977) suggested that the attachment of flagellates to the cuticular lining of the cibarium (prepharynx) interferes with flow detection by sensillae in that part of the gut, so that full engorgement does not occur and probing continues. Killick-Kendrick and Molyneux (1981) suggested that the effect of parasites on sensory input might lead to malfunction of the cibarium and regurgitation of parasites from the pharynx into the skin. As the infective metacyclic forms of leishmanias are found not only in the proboscis but also farther back in the foregut and even in the midgut (Sacks and Perkins, 1985; Lawyer *et al.*, 1987), sensillary blockage could result in enhanced transmission of the parasite.

This idea has been pursued by Molyneux and co-workers (Molyneux *et al.*, 1979; Molyneux, 1980; Jenni *et al.*, 1980; Livesey *et al.*, 1980) with respect to proboscis infections of *Glossina* with the salivarian trypanosomes. Again, infected flies were found to probe more frequently and feed more voraciously than uninfected flies but here a specific relationship was described between trypanosomes and the proboscis mechanoreceptors responsible for detecting the rate of blood flow. Although the large clumps of dividing trypanosomes in the proboscis undoubtedly disturb blood flow through its channel, and this disturbance would be registered by the LC1 and LC2 mechanoreceptors of the labrum and cibarium respectively, there is no clear evidence that it is specifically parasite flagellar attachment to these mechanoreceptors which changes feeding behavior. Thus, *T. brucei* (which may be present in the proboscis lumen and appear entangled with mechanoreceptors in scanning electron micrographs but does not attach to them by its flagella) appears to influence host feeding behavior as well as *T. vivax* and *T. congolense* (which do attach). In the case of the latter two species, it is possible that curvature of the substratum surface near the base of the sensillum encourages flagellar attachment (see Section 5.1.1), but a statistical demonstration of such preference is awaited.

5.2. Parasite Reproduction and Differentiation

The developmental significance of flagellar attachment in the invertebrate host has been studied only with respect to differentiation of the metacyclic stage in the heteroxenous life cycles of trypanosomes.

Using the *in vitro* culture system for the tsetse proboscis stages of *Trypanosoma congolense* developed by Gray *et al.* (1984), Hendry and Vickerman (1988) studied the effect of preventing epimastigote attachment to the wall of the culture flask on division and metacyclogenesis. The attachment of *T. congolense* in subculture occurred within minutes of inoculation and was almost complete within 60 min, but metacyclics did not arise until 10 days afterwards; Ross (1987) has demonstrated that adequate levels of glutamine or proline are necessary for metacyclogenesis in this system. Shaking of cultures at 62 rpm prevented epimastigote attachment, but had no effect on the rate of growth of trypanosomes. Similar results were obtained by growing cultures over a polypropylene substratum to which trypanosome flagella cannot attach. In both series of experiments, however, production of metacyclic trypanosomes was abolished. The lack of metacyclics in shaken cultures was not due to inability of the epimastigotes to transform to metacyclics or to nutritional deprivation, as cessation of shaking was followed by metacyclogenesis. It would, therefore, appear that attachment of epimastigotes is necessary for metacyclic production but not for epimastigote proliferation. Attachment is not an immediate trigger for metacyclogenesis but appears to instigate a series of processes leading to metacyclic production.

In *Trypanosoma cruzi* metacyclogenesis, a similar story is emerging. Metacyclics of *T. cruzi* are produced in liquid cultures with the cessation of exponential growth and up to 90% of trypanosomes may transform (Crane and Dvorak, 1982), suggesting that attachment is unnecessary for this developmental step, yet *in vivo* epimastigotes of this species attach to host rectal chitin and metacyclogenesis occurs while the flagellates are in an attached state (Böker and Schaub, 1984). Development of a chemically defined medium for inducing metacyclic differentiation in multiplying epimastigotes by Contreras *et al.* (1985a,b) has enabled the process to be studied in more detail. Bonaldo *et al.* (1988) found that after transfer to the differentiation medium, epimastigotes attached by their flagella and began to express new 45- to 50-kDa polypeptides. Iodine-labeled surface proteins in the same molecular mass range were recognized by antisera raised against total adhered parasite proteins though the novel surface proteins were not shown to be localized to the flagella. With progress toward the free metacyclic stage (readily recognized in *T. cruzi* by acquisition of the greatly enlarged kinetoplast), expression of these attachment-specific proteins decreased. The authors concluded that expression of the attachment proteins is an essential step in metacyclogenesis.

6. Summary

1. In parasitic flagellates of the order Kinetoplastida, the flagellum may be modified for attachment either to the flagellate's body, to flagella of other flagellates, or to host surfaces during a certain phase of the life cycle. Trypanosome attachment to the vector appears to constitute an essential prelude to differentiation of the vertebrate-infective metacyclic stage, and the same may be true for the leishmanias. Flagellar attachment therefore plays an important part in the epidemiology of major parasitic diseases of man and animals.

2. The flagellar membrane has been poorly studied as a domain independent of the body membrane in kinetoplastids; the two domains appear to share many components active in regulating the host–parasite relationship. Although flagellar movement may be important in cell entry of intracellular kinetoplastids, there is as yet no evidence that specific flagellar membrane components are implicated in host cell recognition and parasite engulfment.

3. Flagellum–body attachment and flagellum–flagellum attachment may be associated with the presence of junctional complexes which bear a morphological resemblance to the desmosomes formed between epithelial cells. These junctions have been poorly studied.

4. Flagellum–host attachment may occur to inert chitinous surfaces of the insect foregut or hindgut, or to living epithelial surfaces. Attachments to inert surfaces are readily reproduced *in vitro* and are characterized by the presence of junctional complexes which morphologically resemble hemidesmosomes. The associated filaments do not appear to correspond to actin or to any of the described intermediate filaments, however, and attachment is strengthened rather than weakened by removal of divalent cations! A novel type of junctional complex would therefore appear to be implicated.

5. Flagellum–host attachments with living microvillar surfaces may also involve hemidesmosome-like junctional complexes, sometimes with damage to the host epithelium. Despite the vector specificity of parasitic flagellates, the specificity of the attachments formed by their flagella is open to doubt.

References

Bailey, C. H., and Brooks, W. M., 1972, Histological observations on larvae of the eye gnat *Hippelates pusio* (Diptera: Chloropidae), infected with the flagellate *Herpetomonas muscarum, J. Invert. Pathol.* **19:**342–353.

Balber, A. E., and Frommel, T. O., 1988, *Trypanosoma brucei gambiense* and *T.b.rhodesiense:* Concanavalin

A binding to the membrane and flagellar pocket of bloodstream and procyclic forms, *J. Protozool.* **35**:214–219.

Banks, K. L., 1978, Binding of *Trypanosoma congolense* to the walls of small blood vessels, *J. Protozool.* **25**:241–245.

Banks, K. L., 1979, The *in vitro* binding of *Trypanosoma congolense* to erythrocytes, *J. Protozool.* **26**:103–108.

Ben Amar, M. F., Pays, A., Tebabi, P., Dero, B., Seebeck, T., Steinert, M., and Pays, E., 1988, Structure and transcription of the actin gene of *Trypanosoma brucei, Mol. Cell. Biol.* **8**:2166–2176.

Benchimol, M., and De Souza, W., 1980, Freeze-fracture study of the plasma membrane of *Leishmania mexicana amazonensis, J. Parasitol.* **66**:941–947.

Bloodgood, R. A., 1975, Ultrastructure of the attachment of *Pyrsonympha* to the hind-gut wall of *Reticulitermes tibialis, J. Insect Physiol.* **21**:391–399.

Böker, C. A., and Schaub, G. A., 1984, Scanning electron microscopic studies of *Trypanosoma cruzi* in the rectum of the vector *Triatoma infestans, Z. Parasitenkd.* **70**:459–469.

Bonaldo, M. C., Souto Padron, T., De Souza, W., and Goldenberg, S., 1988, Cell–substrate adhesion during *Trypanosoma cruzi* differentiation, *J. Cell Biol.* **106**:1349–1358.

Borysenko, J. Z., and Revel, J.-P., 1973, Experimental manipulation of desmosome structure, *J. Anat.* **137**:403–422.

Bray, R. S., and Alexander, J., 1987, *Leishmania* and the macrophage, in: *The Leishmaniases in Biology and Medicine,* Volume I (W. Peters and R. Killick-Kendrick, eds.), Academic Press, New York, pp. 211–233.

Brooker, B. E., 1970, Desmosomes and hemidesmosomes in the flagellate *Crithidia fasciculata, Z. Zellforsch.* **105**:155–166.

Brooker, B. E., 1971a, Flagellar adhesion of *Crithidia fasciculata* to Millipore filters, *Protoplasma* **72**:19–25.

Brooker, B. E., 1971b, Flagellar attachment and detachment of *Crithidia fasciculata* to the gut wall of *Anopheles gambiae, Protoplasma* **73**:191–202.

Brooks, A. S., 1978, Ultrastructure of the flagellar attachment site in three species of trypanosomatids, *Trans. Am. Microsc. Soc.* **97**:287–296.

Brugerolle, G., 1976, Cytologie ultrastructurale, systematique et Evolution des Trichomonadida, *Ann. Stn. Biol. Besse Chandesse* **10**:1–57.

Brugerolle, G., Lom, J., Nohynkova, E., and Joyon, L., 1979, Comparaison et evolution des structures cellulaires chez plusiers especes de bodonides et cryptobiides appartenant aux genres *Bodo, Cryptobia,* et *Trypanoplasma* (Kinetoplastida, Mastigophora), *Protistologica* **15**:197–221.

Brun, R., 1974, Ultrastructur und Zyklus von *Herpetomonas muscarum,* "*Herpetomonas mirabilis*" und *Crithidia luciliae* in *Chrysomyia chloropyga, Acta Trop.* **31**:219–290.

Büngener, W., and Müller, G., 1976, Adhaerenzphänomene bei *Trypanosoma congolense, Tropenmed. Parasitol.* **27**:370–371.

Chang, K. P., 1979, *Leishmania donovani:* Promastigote–macrophage surface interactions *in vitro, Exp. Parasitol.* **48**:175–189.

Chen, L. L., and Haines, T. H., 1976, Flagellar membrane of *Ochromonas danica, J. Biol. Chem.* **251**:1828–1834.

Clark, J. T., and Holberton, D. V., 1988, Triton labile antigens in flagella isolated from *Giardia lamblia, Parasitol. Res.* **74**:415–423.

Contreras, V. T., Morel, C. M., and Goldenberg, S., 1985a, Stage specific gene expression precedes morphological changes during *Trypanosoma cruzi* metacyclogenesis, *Mol. Biochem. Parasitol.* **14**:83–96.

Contreras, V. T., Salles, J. M., Thomas, N., Morel, C. M., and Goldenberg, S., 1985b, *In vitro* differentiation of *Trypanosoma cruzi* under chemically defined conditions, *Mol. Biochem. Parasitol.* **16**:315–327.

Coppens, L., Opperdoes, F. R., Courtoy, P. J., and Baudhuin, P., 1987, Receptor mediated endocytosis in the bloodstream form of *Trypanosoma brucei, J. Protozool.* **34**:465–473.

Cowin, P., and Garrod, D. R., 1983, Antibodies to epithelial desmosomes show wide tissue and species cross-reactivity, *Nature* **302**:148–150.

Crane, M. S. J., and Dvorak, J. A., 1982, *Trypanosoma cruzi:* Spontaneous transformation by a Y strain variant in liquid medium, *Exp. Parasitol.* **54**:87–92.

Cross, G. A. M., 1975, Identification, purification and properties of clone-specific glycoprotein antigens constituting the surface coat of *Trypanosoma brucei, Parasitology* **71**:393–417.

Current, W., 1980, *Cryptobia* sp. in the snail *Triadopsis multilineata* (Say): Fine structure of attached flagellates and their mode of attachment to the spermatheca, *J. Protozool.* **27**:278–287.

De Souza, W., 1984, Cell biology of *Trypanosoma cruzi, Int. Rev. Cytol.* **86:**197–283.

De Souza, W., 1989, The cell surface of trypanosomatids, in: *Progress in Protozoology*, Volume 3 (J. O. Corliss and D. J. Patterson, eds.), Biopress, Bristol, pp. 87–184.

De Souza, W., Martinez-Palomo, A., and Gonzales-Robles, A., 1978, The cell surface of *Trypanosoma cruzi:* Cytochemistry and freeze-fracture, *J. Cell Sci.* **33:**285–299.

De Souza, W., Chavez, B., and Martinez-Palomo, A., 1979, Freeze-fracture study of the cell membrane of *Herpetomonas samuelpessoai, J. Parasitol.* **65:**109–116.

Desser, S. S., 1976, The ultrastructure of the epimastigote stages of *Trypanosoma rotatorium* in the leech *Batracobdella picta, Can. J. Zool.* **54:**1712–1723.

Dollet, M., 1984, Plant diseases caused by flagellate protozoa (*Phytomonas*), *Annu. Rev. Phytopathol.* **22:**115–132.

Duszenko, M., Ivanov, I. E., Ferguson, M. A. J., Plesken, H., and Cross, G. A. M., 1988, Intracellular transport of a variant surface glycoprotein in *Trypanosoma brucei, J. Cell Biol.* **106:**77–86.

Dwyer, D., 1977, *Leishmania donovani:* Surface membrane carbohydrates of promastigotes, *Exp. Parasitol.* **41:**341–358.

Ellis, D. S., Ormerod, W. E., and Lumsden, W. H. R., 1976, Filaments of *Trypanosoma brucei:* Some notes on differences in origin and structure in two strains of *Trypanosoma (Trypanozoon) brucei rhodesiense, Acta Trop.* **33:**151–168.

Ellis, D. S., Evans, D. A., and Stamford, S., 1980, The penetration of the salivary glands of *Rhodnius prolixus* by *Trypanosoma rangeli, Z. Parasitenkd* **62:**63–74.

Erlandsen, S. L., and Feely, D., 1984, Trophozoite motility and the mechanism of attachment, in: *Giardia and Giardiasis: Biology, Pathogenesis and Epidemiology* (S. L. Erlandsen and E. A. Meyer, eds.), Plenum Press, New York, pp. 33–63.

Etges, R., Bouvier, J., and Bordier, C., 1987, The promastigote surface protease of *Leishmania*, in: *Host–Parasite Cellular and Molecular Interactions in Protozoal Infections*, NATO ASI Series H, Volume 11 (K. P. Chang and D. Snary, eds.), Springer-Verlag, Berlin, pp. 165–168.

Evans, D. A., and Ellis, D. S., 1983, Recent observations on the behaviour of certain trypanosomes within their insect hosts, *Adv. Parasitol.* **22:**2–42.

Evans, D. A., Ellis, D. S., and Stamford, S., 1979, Ultrastructural studies of certain aspects of the development of *Trypanosoma congolense* in *Glossina morsitans, J. Protozool.* **26:**557–563.

Farina, M., Attias, M., Souto-Poudron, T., and De Souza, W., 1986, Further studies on the organization of the paraxial rod of trypanosomatids, *J. Protozool.* **33:**352–357.

Farthing, M. J. G., 1985, Receptors and recognition mechanisms in intestinal infection, *Trans. R. Soc. Trop. Med. Hyg.* **79:**569–576.

Fish, W. R., Nelson, R. T., and Hirumi, H., 1987, Cell adhesion in *Trypanosoma in vitro* studies of the interaction of *Trypanosoma vivax* with immobilized organic dyes, *J. Protozool.* **34:**457–464.

Flint, J. E., Schechter, M., Chapman, M. D., and Miles, M. A., 1984, Zymodeme and species specificities of monoclonal antibodies raised against *Trypanosoma cruzi, Trans. R. Soc. Trop. Med. Hyg.* **78:**193–202.

Freymuller, E., and Camargo, E. P., 1981, Ultrastructural differences between species of trypanosomatids with and without symbionts, *J. Protozool.* **28:**175–182.

Gallo, J.-M., and Schrevel, J., 1985, Homologies between paraflagellar rod proteins from trypanosomes and euglenoids revealed by a monoclonal antibody, *Eur. J. Cell Biol.* **36:**163–168.

Gibson, W. C., and Miles, M. A., 1985, Application of new technologies to epidemiology, *Br. Med. Bull.* **41:** 115–121.

Goldstein, S. F., 1974, Isolated, reactivated and laser-irradiated cilia and flagella, in: *Cilia and Flagella* (M. A. Sleigh, ed.), Academic Press, New York, pp. 111–130.

Gray, M. A., Cunningham, I., Gardiner, P. R., Taylor, A. M., and Luckins, A. G., 1981, Cultivation of infective forms of *Trypanosoma congolense* at 28°C from trypanosomes in the proboscis of *Glossina morsitans, Parasitology* **82:**81–95.

Gray, M. A., Ross, C. A., Taylor, A. M., and Luckins, A. G., 1984, *In vitro* cultivation of *Trypanosoma congolense:* The production of infective metacyclic trypanosomes in cultures initiated from cloned stocks, *Acta Trop.* **41:**343–353.

Handman, E., and Goding, J. W., 1985, The *Leishmania* receptor for macrophages is a lipid-containing glycoconjugate, *EMBO J.* **4:**329–336.

Heath, J. P., and Dunn, G. A., 1978, Cell to substratum contacts of chick fibroblasts and their relation to the microfilament system, *J. Cell Sci.* **29:**197–212.

Hendry, K. A. K., 1987, Studies on the flagellar attachment of African trypanosomes, Ph.D. thesis, University of Glasgow.

Hendry, K. A. K., and Vickerman, K., 1988, The requirement for epimastigote attachment during division and metacyclogenesis in *Trypanosoma congolense, Parasitol. Res.* **74:**403–408.

Hoare, C. A., 1949, *Handbook of Medical Protozoology,* Ballière, Tindall & Cox, London.

Hogan, J. C., and Patton, G. L., 1976, Variation in intramembrane components of *Trypanosoma brucei* from intact and x-irradiated rats: A freeze-cleave study, *J. Protozool.* **23:**205–215.

Holberton, D. V., 1974, Attachment of *Giardia*—A hydrodynamic model based on flagellar activity, *J. Exp. Biol.* **60:**207–221.

Hommel, M., and Robertson, E., 1976, *In vitro* attachment of trypanosomes to plastic, *Experientia* **32:**464–466.

Hyams, J. S., 1982, The *Euglena* paraflagellar rod: Structure, relationship to other flagellar components and preliminary biochemical characterization, *J. Cell Sci.* **55:**199–210.

Ismach, R., Cianci, J. P., Caulfield, J. P., Langer, P., and McMahon-Pratt, D., 1989, Flagellar membrane and paraxial rod proteins of *Leishmania:* Characterization employing monoclonal antibodies, *J. Protozool.* in press.

Jackson, P. R., Honigberg, B. M., and Holt, S. C., 1978, Lectin analysis of *Trypanosoma congolense* bloodstream trypomastigote and culture procyclic surface saccharides by agglutination and electron microscopic techniques, *J. Protozool.* **25:**471–481.

Jenni, L., Molyneux, D. H., Livesey, J. L., and Galun, R., 1980, Feeding behaviour of tsetse flies infected with salivarian trypanosomes, *Nature* **283:**383–385.

Jenni, L., Marti, J., Schweizer, J., Betschart, B., Le Page, R. W. F., Wells, J. M., Tait, A., Paindavoine, P., Pays, E., and Steinert, M., 1986, Hybrid formation between African trypanosomes during cyclical transmission, *Nature* **322:**173–175.

Jones, J. C. R., and Goldman, R. D., 1985, Intermediate filaments and the initiation of desmosome assembly, *J. Cell Biol.* **101:**506–517.

Kaminsky, R., Beaudoin, E., and Cunningham, I., 1987, Studies on the development of metacyclic *Trypanosoma bruceis* sspp. cultivated with insect cell lines, *J. Protozool.* **34:**372–377.

Killick-Kendrick, R., 1979, Biology of *Leishmania* in phlebotomine sandflies, in: *Biology of the Kinetoplastida,* Volume 2 (W. H. R. Lumsden and D. A. Evans, eds.), Academic Press, New York, pp. 395–460.

Killick-Kendrick, R., and Molyneux, D. H., 1981, Transmission of leishmaniasis by the bite of phlebotomine sandflies: Possible mechanisms, *Trans. R. Soc. Trop. Med. Hyg.* **75:**152–154.

Killick-Kendrick, R., Molyneux, D. H., and Ashford, R. W., 1974, *Leishmania* in phlebotomid sandflies. I. Modifications of the flagellum associated with attachment to the midgut and oesophageal valve of the sandfly, *Proc. R. Soc. London Ser. B* **187:**409–419.

Killick-Kendrick, R., Leaney, A. J., Ready, P. D., and Molyneux, D. H., 1977, Leishmania in phlebotomid sandflies. IV. The transmission of *Leishmania mexicana amazonensis* by the bite of experimentally-infected *Lutzomyia longipalpis, Proc. R. Soc. London Ser. B* **196:**105–115.

Killick-Kendrick, R., Wallbanks, K. R., Molyneux, D. H., and Lavin, D. R., 1988, The ultrastructure of *Leishmania major* in the foregut and proboscis of *Phlebotomus papatasi, Parasitol. Res.* **74:**586–590.

Lauge, G., and Nishioka, R. S., 1977, Ultrastructural study of the relations between *Leptomonas oncopelti* (Noguchi and Tilden) Protozoa: Trypanosomatidae, and the rectal wall of the adults of *Oncopeltus fasciatus* Dell (Hemiptera: Lygaeidae), *J. Morphol.* **154:**291–305.

Lawyer, P. G., Young, D. G., Butler, J. F., and Akin, D. E., 1987, Development of *Leishmania mexicana* in *Lutzomyia diabolica* and *Lutzomyia shannoni* (Diptera: Psychodidae), *J. Med. Entomol.* **24:**347–355.

Lewis, J. W., and Ball, S. J., 1979, Attachment of the epimastigotes of *Trypanosoma cobitis* (Mitrophanow 1885) to the crop wall of the leech vector *Hemiclepsis marginata, Z. Parasitenkd.* **60:**29–36.

Linder, J. C., and Staehelin, L. A., 1977, Plasma membrane specializations in a trypanosomatid flagellate, *J. Ultrastruct. Res.* **60:**246–262.

Livesey, J. L., Molyneux, D. H., and Jenni, L., 1980, Mechanoreceptor–trypanosome interactions in the labrum of *Glossina:* Fluid mechanics, *Acta Trop.* **37:**151–161.

Lom, J., 1980, *Cryptobia branchialis* Nie from fish gills: Ultrastructural evidence of ectocommensal function, *J. Fish Dis.* **3:**427–436.

Maraghi, S., Mohamed, H. A., Wallbanks, K. R., and Molyneux, D. H., 1987, Scratched plastic as a substrate for trypanosomatid attachment, *Ann. Trop. Med. Parasitol.* **81:**457–458.

Molyneux, D. H., 1969a, The fine structure of the epimastigote forms of *Trypanosoma lewisi* in the rectum of the flea, *Nosopsyllus fasciatus, Parasitology* **59**:55–66.

Molyneux, D. H., 1969b, Intracellular stages of *Trypanosoma lewisi*, in fleas and attempts to find such stages in other trypanosome species, *Parasitology* **59**:737–744.

Molyneux, D. H., 1975, *Trypanosoma (Megatrypanum) melophagium:* Modes of attachment of parasites to midgut, hindgut and rectum of the sheep ked, *Melophagus ovinus, Acta Trop.* **32**:65–74.

Molyneux, D. H., 1980, Host–trypanosome interactions in *Glossina, Insect Sci. Appl.* **1**:39–46.

Molyneux, D. H., 1983, Host–parasite relationships of Trypanosomatidae in vectors, in: *Current Topics in Vector Research*, Volume 1 (K. F. Harris, ed.), Praeger Publications, New York, pp. 117–148.

Molyneux, D. H., and Ashford, R. W., 1975, Observations on a trypanosomatid flagellate in a flea, *Peromyscopsylla silvatica spectabilis, Ann. Parasitol. Hum. Comp.* **50**:265–274.

Molyneux, D. H., and Killick-Kendrick, R., 1987, Morphology, ultrastructure and life-cycles, in: *The Leishmaniases in Biology and Medicine*, Volume 1 (W. Peters and R. Killick-Kendrick, eds.), Academic Press, New York, pp. 121–176.

Molyneux, D. H., Killick-Kendrick, R., and Ashford, R. W., 1975, Leishmania in phlebotomid sandflies. III. The ultrastructure of *Leishmania mexicana amazonensis* in the midgut and pharynx of *Lutzomyia longipalpis, Proc. R. Soc. London Ser. B* **190**:341–357.

Molyneux, D. H., Lavin, D. R., and Elce, B., 1979, A possible relationship between salivarian trypanosomes and *Glossina* labrum mechano-receptors, *Ann. Trop. Med. Parasitol.* **73**:287–290.

Molyneux, D. H., Croft, S. L., and Lavin, P. R., 1981, Studies on the host–parasite relationships of *Leptomonas* species (Protozoa: Kinetoplastida) of Siphonaptera, *J. Nat. Hist.* **15**:395–406.

Molyneux, D. H., Wallbanks, K. R., and Ingram, G. A., 1987, Trypanosomatid–vector interface—*in vitro* studies on parasite substrate interactions, in: *Host–Parasite Cellular and Molecular Interactions in Protozoal Infections*, NATO Series H Volume 11 (K. P. Chang and D. Snary, eds.), Springer-Verlag, Berlin, pp. 387–396.

Mungomba, L. M., Molyneux, D. H.. and Wallbanks, K. R., 1989, Host-parasite relationship of *Trypanosoma corvi* in *Ornithomyia avicularia, Parasitol. Res.* **75**: 167–174.

Nogueira, N., 1983, Host and parasite factors affecting the invasion of mononuclear phagocytes by *Trypanosoma cruzi*, in: *Cytopathology of Parasitic Disease*, Ciba Symposium New Series 90, Pitman, London, pp. 52–73.

Olecnick, J. G., Wolff, R., Nauman, R. K., and McLaughlin, J., 1988, A flagellar pocket membrane fraction from *Trypanosoma brucei rhodesiense:* Immunogold localization and non-variant immunoprotection, *Infect. Immun.* **56**:92–98.

Ouaissi, M. A., 1986, Identification and isolation of *Trypanosoma cruzi* trypomastigote cell surface protein with properties expected of a fibronectin receptor, *Mol. Biochem. Parasitol.* **19**:201–211.

Overton, J., 1975, Experiments with junctions of the adhaerens type, *Curr. Top. Dev. Biol.* **10**:1–34.

Overton, J., 1982, Inhibition of desmosome formation with tunicamycin and with lectin in corneal cell aggregates, *Dev. Biol.* **92**:66–72.

Pays, E., 1988, Expression of variant-specific antigen genes in African trypanosomes, *Biol. Cell.* **64**:121–130.

Pearson, R. D., Wheeler, D. A., Harrison, L. H., and Kay, H. D., 1983, The immunobiology of leishmaniasis, *Rev. Infect. Dis.* **5**:907–927.

Peng, P. L.-M., and Wallace, F. G., 1982, The cysts of *Blastocrithidia triatomae* Cerisola *et al.* 1971, *J. Protozool.* **29**:464–467.

Pereira, M. E. A., Andrade, A. F. B., and Ribeiro, J. M. V., 1981, Lectins of distinct specificity in *Rhodnius prolixus* interact selectively with *Trypanosoma cruzi, Science* **211**:597–600.

Pereira, N. M., De Souza, W., Machado, R. D., and Castro, F. T., 1977, Isolation and properties of flagella of trypanosomatids, *J. Protozool.* **24**:511–514.

Pereira, N. M., Timm, S. L., Da Costa, S. C. G., Rebello, M. A., and De Souza, W., 1978, *Trypanosoma cruzi* isolation and characterization of membrane and flagellar fractions, *Exp. Parasitol.* **46**:225–234.

Petry, K., Baltz, T., and Schottelius, J., 1986, Differentiation of *Trypanosoma cruzi, T.cruzi marinkellei, T.dionisii* and *T.vespertilionis* by monoclonal antibodies, *Acta Trop.* **43**:5–13.

Petry, K., Schottelius, J., and Baltz, T., 1987, Characterization of a 19,000 mol. wt. flagellum-specific protein of *Trypanosoma cruzi, T.dionisii* and *T.vespertilionis* by a monoclonal antibody, *Parasitol. Res.* **73**:180–181.

Pimenta, P. F. P., and De Souza, W., 1987, *Leishmania mexicana*. Distribution of intramembranous particles and filipin–sterol complexes in amastigotes and promastigotes, *Exp. Parasitol.* **63**:117–135.

Pruss, R. M., Mirsky, R., and Raff, M. C., 1981, All classes of intermediate filaments share a common antigenic determinant defined by a monoclonal antibody, *Cell* **27**:419–428.

Reduth, D., and Schaub, G. A., 1988, The ultrastructure of the cysts of *Blastocrithidia triatomae* Cerisola *et al.* 1971, (Trypanosomatidae): A freeze cleave study, *Parasitol. Res.* **74**:301–306.

Reduth, D., Schaub, G. A., and Pudney, M., 1984, Cultivation of *Blastocrithidia triatomae* (Trypanosomatidae) in a cell line of its host *Triatoma infestans* (Reduviidae), *Zentralbl. Bakteriol. Mikrobiol. Hyg.* *[A]* **258**: 383.

Richardson, J. P., Jenni, L., Beecroft, R. P., and Pearson, T. W., 1986, Procyclic tsetse fly midgut forms of African trypanosomes share stage and species-specific surface antigens identified by monoclonal antibodies, *J. Immunol.* **136**:2259–2265.

Roditi, I., Schwarz, H., Pearson, T. W., Beecroft, R. P., Liu, M. K., Richardson, J. P., Buhring, H-J., Pleiss, J., Bulow, R., Williams, R. O., and Overath, P., 1989, Procyclin gene expression and loss of the variant surface glycoprotein during differentiation of *Trypanosoma brucei*, *J. Cell Biol.* **108**: 737–746.

Rogalski, A. A., and Bouck, G. B., 1982, Flagellar surface antigens in *Euglena;* immunological evidence for an external glycoprotein pool and its transfer to the regenerating flagellum, *J. Cell Biol.* **93**:758–766.

Ross, C. A., 1987, *Trypanosoma congolense:* Differentiation to metacyclic trypanosomes in culture depends on the concentration of glutamine or proline, *Acta Trop.* **44**:293–301.

Rowton, E. P., Lushbaugh, W. B., and McGhee, R. B., 1981, Ultrastructure of the flagellar apparatus and attachment of *Herpetomonas ampelophilae* in the gut and Malpighian tubules of *Drosophila melanogaster*, *J. Protozool.* **28**:297–301.

Russell, D. G., and Wilhelm, H., 1986, The involvement of GP63, the major surface glycoprotein in the attachment of *Leishmania* promastigotes to macrophages, *J. Immunol.* **136**:2613–2620.

Russell, D. G., Newsam, R., Palmer, G. C., and Gull, K., 1983, Structural and biochemical characterization of the paraflagellar rod of *Crithidia fasciculata*, *Eur. J. Cell Biol.* **30**:137–143.

Sacks, D. L., and Perkins, P. V., 1985, Development of infective stage *Leishmania* promastigotes within phlebotomine sandflies, *Am. J. Trop. Med. Hyg.* **34**:456–459.

Sacks, D. L., Hieny, S., and Sher, A., 1985, Identification of cell surface carbohydrates and antigenic changes between non-infective and infective developmental stages of *Leishmania major* promastigotes, *J. Immunol.* **135**:564–569.

Schaub, G. A., and Böker, C. A., 1986, Scanning electron microscopic studies of *Blastocrithidia triatomae* (Trypanosomatidae) in the rectum of *Triatoma infestans* (Reduviidae), *J. Protozool.* **33**:266–270.

Schaub, G. A., and Schnitker, A., 1989, Influence of *Blastocrithidia triatomae* (Trypanosomatidae) in the reduviid bug *Triatoma infestans:* Alterations of the Malpighian tubules, *Parasitol. Res.* **75**: 88–97.

Schneider, A., Lutz, H. U., Marugg, R., Gehr, P., and Seebeck, T., 1988, Spectrin-like proteins in the paraflagellar rod structure of *Trypanosoma brucei*, *J. Cell Sci.* **90**:307–315.

Segura, E. L., Paulone, I., Cerisola, J., and Gonzalez-Cappa, S. M., 1976, Experimental Chagas' disease: Protective activity in relation to subcellular fractions of the parasite, *J. Parasitol.* **62**:131–133.

Segura, E. L., Vazquez, C., Bronzina, A., Campos, J. M., Cerisola, J. A., and Gonzalez-Cappa, S. M., 1977, Antigens of the subcellular fractions of *Trypanosoma cruzi*. II. Flagella and membrane fractions, *J. Protozool.* **24**:540–543.

Segura, E. L., Bua, J., Rosenstein de Campini, A., Subias, E., Esteva, M., Moreno, M., and Ruiz, A., 1986, Monoclonal antibodies against the flagellar fraction of epimastigotes of *Trypanosoma cruzi:* Complement-mediated lytic activity against trypomastigotes and passive immunoprotection of mice, *Immunol. Lett.* **13**: 165–171.

Severs, J. N., and Robenek, H., 1983, Detection of microdomains in biomembranes. An appraisal of recent developments in freeze-fracture cytochemistry, *Biochim. Biophys. Acta* **737**:373–408.

Sher, A., and Snary, D., 1982, Specific inhibition of the morphogenesis of *Trypanosoma cruzi* by a monoclonal antibody, *Nature* **300**:639–640.

Skerrow, C. J., 1986, Desmosomal proteins, in: *Biology of the Integument*, Volume 2 (J. Bereiter-Hahn, A. G. Matoltsy, and K. S. Richards, eds.), Springer-Verlag, Berlin, pp. 762–782.

Smith, D. S., Njogu, A. R., Cayer, M., and Jarlfors, U., 1974, Observations on freeze-fractured membranes of a trypanosome, *Tissue Cell* **6**:223–241.

Snary, D., Ferguson, M. A. J., Allen, A. K., Miles, M. A., and Sher, A., 1987, Cell surface glycoproteins of

Trypanosoma cruzi, in: *Host–Parasite Cellular and Molecular Interactions in Protozoal Infections*, NATO ASI Series H, Volume 11 (K. P. Chang and D. Snary, eds.), Springer-Verlag, Berlin, pp. 79–87.

Soares, M. J., and De Souza, W., 1987, The ultrastructure of *Blastocrithidia culicis* as seen in thin sections and freeze-fracture replicas, *Biol. Cell.* **61**:101–108.

Soares, M. J., Brazil, R. P., Tanuri, A., and De Souza, W., 1986, Some ultrastructural aspects of *Crithidia guilhermei* n.sp. isolated from *Phaenicia cuprina* (Diptera: Calliphoridae), *Can. J. Zool.* **64**:2837–2842.

Souto-Padron, T., and De Souza, W., 1983, Freeze-fracture localization of filipin–cholesterol complexes in the plasma membrane of *Trypanosoma cruzi*, *J. Parasitol.* **69**:129–137.

Souto-Padron, T., Goncalves de Lima, V. M. Q., Roitman, I., and De Souza, W., 1980, Fine structure study of *Leptomonas samueli* by the freeze-fracture technique, *Z. Parasitenkd.* **62**:145–157.

Steiger, R., 1973, On the ultrastructure of *Trypanosoma (Trypanozoon) brucei* in the course of its life cycle and some related aspects, *Acta Trop.* **30**:164–168.

Targett, G. A. T., and Viens, P., 1975, The immunological response of CBA mice to *Trypanosoma musculi:* Elimination of the parasite from the blood, *Int. J. Parasitol.* **5**:231–234.

Taylor, A. E. R., and Baker, J. R. (eds.), 1987, *In Vitro Methods for Parasite Cultivation*, Academic Press, New York.

Tetley, L., 1986, Freeze-fracture studies on the surface membranes of pleomorphic bloodstream and *in vitro* transformed procyclic *Trypanosoma brucei*, *Acta Trop.* **43**:307–317.

Tetley, L., and Vickerman, K., 1985, Differentiation in *Trypanosoma brucei:* Host–parasite cell junctions and their persistence during acquisition of the variable antigen coat, *J. Cell Sci.* **74**:1–19.

Tetley, L., Coombs, G. H., and Vickerman, K., 1986, The surface membrane of *Leishmania mexicana mexicana:* Comparison of amastigote and promastigote using freeze-fracture cytochemistry, *Z. Parasitenkd.* **72**:281–292.

Tetley, L., Turner, C. M. R., Barry, J. D., Crowe, J. S., and Vickerman, K., 1987, Onset of expression of the variant surface glycoproteins of *Trypanosoma brucei* in the tsetse fly studied using immunoelectron microscopy, *J. Cell Sci.* **87**:363–372.

Thevenaz, P., and Hecker, H., 1980, Distribution and attachment of *Trypanosoma (Nannomonas) congolense* in the proximal part of the proboscis of *Glossina morsitans morsitans*, *Acta Trop.* **37**:163–173.

Tieszen, K. L., and Molyneux, D. H., 1989, Morphology and host parasite relationships of *Crithidia flexonema* (Trypanosomatidae) in the hindgut and Malpighian tubules of *Gerris odontogaster* (Hemiptera: Gerridae), *J. Parasitol.* **75**: 441–448.

Tieszen, K. L., Heywood, P., and Molyneux, D. H., 1983, Ultrastructure and host–parasite association of *Blastocrithidia gerridis* in the ventriculus of *Gerris odontogaster* (Hemiptera: Gerridae), *Can. J. Zool.* **61**: 1900–1909.

Tieszen, K. L., Molyneux, D. H., and Abdel-Hafez, S. K., 1986, Host–parasite relationships of *Blastocrithidia familiaris* in *Lygaeus pandurus* Scop. (Hemiptera: Lygaeidae), *Parasitology* **92**:1–12.

Tieszen, K. L., Molyneux, D. H., and Abdel-Hafez, S. K., 1989, Host–parasite relationships and cysts of *Leptomonas lygaei* (Trypanosomatidae) in *Lygaeus pandurus* (Hemiptera: Lygaeidae), *Parasitology* **98**: 393–400.

Trinkhaus-Randall, V., and Gipson, I. K., 1984, Role of calcium and calmodulin in hemidesmosome formation *in vitro*, *J. Cell Biol.* **98**:1565–1571.

Turco, S. J., Johnson, C. L., King, D. L., Orlandi, P. A., and Wright, B. L., 1987, The structure, localization and function of the lipophosphoglycan of *Leishmania donovani*, in: *Host–Parasite Cellular and Molecular Interactions in Protozoal Infections*, NATO ASI Series H, Volume 11 (K. P. Chang and D. Snary, eds.), Springer-Verlag, Berlin, pp. 197–201.

Verschueren, H., 1985, Interference reflection microscopy in cell biology: Methodology and applications, *J. Cell Sci.* **75**:279–301.

Vickerman, K., 1969, On the surface coat and flagellar adhesion in trypanosomes, *J. Cell Sci.* **5**:163–194.

Vickerman, K., 1973, The mode of attachment of *Trypanosoma vivax* in the proboscis of the tsetse fly, *Glossina fuscipes*, *J. Protozool.* **20**:394–404.

Vickerman, K., 1974, Antigenic variation in African trypanosomes, in: *Parasites in the Immunized Host: Mechanisms of Survival*, Ciba Symposium New Series 25, Elsevier, Amsterdam, pp. 53–80.

Vickerman, K., 1976, The diversity of the kinetoplastid flagellates, in: *Biology of the Kinetoplastida*, Volume 1 (W. H. R. Lumsden and D. A. Evans, eds.), Academic Press, New York, pp. 1–34.

Vickerman, K., 1985, Developmental cycles and biology of pathogenic trypanosomes, *Br. Med. Bull.* **41**:105–114.

Vickerman, K., 1989, Order Kinetoplastida Honigberg 1963, in: *Handbook of Protoctista* (L. Margulis, J. O. Corliss, M. Melkonian, and D. Chapman, eds.), Jones & Bartlett, Boston, in press.

Vickerman, K., and Barry, J. D., 1982, African trypanosomiasis, in: *Immunology of Parasitic Infections,* 2nd ed. (S. Cohen and K. S. Warren, eds.), Blackwell, Oxford, pp. 204–260.

Vickerman, K., and Preston, T. M., 1976, Comparative cell biology of the kinetoplastid flagellates, in: *Biology of the Kinetoplastida,* Volume 1 (W. H. R. Lumsden and D. A. Evans, eds.), Academic Press, New York, pp. 35–130.

Vickerman, K., and Tetley, L., 1979, Biology and ultrastructure of trypanosomes in relation to pathogenesis, in: *Pathogenicity of Trypanosomes* (G. Losos and A. Chouinard, eds.), IDRC, Ottawa, Canada, pp. 23–31.

Vickerman, K., Tetley, L., Hendry, K. A. K., and Turner, C. M. R., 1988, Biology of African trypanosomes in the tsetse fly, *Biol. Cell.* **64:** 109–119.

Wallbanks, K. R., Molyneux, D. H., and Dirie, M. F., 1989, Chitin derivatives as novel substrates for *Trypanosoma brucei brucei* attachment *in vitro, Acta Trop.* **46:**63–68.

Williams P., 1976, Flagellate infections in cave-dwelling sandflies (Diptera: Psychodidae) in Belize, Central America, *Bull. Entomol. Res.* **65:**615–629.

Witman, G. B., Carlson, K., Berliner, J., and Rosenbaum, J. L., 1972, *Chlamydomonas* flagella. I. Isolation and electrophoretic analysis of microtubules, matrix, membranes and mastigonemes, *J. Cell Biol.* **54:**507–539.

Witman, G. B., Plummer, J., and Sander, G., 1978, *Chlamydomonas* flagellar mutants lacking radial spokes and central tubules, *J. Cell Biol.* **76:**729–742.

Woo, P. T. K., 1987, Cryptobia and cryptobiosis in fishes, *Adv. Parasitol.* **26:**199–237.

Zeledon, R., Alvarenga, N. J., and Schosinsky, K., 1977, Ecology of *Trypanosoma cruzi* in the insect vector, in: *Pan America Health Organisation Scientific Publication No. 347,* Proceedings of an International Symposium 27 June, pp. 59–70.

Zeledon, R., Bolanos, R., and Rojas, M., 1984, Scanning electron microscopy of the final phase of the life cycle of *Trypanosoma cruzi* in the insect vector, *Acta Trop.* **41:**39–43.

Zenian, A., Rowles, P., and Gingell, D., 1979, Scanning electron microscopic study of the uptake of *Leishmania parasites* by macrophages, *J. Cell Sci.* **39:**187–199.

The Sperm Plasma Membrane

A Little More Than Mosaic, a Little Less Than Fluid

Richard A. Cardullo and David E. Wolf

1. Introduction

In this chapter the dynamics of molecular motions in the mammalian sperm plasma membrane will be discussed. Specifically, we will consider the nature of the known or potential forces which control these molecular motions. We will only discuss macromolecules that have been identified as having some physiological role and have been characterized on intact cells using biophysical techniques to gain insight into the organization of the sperm plasma membrane. Information on other cell-surface macromolecules, especially on cilia and flagella, is available in a number of articles (e.g., Brooks, 1985; Vernon *et al.*, 1985, 1987; Olson *et al.*, 1987; Feuchter *et al.*, 1988; Trimmer and Vacquier, 1988).

As we shall see, studies of molecular dynamics on somatic cells have led to a modified fluid mosaic model (Singer and Nicolson, 1972). To paraphrase Hamlet, the mammalian sperm plasma membrane is *a little more than mosaic, a little less than fluid*. This model will thus be our starting point in considering the sperm plasma membrane.

1.1. Why Should the Mammalian Sperm Plasma Membrane Be Given Special Consideration?

Before beginning a discussion of the physical properties of the sperm plasma membrane, it is worth asking why one might *a priori* expect the mammalian sperm plasma membrane to differ from the plasma membrane of other cells. In such a consideration, one must, of course, take care not to ask why the sperm plasma membrane is unique. Clearly, sperm are highly differentiated and specialized cells. However, the same may be said for any of a number of other cell types. What is probably most significant is that sperm

Richard A. Cardullo and David E. Wolf • Worcester Foundation for Experimental Biology, Shrewsbury, Massachusetts 01545.

represent a well-defined model for the study of two important biological processes: membrane–membrane fusion and flagellar or ciliary motion. Two aspects of sperm plasma membrane biology do, however, merit special consideration: the high level of surface regionalization and specialization in this continuous plasma membrane, and the extensive compositional changes that postmeiotic sperm plasma membranes undergo in the absence of significant membrane biosynthesis.

1.2. Regionalization of the Sperm Plasma Membrane

Simplistically, the plasma membrane of sperm is designed for a number of adhesion and fusion events ultimately leading to fertilization. It is indeed impressive to consider the number of cell-surface alterations that the sperm plasma membrane undergoes prior to fertilization. During spermatogenesis, the developing spermatogonia, spermatocytes, and spermatids are in close contact with the surrounding Sertoli cells coincident with the biosynthesis of macromolecules. Subsequent to this, these cells undergo spermiogenesis, which involves impressive morphological transformations leading to the regionalization of various components and the appearance of morphologically distinct regions (i.e., the head, the midpiece, and the principal piece) that can be resolved with a light microscope. Although testicular sperm are released from the Sertoli cell after spermiogenesis (spermiation), extensive surface modifications occur during epididymal maturation and later in the female reproductive tract. The physiological phenomenon known as capacitation enables the sperm, for the first time, to fertilize an ovum. However, fertilization itself requires a number of complex cell-surface interactions (sperm binding to the zona pellucida, acrosome reaction, and membrane–membrane fusion) that necessitate a synergistically controlled sequence of sperm surface alterations to occur. It is therefore possible that the sperm surface is shaped by a number of different mechanisms throughout its development: (1) by its own biosynthetic capabilities prior to meiosis, (2) through direct contact with other cells (e.g., Sertoli cells and ova), and (3) by the extracellular milieu that is provided by the epididymis and the female reproductive tract.

As one might expect, this morphological and functional regionalization is reflected in the distribution of plasma membrane components. Regionalization on sperm has been reported for: lipids (Forrester, 1980; Bearer and Friend, 1982), lectin-binding sites (Nicolson and Yanagamachi, 1974, 1979; Millette, 1977; Koehler, 1978), and sperm-specific antigens (Erickson, 1977; Koo et al., 1977; Tung, 1977; Millette, 1979; Myles et al., 1981; Primakoff and Myles, 1983; Saling and Lakoski, 1985; Lopez and Shur, 1987; O'Rand, 1988). For some examples, see Fig. 1. Several molecules whose functions have been putatively identified with fertilization and the acrosome reaction are localized to the anterior region of the head where the acrosome reaction takes place. While sperm represent one of the clearest examples of the phenomenon of surface regionalization, similar regionalizations and polarizations of surface components are also observed in other highly differentiated cells, such as: epithelia (McNutt and Weinstein, 1973; Staehelin, 1974), muscle (Axelrod et al., 1976b), and early embryos (Johnson, 1981).

According to the fluid mosaic model, the plasma membrane is a quasi-two-dimensional fluid. That is, the molecules of the membrane are free to diffuse in the plane of the membrane. Diffusion results from random thermal motion (Einstein, 1905) and thus, in the absence of external forces, should drive the distribution of diffusing species to homogeneity (Wolf, 1986). A central question thus arises: *How is it that the sperm membrane*

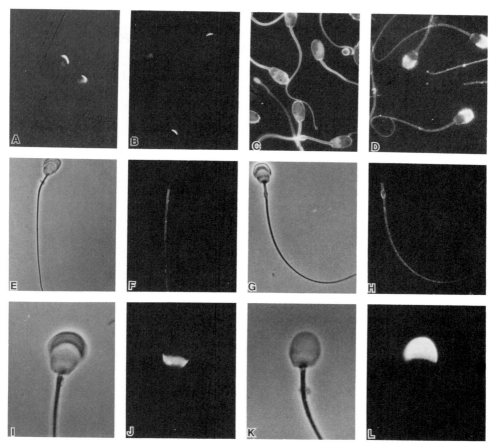

Figure 1. Examples of regionalizations and redistributions of antigens on mammalian sperm. Fluorescence micrographs of: (A) condensing mouse spermatids labeled with anti-M42, (B) caudal epididymal mouse sperm labeled with anti-M42, (C) ejaculated ram sperm labeled with anti-ESA 152, and (D) acrosome-reacted ram sperm labeled with anti-ESA 152. Paired phase-contrast and fluorescence micrographs of: (E, F) caudal epididymal guinea pig sperm labeled with anti-PT-1, (G, H) acrosome-reacted guinea pig sperm labeled with anti-PT-1, (I, J) caudal epididymal guinea pig sperm labeled with anti-PH-20, (K, L) acrosome-reacted guinea pig sperm labeled with anti-PH-20.

overcomes the randomizing effect of diffusion and regionalizes the distribution of its surface components?

1.3. Membrane Modification in the Absence of Macromolecular Biosynthesis

Sperm leaving the rete testis and entering the caput epididymis are largely immobile and incapable of fertilizing ova. They acquire motility and the capacity to fertilize ova only after they come in contact with the luminal fluids of the epididymis (epididymal maturation) and the female reproductive tract, particularly the oviduct (capacitation)

(Voglmayer *et al.*, 1982; Dacheux and Voglmayr, 1983; Russell *et al.*, 1984; Langlais and Roberts, 1985). Posttesticular sperm have little, or no, ability to synthesize plasma membrane lipids or proteins (Lardy and Phillips, 1941; Scott *et al.*, 1967; Premkumar and Bhargava, 1972; Voglmayr, 1975; Bragg and Handel, 1979). Yet the composition of the sperm plasma membrane, both lipid and protein, changes dramatically during maturation and capacitation. The sperm plasma membrane appears to be modified from without as it comes in contact with luminal fluids. *This raises the issue whether sperm membrane proteins are atypical in their mode of membrane insertion which in turn could result in atypical diffusion properties.*

2. Evolving a Model for Membrane Organization and Dynamics

To a large extent the fluid mosaic model of membranes evolved from early studies of patching and capping (Taylor *et al.*, 1971) and of heterokaryon fusion (Frye and Edidin, 1970). In its simplest form, this model envisages the lipid bilayer as being a quasi-two-dimensional fluid solvent in which the membrane proteins are situated, in some cases spanning the bilayer. Because of the fluid nature of the solvent lipid, both the membrane lipids and proteins were thought to be free to diffuse in the plane of the bilayer. Significantly, the popular interpretation of this model does not take into account a variety of "extrinsic" and "intrinsic" forces such as membrane macromolecules interacting with the ectodomain and cytoplasm, molecular interactions within the bilayer, and enthalpically driven processes. Any of these perturbations could easily lead to a highly structured cell surface comprised of microdomains within a cell.

The development of the technique of fluorescence recovery after photobleaching (FRAP) in the early 1970s provided a general and widely applicable approach to testing and refining this model (Peters *et al.*, 1974; Axelrod *et al.*, 1976a; Edidin *et al.*, 1976; Jacobson *et al.*, 1976). In FRAP the membrane molecular species, whose diffusion is to be measured, is tagged *in a non-cross-linking manner* with a fluorophore. Light from an attenuated laser beam is brought to a focus on the membrane so as to illuminate a spot of typically 1-μm radius. The fluorescence signal from this spot is monitored and is found to be essentially constant. The laser is momentarily increased approximately 10,000-fold which irreversibly photobleaches some of the fluorescence within the spot. When the light is returned to the monitoring level, the fluorescence intensity is found to be reduced. Recovery of fluorescence as bleached molecules diffuse out of the spot and unbleached molecules in, is monitored. FRAP provides two measures of molecular diffusibility; the fraction or percent diffusing reflected in the fraction of recovery observed (% R), and the diffusion coefficient (D) reflected in the rate or half time for recovery.

FRAP has to date been applied to a wide variety of molecular species and biological systems. For reviews on this subject the reader is referred to Edidin (1981), Peters (1981), and Wolf and Edidin (1981).

2.1. Lipid Diffusion in Biological Membranes

Early FRAP measurements indicated that in mammalian cell plasma membranes, lipids are completely free to diffuse and exhibit diffusion rates of about 10^{-8} cm^2/sec (Schlessinger *et al.*, 1977a). That is, they can travel a distance of 1 μm in a few tenths of a second. These results supported the concept that the plasma membrane is an isotropic fluid bilayer. More recent and complete data suggest that this is most certainly an oversimplification.

A nondiffusing lipid fraction has been observed on a number of cell types. This, of course, means that not all of the membrane exists in a fluid state. It is well known from biophysical studies on binary lipid mixtures that lipids do not intermix to form homogeneous bilayers, but rather segregate into domains of nonhomogeneous composition and physical state (Shimshick and McConnell, 1973; Mabrey et al., 1978; Stewart et al., 1979). Thus, for instance, one can have coexistent fluid and gel (i.e., solid) phases within the same bilayer. Biological membranes, of course, contain a very large number of distinct lipid species. One would thus expect, on the basis of Gibbs's Phase Rule (e.g., see Moore, 1972), that biological membranes would potentially contain a large number of different types of coexistent lipid domains. Perhaps as a result of this diversity, it has until recently proven difficult to demonstrate the existence of domains in biological membranes. Techniques such as differential scanning calorimetry (DSC) and X-ray diffraction, which were definitive in demonstrating domains in binary lipid mixtures, have generally proven too insensitive to detect potential domains in most homeothermic biological membranes.

Recently, spectroscopic techniques have been developed which exploit the preferential solubility of amphiphilic spectroscopic probes into domains of specific composition (Sklar et al., 1979; Klausner et al., 1980; Klausner and Wolf, 1980; Pringle and Miller, 1979; Williamson et al., 1981, 1982; Wolf et al., 1981a,b; Ethier et al., 1983; Pjura et al., 1984; Packard and Wolf, 1985; Weaver, 1985). As a result, the probe reports on the properties of those specific regions of the membrane rather than on the average or bulk properties of the membrane as a whole. Another approach has been to use a probe which goes into all domains, but which has a particular property (e.g., fluorescence-excited state lifetime) which is distinct in different domains. Use of this latter approach is limited by one's ability to measure the heterogeneity of that spectroscopic property (Klausner et al., 1980; Pjura et al., 1984). Use of these techniques has been extensively reviewed elsewhere (Karnovsky et al., 1982; Wolf, 1988). In general, these experiments point to a bilayer of complex domain organization. *Physiological transformations such as fertilization (Wolf et al., 1981a,b; Wolf, 1983) appear to alter not the bulk membrane properties (e.g., fluidity or viscosity), but rather the ensemble of membrane domains.*

The domain model of membranes may at first seem alien. However, one should consider the fact that if the sole purpose of the lipid bilayer were to provide a bulk homogeneous fluid environment, then there would be no physical need for the diversity of lipids in biological membranes. *Much as the amino acid sequence of proteins ultimately determines their three-dimensional structure, so too may the precise lipid composition determine the ensemble of domains in a membrane.*

One may also conjecture about the potential significance of membrane domains to cell function. Small compositional changes which would have little effect on bulk fluidity could, if localized to a specific domain, have quite profound effects. The domains could potentially act as a lock and key mechanism to activate specific membrane processes. Suppose that two molecules must interact in order for some process to occur, but that interaction is excluded because these molecules are in separate domains. A change in domain organization that resulted in colocalization could thus activate this process. Membrane–membrane fusion events may be localized to domains of specific lipid composition

(Portis *et al.*, 1979). If this is true, then fusion could only occur between specialized domains.

A common feature in many fusion events is an influx of Ca^{2+} ions. Ca^{2+} is known to alter membrane domain organization and to aid in fusion in model systems (Portis *et al.*, 1979; Stewart *et al.*, 1979). A potential mechanistic sequence for fusion is shown in Fig. 2. (1) A lateral reorganization occurs creating fusogenic domains. These would be unstable structures capable of shifting into nonbilayer hexagonal structures. (2) The

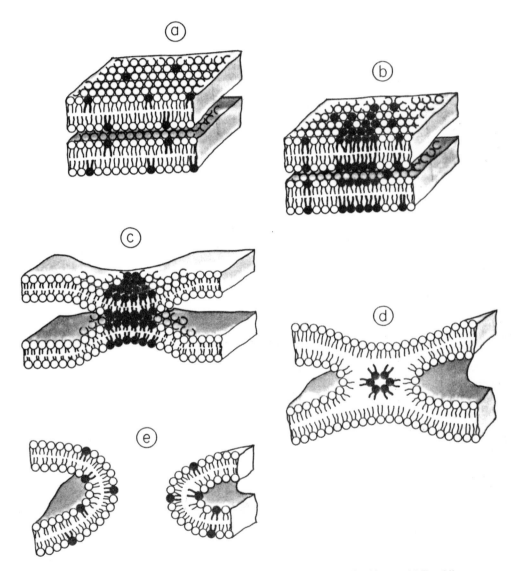

Figure 2. Proposed sequence of membrane-membrane fusion events, as outlined in text. (a) Two bilayers come in close contact. (b) Fusogenic domains within each bilayer and (c) between the bilayers, form. (d) The fusogenic lipids form a hexagonal phase, which connects the monolayers, and (e) the two membranes separate.

membranes move closer together possibly as a result of osmotic stress (Wakelam, 1988; Brocklehurst and Pollard, 1988) to form focal points for fusion. (3) The lipids go from a bilayer to a hexagonal phase, which connects their monolayers. (4) The membranes separate. Ca^{2+} could play a modulating role most likely in either (1) or (3). Stages of this proposed sequence have indeed been observed in studies of model membrane fusion. The focal point stage of fusion is generally transitory. However, the sperm of *Limulus polyphemus* appear to be frozen prior to the acrosome reaction in this dimpled, focal point, state (Tilney, 1985).

2.2. Protein Diffusion in Biological Membranes

Within the context of the fluid mosaic model, the diffusion coefficients of membrane proteins should be determined by the bulk membrane viscosity. If the viscosity is known, the diffusion coefficients of proteins should then be predicted from fluid dynamic theory. Such calculations have been made by Saffmann and Delbruck (1975) modeling the bilayer as a quasi-two-dimensional fluid. The system is quasi-two dimensional because the bilayer, while thin, does have a finite thickness which results in a boundary condition at the two interfaces. In two-dimensional, or near-two-dimensional, systems the diffusion coefficient has a weak logarithmic dependence upon size (molecular weight). As a result, these calculations predict that proteins will diffuse in membranes with nearly the same diffusion coefficients as are observed for lipids.

Early FRAP experiments have indicated that this is clearly not the case. Membrane proteins typically show significant nondiffusing fractions, and those molecules that do diffuse do so typically two or three orders of magnitude slower than lipids (Peters *et al.*, 1974; Axelrod *et al.*, 1976a,b; Edidin *et al.*, 1976; Jacobson *et al.*, 1976).

A protein which spans the bilayer has of course three transverse domains: the ectodomain often containing the amino terminus and the glycosylated regions, the hydrophobic membrane-spanning domain, and the cytoplasmic domain which often contains the carboxyl terminus. Potentially, interactions in any or all of these domains would result in deviations from diffusion properties predicted by the fluid mosaic model.

Studies of patching and capping showed that generally, capping was inhibited by microfilament-disrupting drugs and, in some cases, promoted by microtubule-disrupting agents (Taylor *et al.*, 1971). These results suggested that interactions of the cytoplasmic domain with the cytoskeleton might control protein diffusion. Early FRAP studies of the effect of cytoskeleton-disrupting agents on membrane protein diffusion in general did not support this hypothesis (Schlessinger *et al.*, 1977a,b). Experiments by Webb and colleagues (Wu *et al.*, 1982; Tank *et al.*, 1982) demonstrated that when cells are blebbed, putatively separating the membrane from the underlying cytoskeleton, membrane protein diffusion goes to the fluid dynamic limit. That is, complete recovery with lipidlike diffusion rates is observed. These experimenters point out, however, that there are alternate explanations to these results. For instance, linkages or other interactions with an ectoskeleton could also be severed.

Wolf *et al.* (1977, 1980) have shown that dextran molecules which have been stearoylated so that they insert into membranes behave much like a membrane protein. They exhibit proteinlike diffusion rates and diffusing fractions and can be induced to patch and cap. This capping is sensitive to cytoskeleton-disrupting agents. However, the stearoyl chains are too short to span the bilayer. These results suggest that it is the ectodomain which controls dextran diffusion. Sensitivity to pharmacological agents may result from interaction, through either the ecto- or hydrophobic domain, with other proteins that are specialized for interaction with the cytoskeleton.

Recently, this confusing story has been cleared up using genetically engineered membrane proteins. Edidin and Zuniga (1984) have shown for H2 and Livneh *et al.* (1986) for the EGF receptor that deletion of the cytoplasmic domain does not alter membrane protein diffusion. In addition, Edidin and Wier (1987) and Wier and Edidin (1988) have shown that, upon removal of the glycosylated region, H2 diffusion approaches the fluid dynamic limit.

It has recently been shown that some membrane proteins do not have membrane-spanning peptide sequences, but rather are linked to a phosphatidylinositol (Low and Finean, 1977; Low and Kincaid, 1985; Thomas *et al.*, 1987; Low and Saltieh, 1988). In a sense, these are hybrid molecules which share the properties of both lipid and protein. Such linkages may have particular significance to posttesticular modification of the mammalian sperm surface (Phelps *et al.*, 1988). Only a few of these molecules have been monitored by FRAP (Ishihara *et al.*, 1987; Noda *et al.*, 1987; Phelps *et al.*, 1988) so that no generalities can yet be drawn. It is clear, however, that under certain circumstances, these molecules can indeed diffuse at rates similar to those of lipids.

3. Diffusion on Mammalian Spermatozoa

3.1. Constraints on Sperm Geometry

Before considering the actual results of FRAP experiments on mammalian spermatozoa, it is useful to consider the general results of FRAP measurements and what constraints specific aspects of sperm morphology, size, and structure can *a priori* be expected to play.

Figure 3 shows schematically the broad range of sizes and shapes of spermatozoa from different species. In Table I we estimate the time it would take a typical lipid or protein molecule to diffuse the length of the sperm tail and the head. This estimate is determined from

$$\text{Time} = \text{distance}^2/4D \qquad (1)$$

where D is the diffusion coefficient. Because of the distance^2 dependence of the diffusion correlation time, these times can become quite large especially for molecules with relatively small diffusion coefficients (e.g., membrane proteins). As a result, if diffusion is the only mechanism at work and if there is a change in the constraints to diffusion between different regions (e.g., a barrier to interregional diffusion becomes eliminated), redistribution of some membrane components could take a considerable length of time. As a possible example of such a situation, consider a molecule which is present only in the sperm plasma membrane. Immediately following the acrosome reaction, an exocytotic event which involves fusion between the outer acrosomal membrane and the sperm plasma membrane, one would expect this molecule to be absent from the anterior region of the sperm head. If there are no restrictions to interregional diffusion at the equatorial zone, one would expect with time that these molecules would diffuse into the anterior region of the head. We will discuss below actual examples of such redistributions. If the

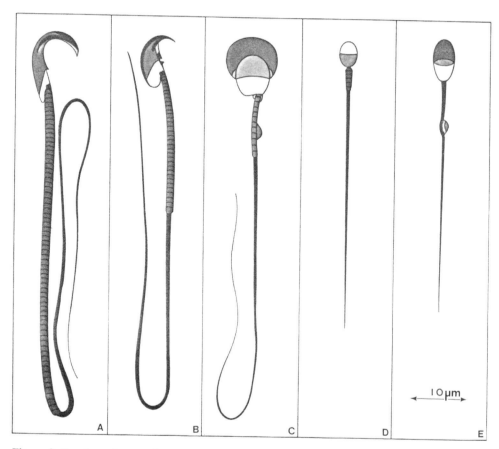

Figure 3. Drawings of mammalian sperm to indicate the range of sperm dimensions. (A) Rat (186 μm), (B) mouse (127 μm), (C) guinea pig (100 μm), (D) human (58 μm), (E) bull (54 μm). Redrawn from Retzius (1906) and taken from Cardullo and Baltz (1988) with permission.

diffusion coefficients can be ascertained from a FRAP experiment, then an estimate of the redistribution time can be obtained from Eq. (1) and Table I.

It is also useful to consider the relative surface areas of the sperm heads and tails. Mammalian spermatozoa show a high degree of diversity in appearance and dimensionality [in general, sperm from different species vary in length from 40 μm to 360 μm (Cummins and Woodall, 1985)]. Some sperm heads are ovoid while others (e.g., rodent sperm) are crescent shaped. Figure 4 and Table II show the distribution of plasma membrane over the three major morphological regions (the head, the midpiece, and the principal piece). The major variation in plasma membrane distribution is over the midpiece. In smaller sperm (<100 μm), only about 10% of the total plasma membrane is localized to the midpiece whereas in larger sperm (> 125 μm) containing crescent-shaped heads, greater than 40% of the plasma membrane is on the midpiece. This increase in midpiece surface area with increasing sperm length results in a smaller fraction overlying

Table I. Time to Diffuse Length of Head and Tail for Sperm from Different Species[a]

Species	Length		Lipid diffusion time[b]		Protein diffusion time[c]	
	Head	Tail	Head	Tail	Head	Tail
Rabbit	9 μm	50 μm	18 sec to 3 min	10 min to 1½ hr	½ hr to 5 hr	15 hr to 1 week
Human	6 μm	52 μm	9 sec to 1½ min	11 min to 2 hr	15 min to 2½ hr	1 day to 1 week
Mouse	8 μm	115 μm	16 sec to 2⅔ min	1 hr to 10 hr	30 min to 5 hr	4 days to 5 weeks
Guinea pig	11 μm	103 μm	30 sec to 5 min	45 min to 7½ hr	1 hr to 8 hr	3 days to 1 month
Syrian hamster	15 μm	177 μm	1 min to 10 min	2 hr to 1 day	1½ hr to 15 hr	1 week to 3 months

[a]Morphometric values are from Cummins and Woodall (1985).
[b]Calculated for $D = 10^{-8}$ cm²/sec to 10^{-9} cm²/sec. Values are rounded off to convenient time units.
[c]Calculated for $D = 10^{-10}$ cm²/sec to 10^{-11} cm²/sec. Values are rounded off to convenient time units.

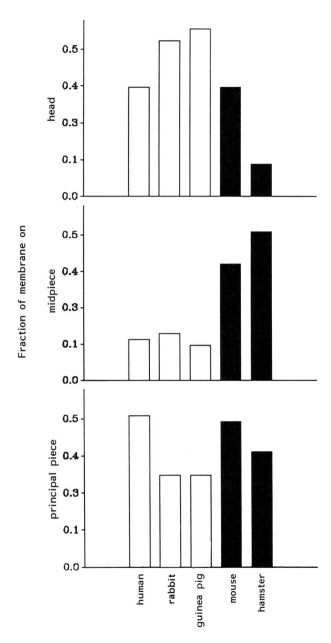

Figure 4. Fraction of surface area in sperm head, midpiece, and principal piece for human, rabbit, guinea pig, mouse, and hamster sperm. In general, as the sperm length increases, the fraction of membrane over the midpiece increases and the fraction over the head decreases. The fraction over the principal piece is more or less constant. Ovoid heads (□); crescent heads (■).

the head ($< 5\%$ in hamster sperm compared with $> 50\%$ in guinea pig sperm). Therefore, on the basis of conservation issues alone there would be a considerable advantage for a larger sperm to isolate a particular molecule on its head whereas this would not be the case for a much smaller sperm. In addition, the larger distances that must be traversed in larger sperm make redistribution of molecules in these sperm prohibitive.

Table II. Estimate of Surface Areas (μm^2) for Head, Midpiece, and Tail of Mammalian Sperm[a]

Species	l_h	w_h	l_{mp}	w_{mp}	l_{pp}	w_{pp}[b]	Head shape[c]	Head area[d]	A_{mp}[e]	A_{pp}[f]	A_t[g]
Rabbit	9	5	9	0.7	41	0.3	O	21π	6.4π	12π	18.8π
Human	6	3.5	4	1.0	48	0.3	O	11π	4π	14π	18π
Mouse	8	3.2	18	1.3	97	0.3	C	7π	24π	29π	53π
Guinea pig	11	9.5	11	1.0	92	0.3	O	52π	11π	27π	38π
Syrian hamster	15	2.5	51	1.0	126	0.3	C	10π	51π	38π	89π

[a]Morphometric values are taken from Cummins and Woodall (1985); length and width units are μm.
[b]Since no values for principal piece are given, we assume a value of 0.3 which we have found for ram sperm.
[c]Head shapes are taken as either ovoid (O) or crescent (C).
[d]For ovoid heads we assume area $= \pi w_h l_h / 2$. For crescent heads we assume area $= (\pi w_h 4)(w_h^2 + l_h^2)^{1/2}$.
[e]For midpiece we assume area $= \pi w_{mp} l_{mp}$.
[f]For principal piece we assume area $= \pi w_{pp} l_{pp}$.
[g]Area of tail = area of midpiece + area of principal piece.

3.2. Lipid Diffusion on Mammalian Spermatozoa

Most FRAP measurements of lipid diffusion employ fluorescent lipid analogues which can be inserted from the aqueous phase either by ethanolic injection or by transfer from vesicles or carrier proteins such as albumin (Wolf and Edidin, 1981). The most useful examples of such analogues are carbocyanine dyes (Sims *et al.*, 1974; Wolf, 1988) and NBD-labeled phospholipids (Struck and Pagano, 1980; Nichols and Pagano, 1981; Pagano and Sleight, 1985). An interesting alternative approach has been developed by Dr. R. Pagano and colleagues, who expose cells to NBD-labeled phosphatidic acid and then allow the cell to synthesize fluorescent phospholipids from this precursor (Pagano and Sleight, 1985). In our hands, the carbocyanine dyes have been most useful for the study of mammalian spermatozoa, particularly 1,1'-dihexadecyl-3,3,3',3'-tetramethylindocarbocyanine (C_{16}diI). Mammalian sperm are best handled in medium containing some albumin, which tends to remove NBD-labeled phospholipids. In addition, our preliminary results indicate that sperm do not convert NBD phosphatidic acid beyond diacylglycerol. This is consistent with the sperm having surface phosphatases but only limited ability to synthesize phospholipids.

Figure 5 shows space-filling models of C_{18}diI and phosphatidylcholine. Table III gives the diffusion coefficients and diffusing fractions for a variety of fluorescent lipid analogues in the anterior head region of ejaculated ram spermatozoa (Wolf and Voglmayr, 1984). In Fig. 6 we compare the diffusion of C_{16}diI on ejaculated ram and human spermatozoa and on cauda epididymal mouse spermatozoa (Wolf and Voglmayr, 1984; Wolf *et al.*, 1986a,b,c; Wolf and Tanphaichitr, unpublished results). In Fig. 6 we also consider for ram and mouse spermatozoa the diffusion on the different morphological regions. Three conclusions can be drawn from these data:

1. The rates of diffusion of lipids on mammalian sperm are similar to those observed on other mammalian cell types.

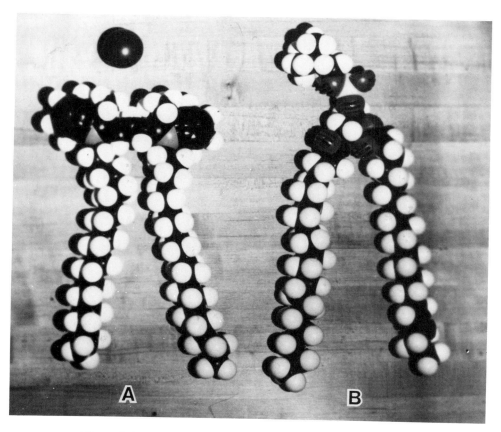

Figure 5. Space-filling models of (A) $C_{18}diI$ and (B) phosphatidylcholine.

Table III. Diffusion of Fluorescent Lipid Analogues on the Anterior Region of Ejaculated Ram Sperm Heads[a,b]

Analogue	D ($\times$ 10^9 sec/cm^2)	% R
$C_{10}diI$	9.3 $\pm$ 1.3 (18)	65 $\pm$ 2 (18)
$C_{12}diI$	4.65 $\pm$ 0.40 (20)	65 $\pm$ 2 (20)
$C_{14}diI$	3.56 $\pm$ 0.28 (40)	63 $\pm$ 2 (40)
$C_{16}diI$	4.73 $\pm$ 0.71 (134)	62 $\pm$ 1 (134)
$C_{18}diI$	4.68 $\pm$ 0.49 (10)	75 $\pm$ 3 (10)
NBD-PE	6.0 $\pm$ 1.1 (15)	58 $\pm$ 2 (15)
NBD-PC(6)	3.28 $\pm$ 0.32 (10)	64 $\pm$ 2 (10)

[a]From Wolf *et al.* (1988).
[b]Values are mean $\pm$ S.E.M. The number of measurements is given in parentheses.

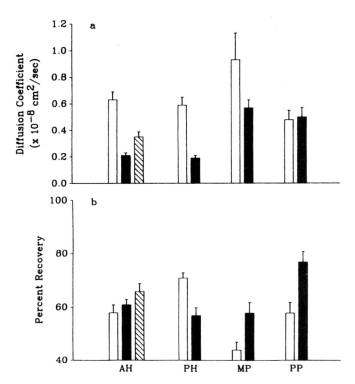

Figure 6. Diffusibility of C_{16}diI on the different morphological regions of ram (□) (Wolf and Voglmayr, 1984), mouse (■) (Wolf *et al.*, 1986b), and human (▨) sperm (Wolf and Tanphaichitr, unpublished results). (a) Diffusion coefficient; (b) percent recovery. AH, anterior head; PH, posterior head; MP, midpiece; PP, principal piece.

2. Unlike the lipids of somatic cells, a significant fraction of mammalian sperm lipid is not diffusing.
3. There are significant differences in lipid diffusibility, both diffusion rate and fraction, between the morphologically distinct regions of the sperm surface.

In this context it is also interesting to consider the labeling patterns of mouse and ram sperm with C_{16}diI (Fig. 7). We see that in both of these sperm types the lipid probe distribution is regionalized. This regionalization of both distribution and diffusibility reflects the properties of the outer leaflet of the sperm plasma membrane, since we have shown using membrane-impermeable quenching agents that the probe is confined to this leaflet (Wolf, 1985).

3.3. Changes in Sperm Plasma Membrane Lipid Diffusibility during Spermatogenesis, Maturation, and Capacitation

In Fig. 8 we compare C_{16}diI diffusion on the different regions of mouse spermatozoa as a function of developmental stage (Wolf *et al.*, 1986a,b,c). That is, we consider the effects of spermatogenesis, epididymal maturation, hyperactivation, and capacitation

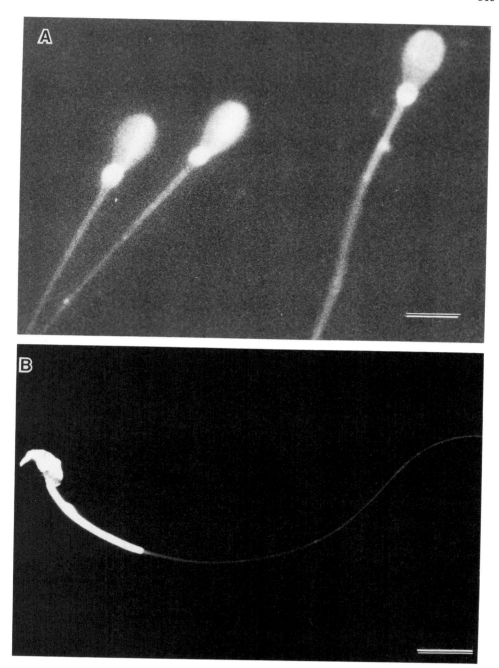

Figure 7. Fluorescent labeling of (A) testicular ram spermatozoa and (B) caudal epididymal mouse spermatozoa with C_{16}diI. Bars = 10 μm.

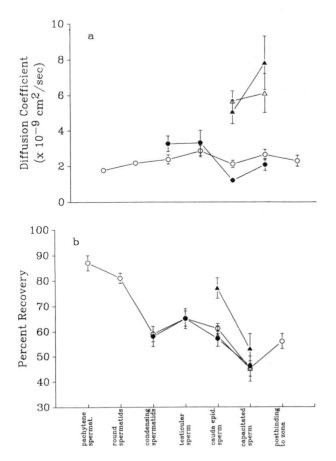

Figure 8. Diffusion of $C_{16}diI$ on the different regions of mouse sperm and spermatids as a function of developmental stage. (a) Diffusion coefficient; (b) percent recovery. Data from Wolf *et al.* (1986b,c). AH (O–O), anterior head; PH (●–●), posterior head; MP (△–△), midpiece; PP (▲–▲), principal piece.

upon diffusion coefficient, diffusing fraction, and the regionalization of diffusibility. From these data, several conclusions may be drawn:

1. In tetraploid pachytene spermatocytes, the final premeiotic phase in spermatogenesis, the diffusing lipid fraction is similar to that observed on most somatic cells. In haploid round spermatids, the first postmeiotic phase, the diffusing fraction decreases somewhat. However, with the development of sperm shape during spermiogenesis, large nondiffusing fractions observed on mature spermatozoa develop. Indeed, as soon as the anterior and posterior regions of the head become distinguishable, at stage 10 or 11, both the diffusing fractions and the diffusing rates of these fractions are statistically identical with those observed on testicular spermatozoa. In Table IV, we consider the diffusibility on the round structures associated with spermatogenesis (residual bodies and cytoplasmic droplets). These data demonstrate that the development of nondiffusing lipid concurrent with changing cell shape is not an artifact of the cell no longer being round. Rather, this change in diffusibility reflects a stage-specific phenomenon.

2. Diffusibility, both diffusion coefficient and diffusing fraction, can change with sperm development: spermiogenesis, epididymal maturation, hyperactivation, and capaci-

Table IV. C_{16}diI Diffusion in the Plasma Membrane of Round Anucleate Structures of Mouse Spermatogenic Cells[a,b]

Structure	D ($\times$ 10^9 sec/cm²)	Percent R
Residual bodies	2.41 ± 0.30	54 ± 4
	(23)	(23)
Cytoplasmic droplets of condensing spermatids	5.7 ± 1.7	64 ± 5
	(13)	(13)
Cytoplasmic droplets of testicular spermatozoa	2.3 ± 0.8	70 ± 17
	(3)	(3)

[a]Data from Wolf *et al.* (1986c).
[b]Values are mean ± S.E.M. The number of measurements is given in parentheses.

tation. Just as these properties are regionalized, so too can the effects of these transformations be regionalized with some regions changing more than others. In the mouse, for instance, if we compare diffusion in the mid- and principal pieces of the tail on hyperactivated-capacitated sperm to that on control sperm (maintained in noncapacitating media for equivalent periods of time), we see that C_{16}diI diffusion becomes considerably increased. While it may be tempting to suggest that this represents an increase in membrane "fluidity" necessary for the greater curvatures of the hyperactivated sperm waveform, it is much more likely that this result is indicative of subtler changes in the ensemble of membrane events. Since hyperactivation in demembranated spermatozoa can be induced by altering ionic and AMP concentrations (Witman, this volume), our results suggest that one probable role of the sperm tail plasma membrane in hyperactivation is to regulate the ionic and AMP concentrations of the tail cytoplasm. In this context, one might hypothesize that changes in tail membrane domains alter the activity of either membrane channels or cyclases. As is shown in Fig. 9, dramatic changes in C_{16}diI diffusion are also

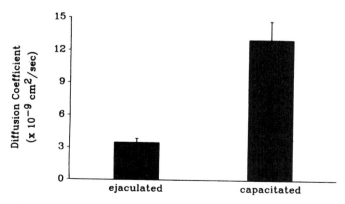

Figure 9. Effect of capacitation on C_{16}diI diffusion coefficient in the anterior region of the human sperm head. Error bars represent the standard deviation of at least four determinations. Unpublished data of Wolf and Tanphaichitr.

observed in the human sperm anterior region of the head plasma membrane with capacitation (Wolf and Tanphaichitr, unpublished results). Again, we hypothesize that such changes are indicative of alterations in membrane domain organization preparatory to fusion events leading to fertilization.

3.4. Causes of Nondiffusing Lipid and the Question of Lipid Domains

These results leave us with at least two pressing questions: (1) What is the cause of the nondiffusing lipid fraction in the plasma membrane of mammalian spermatozoa? (2) Can one definitively demonstrate the existence of lipid domains in these membranes? We believe in fact that these two questions are interconnected.

Several possible explanations of the nondiffusing lipids come to mind:

1. They might be an artifact of membrane oxidation.
2. They might be due to protein–lipid interactions.
3. They might be due to interactions of the membrane with the cytoskeleton.
4. They might be due to lipid domains.

The results of experiments designed to address this question are summarized in Table V. Most significantly, when one prepares plasma membrane fractions from the anterior region of the ram sperm head, extracts lipids from these fractions, and measures diffusion of C_{16}diI in membranes reconstituted from these extracts, nondiffusing lipid fractions are still observed (Wolf et al., 1988). Figure 10 shows recovery curves observed in model membranes from pure lipids. In homogeneous fluid membranes prepared from dilaurylphosphatidylcholine, we observe complete rapid recovery. In homogeneous gel-phase membranes, prepared from dipalmitoylphosphatidylcholine, we observe no recovery. In membranes made from a one-to-one mixture of these two lipids, known to have coexistent fluid and gel domains, only 50% recovery is observed. Our data on reconstituted sperm membrane lipids demonstrate that nondiffusing lipid can result, in these membranes, from purely lipid factors or interactions. Quite probably, these interactions are the same as those which result in lipid domains in model systems.

Recently, we have been able to demonstrate the existence of lipid domains in both

Table V. Experiments Investigating the Cause of Nondiffusing Lipid Fraction in Ejaculated Ram Spermatozoa[a]

Control condition[b]	Experimental condition	D		Percent R	
		Control	Experimental	Control	Experimental
Oxygen-free AH	Oxygen-enriched AH	6.1 ± 0.5 (42)	4.9 ± 0.5 (42)	72 ± 1 (42)	78 ± 1 (42)
Untreated AH	Pronase-treated AH	12.3 ± 0.7 (39)	11.4 ± 0.1 (22)	73 ± 1 (39)	75 ± 2 (22)
Untreated T	Blebbed T	5.2 ± 1.4 (31)	5.0 ± 0.6 (31)	60 ± 3 (31)	64 ± 2 (31)
AH	Bilayers from lipid extracts of AH	6.3 ± 0.6 (24)	9.8 ± 0.6 (70)	58 ± 3 (24)	74 ± 2 (70)

[a]Data from Wolf et al. (1988).
[b]AH, anterior region of the head; T, tail.

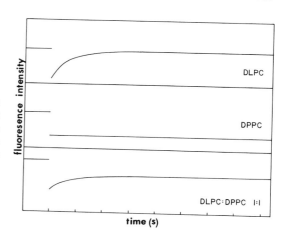

Figure 10. Recovery curves showing diffusion of C_{16}diI in model membranes. Complete rapid recovery is observed in homogeneous dilaurylphosphatidylcholine (DLPC) fluid-phase membranes. No recovery is observed in homogeneous dipalmitoylphosphatidylcholine (DPPC) gel-phase membranes. Approximately 50% recovery is observed in mixed-phase DLPC : DPPC (1 : 1) membranes. From Wolf (1988) with permission.

the intact and reconstituted plasma membranes of the anterior region on the ram sperm head using differential scanning calorimetry (Wolf *et al.*, 1989b). Unlike the case for mammalian somatic cell plasma membranes where calorimetry gives little or no evidence of lipid domains, dramatic phase transitions are observed in these membranes. The energies of these transitions are consistent with those of fluid-to-gel transitions observed in pure lipid membranes. Furthermore, the membrane clearly exists in a state of mixed phases at physiological temperatures. It also appears to be the case that the domain organization is different at testicular, epididymal, and postejaculatory (female reproductive tract) temperatures. It has been previously demonstrated that the decreased temperatures during spermatogenesis and epididymal maturation maintain sperm viability by lowering the rates of intermediary metabolism (Cardullo, 1985; Djakiew and Cardullo, 1986) and lipid peroxidation (Alvarez and Storey, 1986) during maturation and storage. Hence, the decreasing temperature during epididymal maturation may act as a driving force for reorganizing the sperm membrane as well as optimizing conditions for sperm storage. These possibilities are of course not mutually exclusive. Upon ejaculation and entry into the higher temperatures of the female reproductive tract, membrane organization can again be altered concomitant with an increase in metabolism necessary for motility and fertilization. *That is, temperature may play an activating role in mammalian sperm cells.* As we have indicated, immobile lipid fractions have often been detected in poikilothermic cells (Wolf *et al.*, 1981a; Weaver, 1985; Treistman *et al.*, 1987). The sperm cell may in a sense be said to be a *poikilotherm among homeotherms.*

3.5. Protein Diffusion on Mammalian Sperm

We turn next to consider protein diffusion on mammalian spermatozoa. Our ultimate goal is to understand the mechanism whereby mammalian sperm organize their surfaces, and how this organization and changes in this organization during maturation, capacitation, and the acrosome reaction result in fertilization. Since to date only a few membrane proteins have been sufficiently characterized in either their diffusibility or physiological role, it is too early to draw generalities. We shall begin therefore by considering separately some of the results which have been obtained.

3.5.1. The Immobile Mouse Sperm Head: M42, P220, and Less Specific Probes of Membrane Protein Diffusion. We have considered the diffusibility of two mouse sperm head antigens, M42 and P220, which have been characterized by the laboratory of Dr. P. Saling. The results of preliminary experiments with these antigens are given in Table VI. While characterization of P220 has just begun, considerable characterization of M42 has been done. This antigen is a membrane protein which is apparently posttranslationally modified during spermatogenesis and epididymal maturation. Prior to spermatogenesis, the antigen displays a molecular mass greater than 300 kDa and in epididymal spermatozoa the protein is present as a doublet of 240/270 kDa in the caput, 235/260 kDa in the corpus, and 220/240 kDa in the cauda (Lakoski *et al.*, 1988). Given that biosynthetic activity of proteins is stopped soon after meiosis, it is possible that these posttranslational modifications are regulated within the milieu of the epididymis itself. M42 can be detected homogeneously over the surface of pre-acrosomal-phase round spermatids, and becomes localized to the plasma membrane over the acrosome during the acrosomal phase (unpublished observations). On caudal epididymal sperm, it is localized to a sharp crescent over the acrosome (Fig. 1). Antibodies against this antigen block the zona and ZP3-induced acrosome reaction, but not the ionophore-induced acrosome reaction (Leyton *et al.*, 1989). In addition, antibodies against M42 antigen block fertilization both *in vitro* and *in vivo* (Saling *et al.*, 1986).

On caudal epididymal sperm, P220 is generally localized to the post acrosomal region of the sperm head plasma membrane. A subpopulation of sperm also show P220 in the anterior region of the head.

As seen in Table VI, neither M42 nor P220 antibodies diffuse in the caudal epididymal mouse sperm head (Wolf *et al.*, 1989a). For comparison purposes, we have also considered the diffusion of less specific probes of protein diffusion, fluorescent succinylated Con A and fluorescent succinylated wheat germ agglutinin. For these probes, only very slight recovery is observed. It may in fact be the case that what little diffusion is observed for these fluorescent lectins is a result of glycolipid rather than glycoprotein diffusion.

While these results are too limited to be conclusive in the matter, they suggest the possibility that little protein diffusion is present in the plasma membrane of the mouse sperm head.

3.5.2. A Maturation-Dependent Ram Sperm Protein: ESA 152. ESA 152 is a ram sperm antigen present on ejaculated but not testicular ram spermatozoa (Wolf *et al.*, 1986b). It is a glycosylated integral membrane protein of 18 kDa (Weaver *et al.*, 1989). ESA 152 is distributed to all regions of the ram sperm plasma membrane (Fig. 1). This distribution is somewhat regionalized, however, in that there is less ESA 152 over the anterior region of the ram sperm head than over the posterior region of the head. This "intrinsic" distribution is what is observed when one uses either fixed sperm or unfixed sperm directly labeled with a fluorescent Fab fragment against ESA 152. If an intact antibody followed by a fluorescent Fab fragment against the first antibody is used, then one observes that this antigen is now heavily present on the anterior region of the head and largely excluded from the posterior region (see Fig. 1). Preliminary results suggest that this redistribution is not a capping of antigen but a redistribution following the acrosome reaction. It thus appears that ESA 152 antibodies may trigger the acrosome reaction in ram

Table VI. Plasma Membrane Protein Diffusion on Mammalian Sperm

Protein probe	Sperm	AH		PH		M		T	
		D^a	Percent R	D	Percent R	D	Percent R	D	Percent R
Rhodamine succinylated Con A[b]	Mouse testicular	6.7 ± 1.4	10 ± 1	25 ± 10	8 ± 1	—	—	—	—
WGA[b]	Mouse testicular	4.6 ± 1.3	10 ± 1	12 ± 4	9 ± 1	—	—	—	—
M42	Mouse caudal	—	0	—	0	—	—	—	—
P220	Mouse caudal	—	0	—	0	—	—	—	—
	Mouse caudal capacitated	—		—	—	—	—	—	—
WGA[b]	Mouse caudal	6.7 ± 2.0	5 ± 1	3.3 ± 0.7	8 ± 1	—	—	—	—
ESA 152 Fab[c]	Ram ejaculated	1.3 ± 0.2	50 ± 3	1.0 ± 0.2	51 ± 3	7.0 ± 1.6	28 ± 3	2.6 ± 0.2	54 ± 3
ESA 152 intact[c]	Ram ejaculated acrosome reacted?	4.1 ± 0.6	61 ± 2	—	—	2.0 ± 0.2	50 ± 3	2.7 ± 0.3	66 ± 3
PH-20[d]	Guinea pig testicular	—	—	0.019 ± 0.003	72 ± 7	—	—	—	—
PH-20[e]	Guinea pig caudal	—	—	0.18 ± 0.05	73 ± 3	—	—	—	—
	Guinea pig caudal acrosome reacted	—	—	0.18 ± 0.05	73 ± 3	—	—	—	—
PT-1[f]	Guinea pig	45 ± 2	78 ± 9	—	—	—	—	2.5 > 90%	

[a]Diffusion coefficients given as mean ± S.E.M. × 10⁹ sec/cm². [d]Phelps and Myles (1987); [c]Cowan et al. (1986); [f]Myles et al. (1984).
[b]Wolf et al. (1989a); [c]Wolf et al. (1986b);

sperm. The identical anterior head distribution is observed on sperm whose acrosome reactions are induced by ionophore. The results of FRAP measurements of ESA 152 diffusibility are summarized in Table VI. In contrast to most diffusion coefficients on somatic cells, ESA 152 shows lipidlike diffusion coefficients. Significant nondiffusing fractions are also observed on all regions of the sperm surface.

3.5.3. A Freely Diffusing Regionalized Guinea Pig Sperm Antigen: PT-1.
PT-1 is a guinea pig plasma membrane antigen which is confined to the principal piece of the tail in caudal epididymal sperm (Primakoff and Myles, 1983) (Fig. 1). As shown in Table VI, it is nearly completely free to diffuse within this region (> 90% recovery) and in common with ESA 152 has a "lipidlike" diffusion coefficient (Myles *et al.*, 1984). Induction of the acrosome reaction with ionophore causes PT-1 to redistribute into the midpiece (see Fig. 1). This redistribution can, however, be prevented by immobilizing the sperm in agarose.

The results with PT-1 lead to two important conclusions:

1. That regionalization can occur without immobilizing the antigen. That is, a molecule can be completely confined to a region of the sperm surface, and yet be completely free to diffuse within that region. The results with PT-1 suggest that in this case, regionalization might be controlled by a barrier to interregional diffusion, which can be relaxed when the acrosome reaction is fired on motile sperm.
2. That motility can be required to cause redistributions

3.5.4. A Redistributing Guinea Pig Sperm Antigen: PH-20.
PH-20 is a guinea pig sperm plasma membrane antigen which on caudal epididymal sperm is localized to the posterior region of the sperm head (Myles and Primakoff, 1984). Antibodies against PH-20 block the interaction of acrosome-reacted but not acrosome-intact sperm with the zona pellucida (Myles *et al.*, 1987). In addition to being present in the plasma membrane, PH-20 is also found in the inner acrosomal membrane. Two forms of PH-20 have been detected; the major form has a molecular mass of 66 kDa and the minor form, 56 kDa. The distribution of these two forms between the acrosomal and plasma membranes has yet to be determined. The appearance of PH-20 during spermatogenesis has been studied by Phelps and Myles (1987) PH-20 is first detected in the Golgi apparatus of round spermatids at the start of spermiogenesis. It next appears in the developing acrosome. Subsequent to this, a second population of PH-20 appears to be inserted into the plasma membrane directly. Initially, PH-20 distribution in both the acrosomal and plasma membranes is homogeneous. Subsequently, PH-20 becomes localized to the inner acrosomal membrane, and as a result of epididymal maturation, the plasma membrane population becomes confined to the posterior head. Following the acrosome reaction, the plasma membrane PH-20 migrates into the anterior region of the head, and thus joins the population of PH-20 which was initially localized to that region (Primakoff *et al.*, 1985; Cowan *et al.*, 1986) (see Fig. 1). It is interesting that PH-20 goes from being sharply confined to the posterior region of the head to being sharply confined to the anterior region.

The results of FRAP measurements of PH-20 diffusibility are given in Table VI for testicular and caudal epididymal sperm (Myles *et al.*, 1984; Cowan *et al.*, 1986, 1987). On the posterior region of the head of caudal epididymal sperm, PH-20 shows diffusion coefficients and mobile fractions that are typical of proteins in somatic cell plasma membranes ($D = 1.8 \times 10^{-10}$ cm^2/sec, % $R = 73$). Upon migrating into the anterior region of the head (i.e., into what was previously the inner acrosomal membrane) PH-20 exhibits "lipidlike" diffusion properties ($D = 10^{-9}$ cm^2/sec, % $R = 78$). In contrast, diffusibility of PH-20 on testicular sperm is more severely restricted with diffusion coefficients on the order of 10^{-11} cm^2/sec. These results are particularly interesting in light of the recent demonstration (Phelps *et al.*, 1988) that PH-20 is a phosphatidylinositol-linked membrane protein. The obvious question is; how is the diffusibility of PH-20 regulated in response to physiological transformation and ultimately how is this response significant to the role played by PH-20 in fertilization?

4. Mechanisms of Membrane Regionalization

Even with the relatively modest group of sperm membrane macromolecules which have been considered to date, it is clear that sperm use a variety of mechanisms to regionalize and specialize their surfaces (Wolf, 1986).

4.1. Regionalization by Immobilization

Antigens such as M42 and P220 as well as other mouse sperm head components may well be localized by virtue of their inability to diffuse. It should be noted that the fact that diffusion does not take place over micrometer distances, as measured by FRAP, does not *a priori* exclude the possibility that diffusion might take place over shorter, submicrometer distances. It is possible, for instance, that these molecules could be corralled in a reticular network, free to diffuse within a corral but not across the corral fences. It is also, of course, possible that these macromolecules are genuinely immobilized by virtue of being tied to some cytoskeletal element or by being confined to a gel-phase lipid domain, which in turn is anchored to the cytoskeleton.

4.2. Regionalization by Diffusional Barriers

Fencing or corralling of macromolecules could also occur on a larger scale. That is, barriers to interregional diffusion could also exist. This appears, for instance, to be the case for PT-1 (Primakoff and Myles, 1983; Myles *et al.*, 1984) which, while completely free to diffuse in the principal piece of the tail, cannot cross over to the midpiece. A likely candidate for such barriers are intramembranous particle arrays which often appear at the interregional connection points in sperm (Friend, 1982). Often these barriers are not complete since channels across them appear to exist. This raises the possibility that the barrier could act much like a molecular sieve (e.g., a dialysis bag), allowing only macromolecules below a certain size to cross. Figure 11 shows an experiment designed to test whether lipids can cross such a barrier. Here a sperm was labeled with C_{16}diI and then all of the fluorescence from the midpiece was bleached away. Recovery from both the head and tail was observed. The rate of this recovery is consistent with that predicted from diffusion coefficients. However, it is not clear whether, as one might expect, the barriers offer any selective resistance to free interregional diffusion.

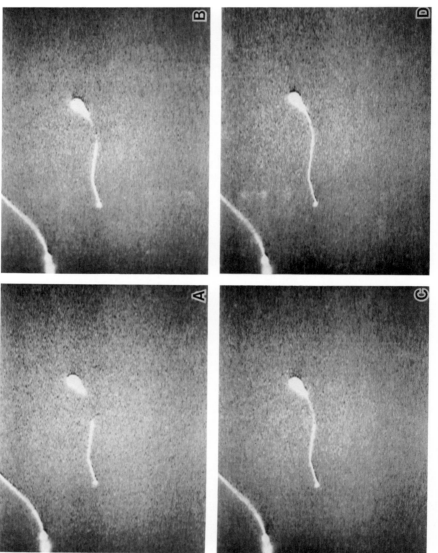

Figure 11. A photobleaching experiment showing interregional diffusion of C_{16}diI in the ram sperm plasma membrane. Fluorescence micrographs were taken with a SIT camera. The entire fluorescence in the midpiece is bleached away and recovery from the head and principal piece of the tail is observed. Time = (A) 0, (B) 10, (C) 20, (D) 30 sec.

4.3. Regionalization Due to Selective Solubility

It is also important to point out that a barrier need not be a physical structure in the same sense as a fence or array of intramembranous particles. It is, for instance, possible that we are dealing with very large-scale lipid regionalizations resulting in the exclusion of specific proteins by virtue of their solubility preference between these domains. For instance, as a result of lateral phase segregations the lipids of the head might be quite different from those of the midpiece, so that a given protein might be soluble in only the head membrane lipids. In this context it is also possible to conjecture about a situation where a protein might be soluble in, say, both the anterior region and the posterior region of the head, but excluded from one or the other by virtue of its being excluded from the equatorial region.

In physical chemistry terms, this issue of selective solubility is essentially one of chemical potential. *It is important to note that in dealing with the issue of sperm surface regionalization, one is potentially dealing with a highly interactive system. Once you regionalize a single component, by whatever mechanism, you potentially regionalize, to a greater or lesser extent, all of the membrane components by virtue of the cascade of molecular interactions.*

5. How Does the Sperm Become Regionalized?

Related to the question of how the mammalian sperm maintains its surface regionalization is how it became regionalized in the first place. We have discussed the fact that, if one had localized insertion, then even in the presence of free diffusion, some level of regionalization would persist for a considerable length of time.

We have also discussed examples of sperm membrane components which start off homogeneously distributed and subsequently become regionalized. Either material is selectively removed from specific regions or the system redistributes in response to changes in the molecular interactions. In this context, it is useful to consider the problem of diffusion to a trap (Chao *et al.*, 1981). Suppose, for instance, that a system is initially homogeneously distributed and free diffusing. At some time a set of immobilizing receptor sites (e.g., cytoskeletal linkages) appear. Whenever a molecule attaches to one of these sites, it becomes trapped there. Eventually, the entire system will be trapped and its regionalization will reflect the distribution of these traps. The time needed to trap the system can again be estimated from Eq. (1). This may be particularly relevant to antigens localized to the anterior head of mouse sperm (e.g., PH-20 antigen, M42 antigen, and galactosyltransferase) which may be associated with cytoskeletal components that coalesce to the anterior head late in spermatogenesis (Phelps and Myles, 1987; Scully *et al.*, 1987).

6. Mechanisms of Redistribution

Thermodynamically speaking, the processes of macromolecular regionalization and redistribution reflect a delicate balance between entropically driven and enthalpically driven processes. The dominant factor is ultimately determined by some physiological stimulus that results in a redistribution of populations of macromolecules, be they lipids or proteins. Some redistributions, that of PT-1 for instance, require motility and metabolic energy to occur. Other redistributions, to first order, are consistent with redistribution by diffusion.

One typically thinks of diffusion as being entropically driven. That is, if one localizes a solute to one side of a partition and then removes that partition, the molecules move to maximize entropy and become homogeneously distributed on both sides. However, if there is a solubility preference for the solute on one side or the other, then the final distribution reflects this preference. In addition, the redistribution can become enthalpically driven, either speeded up or slowed down, by interactions of the solute with the solvents on the different sides of the barrier. In mathematical terms one usually thinks of the diffusion coefficient as being given by

$$D = kT/f \tag{2}$$

where f is the frictional coefficient. However, when the system becomes interactive, one must consider the enthalpy or chemical activity, a. Under this constraint, Eq. (2) becomes

$$D = (kT/f)(1 + C \, d\ln a/dC) \tag{3}$$

where C is the concentration (Van Holde, 1971).

A further level of complexity is to recognize that redistributions are potentially nonequilibrium. Upon removing an interregional barrier, it is not just a single membrane component which is going to exchange (McCall and Douglass, 1967; Ryan et al., 1989). Indeed, not only the proteins but the lipids will redistribute to establish a new equilibrium including a new ensemble of domains. *Until this new distribution is established, the intermixing will represent a nonequilibrium situation. That is, the mixing will not necessarily be governed by the equilibrium diffusion rates which one would measure either in the equilibrium state before the barrier is removed or in the equilibrium state which is established after the barrier is removed.*

7. Summary

We have thus found that the sperm plasma membrane is indeed "a little more than mosaic, a little less than fluid." Unlike most mammalian somatic cells, a significant fraction of mammalian sperm plasma membrane lipid is not free to diffuse. In addition, despite the fact that the sperm is covered by a continuous plasma membrane lipid, diffusibility is different over the morphologically distinct regions of the sperm surface. The physiological transformations of the sperm surface which result in the acrosome reaction and ultimately in fertilization, in general lead to an increase both in immobile lipid and in the diffusion rate of the mobile fraction. However, these transformations and the way in which they control sperm function, are more accurately represented as a change in the ensemble of membrane domains than as a change in bulk membrane fluidity.

The diffusibility of many sperm plasma membrane proteins is also quite different from the typical somatic cell plasma membrane protein. Some sperm plasma membrane proteins have been reported to be completely free to diffuse and to diffuse at the fluid dynamic limit.

Initial evidence suggests that sperm use a variety of mechanisms to regionalize the distribution of their surface components. These appear to include: lateral phase segregations, immobilization, and barriers to interregional diffusion. These regionalizations are ultimately related to the changing environments and physiological stimuli that are unique to the processes of spermatogenesis, epididymal maturation, and fertilization.

ACKNOWLEDGMENTS. We are grateful to the other members of our laboratory: K. M. Bocian, C. A. McKinnon, and R. M. Mungovan for excellent technical assistance both in the laboratory and in the preparation of the manuscript. We are also indebted to Dr. Patricia Saling (Duke University Medical School) for providing us with antibodies against M42 and P220 as well as for many stimulating discussions. We thank the laboratory of Dr. Grant Fairbanks (Worcester Foundation), and especially Dr. Frances Weaver, for providing us with the ESA 152 antibody against ram sperm. Dr. Diana Myles (University of Connecticut Health Sciences Center) was gracious in providing us with photographs of both PT-1 and PH-20 antibodies on guinea pig sperm. Finally, we thank Donna Dangott for her artistic skill in generating the drawings for the manuscript. Work from this laboratory is supported by NIH Grants HD 17377 and HD 23294 and by grants from the A. W. Mellon and Whittaker Foundations. R.A.C. is an NIH National Research Service Award Recipient (HD 07312).

References

Alvarez, J. G., and Storey, B. T., 1986, Spontaneous lipid peroxidation in rabbit and mouse epididymal spermatozoa: dependence of rate on temperature and oxygen concentration, *Biol. Reprod.* **32**:342–351.

Axelrod, D., Koppel, D. E., Schlessinger, J., Elson, E., and Webb, W. W., 1976a, Mobility measurement by analysis of fluorescence photobleaching recovery kinetics, *Biophys. J.* **16**:1055–1069.

Axelrod, D., Ravdin, P., Koppel, D. E., Schlessinger, J., Webb, W. W., Elson, E. L., and Podleski, T. R., 1976b, Lateral motion of fluorescently labeled acetylcholine receptors in membranes of developing muscle fibers, *Proc. Natl. Acad. Sci. USA* **73**:4594–4598.

Bearer, E. L., and Friend, D. S., 1982, Modifications of anionic-lipid domains preceding membrane fusion in guinea pig sperm, *J. Cell Biol.* **92**:604–615.

Bragg, P. W., and Handel, M. A., 1979, Protein synthesis in mouse spermatozoa, *Biol. Reprod.* **20**:333–337.

Brocklehurst, K. W., and Pollard, H. B., 1988, Osmotic effects in membrane fusion during exocytosis, *Curr. Top. Membr. Transp.* **32**:203–225.

Brooks, D. E., 1985, Characterization of a 22 kDa protein with widespread tissue distribution but which is uniquely present in secretions of the testis and epididymis and on the surface of spermatozoa, *Biochim. Biophys. Acta* **841**:59–70.

Cardullo, R. A., 1985, Oxygen metabolism, motility, and the maintenance of viability of caudal epididymal rat sperm, Ph. D. dissertation, The Johns Hopkins University, Baltimore.

Cardullo, R. A., and Baltz, J. M., 1988, A relationship between metabolism, motility, and sperm dimensions, *Biophys. J.* **53**:371a.

Chao, N.-m., Young, S. H., and Poo, M.-M., 1981, Localization of cell membrane components by surface diffusion into a "trap," *Biophys. J.* **36**:139–153.

Cowan, A. E., Primakoff, P., and Myles, D. G., 1986, Sperm exocytosis increases the amount of PH-20 antigen on the surface of guinea pig sperm, *J. Cell Biol.* **103**:1289–1297.

Cowan, A. E., Myles, D. G., and Koppel, D. E., 1987, Lateral diffusion of the PH-20 protein on guinea pig sperm: Evidence that barriers to diffusion maintain plasma membrane domains in mammalian sperm, *J. Cell Biol.* **104:**917–923.

Cummins, J. M., and Woodall, P. F., 1985, On mammalian sperm dimensions, *J. Reprod. Fertil.* **75:**153–175.

Dacheux, J. L., and Voglmayr, J., 1983, Sequence of specific cell surface differentiation and its relationship to exogenous fluid proteins in the ram epididymis, *Biol. Reprod.* **29:**1033–1046.

Djakiew, D., and Cardullo, R., 1986, Lower temperature of the cauda epididymidis facilities the storage of sperm by enhancing oxygen availability, *Gamete Res.* **15:**237–245.

Edidin, M., 1981, Molecular motions and membrane organization and function, in: *Comprehensive Biochemistry* (J. B. Finean and R. H. Mitchell, eds.), Elsevier/North-Holland, Amsterdam, pp. 37–82.

Edidin, M., and Wier, M. L., 1987, Lateral diffusion of class I MHC antigens depends upon the extent of their glycosylation, *J. Cell Biol.* **105:**304a.

Edidin, M., and Zuniga, M., 1984, Lateral diffusion of wild type and mutant L^d antigens in L cells, *J. Cell Biol.* **99:**2333–2335.

Edidin, M., Zagyansky, Y., and Lardner, T. J., 1976, Measurement of membrane lateral diffusion in single cells, *Science* **191:**466–468.

Einstein, A., 1905, On the movement of small particles in a stationary liquid demanded by the molecular theory of heat, *Ann. Phys.* **17:**549–561.

Erickson, R. R., 1977, Differentiation and other alloantigens of spermatozoa, in: *Immunobiology of Gametes* (M. Edidin and M. H. Johnson, eds.), Cambridge University Press, London, pp. 85–114.

Ethier, M. F., Wolf, D. E., and Melchior, D., 1982, A calorimetric investigation of the phase partitioning of the fluorescent carbocyanine probes in phosphatidylcholine bilayers, *Biochemistry* **21:**1178–1182.

Feuchter, F. A., Tabet, A. J., and Green, M. F., 1988, Maturation antigen of the mouse sperm flagellum. I. Analysis of its secretion, association with sperm, and function, *Am. J. Anat.* **181:**67–76.

Forrester, I., 1980, Effects of digitonin and polymixin B on plasma membrane of ram spermatozoa—An EM study, *Arch. Androl.* **4:**195–204.

Friend, D. S., 1982, Plasma membrane diversity in a highly polarized cell, *J. Cell Biol.* **93:**243–249.

Frye, L. D., and Edidin, M., 1970, The rapid intermixing of cell surface antigens after formation of mouse–human heterokaryons, *J. Cell Sci.* **7:**319–335.

Ishihara, A., Hou, Y., and Jacobson, K., 1987, The Thy-1 antigen exhibits rapid lateral diffusion in the plasma membrane of rodent lymphoid cells and fibroblasts, *Proc. Natl. Acad. Sci. USA* **84:**1290–1293.

Jacobson, K., Wu, E.-S., and Poste, G., 1976, Measurement of the translational mobility of concanavalin A in glycerol–saline solutions and on the cell surface by fluorescence recovery after photobleaching, *Biochim. Biophys. Acta* **423:**215–222.

Johnson, M. H., 1981, Membrane events associated with the generation of a blastocyst, *Int. Rev. Cytol. Suppl.* **12:**1–37.

Karnovsky, M. J., Kleinfeld, A. M., Hoover, R. L., and Klausner, R. D., 1982, The concept of lipid domains in membranes, *J. Cell Biol.* **94:**1–6.

Klausner, R. D., and Wolf, D. E., 1980, Selectivity of fluorescent lipid analogs for lipid domains, *Biochemistry* **19:**6199–6203.

Klausner, R. D., Kleinfeld, A. M., Hoover, R. L., and Karnovsky, M. J., 1980, Lipid domains in membranes: Evidence derived from structural perturbations induced by free fatty acids and life time heterogeneity analysis, *J. Biol. Chem.* **255:**1285–1295.

Koehler, J. K., 1978, The mammalian sperm surface: Studies with specific labeling techniques, *Int. Rev. Cytol.* **54:**73–108.

Koo, G. C., Boyse, E. A., and Wachtel, S. S., 1977, Immunogenic techniques and approaches in the study of sperm and testicular cell surface antigens, in: *Immunobiology of Gametes* (M. Edidin and M. H. Johnson, eds.), Cambridge University Press, London, pp. 73–84.

Lakoski, K. A., Carron, C. P., Cabot, C. L., and Saling, P. M., 1988, Epididymal maturation and the acrosome reaction in mouse sperm: Response to zona pellucida develops coincident with modification of M42 antigen, *Biol. Reprod.* **38:**221–233.

Langlais, J., and Roberts, K. D., 1985, A molecular membrane model of sperm capacitation and the acrosome reaction of mammalian spermatozoa, *Gamete Res.* **12:**183–224.

Lardy, H. A., and Phillips, P. H., 1941, The interrelation of oxidative and glycolytic processes as sources of energy for bull spermatozoa, *Am. J. Physiol.* **133:**602–609.

Leyton, L., Robinson, A., and Saling, P. M., 1989, Relationship between the M42 antigen of mouse sperm and the acrosome reaction induced by ZP3, *Dev. Biol.* **132:** 174–178.

Livneh, E., Benveniste, M., Prywes, R., Felder, S., Kam, Z., and Schlessinger, J., 1986, Large deletions in the cytoplasmic kinase domain of the epidermal growth factor receptor do not affect its lateral mobility, *J. Cell Biol.* **103:**327–331.

Lopez, L. C., and Shur, B. D., 1987, Redistribution of mouse sperm galactosyltransferase after the acrosome reaction, *J. Cell Biol.* **105:**1663–1670.

Low, M. G., and Finean, J. B., 1977, Nonlytic release of acetylcholine esterase from erythrocytes by a phosphatidylinositol specific phospholipase C, *FEBS Lett.* **82:**143–146.

Low, M. G., and Kincade, P. W., 1985, Phosphatidylinositol is the membrane-anchoring domain of the Thy-1 glycoprotein, *Nature* **318:**62–64.

Low, M. G., and Saltieh, A. R., 1988, Structural and functional roles of glycosylphosphatidylinositol in membranes, *Science* **239:**268–275.

Mabrey, S., Mateo, P. L., and Sturtevant, J. M., 1978, High-sensitivity scanning calorimetric study of mixtures of cholesterol with dimyristoyl- and dipalmitoylphosphatidylcholines, *Biochemistry* **17:**2464–2468.

McCall, D. W., and Douglass, D. C., 1967, Diffusion in binary solutions, *J. Phys. Chem.* **71:**987–991.

McNutt, N. S., and Weinstein, R. S., 1973, Membrane ultrastructure at mammalian intracellular junctions, *Prog. Biophys. Mol. Biol.* **26:**45–101.

Millette, C. F., 1977, Distribution and mobility of lectin binding sites on mammalian spermatozoa, in: *Immunobiology of Gametes* (M. Edidin and M. H. Johnson, eds.), Cambridge University Press, London pp. 51–71.

Millette, C. F., 1979, Appearance and partitioning of plasma membrane antigens during mouse spermatogenesis, in: *The Spermatozoon* (D. W. Fawcett and J. M. Bedford, eds.), Urban & Schwarzenberg, Munich, pp. 177–186.

Moore, W. J., 1972, *Physical Chemistry,* 2nd ed., Prentice–Hall, Englewood Cliffs, N.J., pp. 206–209.

Myles, D. G., and Primakoff, P., 1984, Localized surface antigens of guinea pig sperm migrate to new regions prior to fertilization, *J. Cell Biol.* **99:**1634–1641.

Myles, D. G., Primakoff, P., and Bellve, A. R., 1981, Surface domain of the guinea pig sperm defined with monoclonal antibodies, *Cell* **23:**433–439.

Myles, D. G., Primakoff, P., and Koppel, D. E., 1984, A localized surface protein of guinea pig sperm exhibits free diffusion in its domain, *J. Cell Biol.* **98:**1905–1909.

Myles, D. G., Hyatt, H., and Primakoff, P., 1987, Binding of both acrosome-intact and acrosome-reacted guinea pig sperm to the zona pellucida during in vitro fertilization, *Dev. Biol.* **121:**559–567.

Nichols, J. W., and Pagano, R. E., 1981, Kinetics of soluble lipid monomer diffusion between vesicles, *Biochemistry* **20:**2783–2789.

Nicolson, G. L., and Yanagamachi, R., 1974, Mobility and the restriction of mobility of plasma membrane lectin-binding components, *Science* **184:**1294–1296.

Nicolson, G. L., and Yanagamachi, R., 1979, Cell surface changes associated with the epididymal maturation of mammalian spermatozoa, in: *The Spermatozoon* (D. W. Fawcett and J. M. Bedford, eds.), Urban & Schwarzenberg, Munich, pp. 187–194.

Noda, M., Yoon, K., Rodan, G. A., and Koppel, D. E., 1987, High lateral mobility of endogenous and transfected alkaline phosphatase: A phosphatidylinositol-anchored membrane protein, *J. Cell Biol.* **105:** 1671–1677.

Olson, G. E., Lifsics, M. R., Winfrey, V. P., and Rifkin, J. M., 1987, Modification of the rat sperm flagellar membrane during maturation in the epididymis, *J. Androl.* **8:**129–147.

O'Rand, M. G., 1988, Sperm–egg recognition and barriers to interspecies fertilization, *Gamete Res.* **19:**315–328.

Packard, B. S., and Wolf, D. E., 1985, Fluorescence properties of dialkyl indocarbocyanine dyes in phospholipid membranes, *Biochemistry* **24:**5176–5181.

Pagano, R. E., and Sleight, R. G., 1985, Defining lipid transport pathways in animal cells, *Science* **229:**1051–1057.

Peters, R., 1981, Translational diffusion in the plasma membrane of single cells as studied by fluorescence microphotolysis, *Cell Biol. Int. Rep.* **5:**733–760.

Peters, R., Peters, J., Tews, K. H., and Bahr, W., 1974, A microfluorometric study of translational diffusion in erythrocyte membranes, *Biochim. Biophys. Acta* **367:**282–294.

Phelps, B. M., and Myles, D. G., 1987, The guinea pig sperm plasma membrane protein, PH-20, reaches the surface via two transport pathways and becomes localized to a domain after an initial uniform distribution, *Dev. Biol.* **123**:63–72.

Phelps, B. M., Primakoff, P., Koppel, D. E., Low, M. G., and Myles, D. G., 1988, Restricted lateral diffusion of PH-20, a PI-anchored sperm membrane protein, *Science* **240**:1780–1782.

Pjura, W. J., Kleinfeld, A. M., and Karnovsky, M. J., 1984, Partition of fatty acids and fluorescent fatty acids into membranes, *Biochemistry* **23**:2039–2043.

Portis, A., Newton, C., Pangborn, W., and Papahadjopoulos, D., 1979, Studies on the mechanism of membrane fusion—Evidence for an intermembrane Ca^{++}–phospholipid complex, synergism with Mg^{++}, and inhibition by spectrin, *Biochemistry* **18**:780–790.

Premkumar, E., and Bhargava, P. M., 1972, Transcription and translation in bovine spermatozoa, *Nature New Biol.* **240**:139–143.

Primakoff, P., and Myles, D. G., 1983, A map of the guinea pig sperm surface constructed with monoclonal antibodies, *Dev. Biol.* **98**:417–428.

Primakoff, P., Hyatt, H., and Myles, D. G., 1985, A role for the migrating sperm surface antigen PH-20 in guinea pig sperm binding to the egg zona pellucida, *J. Cell Biol.* **101**:2239–2244.

Pringle, M. J., and Miller, K. W., 1979, Differential effects on phospholipid phase transitions produced by structurally related long chain alcohols, *Biochemistry* **18**:3314–3320.

Retzius, G., 1906, *Biologische Untersuchungen,* Volume XIII, Jena, Stockholm.

Russell, L. D., Peterson, R. N., Hunt, W., and Strack, L. E., 1984, Posttesticular surface modifications and contributions of reproductive tract fluids to the surface polypeptide composition of boar spermatozoa, *Biol. Reprod.* **30**:959–978.

Ryan, T., Myers, J., and Webb, W. W., 1989, Molecular interaction on the cell surface revealed by electrophoresis, *Biol. Bull. (Woods Hole, Mass.)* **176**: 164–169.

Saffmann, P. G., and Delbruck, M., 1975, Brownian motion in biological membranes, *Proc. Natl. Acad. Sci. USA* **72**:3111–3113.

Saling, P. M., and Lakoski, K. A., 1985, Mouse sperm antigens that participate in fertilization. II. Inhibition of sperm penetration through the zona pellucida using monoclonal antibodies, *Biol. Reprod.* **33**:527–536.

Saling, P. M., Martin, P. C., and Waibel, R., 1986, Contraceptive effect of two anti-sperm monoclonal antibodies administered singly and in combination in the mouse, in: *Immunological Approaches to Contraception and Fertility* (G. B. Talwar, ed.), Plenum Press, New York, pp. 191–199.

Schlessinger, J., Axelrod, D., Koppel, D. E., Webb, W. W., and Elson, E. L., 1977a, Lateral transport of a lipid probe and labeled proteins on a cell membrane, *Science* **195**:307–309.

Schlessinger, J., Barak, L. S., Hammes, G. G., Yamada, K. M., Pastan, I., Webb, W. W., and Elson, E. L., 1977b, Mobility and distribution of a cell surface glycoprotein and its interaction with other membrane components, *Proc. Natl. Acad. Sci. USA* **74**:2909–2913.

Scott, T. W., Voglmayr, J. K., and Setchell, B. P., 1967, Lipid composition and metabolism in testicular and ejaculated ram spermatozoa, *Biochem. J.* **102**:456–461.

Scully, N. F., Shaper, J. H., and Shur, B. D., 1987, Spatial and temporal expression of cell surface galactosyltransferase during mouse spermatogenesis and epididymal maturation, *Dev. Biol.* **124**:111–124.

Shimshick, E. J., and McConnell, H. M., 1973, Lateral phase separation in phospholipid membranes, *Biochemistry* **12**:2351–2360.

Sims, P. J., Waggoner, A. S., Wang, C.-H., and Hoffman, J. F., 1974, Studies on the mechanism by which cyanine dyes measure membrane potential in red blood cells and phosphatidylcholine vesicles, *Biochemistry* **13**:3315–3330.

Singer, S. J., and Nicolson, G. L., 1972, The fluid mosaic model of the structure of cell membranes, *Science* **175**:720–731.

Sklar, L. A., Miljanich, G. P., and Dratz, E. A., 1979, Phospholipid lateral phase separation and the partition of cis-parinaric and trans-parinaric acid among aqueous solid lipid and fluid lipid phases, *Biochemistry* **18**: 1707–1716.

Staehelin, L. A., 1974, Structure and function of intercellular junctions, *Int. Rev. Cytol.* **39**:191–283.

Stewart, T. P., Hui, S. W., Portis, A. R., Jr., and Papahadjopoulos, D., 1979, Complex phase mixing of phosphatidylcholine and phosphatidylserine in multilamellar membrane vesicles. *Biochim. Biophys. Acta* **556**:1–16.

Struck, D. K., and Pagano, R. E., 1980, Insertion of fluorescent phospholipids into the plasma membrane of a mammalian cell, *J. Biol. Chem.* **255**:5404–5410.

Tank, D. W., Wu, E.-S., and Webb, W. W., 1982, Enhanced molecular diffusibility in muscle membrane blebs: Release of lateral constraints, *J. Cell Biol.* **92**:207–212.

Taylor, R. B., Duffus, W. P. H., Raff, M. C., and dePetris, S., 1971, Redistribution and pinocytosis of lymphocyte surface immunoglobulin molecules induced by anti-immunoglobulin antibody, *Nature New Biol.* **233**:225–229.

Thomas, J., Webb, W., Davitz, M. A., and Nussenzweig, V., 1987, Decay accelerating factor diffuses rapidly on HeLa$_{AE}$ cell surfaces, *Biophys. J.* **51**:522a.

Tilney, L. G., 1985, The acrosomal reaction, in: *Biology of Fertilization*, Volume 2 (C. Metz and A. Monroy, eds.), Academic Press, New York, pp. 156–213.

Treistman, S. N., Moynihan, M. M., and Wolf, D. E., 1987, Effects of temperature and alcohols on lipid lateral diffusibility in Aplysia neurons, *Biochim. Biophys. Acta* **898**:109–120.

Trimmer, J. S., and Vacquier, V. D., 1988, Monoclonal antibodies induce the translocation, patching, and shedding of surface antigens of sea urchin spermatozoa, *Exp. Cell Res.* **175**:37–51.

Tung, K. S. K., 1977, The nature of antigens and pathogenetic mechanisms in autoimmunity to sperm, in: *Immunobiology of Gametes* (M. Edidin and M. H. Johnson, eds.), Cambridge University Press, London, pp. 157–185.

Van Holde, K. E., 1971, *Physical Biochemistry*, Prentice-Hall, Englewood Cliffs, N.J., p. 86.

Vernon, R. B., Hamilton, M. S., and Eddy, E. M., 1985, Effects of in vivo and in vitro fertilization environments on the expression of a surface antigen of the mouse sperm tail, *Biol. Reprod.* **32**:669–680.

Vernon, R. B., Muller, C. S., and Eddy, E. M., 1987, Further characterization of a secreted epididymal glycoprotein in mice that binds to sperm tails, *J. Androl.* **8**:123–128.

Voglmayr, J. K., 1975, Metabolic changes in spermatozoa during epididymal transit, in: *Handbook of Physiology*, Volume 5 (R. O. Greep and D. W. Hamilton, eds.), American Physiological Society, Washington, D.C., pp. 437–451.

Voglmayr, J. K., Fairbanks, G., Vespa, D. B., and Colella, J. R., 1982, Studies on mechanisms of surface modification in ram spermatozoa during the final stages of differentiation, *Biol. Reprod.* **26**:483–500.

Wakelam, M. J. O., 1988, Myoblast fusion—A mechanistic analysis, *Curr. Top. Membr. Transp.* **32**:87–112.

Weaver, F. E., 1985, Studies on the effects of temperature and membrane composition on the organization of eukaryotic cell membranes, Ph.D. dissertation, The Johns Hopkins University, Baltimore.

Weaver, F. E., Gaffney, K. J., Dino, J. E., Lewis, R. G., and Fairbanks, G., 1989, Origin and properties of ESA 152, a maturation dependent ram sperm surface antigen, *J. Cell Biol.* **107**: 165a.

Wier, M., and Edidin, M., 1988, Constraint of the translational diffusion of a membrane glycoprotein by its external domains, *Science* **242**:412–414.

Williamson, P. L., Massey, W. A., Phelps, B. M., and Schlegel, R. A., 1981, Membrane phase state and the rearrangement of hematopoietic cell surface receptors, *Mol. Cell Biol.* **1**:128–135.

Williamson, P. L., Bateman, J., Kozarsky, K., Mattocks, K., Hemanowicz, N., Choe, H.-R., and Schlegel, R. A., 1982, Involvement of spectrin in the maintenance of phase-state asymmetry in the erythrocyte membrane, *Cell* **30**:725–733.

Wolf, D. E., 1983, The plasma membrane in early embryogenesis, in: *Development in Mammals*, Volume 5 (M. H. Johnson, ed.), Elsevier/North-Holland, Amsterdam, pp. 187–208.

Wolf, D. E., 1985, Determination of the sidedness of carbocyanine dye labeling of membranes, *Biochemistry* **24**:582–586.

Wolf, D. E., 1986, Corralling the drunken beggar, *Bioessays* **6**:116–121.

Wolf, D. E., 1987a, Probing the lateral organization and dynamics of membranes, in: *Spectroscopic Membrane Probes* (L. Loew, ed.), CRC Press, Boca Raton, Fla., pp. 193–220.

Wolf, D. E., 1988, Diffusion and the control of membrane regionalization, in: *Cell Biology of the Testis and Epididymis* (M.-C. Orgebin-Crist and B.J. Danzo, eds.), New York Academy of Sciences, New York, pp. 247–261.

Wolf, D. E., and Edidin, M., 1981, Diffusion and mobility of molecules in surface membranes, *Tech. Cell. Physiol.* **P105**:1–14.

Wolf, D. E., and Voglmayr, J. K., 1984, Diffusion and regionalization in membranes of maturing ram spermatozoa, *J. Cell Biol.* **98**:1678–1684.

Wolf, D. E., Schlessinger, J., Elson, E. L., Webb, W. W., Blumenthal, R., and Henkart, P., 1977, Diffusion and patching of macromolecules on planar lipid bilayer membranes, *Biochemistry* **16**:3476–3483.

Wolf, D. E., Henkart, P., and Webb, W. W., 1980, Diffusion, patching, and capping of stearoylated dextrans on 3T3 cell plasma membranes, *Biochemistry* **19**:3893–3904.

Wolf, D. E., Edidin, M., and Handyside, A. H., 1981a, Changes in the organization of the mouse egg plasma membrane upon fertilization and first cleavage: Indications from the lateral diffusion rates of fluorescent lipid analogs, *Dev. Biol.* **85**:195–198.

Wolf, D. E., Kinsey, W., Lennarz, W., and Edidin, M., 1981b, Changes in the organization of the sea urchin egg plasma membrane upon fertilization: Indications from the lateral diffusion rates of lipid-soluble fluorescent dyes, *Dev. Biol.* **81**:133–138.

Wolf, D. E., Hagopian, S. S., and Ishijima, S., 1986a, Changes in sperm plasma membrane lipid diffusibility following hyperactivation during in vitro capacitation in the mouse, *J. Cell Biol.* **102**:1372–1377.

Wolf, D. E., Hagopian, S., Lewis, R., Voglmayr, J. K., and Fairbanks, G., 1986b, Lateral regionalization and diffusion of a maturation dependent antigen in the ram sperm plasma membrane, *J. Cell Biol.* **102**:1826–1831.

Wolf, D. E., Scott, B. K., and Millette, C. F., 1986c, The development of regionalized diffusibility in the germ cell plasma membrane during spermatogenesis in the mouse, *J. Cell Biol.* **103**:1745–1750.

Wolf, D. E., Lipscomb, A. C., and Maynard, V. M., 1988, Causes of nondiffusing lipid in the plasma membrane of mammalian spermatozoa, *Biochemistry* **27**:860–865.

Wolf, D. E., Cardullo, R. A., McKinnon, C. M., Lakoski, K. A., and Saling, P. M., 1989a, Diffusion of proteins in the head plasma membrane of mouse spermatozoa, *J. Cell Biol.* in press.

Wolf, D. E., Maynard, V. M., McKinnon, C. M., and Wolf, D. E., 1989b, Lipid domains in the ram sperm plasma membranes demonstrated by differential scanning calorimetry, submitted for publication.

Wu, E.-S., Tank, D. W., and Webb, W. W., 1982, Unconstrained lateral diffusion of concanavalin A receptors of bulbous lymphocytes, *Proc. Natl. Acad. Sci. USA* **79**:4962–4966.

Structure and Assembly of the Oviduct Ciliary Membrane

Bernadette Chailley, Emmanuelle Boisvieux-Ulrich, and Daniel Sandoz

1. Introduction

For species whose fertilization is internal, the oviduct is a part of the female genital tract which is the site for a number of different steps required for reproduction, including gamete transport, fertilization, and early embryonic development. In mammals the two oviducts (or Fallopian tubes) are connected to the uterus (for review see Hafez and Blandau, 1969). They can be subdivided in three segments:

1. The fimbria, which embraces the ovary and collects the ovum with its cumulus, has an epithelial surface which is entirely ciliated. Only some stem cells located in the basal part of the epithelium are unciliated.
2. The ampulla, in which fertilization takes place, possesses fewer ciliated cells interspersed with mucous and stem cells. Together, the fimbria and ampulla form the infundibulum.
3. The isthmus, joining the uterus, is the site where segmentation of the embryo begins. It displays only rare ciliated cells.

In the bird oviduct, only the left ovary and oviduct develop, the right gonad and oviduct remaining rudimentary (for review see Romanoff and Romanoff, 1949). The oviduct of the laying bird is very long (30 cm in hen, 20 cm in quail). It is subdivided into:

1. The fimbria and the ampulla forming the infundibulum, whose functions are similar to those indicated for mammals.
2. The magnum which synthesizes the egg white and in which segmentation begins.
3. The isthmus which secretes the shell membranes.
4. The shell gland or so-called uterus in which the calcareous shell is deposited.
5. The vagina with glands, which stores spermatozoa.

Bernadette Chailley, Emmanuelle Boisvieux-Ulrich, and Daniel Sandoz • Centre de Biologie Cellulaire C.N.R.S., 94205 Ivry sur Seine Cedex, France.

All segments of the bird oviduct are ciliated; the ciliated cells are often interspersed with mucous cells forming a mucociliary epithelium (Aitken and Johnston, 1963; Fertuck and Newstead, 1970; Wyburn *et al.*, 1970; Sandoz *et al.*, 1971). In reptiles a great variation of organization is observed according to the viviparous, oviparous, or ovoviviparous species, but in all cases the oviduct epithelium possesses ciliated cells. In lower vertebrates (Boisseau, 1973a) the oviduct is also ciliated. Thus, in all vertebrates the epithelium of the female genital tract is ciliated. Cilia beat in metachronal waves from the fimbria to the isthmus. As in the respiratory tract, mucous cells are often associated with ciliated cells.

The oviduct ciliated cells have been studied primarily in mammals, such as mouse (Dirksen, 1971, 1974), rabbit (Anderson and Hein, 1977), monkey (Brenner, 1969a,b; Anderson and Brenner, 1971), or after biopsies of collected segments from human Fallopian tubes (for review see Brenner, 1969a). Many studies have also been devoted to the long bird oviduct which provides numerous ciliated cells allowing biochemical analyses (Anderson, 1974; Klotz *et al.*, 1986a,b).

The oviduct ciliated cells are 20 to 30 μm high. The ovoid or lobed nucleus is located in the apical half of the cell. Mitochondria are numerous in the apical region of the cell and provide the ATP necessary for ciliary beating. A highly developed cytoskeleton including microfilaments, labile microtubules, and cytokeratin filaments is organized in the apex of the cell (for review see Sandoz *et al.*, 1988). This complex cytoskeleton is connected to 100 or 200 basal bodies which are aligned under the cell surface and on which the axonemes are polymerized. The cilia are about 7 μm long and 0.25 μm in diameter. They are interspersed with numerous, sometimes branched microvilli (Sandoz *et al.*, 1971; Rumery *et al.*, 1978) which are about 2μm long and 0.1 μm in diameter. The apical plasma membrane surface is subdivided into three distinct domains roughly distributed as follows: about 80% ciliary membrane and about 20% microvillous membrane, with a small residual portion forming the interciliary membrane. These three domains can be distinguished by their morphological and cytochemical properties and probably by their physiological functions too. This chapter will be devoted mainly to the ciliary membrane.

No biochemical studies have been performed on the ciliary membrane of the oviduct thus far because deciliation does not occur as easily in vertebrate cells as in protozoa (for review see Adoutte *et al.*, 1980; Witman, 1986). Our knowledge of the ciliary membrane comes essentially from morphological and cytochemical studies.

2. Organization of the Ciliary Membrane

The ciliary membrane is an extension of the apical plasma membrane which covers the axoneme. Ultrastructural and cytochemical studies have shown that the ciliary membrane has unique properties and must be considered as a specialized domain of the apical plasma membrane.

2.1. Ultrastructural Data

Freeze-fracture studies have revealed that the organization of the ciliary membrane differs from that of the other plasma membrane domains by the presence of intramembrane particles (IMPs) grouped at the base of the organelle; microvillous and interciliary membranes display only randomly dispersed IMPs (Fig. 1). In fact, the ciliary membrane possesses two major domains.

2.1.1. The Ciliary Necklace. At the base of the cilium, the IMPs are grouped in five to seven rows (Dirksen *et al.*, 1971; Boisvieux-Ulrich *et al.*, 1977) forming the ciliary necklace (Gilula and Satir, 1972). The ciliary necklace of oviduct cilia displays the largest number of strands among the motile cilia (Menco, 1980). The sinuous rows are

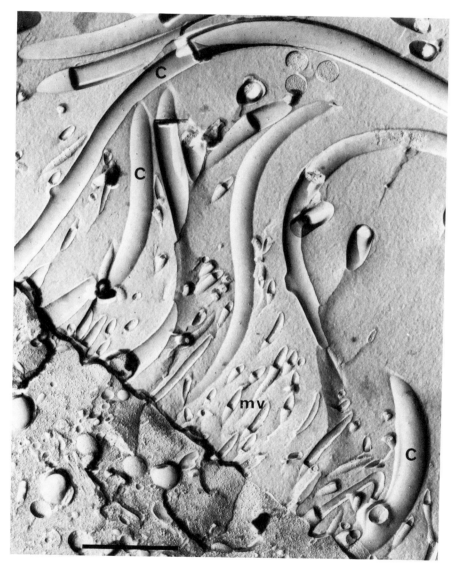

Figure 1. Freeze-fracture replica of quail oviduct ciliated cell. Between cilia (C) curved in the recovery stroke are interspersed numerous microvilli (mv). Bar = 1 μm.

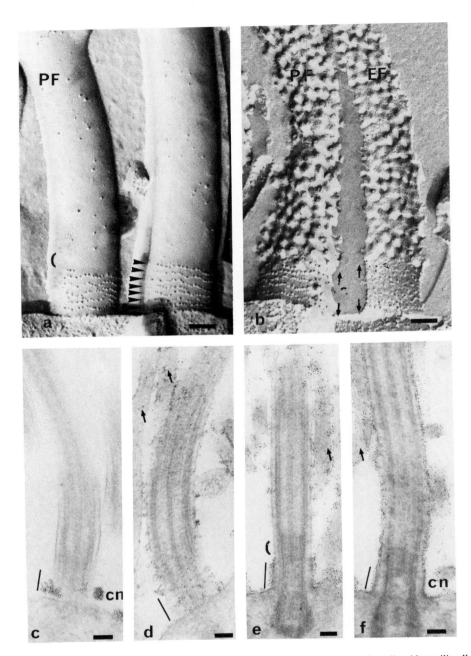

Figure 2. Proximal part of the ciliary membrane. (a) Replica of proximal part of quail oviduct cilia allows visualization of the protoplasmic fracture face (PF) of the membrane. The sinuous strands of intramembrane particles (IMPs) form the ciliary necklace (arrowheads). Just above the ciliary necklace, the shaft membrane is devoid of IMPs for a distance of 0.1 μm (bracket). The remainder of the shaft membrane displays scattered IMPs. (b) Replica after filipin–glutaraldehyde fixation. Both protoplasmic (PF) and exoplasmic (EF) fracture faces of the ciliary membrane are visible. The shaft membrane shows a waffled aspect due to the formation of filipin–cholesterol complexes while the ciliary necklace area (cn) is almost devoid of these membrane altera-tions. (c, d) Cationized ferritin labeling (1 mg/ml). The pattern of the ferritin molecules differs depending on whether the labeling takes place before (c) or after (d) glutaraldehyde fixation. (c) Ferritin molecules are

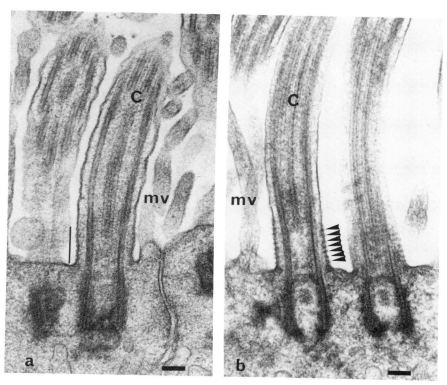

Figure 3. Transmission electron micrographs of quail cilia from (a) trachea and (b) oviduct. At the base of the organelles, the characteristic ciliary necklace is visualized as six or seven parallel rows of beads (arrowheads) in both cilia (C). However, the bead structures appear more distinct in the oviduct cilium than in the tracheal one. mv, microvilli. Bars = 0.1 μm.

parallel and regularly spaced (about 26 to 30 nm). Occasionally, two rows can fuse always forming a continuous circle (Fig. 2a). The large IMPs (10 nm in diameter) of each row are very close to each other and there are about 70 IMPs per row. Thus, the ciliary necklace could be formed by about 400 to 500 IMPs. The necklace IMPs can be visualized in thin sections by conventional electron microscopy because they are larger than the membrane thickness (Anderson and Hein, 1977; Boisvieux-Ulrich *et al.*, 1977). The ciliary necklace is a rare example of a clear visualization of IMPs in thin section, providing a good model for studying the relationships between cytoskeleton, IMPs, and lectin receptors. In quail, ciliary necklace IMPs are more visible in the oviduct than in the trachea (Fig. 3), indicat-

aggregated in a massive strand surrounding the necklace (cn) while the shaft membrane is devoid of labeling. (d) Ferritin molecules are regularly distributed on the necklace area and scattered all along the shaft membrane. Microvilli are also decorated (arrows). (e) Con A receptor detection. A regular deposit of glycosylated ferritin molecules is mainly observed on the necklace area. A membrane segment 0.1 μm high is without labeling (bracket). Labeling in cluster is scattered on the shaft membrane. Microvilli also show clusters (arrow). (f) WGA receptor detection. A dense deposit of glycosylated ferritin molecules is seen all along the ciliary shaft membrane; there is less label on the necklace area. Microvilli are irregularly labeled (arrow). Bars = 0.1 μm.

ing that in the same species ciliary necklace IMPs can differ according to the organ. The IMPs probably correspond to large protein complexes. The ciliary necklace extends to a height of 0.25 μm corresponding to the 9 + 0 transitional zone between the 9 triplet–basal body and the 9 + 2 axoneme.

2.1.2. The Shaft Membrane. Above the collar region the entire shaft membrane presents a smooth aspect on replicas (Fig. 2a). Only a few IMPs are observed, generally dispersed but sometimes grouped in incomplete strands. Like the necklace IMPs, they are easily spotted in thin sections. In all vertebrate motile cilia, only a few particles have been found in the shaft membrane in contrast to the variability noted in ciliated protozoa (Bardele, 1981).

One can note two minor subdomains in the proximal part of the shaft membrane. A first smooth subdomain extends 0.1 μm above the necklace (Sandoz *et al.*, 1979) (Fig. 2a). In this area the ciliary membrane is close to the axonemal doublets without visible links between microtubules and membrane but microtubules are surrounded by a dense material inside and outside the axoneme. This domain, which has been observed also in the ciliary membrane of other organs (Gilula and Satir, 1972; Chailley *et al.*, 1981), is continued by the second minor domain, which is characterized by incomplete IMP rows of unequal length. These IMPs have the same characteristics as the ciliary necklace IMPs (Boisvieux-Ulrich *et al.*, 1977). This domain, which is about 0.5 μm high, does not have the homogeneous pattern of the ciliary necklace. Even in the same sample the number of particles is quite variable from one cilium to another. In the quail oviduct shaft, IMPs are more numerous in the infundibulum than in the magnum (Sandoz *et al.*, 1988).

Although no fibrillar link is visible on thin sections, some connections seem to be present between the ciliary membrane and axonemal doublet 1, since in the quail oviduct, this doublet remains associated with the membrane in isolated cilia whose base is folded during isolation (Chailley *et al.*, 1986). The shaft membrane probably contains only a small number of protein molecules which are mainly grouped in IMP arrays within the basal third of the shaft.

2.1.3. The Ciliary Tip. The ciliary tip is characterized by a cap inside the cilium (see Dentler, this volume) and outside by the long process (22 to 30 nm) of a special glycocalyx forming the tip crown. Such a crown, described in vertebrate oviduct cilia (Dirksen and Satir, 1972; Anderson and Hein, 1977; Sandoz *et al.*, 1979), was also found on tracheal cilia (see Dentler, this volume). Curiously, it seems to be absent from primate oviduct cilia (Blandau *et al.*, 1979) whereas it is present in human oviduct (unpublished observations). In freeze-fracture no IMP organization is detected in the tip membrane (Fig. 4a,b).

2.1.4. The Interciliary Membrane. The interciliary membrane shows some differentiation near the cilia. Just below the ciliary necklace a small domain is devoid of particles. Around the site of emergence of the cilium, some arrays of IMPs have been described (Boisvieux-Ulrich *et al.*, 1977; Weiss *et al.*, 1977; Menco, 1980) and probably correspond to the site of fixation of the anchoring fibers to the membrane.

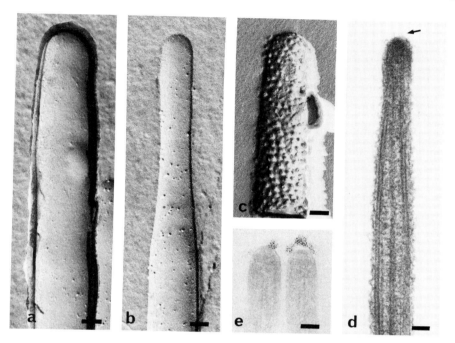

Figure 4. Ciliary tip membrane of growing and mature cilia. (a) Replica of a growing cilium with a constant diameter and a rounded tip whose membrane is without IMP organization. (b) Replica of a growing cilium end once the axonemal wall is composed of peripheral singlets, the B microtubules being shorter than the A microtubules. The ciliary extremity gets thinner and the tip membrane shows no IMP organization. (c) Replica of a growing cilium showing the filipin–cholesterol complexes which are less numerous at the rounded tip. (d) Thin section allows visualization of the membrane alterations following the formation of filipin–cholesterol complexes which are less numerous at the thinner tip. Once the cilium extremity gets thinner, the crown glycocalyx and cap structure differentiations begin to form (arrow). (e) Cationized ferritin labeling (1 mg/ml) before fixation observed on thin section. The ciliary tip of a mature cilium displays a large decoration on the crown glycocalyx. Bars = 0.1 μm.

2.2. Cytochemical Data

2.2.1. Glycoconjugates.
Different cytochemical techniques detecting either anionic groups or lectin receptors allow the visualization of surface glycoconjugates. The entire ciliary shaft membrane is rich in anionic charges, as demonstrated by the uniform labeling obtained with ruthenium red (RR) and colloidal iron hydroxide (CIH) at low pH (Bois-vieux-Ulrich *et al.*, 1977; Sandoz *et al.*, 1979; Odor and Blandau, 1988). In the necklace region, RR and CIH deposits form parallel lines similar to the necklace design. The labeling is suppressed when samples are pretreated with *Vibrio cholerae* neuraminidase, which indicates that the ciliary membrane is rich in sialyl residues.

Labeling with cationized ferritin (CF) introduced by Danon *et al.* (1972) allows the labeling of cells at physiological pH and can be used before or after aldehyde fixation. Without fixation, CF patches mainly label the ciliary necklace area and the tip crown in rabbit oviduct (Anderson and Hein, 1977) or in quail oviduct (Figs. 2c, 4e), whereas after glutaraldehyde fixation a monolayer of CF covers the necklace and the tip crown. Numerous spots of label are also observed along the shaft membrane (Fig. 2d). These results differ from those obtained by Moller *et al.* (1981) in rat trachea which showed a continuous CF distribution over the entire ciliary surface with or without fixation. The reasons for the differences in CF labeling observed before and after fixation have been discussed by several authors. On one hand, glutaraldehyde fixation has been shown to increase the negative charges on the cell surface (Vassar *et al.*, 1972). On the other hand, free aldehyde radicals introduced by glutaraldehyde can adsorb ferritin molecules; finally, fixation can make anionic residues more accessible to CF which is a large marker molecule (Temmink, 1979).

The oviduct ciliary membrane reacts weakly after Thiery's technique (1967) (Sandoz *et al.*, 1971; Schulte *et al.*, 1985), whereas other vertebrate cilia are strongly stained (Chailley *et al.*, 1981), which indicates that the ciliary glycocalyx varies greatly depending on the species and organ examined.

Lectins known for their specific affinity for sugar residues have been used in light and electron microscopy in order to characterize the plasma membrane of oviduct cells and also the secretory products. Histochemical studies performed on mouse oviduct (Lee *et al.*, 1983; Wu *et al.*, 1983) provide little information on ciliated cell membranes. Wheat germ agglutinin (WGA), which binds to sialyl and *N*-acetylglucosaminic residues, labels the apical membrane of rabbit ciliated cells (Menghi *et al.*, 1985). Terminal sialic acid-β-galactose disaccharide was shown to be present in human Fallopian tube (Schulte *et al.*, 1985). Cilia from other tissues, such as trachea (Schulte and Spicer, 1985) and ductuli efferentes (Burkett *et al.*, 1987), also were labeled by lectins, revealing the presence of *N*-acetylgalactosamine and *N*-acetylneuraminic acid residues. However, the light microscope does not allow a precise localization of glycosyl residues.

In contrast, using sequential application of lectin and glycosylated ferritin, Sandoz *et al.*, (1979) have shown that the necklace IMPs bind Con A revealing α-D-mannosyl residues (Fig. 2e) and WGA (Fig. 2f). Prior digestion by neuraminidase does not prevent WGA binding, indicating the presence of *N*-acetylglucosamine in the ciliary necklace IMPs. In the necklace of connecting cilia of photoreceptors, Horst *et al.*, (1987) showed labeling with succinylated WGA suggesting a predominance of *N*-acetylglucosamine in necklace glycoconjugates. No α- or β-D-galactose residues were detected in the necklace glycoproteins using *Ricinus communis* agglutinin I (RCA I) or peanut agglutinin (PNA) (Sandoz *et al.*, 1979).

All along the ciliary shaft membrane, WGA receptors are uniformly distributed while Con A receptors are restricted to certain spots corresponding to the IMP distribution. At the ciliary tip, Con A, WGA, and RCA I bind to the tip crown, indicating respectively the presence of α-D-mannosyl, *N*-acetylglucosaminyl and/or sialyl, and D-galactosyl residues.

2.2.2. Cholesterol. As a probe for free cholesterol in membrane, filipin, a polyene antibiotic forming complexes with membrane cholesterol, was introduced in electron microscopy by Elias *et al.* (1979). These filipin–cholesterol complexes (FC) are visible after freeze-fracture (Fig. 2b) or in thin sections (Fig. 4d). The ciliary membrane of quail oviduct is rich in cholesterol except in the ciliary necklace (Chailley and Boisvieux-Ulrich, 1985) as previously shown for tracheal cilia (Montesano, 1979). At the cilium tip the FC are less numerous, suggesting that the tip membrane contains less cholesterol (Chailley and Boisvieux-Ulrich, 1985) (Fig. 4c,d). The ciliary shaft membrane contains more cholesterol than microvillous and interciliary membranes. Knowing the role of cholesterol in membrane fluidity, one can suggest that the high cholesterol concentration in the shaft membrane increases its mechanical stability restricting the planar diffusion of molecules.

2.2.3. Lipid Extraction. Whereas the lipid composition of the protozoan ciliary membrane has been studied by biochemical methods (Thompson and Nozawa, 1977; Andrews and Nelson, 1979; Rhoads and Kaneshiro, 1979), lipids have not been identified in oviduct cilia because of the difficulty of isolating vertebrate cilia with their membranes.

Different extractions indicate that the shaft membrane contains a large amount of lipids. The shaft membrane is easily solubilized with Triton X-100 (Anderson, 1974). The ciliary necklace is resistant to extraction and the ciliary tip is less easily extracted than the shaft membrane.

2.2.4. Detection of Enzyme Activity and Immunocytochemistry. ATPase activity has been demonstrated in apical plasma membrane of oviduct ciliated cells and especially in the ciliary membrane when cells were treated with estrogen (Nayak, 1972; Nayak and Wu, 1975). Immunologically, the presence of calmodulin along the ciliary membrane and especially in the necklace area has been demonstrated in tracheal cilia (Gordon *et al.*, 1982). No spectrin, a membrane-associated protein, has been revealed in ciliary membrane while this protein has been shown to be present in the lateral plasma membrane and the apical cytoskeleton of ciliated cells (Chailley *et al.*, 1989). Biochemically, tubulin is found to be an integral protein component of the molluscan gill ciliary membrane (Stephens, 1977, 1986). This tubulin differs from that of the axoneme as suggested by amino acid composition and differential detergent binding (for review see Stephens, this volume).

2.2.5. Conclusions of Cytochemical Studies. Extraction of lipids including glycolipids by chloroform/methanol treatment of glutaraldehyde-fixed tissue abolishes the binding of WGA on the shaft membrane, strongly suggesting that the shaft membrane contains a large amount of glycolipids (Sandoz *et al.*, 1979). Since the CIH labeling is abolished by neuraminidase digestion, it may be concluded that most of these glycolipids are sialolipids. On the contrary, digestion of fixed tissue by proteases does not modify WGA labeling of the shaft membrane. Con A, which labels mainly α-D-mannosyl residues, localizes glycoproteins since mannosyl residues are not found in membrane glycolipids.

In conclusion, the different labeling studies clearly indicate that the IMPs of the shaft membrane and the ciliary necklace are glycoproteins with *N*-acetylglucosaminyl, α-D-mannosyl, and sialyl residues. These glycoproteins, in analogy with the biochemical results on the ciliary membranes of photoreceptors, could have molecular masses above 400 kDa and even as high as 600 kDa (Horst *et al.*, 1987). The tip crown is composed of glycoproteins which do not form IMPs. These glycoproteins contain *N*-acetylglucosaminyl, α-D-mannosyl, galactosyl, and sialyl residues. The shaft membrane appears to be poor in proteins and rich in sialolipids and cholesterol (Fig. 5).

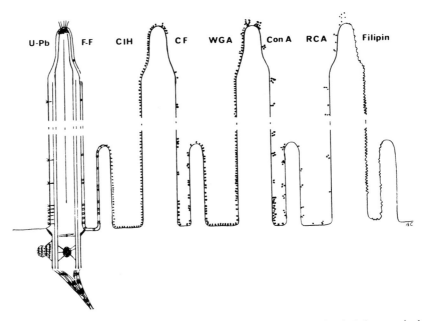

Figure 5. Schematic representations summarizing the ultrastructural and cytochemical data acquired on the ciliary, microvillous, and interciliary membranes. Thin section contrasted by uranyl citrate and lead citrate (U-Pb). Freeze-fracture (F-F) image revealing the intramembrane particles (IMPs). Labeling by colloidal iron hydroxide (CIH) at low pH and by cationized ferritin (CF) at neutral pH on fixed tissues revealing anionic sites of sialyl residues. Labeling by lectins such as wheat germ agglutinin (WGA) revealing *N*-acetylglucosaminyl and sialyl residues, Con A revealing α-D-mannose, *Ricinus communis* agglutinin I (RCA) revealing D-galactose. Labeling with filipin, an antibiotic inducing complex formation with free membrane cholesterol.

2.3. Ciliary Membrane–Cytoskeleton Relationships

In order to stick out from the cell surface, the microtubular structures must be strongly bound to the membrane. The ciliary microtubular skeleton is linked to the membrane at several levels (for review see Dentler, 1981, 1987).

Basal bodies dock with the membrane through anchoring fibers also named transitional fibers (Gibbons, 1961) or alar sheets (Anderson, 1972). These transitional fibers, whose nature has not been characterized, bind to IMP arrays regularly arranged in the interciliary membrane around the ciliary base in quail oviduct (Boisvieux-Ulrich *et al.*, 1977) as well as in *Chlamydomonas* (Weiss *et al.*, 1977). As shown by Gilula and Satir (1972), numerous fibrous links connect the ciliary necklace IMPs and the microtubular doublets and these authors proposed that the connections are arranged like champagne glasses in protozoan and invertebrate cilia. In oviduct, the fibrillar links appear as Y-shaped in cross sections (Anderson, 1974; Boisvieux-Ulrich *et al.*, 1977) but the three-dimensional organization has not been reconstructed yet. In the shaft all IMPs are also connected to the axonemal doublets by linear links which are shorter than the Y-shaped links of the necklace.

In ciliated protozoa, Dentler *et al.* (1980) suggested that peculiar dynein arms could connect axonemal tubulins and membrane tubulin molecules. Although such peripheral arms are present in oviduct cilia (Sandoz *et al.*, 1988), their molecular nature remains unknown. After lipid (and glycolipid) extraction, the lectin-labeled glycoproteins remain connected to the axonemal doublets by fibrillar links in oviduct (Sandoz *et al.*, 1979) as well as in the connecting cilium of photoreceptors (Horst *et al.*, 1987).

At the ciliary tip, the fibrils of the glycocalyx are very probably connected to the internal cap via transmembrane proteins following observations by Dentler and Le Cluyse (1982) and Le Cluyse and Dentler (1984). This cap is also associated with the plug structures which attach to the peripheral single microtubules (see Dentler, this volume).

3. Assembly of the Ciliary Membrane

The differentiation of committed oviduct cells is controlled by steroid hormones. Estrogen induces mitosis and ciliogenesis while progesterone stimulates the secretory process in birds (Kohler *et al.*, 1969; Oka and Schimke, 1969; Sandoz *et al.*, 1976; Pageaux *et al.*, 1986) as well as in mammals (Allen, 1928; Brenner, 1969a; Verhage *et al.*, 1973b, 1979; Jansen, 1980). Ciliogenesis is asynchronous all along the tract and is randomly dispersed or grouped in cell islets. Ciliated cells are the result of a terminal differentiation of postmitotic cells (see Lemullois *et al.*, 1988). They do not replicate their DNA (Conti *et al.*, 1981) and do not divide. They are always interspersed by committed cells or, in bird oviduct, by mucous cells which can divide and transdifferentiate into ciliated cells (Sandoz and Boisvieux-Ulrich, 1976).

3.1. Cytoplasmic Events

Ciliogenesis is preceded by microvillogenesis (Dirksen, 1974; Verhage *et al.*, 1979; Chailley *et al.*, 1982; Chailley and Boisvieux-Ulrich, 1985). Microvilli do not seem to develop in preorganized membrane domains (Chailley and Boisvieux-Ulrich, 1985). After mitosis the centrioles of the diplosome leave their position in the Golgi area and move toward the apical membrane. One of the centrioles generates a primary cilium without central tubules. Such primary cilia are observed by scanning electron microscopy at the surface of cells from immature oviductal tissues (Odor and Blandau, 1985) but are characteristic of quiescent cells (Roth *et al.*, 1988). As described by Anderson and Brenner (1971), some centrioles are generated around the centrioles of the diplosome according to the *centriolar pathway*. However, most of the centrioles are generated near the nucleus without contact with the centrioles of the diplosome according to the *acentriolar pathway* (Brenner, 1969a; Dirksen, 1971; Anderson and Brenner, 1971; Boisseau, 1973b; Verhage *et al.*, 1973b; Nayak *et al.*, 1976; for review see Lemullois *et al.*, 1988).

After centriologenesis, centrioles move toward the apical cell membrane. During the migration they acquire the anchoring fibers at their distal pole and the basal foot on one side, becoming basal bodies or kinetosomes (for review see Lemullois *et al.*, 1988). In some cases, basal bodies attach directly to the apical plasma membrane through anchoring fibers (Anderson and Brenner, 1971; Chailley *et al.*, 1982). In other cases, cytoplasmic vesicles attach to the migrating basal bodies through the anchoring fibers according to two patterns. In the first one, some small vesicles (about 0.1 μm in diameter) can bind

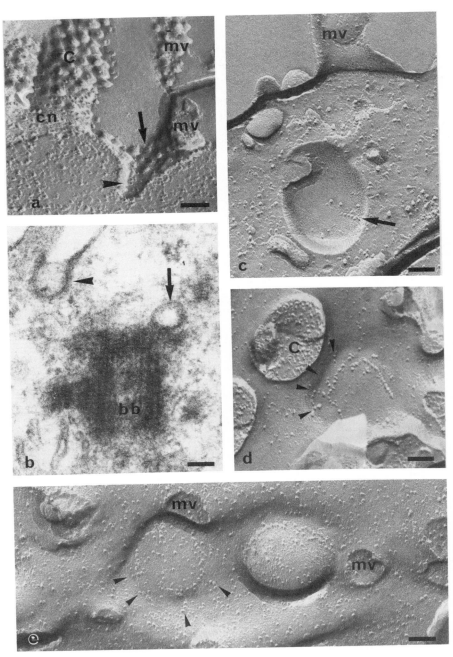

Figure 6. Apical surface organization during ciliogenesis. (a) Replica prepared after filipin–glutaraldehyde fixation. A pit (arrowhead) is observed near a microvillus (mv) and a growing cilium (C) whose ciliary necklace (cn) is at this stage formed by only three IMP strands. The pit membrane displays some IMPs but no filipin–cholesterol complexes which are present on ciliary shaft, microvillous, and interciliary (arrow) membranes. (b) Transmission electron micrograph of a migrating basal body (bb) with an associated vesicle (arrow). The vesicle

individually to the anchoring fibers arising from centriolar triplets (Fig. 6b). The basal body-vesicle association is then asymmetrical (Chailley *et al.*, 1982). In the second case, one basal body binds to a large cup-shaped vesicle which covers all of the distal part of the basal body as is the case in rat trachea (Sorokin, 1968) and in lamprey (Youson, 1982). In both cases, these vesicles are characterized by the presence of large IMPs (Fig. 6c) and they fuse with the apical plasma membrane through exocytosis. The cytoplasmic side of IMPs could be considered as receptors for the transitional fibers of basal bodies. Such presumptive receptors have not yet been identified. They could also be present in the lateral membrane, since basal bodies occasionally bind to the lateral membrane, thus generating cilia (Sandoz and Boisvieux-Ulrich, 1976; Chailley *et al.*, 1982). After treatment of ciliogenic cells with benzodiazepines such as diazepam or medazepam, the number of basal bodies docking with the lateral membrane increases (Boisvieux-Ulrich *et al.*, 1987).

In oviduct cells the docking of basal bodies with the plasma membrane does not seem to be controlled by a precise pattern as described in ciliated protozoa (Hufnagel, 1983). They often dock near and sometimes just under a microvillus (Chailley *et al.*, 1982; Lemullois *et al.*, 1988).

3.2. Plasma Membrane Events

The anchoring of the basal bodies to the plasma membrane induces or at least expresses membrane modifications which are visualized in freeze-fracture by a circular array and some radial arrays of IMPs (Chailley *et al.*, 1982) (Fig. 6d,e). Such IMP rings have also been observed in the plasma membrane of other ciliogenic cells such as thymic cells (Cordier and Haumont, 1979) and respiratory ciliated cells (Menco, 1980; Carson *et al.*, 1981). Ciliogenesis can be distinguished from deciliation by the organization of the IMPs and the presence of microvilli in ciliogenic cells. Some figures described as ciliogenesis were clearly figures of deciliation (Carson *et al.*, 1985) or at least figures of reciliation as shown in ciliated protozoa (Satir *et al.*, 1976).

The necklace forms progressively during ciliary growth (Chailley *et al.*, 1982). Only one IMP row is in place until the cilium buds out and grows to about 1 μm in length. Then, the other necklace rows appear progressively from the interciliary membrane in which they are probably integrated by an exocytosis process. They are added under the first row. At this stage, the IMPs are already the characteristic necklace glycoproteins as shown by their Con A affinity (Chailley *et al.*, 1982) (Fig. 7d–f) and the exclusion of cholesterol from their environment (Chailley and Boisvieux-Ulrich, 1985) (Fig. 7b). The ciliary shaft of the growing cilium is rich in IMPs, glycoproteins (Chailley *et al.*, 1982),

---

membrane shows a dense segment with beads in area of close association with the radial fibers developed at the distal end of basal body. Arrowhead indicates a pit. (c) Replica of a PF vesicle membrane showing IMP alignment (arrow) and large IMPs in the plasma membrane of a cell undergoing ciliogenesis. (d) Replica of PF apical plasma membrane on which the aligned IMPs organize as a circular arc in a smooth membrane depression corresponding to the site of docking of a basal body underneath the membrane. IMP radial arrays (arrowheads) arise from the IMP arc. C, fractured cilium. (e) Replica of PF plasma membrane. On the left, a ring with IMP radial arrays (arrowheads) evokes the docking of a basal body. On the right, a new cilium is budding with the first IMP row of the ciliary necklace. mv, fractured microvilli. (b, e) Reprinted from Chailley *et al.* (1982) with permission. Bars = 0.1 μm.

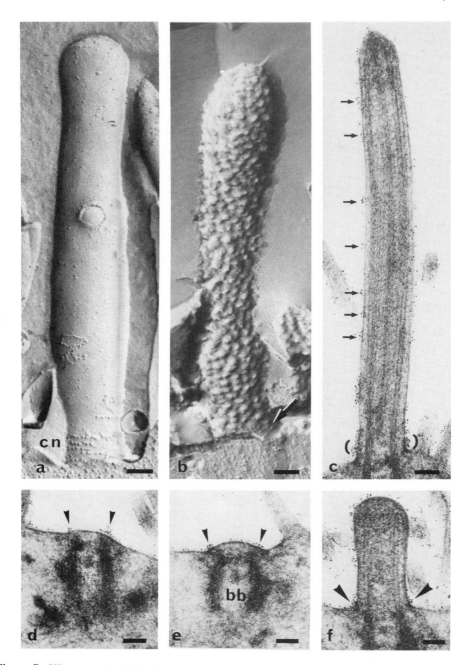

Figure 7. Ciliary growth. (a) Replica of a short cilium showing numerous scattered IMPs and a few IMP strands. The necklace (cn) is formed of three IMP rows. (b) Replica after filipin–glutaraldehyde fixation. The short cilium membrane shows an entirely waffled aspect because of the formation of filipin–cholesterol complexes, except in the region of the ciliary necklace, which has only one or two IMP rows (arrow). (c–f) Con A receptor detection. (c) In a growing cilium, glycosylated ferritin molecules visualize numerous Con A receptors

and cholesterol (Chailley and Boisvieux-Ulrich, 1985) (Fig. 7a–c). Thus, progressive development of the necklace forms a boundary between the shaft domain and the interciliary domain controlling the planar diffusion of molecules across the IMP rows. This diffusion barrier allows the unique properties of the ciliary membrane to be established and maintained. The glycoprotein content and IMP density decrease in the shaft membrane during elongation. The growth of the membrane is simultaneous with the axonemal elongation. Drugs, such as colchicine or nocodazole which stop axoneme polymerization, stop cilium elongation but do not prevent the complete ciliary necklace from being constructed (Boisvieux-Ulrich *et al.*, 1989b).

The origin of the ciliary membrane is not clearly established. At the beginning of differentiation of the ciliated cells, the Con A receptors decrease whereas the WGA receptors increase in the apical plasma membrane (unpublished results) suggesting changes in membrane glycoproteins according to biochemical data (De Rosa and Lucas, 1982). The apical membrane is also enriched in cholesterol. The total apical surface area increases rapidly through microvillogenesis and ciliogenesis; therefore, a supply of new membrane is necessary. The Golgi apparatus is well developed in ciliogenic cells and numerous vesicles are observed in the apical cytoplasm. However, the formation of endocytotic vesicles (Fig. 6a) and the presence of numerous multivesicular bodies indicate a recycling of the apical plasma membrane.

The microvillous domain which displays more IMPs and less cholesterol than the surrounding plasma membrane differentiates first without any visible diffusion barrier (Chailley and Boisvieux-Ulrich, 1985). The presumptive receptors for anchoring fibers which form IMP arrays could be integrated into the plasma membrane before the centriole migration or simultaneously with it. In the latter case, the vesicle carrying the receptors can bind to the migrating basal bodies. The necklace glycoproteins come probably through Golgi vesicles; the docking of basal bodies with the membrane could induce their assembly. However, only a study using antibodies against these IMPs could definitely show the pathway of these glycoproteins. Numerous cholesterol-rich vesicles probably supply the ciliary membrane.

The growing cilium has a constant diameter and ends in a round tip until the cilium has reached a certain length (about 4.5 μm) and the B microtubules have terminated their growth. Unlike the situation observed in amphibian palate cilia in which the ciliary tip glycocalyx appears early (Portman *et al.*, 1987), the ciliary crown glycocalyx of oviduct cilia appears only on fully (Dirksen and Satir, 1972) or nearly fully grown cilia (Chailley *et al.*, 1982). At this step, the cap structure is differentiated and the cap–membrane–crown attachment is present without preventing the last bit of ciliary growth (Fig. 8). Such a cap structure could stabilize the ciliary tips and control the terminal tubulin assembly. The short cilia (about 1 μm high) observed after colchicine or nocodazole action never display the crown of the ciliary tip while they differentiate a complete ciliary necklace (Boisvieux-Ulrich *et al.*, 1989b). The membrane of the ciliary tip in the growing cilium as

---

in the necklace area (bracket) and dispersed ones on the shaft membrane (arrows). (d, e) A few Con A receptors are already detected on the new bulge membrane (arrowheads) in the sites corresponding to the docking of a basal body (bb) which is located underneath the plasma membrane. (f) In a short cilium, the necklace, composed of one IMP row, displays Con A receptors (arrowheads). Con A receptors are also present on the newly formed shaft membrane. From Chailley *et al.* (1982). Bars = 0.1 μm.

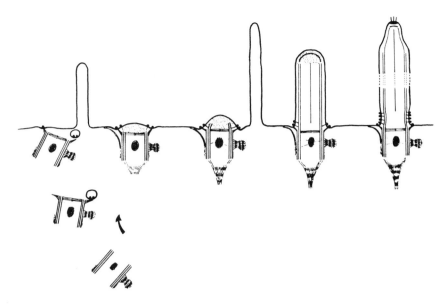

Figure 8. Schematic representation summarizing the sequence of events in ciliogenesis of quail oviduct. Modified from Chailley *et al.* (1982).

well as in the mature cilium is poor in cholesterol and IMPs, suggesting a higher fluidity favoring membrane elongation. During elongation, the cilia remain perpendicular to the cell surface and are unable to move until they reach a length of about 3 μm. The longer cilia bend slightly and the characteristic three-dimensional beating occurs simultaneously with the reorientation of all the basal bodies in the cell (Frisch and Farbman, 1968; Boisvieux-Ulrich *et al.*, 1985).

Ciliogenesis occurs asynchronously in the same cells giving them a daisy-like appearance in scanning electron microscopy since the peripheral cilia grow before the central ones (Dirksen, 1974; Rumery *et al.*, 1978). The same asynchrony is noted for the preceding microvillus development. During elongation of cilia, a concomitant elongation of microvilli can be seen (Verhage *et al.*, 1973a; Chailley and Boisvieux-Ulrich, 1985). Although the microvilli are arranged around the cilia in the oviduct, they do not exhibit a regular pattern as is the case in other ciliated epithelia (Reed *et al.*, 1984; Inoue and Hogg, 1977).

4. Deciliation

In mammals, the oviduct ciliated cells are sensitive to the ovarian steroid level in plasma (Verhage *et al.*, 1979; Jansen, 1980) and thus to the stage of the reproductive cycle (annual or menstrual cycle and pregnancy). Ciliated cells can deciliate during the midluteal phase but maintain the potential to reciliate (Brenner, 1969b; Verhage *et al.*,

1973a; West *et al.*, 1977; Rumery *et al.*, 1978). However, it was noted that the epithelial cells from the human fimbria and ampulla appear more resistant to deciliation than those from subhuman primates (Brenner and West, 1975). A copper intrauterine device used as contraceptive induces a deciliation of ampulla but only on patients who wore the copper device for more than 2 years (Eibschitz *et al.*, 1986).

In quail oviduct, deciliation occurs *in vivo* a few (3–6) weeks after ovariectomy of laying birds and is accelerated by progesterone injections, but is never synchronous (Boisvieux-Ulrich *et al.*, 1980). Different methods of deciliation were observed. Exceptionally a retraction of cilia into the cytoplasm has been found because of alterations of the membrane–axoneme connections. More often a scission of individual cilia occurs at the level of the ciliary necklace. A prior fusion of the ciliary membrane shaft producing polycilia is also observed. The ciliary necklace membrane does not fuse as has been described during the formation of the macrocilia of *Beroë* (Tamm and Tamm, 1988). In quail oviduct, the scission of polycilia from the cell body is achieved at the level of the ciliary necklace after depolymerization (or proteolysis) of the microtubules of the transition zone. After constriction of the ciliary membrane, part of the IMPs remain associated with the ciliary membrane while the rest remain with the cell membrane (Fig. 9a) This process is comparable with that previously described in protozoa (Satir *et al.*, 1976) but the level of scission differs since in the ciliated protozoa all the necklace IMPs remain associated with the plasma membrane.

In ciliated epithelium, the deciliation process usually is accompanied by microvillous vesiculation resulting in a smooth apical surface showing only membrane stubs with underlying basal bodies linked to the membrane (Boisvieux-Ulrich *et al.*, 1980). The stub membrane can be reorganized: the necklace IMPs lose their circular arrangement and scatter on the stub surface (Fig. 9b–d). Then cholesterol diffuses into the stub membrane (Fig. 9c). Although reciliation can occur during the mammalian reproductive cycle (Brenner, 1969b; Rumery *et al.*, 1978), the membrane changes accompanying reciliation have not been followed, as has been done in the protozoa (Satir *et al.*, 1976; Williams, 1983; for review see Lefebvre and Rosenbaum, 1986).

Partial deciliation of the oviduct can be experimentally induced following a strong mechanical stirring (Chailley *et al.*, 1986). On the contrary, all the methods used for the induction of deciliation in protozoa (calcium, pH shock, dibucaine) (Bloodgood, 1974; Satir *et al.*, 1976; Witman, 1986) are ineffective in the quail oviduct. In the vertebrate oviduct a massive deciliation is obtained after Ca^{2+} shock only when the ciliary membrane is extracted as described by Anderson (1974) and Torres *et al.* (1977) and in porcine trachea by Hastie *et al.* (1986). The properties of the vertebrate ciliary membrane could be responsible for the resistance to deciliation methods rather than differences in the microtubules. This resistance to deciliation is responsible for the lack of data on the biochemical composition of the vertebrate ciliary membrane.

5. Functions of Ciliary Membrane in Oviduct

Two different aspects will be discussed: the functions of the ciliary membrane and the functions of cilia in the oviduct.

Since demembranated cilia can be reactivated in the presence of ATP and Mg^{2+}

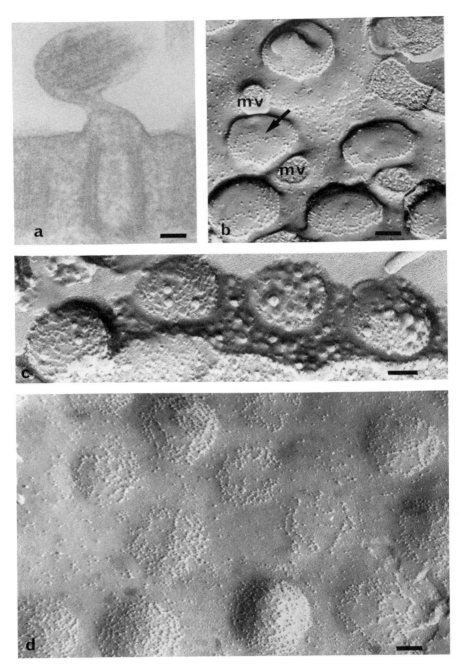

Figure 9. Deciliation. (a) Thin section (uncontrasted) showing the cilium scission which occurs in the ciliary necklace area by constriction of the membrane between the necklace rows. (b) Replica of an almost deciliated cell whose PF surface is covered with rounded and flat stubs. They bear one to three IMP rows which are beginning to disorganize (arrow). mv, microvilli. (c) Replica prepared after filipin–glutaraldehyde fixation. PF

(Gibbons and Gibbons, 1972; Torres *et al.,* 1977; Dirksen and Zeira, 1981), the ciliary membrane has been considered to play a minor role in ciliary motility. As shown in newt lung (Hard and Weaver, 1983), the metachronal wave beating is maintained in demembranated epithelium. However, it is now known that the ionic environment regulates ciliary beating. Although this regulation is less important in ciliated epithelia than in protozoa in which the direction of active stroke can be reversed (Naitoh and Kaneko, 1972; Preston and Saimi, this volume), the frequency and amplitude of the ciliary beat have been shown to be sensitive to Ca^{2+} in respiratory epithelia (Girard and Kennedy, 1986; Sanderson and Dirksen, 1986). The necklace glycoproteins could be implicated in the Ca^{2+} influx. The necklace is known to fix calcium in protozoan cilia (Plattner, 1975; Fisher *et al.,* 1976). Two functions have been suggested for necklace glycoprotein complexes: either as Ca^{2+} pumps or as Ca^{2+} channels (for review, see Satir, 1980). However, the direction of Ca^{2+} transport has not been demonstrated thus far. Ionic movements across the ciliary membrane are highly probable. The isolation of necklace glycoproteins and their integration into liposomes could allow the functional characterization of these proteins. Membrane-associated calmodulin (Gordon *et al.,* 1982) could be involved in the Ca^{2+} modulation of ciliary beating (Schultz *et al.,* 1983; for review see Andrivon, 1988).

Another obvious function of the IMPs in the ciliary and interciliary membrane is the binding of cilia and basal bodies to the plasma membrane. Furthermore, ciliary necklace IMPs may act as a selective diffusion barrier between the intraciliary membrane domain and the shaft membrane domain. Such control of the flux of membrane components is possible because IMPs are themselves linked to axonemal microtubules. Immobilization of IMPs by their link to cytoskeletal elements allows the great variation in IMP pattern that has been described in protist cilia and flagella (Bardele, 1981). A function of diffusion barrier for IMP strands was first proposed by Friend and Fawcett (1974) for the maintenance of the numerous membrane domains observed in spermatozoa. As already mentioned by Dentler (1981), the ciliary necklace is not a tight barrier since there is an enormous flow of lipids and proteins through the highly developed necklace (40 strands) of the connecting cilia of the rod photoreceptor. However, it could act as a *selective* barrier or a filter allowing the polarized crossing of the ciliary necklace only to the ciliary membrane components. The high content of cholesterol in the ciliary membrane may induce a characteristic mechanical membrane behavior during the ciliary beat, the cholesterol-rich membrane being more rigid than other areas of membrane. It may also prevent any endocytosis along the shaft (Anderson and Hein, 1977), whereas this process is highly developed elsewhere in oviduct ciliated cells (Parr *et al.,* 1988).

In oviduct ciliated cells, the height of the ciliary necklace is strictly related to the height of the 9 + 0 transitional zone, the basal end of the central pair microtubules always correlating with the level of the last strand of the necklace (Sandoz *et al.,* 1988). In the connecting cilium of immotile sensory cilia, the necklace is very well developed (Röhlich,

---

stub membrane is more or less altered by filipin–cholesterol complexes according to the IMP number and the state of disorganization of the IMP rows. (d) Replica of an entirely deciliated cell. The PF membrane displays numerous patterns of deciliation. Stubs are either still elevated (numerous IMPs) or integrated into the cell surface (few IMPs). IMP rows are disorganized and the microvilli have completely disappeared. (b, d) Reprinted from Boisvieux-Ulrich *et al.* (1980) with permission. Bars = 0.1 μm.

1975; Menco, 1980; Horst *et al.,* 1987). Thus, the development of the necklace cannot be correlated with ciliary movement.

The association of basal bodies with the plasma membrane appears to be a signal for the induction of tubulin polymerization onto the distal pole of basal bodies (Chailley *et al.,* 1982). This induction effect appears essential even when tubulin polymerization is stimulated by taxol (Boisvieux-Ulrich *et al.,* 1989a).

In trachea ciliated cells, many infectious agents which bind to the ciliary membrane induce deciliation and disorganization of necklace glycoproteins (Collier *et al.,* 1977; Carson *et al.,* 1979; Tuomanen, this volume). These pathogenic agents primarily act on the ciliary beat, suggesting that the ciliary membrane is implicated in information transfer and may also be directly involved in the regulation of ciliary beating (Andrivon, 1988). In human oviduct, many infectious agents, such as *Chlamydia,* induce ciliated cell alterations (Boisvieux-Ulrich *et al.,* unpublished observations). Such agents are responsible for a decrease in fertility (De Cherney and Laufer, 1986), demonstrating the importance of cilia in the reproductive process.

In the oviduct, the active stroke of the ciliary beat is directed from the ovary toward the uterus. The ciliary activity is probably involved in the transport of oviduct secretion, avoiding the accumulation of secretory product in the lumen and the development of germs. The ciliary beating of the fimbria seems to act in the capture of the ovum, surrounded by the cumulus, in two ways: first it induces a fluid movement which drives the ovum into the oviduct; then the negative charges of the ciliary tip crown directly interact with the cumulus and propel the ovum into the oviduct. When the anionic charges of the glycocalyx crown are neutralized by polycation binding, the ovum transport is inhibited (Norwood *et al.,* 1978; Norwood and Anderson, 1980). It may be assumed that the numerous negative charges of the tip crown (Sandoz *et al.,* 1979) increase both the cumulus–cilia relationships and the ovum–cilia relationships. In rabbits, experimental inversion of an oviduct segment in which the cilia keep their original active stroke for several weeks prevents the ovum progression through the reverse segment (Eddy *et al.,* 1982).

The cause of ectopic pregnancy in human females remains obscure despite many investigations. However, the authors speculate that loss of ciliated cells from the ampulla may decrease the efficiency of ovum uptake and transport, leading to a delay in the ovum's entering in the isthmus and subsequently to the uterine cavity (for review see Fredericks, 1986). The delay could result in tubal implantation. Deciliation may also explain the increased incidence of ectopic pregnancy after tuboplasty.

In the human immotile cilia syndrome, contradictory results were reported about female fertility. At the beginning of the exploration, it was believed that female fertility was unaffected by the ciliary immotility (Afzelius *et al.,* 1978). After the examination of a larger number of patients, it became clear that a ciliary dismotility involves a risk of infertility (Afzelius and Eliasson, 1983; Afzelius, 1985). However, no increase of ectopic pregnancy was observed.

Thus, cilia have important functions in ovum uptake and transport and also in the transport of oviduct secretion. However, peristaltic mechanisms may partially compensate for a role of normal ciliary activity in the oviducts. Some authors have suggested a dominant role for tubal smooth muscle contraction in fluid movements (Bourdage and Halbert, 1988). In fact, we have to consider the possibility that ciliary beat may be implicated in the movement of fluid (Leese, 1988) carrying the ovum near the ciliated epithelium, while the smooth muscle contraction may help the progression of spermatozoa in the middle of the oviduct toward the fertilization site.

References

Adoutte, A., Ramanathan, R., Lewis, R. M., Dute, R. R., Ling, K. Y., Kung, C., and Nelson, D. L., 1980, Biochemical studies of the excitable membrane of *Paramecium tetraurelia*. III. Proteins of cilia and ciliary membranes, *J. Cell Biol.* **84**:717–738.

Afzelius, B. A., 1985, The immotile-cilia syndrome: A microtubule-associated defect, *CRC Rev. Biochem.* **19**: 63–87.

Afzelius, B. A., and Eliasson, R., 1983, Male and female infertility problems in the immotile-cilia syndrome, *Eur. J. Respir. Dis.* **64**(Suppl. 127):144.

Afzelius, B. A., Camner, P., and Mossberg, B., 1978, On the function of cilia in the female reproductive tract, *Fertil. Steril.* **29**:72–74.

Aitken, R. N. C., and Johnston, H. S., 1963, Observations on the fine structure of the infundibulum of the avian oviduct, *J. Anat.* **97**:87–99.

Allen, E., 1928, Reactions of immature monkey (*Macacus rhesus*) to injections of ovarian hormone, *J. Morphol.* **46**:479–520.

Anderson, R. G. W., 1972, The three-dimensional structure of the basal body from the rhesus monkey oviduct, *J. Cell Biol.* **54**:246–265.

Anderson, R. G. W., 1974, Isolation of ciliated or unciliated basal bodies from the rabbit oviduct, *J. Cell Biol.* **60**:393–404.

Anderson, R. G. W., and Brenner, R. M., 1971, The formation of basal bodies (centrioles) in the rhesus monkey oviduct, *J. Cell Biol.* **50**:10–34.

Anderson, R. G. W., and Hein, C. E., 1977, Distribution of anionic sites on the oviduct ciliary membrane, *J. Cell Biol.* **72**:482–492.

Andrews, D., and Nelson, D. L., 1979, Biochemical studies of the excitable membrane of *Paramecium tetraurelia*. II. Phospholipids of ciliary and other membranes, *Biochim. Biophys. Acta* **550**:174–187.

Andrivon, C., 1988, Membrane control of the ciliary movement, *Biol. Cell.* **63**:133–142.

Bardele, C. F., 1981, Functional and phylogenetic aspects of the ciliary membrane: A comparative freeze-fracture study, *BioSystems* **14**:403–421.

Blandau, R. J., Bourdage, R., and Halbert, S., 1979, Tubal transport, in: *The Biology of the Fluids of the Female Genital Tract* (F. K. Beller and G. F. B. Schumacher, eds.), Elsevier/North-Holland, Amsterdam, pp. 319–333.

Bloodgood, R. A., 1974, Resorption of organelles containing microtubules, *Cytobios* **9**:143–161.

Boisseau, C., 1973a, Etude ultrastructurale de l'oviducte du triton *Pleurodeles waltlii* Michah. I. Ultrastructure des cellules épithéliales de l'oviducte moyen différencié, *J. Microsc. (Paris)* **18**:341–358.

Boisseau, C., 1973b, Etude ultrastructurale de l'oviducte du triton *Pleurodeles waltlii* Michah. II. Morphogénèse des glandes et différenciation des cellules épithéliales de l'oviducte moyen, *J. Microsc. (Paris)* **18**:359–382.

Boisvieux-Ulrich, E., Sandoz, D., and Chailley, B., 1977, A freeze-fracture and thin section study of the ciliary necklace in quail oviduct, *Biol. Cell.* **30**:245–252.

Boisvieux-Ulrich, E., Sandoz, D., and Chailley, B., 1980, A thin section and freeze-fracture study of deciliation in bird oviduct, *Biol. Cell.* **37**:261–268.

Boisvieux-Ulrich, E., Lainé, M. C., and Sandoz, D., 1985, The orientation of ciliary basal bodies in quail oviduct is related to the ciliary beating cycle commencement, *Biol. Cell.* **55**:147–150.

Boisvieux-Ulrich, E., Lainé, M. C., and Sandoz, D., 1987, *In vitro* effects of benzodiazepines on ciliogenesis in the quail oviduct, *Cell Motil. Cytoskel.* **8**:333–344.

Boisvieux-Ulrich, E., Lainé, M. C., and Sandoz, D., 1989a, *In vitro* effects of taxol on ciliogenesis in quail oviduct, *J. Cell Sci.* **92**:9–20.

Boisvieux-Ulrich, E., Lainé, M. C., and Sandoz, D., 1989b, *In vitro* effects of colchicine and nocodazole on ciliogenesis in quail oviduct, *Biol. Cell.* **67** (1), in press.

Bourdage, R. J., and Halbert, S. A., 1988, Distribution of embryos and 500 μm microspheres in the rabbit oviduct: Controls for acute motion analysis during transport, *Biol. Reprod.* **38:**282–291.

Brenner, R. M., 1969a, The biology of oviductal cilia, in: *The Mammalian Oviduct* (E. S. E. Hafez and R. J. Blandau, eds.), University of Chicago Press, Chicago, pp. 203–229.

Brenner, R. M., 1969b, Renewal of oviduct cilia during the menstrual cycle of the rhesus monkey, *Fertil Steril.* **20:**599–611.

Brenner, R. M., and West, N. B., 1975, Hormonal regulation of the reproductive tract in female mammals, *Annu. Rev. Physiol.* **37:**273–302.

Burkett, B. N., Schulte, B. A., and Spicer, S. S., 1987, Histochemical evaluation of glycoconjugates in the male reproductive tract with lectin–horseradish peroxidase conjugates. II: Staining of ciliated cells, basal cells, flask cells, and clear cells in the mouse, *Am. J. Anat.* **178:**23–29.

Carson, J. L., Collier, A. M., and Clyde, W. A., 1979, Ciliary membrane alterations occurring in experimental *Mycoplasma pneumoniae* infection, *Science* **206:**349–351.

Carson, J. L., Collier, A. M., Knowles, M. R., Boucher, R. C., and Rose, J. G., 1981, Morphometric aspects of ciliary distribution and ciliogenesis in human nasal epithelium, *Proc. Natl. Acad. Sci. USA* **78:**6996–6999.

Carson, J. L., Collier, A. M., Knowles, M. R., and Boucher, R. C., 1985, Ultrastructural characterization of epithelial cell membranes in normal human conducting airway epithelium: A freeze-fracture study, *Am. J. Anat.* **173:**257–268.

Chailley, B., and Boisvieux-Ulrich, E., 1985, Detection of plasma membrane cholesterol by filipin during microvillogenesis and ciliogenesis in quail oviduct, *J. Histochem. Cytochem.* **33:**1–10.

Chailley, B., N'Diaye, A., Boisvieux-Ulrich, E., and Sandoz, D., 1981, Comparative study of the distribution of fuzzy coat, lectin receptors, and intramembrane particles of the ciliary membrane, *Eur. J. Cell Biol.* **25:**300–307.

Chailley, B., Boisvieux-Ulrich, E., and Sandoz, D., 1982, Ciliary membrane events during ciliogenesis in quail oviduct, *Biol. Cell.* **46:**51–64.

Chailley, B., Bork, K., Gounon, P., and Sandoz, D., 1986, Immunological detection of actin in isolated cilia from quail oviduct, *Biol. Cell.* **58:**43–52.

Chailley, B., Frappier, T., Regnouf, F., and Lainé, M. C., 1989, Immunological detection of spectrin during differentiation and in mature ciliated cells from quail oviduct, *J. Cell Sci.* **93:** 683–690.

Collier, A. M., Peterson, L. P., and Baseman, J. B., 1977, Pathogenesis of infection with *Bordetella pertussis* in hamster tracheal organ culture, *J. Infect. Dis.* **136:**S196–S203.

Conti, C. J., Conner, E. A., Gimenez-Conti, I. B., Silverberg, S. G., and Gerschenson, L. E., 1981, Regulation of ciliogenesis and proliferation of uterine epithelium by 20α-hydroxy-pregn-4-en-3-one administration and withdrawal in ovariectomized rabbits, *Biol. Reprod.* **24:**903–911.

Cordier, A. C., and Haumont, S., 1979, Origin of necklace particles in thymic ciliating cells, *Am. J. Anat.* **156:** 91–97.

Danon, D., Goldstein, L., Marikovsky, Y., and Skutelsky, E., 1972, Use of cationized ferritin as a label of negative charges on cell surfaces, *J. Ultrastruct. Res.* **38:**500–510.

De Cherney, A. H., and Laufer, N., 1986, Pelvic infection and infertility: Mechanisms of action, in: *The Fallopian Tube: Basic Studies and Clinical Contributions* (A. M. Siegler, ed.), Futura Publishing, New York, pp. 201–209.

Dentler, W. L., 1981, Microtubule–membrane interactions in cilia and flagella, *Int. Rev. Cytol.* **72:**1–47.

Dentler, W. L., 1987, Cilia and flagella, *Int. Rev. Cytol. Suppl.* **17:**391–456.

Dentler, W. L., and Le Cluyse, E. L., 1982, Microtubule capping structures at the tips of tracheal cilia: Evidence for their firm attachment during ciliary bend formation and the restriction of microtubule sliding, *Cell Motil.* **2:**549–573.

Dentler, W. L., Pratt, M. H., and Stephens, R. E., 1980, Microtubule–membrane interaction in cilia. II. Photochemical cross-linking of bridge structures and the identification of a membrane-associated dynein-like ATPase, *J. Cell Biol.* **84:**381–403.

De Rosa, P. A., and Lucas, J. J., 1982, Estrogen-induced changes in chick oviduct membrane glycoproteins, *J. Biol. Chem.* **257:**1017–1024.

Dirksen, E. R., 1971, Centriole morphogenesis in developing ciliated epithelium of the mouse oviduct, *J. Cell Biol.* **51:**286–302.

Dirksen, E. R., 1974, Ciliogenesis in the mouse oviduct: A scanning electron microscope study, *J. Cell Biol.* **62:**899–904.

Dirksen, E. R., and Satir, P., 1972, Ciliary activity in the mouse oviduct as studied by transmission and scanning electron microscopy, *Tissue Cell* **4:**389–404.

Dirksen, E. R., and Zeira, M., 1981, Microtubule sliding in cilia of the rabbit trachea and oviduct, *Cell Motil.* **1:** 247–260.

Dirksen, E. R., Gilula, N. B., Davidson, L., Schooley, C., Satir, B., and Satir, P., 1971, New aspects of cilia structure, *Anat. Rec.* **169:**464.

Eddy, C. A., Archer, D. R., and Pauerstein, C. J., 1982, Failure of cilia to reprogram following segmental ampullary reversal of rabbit oviduct, *Experientia* **38:**104–105.

Eibschitz, I., Sharf, M., and de Vries, K., 1986, The effect of an IUD on tubal pathophysiology, in: *The Fallopian Tube: Basic Studies and Clinical Contributions* (A. M. Siegler, ed.), Futura Publishing, New York, pp. 239–246.

Elias, P. M., Friend, D. S., and Goerke, J., 1979, Membrane sterol heterogeneity. Freeze-fracture detection with saponins and filipin, *J. Histochem. Cytochem.* **27:**1247–1260.

Fertuck, H. C., and Newstead, J. D., 1970, Fine structural observations on magnum mucosa in quail and hen oviducts, *Z. Zellforsch. Mikrosk. Anat.* **103:**447–459.

Fisher, G., Kaneshiro, E. S., and Peters, P. D., 1976, Divalent cation affinity sites in *Paramecium aurelia*, *J. Cell Biol.* **69:**429–442.

Fredericks, C. M., 1986, Morphological and functional aspects of the oviductal epithelium, in: *The Fallopian Tube: Basic Studies and Clinical Contributions* (A. M. Siegler, ed.), Futura Publishing, New York, pp. 67–80.

Friend, D. S., and Fawcett, D. W., 1974, Membrane differentiations in freeze-fractured mammalian sperm, *J. Cell Biol.* **63:**641–664.

Frisch, D., and Farbman, A., 1968, Development of order during ciliogenesis, *Anat. Rec.* **162:**221–232.

Gibbons, B. H., and Gibbons, I. R., 1972, Flagellar movement and adenosine triphosphatase activity in sea urchin sperm extracted with Triton X-100, *J. Cell Biol.* **54:**75–97.

Gibbons, I. R., 1961, The relationship between the fine structure and direction of beat in gill cilia of a lamellibranch mollusc, *J. Biophys. Biochem. Cytol.* **11:**179–205.

Gilula, N. B., and Satir, P., 1972, The ciliary necklace. A ciliary membrane specialization, *J. Cell Biol.* **53:** 494–509.

Girard, P. G., and Kennedy, J. R., 1986, Calcium regulation of ciliary activity in rabbit tracheal epithelial explants and outgrowth, *Eur. J. Cell Biol.* **40:**203–209.

Gordon, R. E., Williams, K. B., and Puszkin, S., 1982, Immune localization of calmodulin in the ciliated cells of hamster tracheal epithelium, *J. Cell Biol.* **95:**57–63.

Hafez, E. S. E., and Blandau, R. J. (eds.), 1969, *The Mammalian Oviduct*, University of Chicago Press, Chicago.

Hard, R., and Weaver, A., 1983, Newt lungs: A versatile system for studying mucociliary transport, *Tissue Cell* **15:**217–226.

Hastie, A. T., Dicker, D. T., Hingley, S. T., Kueppers, F., Higgins, M. L., and Weinbaum, G., 1986, Isolation of cilia from porcine tracheal epithelium and extraction of dynein arms, *Cell Motil. Cytoskel.* **6:** 25–34.

Horst, C. J., Forestner, D. M., and Besharse, J. C., 1987, Cytoskeletal–membrane interactions: A stable interaction between cell surface glycoconjugates and doublet microtubules of the photoreceptor connecting cilium, *J. Cell Biol.* **105:**2973–2987.

Hufnagel, L. A., 1983, Freeze-fracture analysis of membrane events during early neogenesis of cilia in *Tetrahymena:* Changes in fairy-ring morphology and membrane topography, *J. Cell Sci.* **60:**137–156.

Inoue, S., and Hogg, J. C., 1977, Freeze-etch study of the tracheal epithelium of normal guinea pigs with particular reference to intercellular junctions, *J. Ultrastruct. Res.* **61:**89–99.

Jansen, R. P. S., 1980, Cyclic changes in the human fallopian tube isthmus and their functional importance, *Am. J. Obstet. Gynecol.* **136:**292–308.

Klotz, C., Bordes, N., Lainé, M. C., Sandoz, D., and Bornens, M., 1986a, A protein of 175,000 daltons associated with striated rootlets in ciliated epithelia, as revealed by a monoclonal antibody, *Cell Motil. Cytoskel.* **6:**56–67.

Klotz, C., Bordes, N., Lainé, M. C., Sandoz, D., and Bornens, M., 1986b, Myosin at the apical pole of ciliated epithelial cells as revealed by a monoclonal antibody, *J. Cell Biol.* **103**:613–619.

Kohler, P. O., Grimley, P. M., and O'Malley, B. W., 1969, Estrogen induced cytodifferentiation of the ovalbumin-secreting glands of the chick oviduct, *J. Cell Biol.* **40**:8–27.

Le Cluyse, E. L., and Dentler, W. L., 1984, Asymmetrical microtubule capping structures in frog palate cilia, *J. Ultrastruct. Res.* **86**:75–85.

Lee, M. C., Wu, T. C., Wan, Y. J., and Damjanov, I., 1983, Pregnancy-related changes in the mouse oviduct and uterus revealed by differential binding of fluoresceinated lectins, *Histochemistry* **79**:365–375.

Leese, H. J., 1988, The formation and function of oviduct fluid, *J. Reprod. Fertil.* **82**:843–856.

Lefebvre, P. A., and Rosenbaum, J. L., 1986, Regulation of the synthesis and assembly of ciliary and flagellar proteins during regeneration, *Annu. Rev. Cell Biol.* **2**:517–546.

Lemullois, M., Boisvieux-Ulrich, E., Lainé, M. C., and Sandoz, D., 1988, Development and functions of the cytoskeleton during ciliogenesis in metazoa, *Biol. Cell.* **63**:195–208.

Menco, B. P. M., 1980, Qualitative and quantitative freeze-fracture studies on olfactory and respiratory epithelial surfaces of frog, ox, rat and dog. IV. Ciliogenesis and ciliary necklaces (including high-voltage observations), *Cell Tissue Res.* **212**:1–16.

Menghi, G., Bondi, A. M., and Materazzi, G., 1985, Distribution of lectin binding sites in rabbit oviduct, *Anat. Rec.* **211**:279–284.

Moller, P. C., Chang, J. P., and Partridge, L. R., 1981, The distribution of cationized ferritin receptors on ciliated epithelial cells of rat trachea, *Tissue Cell* **13**:731–737.

Montesano, R., 1979, Inhomogeneous distribution of filipin–sterol complexes in the ciliary membrane of rat tracheal epithelium, *Am. J. Anat.* **156**:139–145.

Naitoh, Y., and Kaneko, H., 1972, Reactivated Triton-extracted models of *Paramecium:* Modification of ciliary movement by calcium ions, *Science* **176**:523–524.

Nayak, R. K., 1972, Ultrastructural localization of phosphatase activities in the bovine, porcine and rabbit oviduct, *J. Histochem. Cytochem.* **20**:841–842.

Nayak, R. K., and Wu, A. S. H., 1975, Fine structural localization of adenosine tri-phosphatase in the epithelium of the rabbit oviduct, *J. Anim. Sci.* **41**:1077–1082.

Nayak, R. K., Zimmerman, D. R., and Albert, E. N., 1976, Electron microscopic studies of estrogen-induced ciliogenesis and secretion in uterine tube of the gilt, *Am. J. Vet. Res.* **37**:189–197.

Norwood, J. T., and Anderson, R. G. W., 1980, Evidence that adhesive sites on the tips of oviduct cilia membranes are required for ovum pickup in situ, *Biol. Reprod.* **23**:788–791.

Norwood, J. T., Hein, C. E., Halbert, S. A., and Anderson, R. G. W., 1978, Polycationic macromolecules inhibit cilia-mediated ovum transport in the rabbit oviduct, *Proc. Natl. Acad. Sci. USA* **75**:4413–4416.

Odor, D. L., and Blandau, R. J., 1985, Observations on the solitary cilium of rabbit oviductal epithelium: Its motility and ultrastructure, *Am. J. Anat.* **174**:437–453.

Odor, D. L., and Blandau, R. J., 1988, Light and electron microscopic observation on the cervical epithelium of the rabbit. I, *Am. J. Anat.* **181**:289–319.

Oka, T., and Schimke, R. T., 1969, Interaction of estrogen and progesterone in chick oviduct development. II. Effect of estrogen and progesterone on tubular gland cell function, *J. Cell Biol.* **43**:123–137.

Pageaux, J. F., Laugier, C., Pal, D., D'Almeida, M. A., Sandoz, D., and Pacheco, H., 1986, Magnum morphogenesis during the natural development of the quail oviduct: Analysis of egg white proteins and progesterone receptor concentration, *Biol. Reprod.* **35**:657–666.

Parr, E. L., Tung, H. N., and Parr, M. B., 1988, Endocytosis in the epithelium of the mouse oviduct, *Am. J. Anat.* **181**:393–400.

Plattner, H., 1975, Ciliary granule plaques, membrane intercalated particle aggregates associated with Ca^{++} binding sites in *Paramecium, J. Cell Sci.* **18**:257–269.

Portman, R. W., Le Cluyse, E. L., and Dentler, W. L., 1987, Development of microtubule capping structures in ciliated epithelial cells, *J. Cell Sci.* **87**:85–94.

Reed, W., Avolio, J., and Satir, P., 1984, The cytoskeleton of the apical border of the lateral cells of freshwater mussel gill: Structural integration of microtubule and actin filament-based organelles, *J. Cell Sci.* **68**:1–33.

Rhoads, D. E., and Kaneshiro, E. S., 1979, Characterization of phospholipids from *Paramecium tetraurelia* cells and cilia, *J. Protozool.* **26**:329–338.

Röhlich, P., 1975, The sensory cilium of retinal rods is analogous to the transitional zone of motile cilia, *Cell Tissue Res.* **161**:421–430.

Romanoff, A. L., and Romanoff, A. J. (eds.), 1949, *The Avian Egg.* Wiley/Chapman & Hall, New York/London, pp. 174–252.

Roth, K. E., Rieder, C. L., and Bowser, S. S., 1988, Flexible-substratum technique for viewing cells from the side: Some *in vivo* properties of primary (9 + 0) cilia in cultured kidney epithelia, *J. Cell Sci.* **89**:457–466.

Rumery, R. E., Gaddum-Rosse, P., Blandau, R. J., and Odor, D. L., 1978, Cyclic changes in ciliation of the oviductal epithelium in the pig-tailed macaque (*Macaca nemestrina*), *Am. J. Anat.* **153**:345–365.

Sanderson, M. J., and Dirksen, E. R., 1986, Mechanosensitivity of cultured ciliated cells from the mammalian respiratory tract: Implications for the regulation of mucociliary transport, *Proc. Natl. Acad. Sci. USA* **83**:7302–7306.

Sandoz, D., and Boisvieux-Ulrich, E., 1976, Ciliogénèse dans les cellules à mucus de l'oviducte de caille. I. Etude ultrastructurale chez la caille en ponte, *J. Cell Biol.* **71**:449–459.

Sandoz, D., Ulrich, E., and Brard, E., 1971, Etude des ultrastructures du magnum des oiseaux. I. Evolution au cours du cycle de ponte chez la poule *Gallus domesticus, J. Microsc. (Paris)* **11**:371–400.

Sandoz, D., Boisvieux-Ulrich, E., Laugier, C., and Brard, E., 1976, Ciliogénèse dans les cellules à mucus de l'oviducte de caille. II. Contrôle hormonal, *J. Cell Biol.* **71**:460–471.

Sandoz, D., Boisvieux-Ulrich, E., and Chailley, B., 1979, Relationships between intramembrane particles and glycoconjugates in the ciliary membrane of the quail oviduct, *Biol. Cell.* **36**:267–280.

Sandoz, D., Chailley, B., Boisvieux-Ulrich, E., Lemullois, M., Lainé, M. C., and Bautista-Harris, G., 1988, Organization and functions of cytoskeleton in metazoan ciliated cells, *Biol. Cell.* **63**:183–193.

Satir, B. H., 1980, The role of local design in membranes, in: *Membrane–Membrane Interactions* (N. B. Gilula, ed.), Raven Press, New York, pp. 45–58.

Satir, B., Sale, W. S., and Satir, P., 1976, Membrane renewal after dibucaine deciliation of *Tetrahymena.* Freeze-fracture technique, cilia, membrane structure, *Exp. Cell Res.* **97**:83–91.

Schulte, B. A., and Spicer, S. S., 1985, Histochemical methods for characterizing secretory and cell surface sialoglycoconjugates, *J. Histochem. Cytochem.* **33**:427–438.

Schulte, B. A., Rao, K. P. P., Kreutner, A., Thomopoulos, G. N., and Spicer, S. S., 1985, Histochemical examination of glycoconjugates of epithelial cells in the human Fallopian tube, *Lab. Invest.* **52**:207–219.

Schultz, J. E., Schönefeld, U., and Klumpp, S., 1983, Calcium/calmodulin-regulated guanylate cyclase and calcium-permeability in the ciliary membrane from *Tetrahymena, Eur. J. Biochem.* **137**:89–94.

Sorokin, I., 1968, Reconstruction of centriole formation and ciliogenesis in mammalian lungs, *J. Cell Sci.* **3**:207–230.

Stephens, R. E., 1977, Major membrane protein differences in cilia and flagella: Evidence for a membrane-associated tubulin, *Biochemistry* **16**:2047–2058.

Stephens, R. E., 1986, Membrane tubulin, *Biol. Cell.* **57**:95–110.

Tamm, S. L., and Tamm, S., 1988, Development of macrociliary cells in *Beroë.* II. Formation of macrocilia, *J. Cell Sci.* **89**:81–95.

Temmink, J. H. M., 1979, Application of cytochemical methods to electron microscope investigations of cell surface receptors, *Biol. Cell.* **36**:227–236.

Thiery, J. P., 1967, Mise en évidence des polysaccharides sur coupes fines en microscopie électronique, *J. Microsc. (Paris)* **6**:987–1018.

Thompson, G. A., and Nozawa, Y., 1977, *Tetrahymena:* A system for studying dynamic membrane alterations within the eukaryotic cell, *Biochim. Biophys. Acta.* **472**:55–92.

Torres, L. D., Renaud, F. L., and Portocarrero, C., 1977, Studies on reactivated cilia, II. Reactivation of ciliated cortices from the oviduct of *Anolis cristatellus, Exp. Cell Res.* **108**:311–320.

Vassar, P. S., Hards, J. M., Brooks, D. E., Hagenberger, B., and Seaman, G. V. F., 1972, Physicochemical effects of aldehydes on the human erythrocyte, *J. Cell Biol.* **53**:809–818.

Verhage, H. G., Abel, J. H., Tietz, W. J., and Barrau, M. D., 1973a, Development and maintenance of the oviductal epithelium during the estrous cycle of the bitch, *Biol. Reprod.* **9**:460–474.

Verhage, H. G., Abel, J. H., Tietz, W. J., and Barrau, M. D., 1973b, Estrogen-induced differentiation of the oviductal epithelium in prepubertal dogs, *Biol. Reprod.* **9**:475–488.

Verhage, H. G., Bareither, M. L., Jaffe, R. C., and Akbar, M., 1979, Cyclic changes in ciliation, secretion and cell height of the oviductal epithelium in women, *Am. J. Anat.* **156**:505–522.

Weiss, R. L., Goodenough, D. A., and Goodenough, U. W., 1977, Membrane particle arrays associated with the basal body and with contractile vacuole secretion in *Chlamydomonas, J. Cell Biol.* **72**:133–143.

West, N. B., Verhage, H. G., and Brenner, R. M., 1977, Changes in nuclear estradiol receptor and cell structure during estrous cycles and pregnancy in the oviduct and uterus of cats, *Biol. Reprod.* **17:**138–143.

Williams, N. E., 1983, Surface membrane regeneration in deciliated *Tetrahymena, J. Cell Sci.* **62:**407–417.

Witman, G. B., 1986, Isolation of *Chlamydomonas* flagella and flagellar axonemes, *Methods Enzymol.* **134:** 280–290.

Wu, T. C., Wan, Y. J., and Damjanov, I., 1983, Distribution of *Bandeiraea simplicifolia* lectin binding sites in the genital organs of female and male mice, *Histochemistry* **77:**233–241.

Wyburn, G. M., Johnston, H. S., Draper, M. H., and Davidson, M. F., 1970, The fine structure of the infundibulum and magnum of the oviduct of *Gallus domesticus, Q. J. Exp. Physiol.* **55:**213–232.

Youson, J. H., 1982, Replication of basal bodies and ciliogenesis in a ciliated epithelium of the lamprey, *Cell Tissue Res.* **223:**255–266.

The Surface of Mammalian Respiratory Cilia
Interactions between Cilia and Respiratory Pathogens

Elaine Tuomanen

1. Introduction

Ciliated cells are the most numerous (30–65%) of the eight cell types on the human respiratory epithelium (Jeffrey and Reid, 1975; Plopper *et al.*, 1983). These cells are restricted to the central airways: the trachea, where they form an almost continuous cover, and the primary, secondary, and tertiary bronchioles, where the proportion of other cells increases. Thus, the surface of the central airways can be viewed as a continuous carpet of cilia with intervening islands of nonciliated cells (e.g., goblet or serous cells). Infection of the ciliated respiratory epithelium (i.e., bronchitis) is clearly distinguishable pathologically from infection of the nonciliated terminal alveoli (i.e., pneumonia). Each ciliated cell extends a tuft of approximately 200 cilia from one face of the cell body (5-μm cell diameter, density of cilia 6–8 μm^{-2}) into the airway lumen (Sleigh *et al.*, 1988). Human cilia measure 6 μm in length in larger airways and 5 μm in smaller bronchioles (Sleigh *et al.*, 1988). Such shortening of cilia toward the lung periphery has also been documented in mouse and rat (Jeffrey and Reid, 1975).

The cilium is anchored to the cell by a basal body with a basal foot, short striated rootlets, and attached cytoplasmic microtubules (Sleigh *et al.*, 1988). Basal feet are aligned on a single cell and thus the effective ciliary strokes have a common orientation (Sleigh *et al.*, 1988). The ciliary carpet is overlaid by a patchy gel-like mucous blanket which is slowly propelled by the ciliary beat from the alveoli toward the trachea (ciliary beat frequency in human is 14 Hz; Low *et al.*, 1984). Nonspecific (nonimmune) clearance of pathogens from the lung is contingent on entrapment of the pathogen (or particles in the range of 0.1 to 1.0 μm which reach the lower airways in aerosols) in the moving mucous blanket.

Elaine Tuomanen • The Rockefeller University, New York, New York 10021.

The surface of the mammalian respiratory cilium is a highly complex, organized membrane bilayer coated with a complex surface-associated glycocalyx derived from the copious secretions of neighboring nonciliated cells. The differentiation of the ciliary surface is remarkable, presenting an intricate arrangement of surface-exposed ligands. The sine qua non of infection of this naturally sterile site is anchoring or adherence of the pathogen to the ciliated epithelium, thereby protecting the pathogen from clearance driven by the constant ciliary beat. In comparison to the multiplicity of pathogens causing pneumonia, it is the rare pathogen that has evolved the highly specific adhesin structures on its surface needed to attach to and thereby infect cilia. It is the aim of this review to present a current description of the mammalian respiratory ciliary surface in terms of membrane structure and biochemical composition with an emphasis on studies elucidating features of human respiratory cilia. Since information on the composition of the human ciliary surface is limited, it is necessary to draw on widely diverse studies in animals and humans in order to construct a composite picture of a representative cilium. This ciliary map will then be translated to the process of disease by examination of the host–pathogen interactions which govern infections on the ciliated respiratory epithelium. Although the number of cilium-directed pathogens is few (certainly less than ten are characterized), virtually every person experiences many of these infections, be they as mild as the common cold or as severe as whooping cough.

2. The Structure and Cytochemistry of the Surface of Respiratory Cilia

Knowledge of the structure of the respiratory ciliary membrane of humans is only rudimentary at the current time. Much of the structure is inferred from data gathered in mammals and amphibia where some systematic studies have been carried out. This surprising information gap arises because most studies have focused on the axoneme, the 9 + 2 filament array which forms the core of the cilium and is responsible for ciliary motility (for review see Gibbons, 1981). In order to study the axoneme, the ciliary membrane is systematically destroyed by detergents, leaving no opportunity to investigate membrane structure or function. This discussion will be directed at summarizing the current view of the emerging complex structure of the surface of the human respiratory ciliary membrane.

2.1. Structure

The mammalian respiratory cilium is characterized by several highly differentiated structures on its surface (Fig. 1). In general, these structures are features of all cilia. However, some differences in structure do occur between cilia from different species and between cilia from different body sites. Species-specific differences are exemplified by the granule plaque found at the base of *Paramecium* but not mammalian cilia (Plattner, 1975). Structural differences specific to body site are common, e.g., membrane particle density differs between nasal, tracheal, and oviductal cilia within one species, but is similar for tracheal cilia from dog, ox, rat, and frog (Menco, 1980). The composite cilium shown in Fig. 1 is derived from many sources and the reader is referred to the text for the specific situation in which a feature of the surface has been observed.

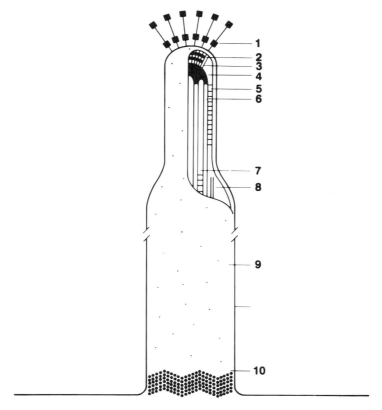

Figure 1. Schematic diagram of the structure of a human respiratory cilium. 1: ciliary crown with six bristles; 2: multilayered plate; 3: dense filaments connecting crown and cap; 4: cap; 5: lateral spokes; 6: internal microtubule; 7: radial spokes; 8: external microtubule; 9: cell membrane; 10: ciliary necklace. Adapted from Foliquet and Puchelle (1986).

2.1.1. Crown. At the tip of the cilium is a tuft of fibrils termed the ciliary crown. This structure was first observed in mouse oviduct (Dirksen and Satir, 1972) and subsequently in thymic cysts of nude mice (Cordier, 1975), and the mammalian trachea (Kuhn and Engleman, 1978). A related structure exists on cilia of the oral membranelle of *Tetrahymena pyriformis* (Kuhn and Engleman, 1978). However, in contrast to protozoa, the crown of mammalian avian, and amphibian epithelial cells is distinguished by 4–6 bristles approximately 35 μm long which extend from the cap–plug complex at the terminus of the axoneme through the plasma membrane into the periciliary space (Kuhn and Engleman, 1978; Dentler and LeCluyse, 1982a; Dentler, 1987, this volume). The tips of the coronal bristles extend into the gel-like mucous blanket where they are thought to force mucous translocation during the ciliary beat. The crown is present only on fully developed cilia, suggesting it may function as a cap to prevent further ciliary elongation (Kuhn and Engleman, 1978; Dentler and LeCluyse, 1982b).

The core of the coronal fibrile is said to be composed of a nonsialylated protein based on the observations that the crown is removed completely by Pronase and incompletely by trypsin and remains intact following treatment with neuraminidase (Anderson and Hein, 1977). Phosphotungstic acid exhibits only weak affinity for the crown, suggesting the core protein is not rich in lysine or arginine (Kuhn and Engleman, 1978).

2.1.2. Shaft. The composition of the membrane of the ciliary shaft differs by species. The most work has been done on cilia from scallop gill, *Tetrahymena,* and *Paramecium.* According to Stephens (1977, 1983), the scallop gill ciliary membrane (Triton X-100-soluble fraction) is 45% protein of which the major component is a 55-kDa cilium-specific, glycosylated, hydrophobically unique tubulin. The tubulin is in a tubulin–lipid–protein complex associated with a dynein-like membrane-associated ATPase (Stephens, 1985, this volume). In contrast, when the ciliary membranes of *Paramecium, Tetrahymena,* or *Chlamydomonas* are purified (in the absence of detergent), tubulin is absent or only a minor membrane component and the majority of the approximately 70 membrane-associated proteins are acidic ranging in size from 15 to 150 kDa (Adoutte *et al.,* 1980; Monk *et al.,* 1983).

Only a few studies have attempted to determine the ciliary membrane composition of higher animals. In the quail oviduct and rat tracheal epithelium, the membrane along the ciliary shaft has been shown to be rich in cholesterol (Chailley and Boisvieux-Ulrich, 1985; Montesano, 1979). Chen and Lancet (1984) have determined the protein profile of frog nasal and respiratory ciliary membranes. Respiratory cilia contain three- to tenfold less membrane-associated proteins than those in the nose. Four major unidentified polypeptides are common to ciliary membranes from both sites while seven are unique to the nose, including a form of tubulin. Immunoelectron microscopy of rabbit tracheal explants using gold- or ferroisothiocyanate-conjugated antitubulin antibody failed to detect tubulin-positive staining on untreated membranes (Moller *et al.,* 1983). If Triton X-100 permeabilization of the membrane was carried out, then axonemes and basal bodies bound antibody. By this criterion, rabbit tracheal ciliary membranes do not appear to contain an antibody-accessible form of tubulin. Taken together, these studies suggest that mammalian ciliary membranes present a unique, heterogeneous array of surface proteins, some structural (probably not tubulin) and some capable of binding calcium [as described in *Tetrahymena* (Hirano and Watanabe, 1985)]. Some of these proteins are contained in randomly dispersed intramembranous particles visible on freeze-fracture (Menco, 1980). In amphibia, the particles have been found to contain mannosylated transmembrane glycoproteins and may serve to bind internal cytoplasmic structures to the membrane as well as to anchor glycocalyx to the ciliary surface (Chailley *et al.,* 1981).

Lectin binding to isolated proteins and to whole cilia has been used to demonstrate glycosylation of the ciliary membrane. Frog respiratory cilia do not bind Con A (affinity for D-mannosyl and D-glucosyl residues) or wheat germ agglutinin (WGA) (affinity for *N*-acetyl-D-glucosamine); nasal cilia bound both lectins (Chen and Lancet, 1984) (Fig. 2). Chen and Lancet (1984) state that unpublished evidence indicates mammalian ciliary membranes are similar to those of the frog. However, Chailley *et al.,* (1981) have demonstrated that WGA binds randomly on the entire ciliary surface of amphibian and quail oviducts, indicating the presence of *N*-acetylglucosamine. In areas where membrane-associated particles are absent, such glycosylation is thought to be linked to lipids (Chailley *et al.,* 1981; Sandoz *et al.,* 1979).

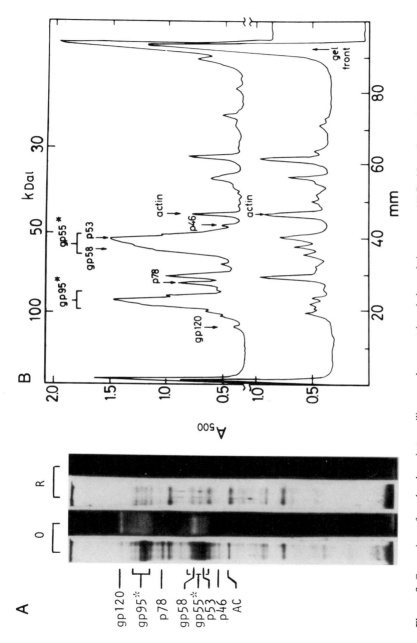

Figure 2. Comparison of nasal and respiratory ciliary membrane protein and glycosylation pattern. (A) Positive: silver-stained gel electrophoretic patterns of membrane glycoproteins (gp) from frog olfactory (O) and respiratory (R) cilia (AC, actin). Negative: autoradiography of [^{125}I]-Con A binding to the same gel lanes. (B) Densitometric scans of olfactory (upper) and respiratory (lower) silver-stained lanes of A. Reprinted with permission from Chen and Lancet (1984).

2.1.3. Necklace. The ciliary necklace is formed by an undulating band of membranous particles in two to six parallel rows at the base of the ciliary shaft (Gilula and Satir, 1972). It has been suggested that the necklace may function as a timing device for the ciliary beat by regulating localized membrane permeability (Gilula and Satir, 1982). The necklace may also serve to maintain the regionalization of the composition of the ciliary surface glycocalyx (Spicer *et al.*, 1983). In the murine oviduct, the composition of the particles of this structure differs from the ciliary crown in that they are Pronase stable but, like the crown, they are sialic acid-free (neuraminidase stable) (Anderson and Hein, 1977). Con A and WGA bind to the membrane-particle-rich region of the necklace, suggesting the presence of glycoproteins containing mannose and *N*-acetylglucosamine (Chailley *et al.*, 1981). In contrast to the membrane of the ciliary shaft, the necklace appears not to contain free cholesterol (Chailley and Boisvieux-Ulrich, 1985).

2.2. Chemistry

Several probes involving electron microscopy have been used to determine the chemical nature of the surface of the ciliary membrane. The distribution of binding of lectins (carbohydrate-binding proteins) has suggested that a variety of carbohydrates are important components of the ciliary surface. The binding of ruthenium red, ferritin, and poly-L-lysine has been used to map anionic sites on the surface (Tuomanen and Hendley, 1983; Anderson and Hein, 1977). Such agents bind both to the extracellular domains of integral membrane components and to components loosely associated with the membrane which is continuously bathed in carbohydrate-rich mucous, serous secretions, and periciliary fluid. These fluids contribute components which, along with integral membrane proteins, form a fuzzy carbohydrate-rich coat associated with the plasma membrane on the surface of human (Tuomanen and Hendley, 1983), amphibian (Chailley *et al.*, 1981), and *Paramecium* (Adoutte *et al.*, 1980) cilia. This glycocalyx is the true functional surface of the cilium. Its composition differs depending on the region of the cilium and it is chemically distinct from microvillar and ciliated epithelial cell body surfaces. These differences determine the sites where exogenous ligands, particularly those present on pathogens, first establish contact with the target cilium.

Over the past decade, mapping studies directed at the chemical nature of the differentiated ciliary surface have begun to appear. These seminal studies will be reviewed in detail since they form the basis for understanding why there exists a specialized topography of infectious diseases of the cilium. Figure 3 is a composite of currently available information and represents a hypothetical mammalian respiratory ciliary surface map.

2.2.1. Sites of Surface Charge. In 1977, Anderson and Hein began the study of the biochemistry of the ciliary surface by examining the binding of cationic ferritin to the surface of mouse oviductal cilia using transmission electron microscopy. The anionic sites thus disclosed were concentrated at the ciliary crown and necklace, with significantly fewer anionic sites along the ciliary shaft. Binding of ferritin to all sites was blocked by cationic poly-L-lysine. Ferritin binding to the tip and necklace was sensitive to trypsin, Pronase, or neuraminidase treatment of the ciliated cell, indicating that the anionic sites were glycoproteins containing sialic acid. Thus, although the ciliary necklace does not

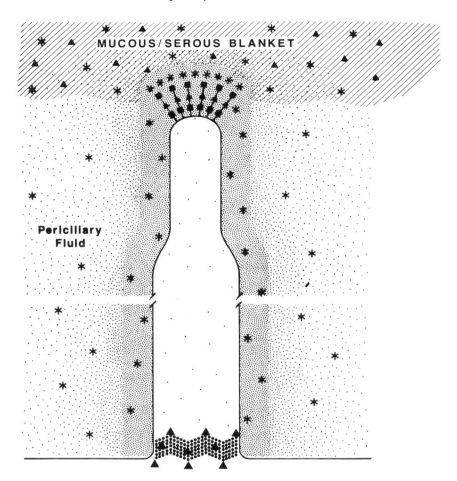

Figure 3. Schematic representation of surface chemistry of a human respiratory cilium. Heavy dots adjacent to ciliary membrane represent glycocalyx enriched with anionic carbohydrates. ▲, anionic sialylated glycoconjugates on bristles and in mucous blanket; ∗, sulfated glycoconjugates enriched in glycocalyx and mucous blanket. Compiled from Spicer *et al.* (1983), Mazzuca *et al.* (1982), Anderson and Hein (1977), Tuomanen and Hendley (1983), and Kuhn and Engleman (1978).

contain sialylated glycoprotein as a structural component, it appears to have such moieties within its surface-associated glycocalyx.

2.2.2. Surface Glycoconjugates. Ruthenium red, an electron microscopic stain for anionic carbohydrates, stains a fuzzy layer extending from the tip, all along the shaft, down to the necklace of cilia from humans (Tuomanen and Hendley, 1983), rabbits, rats, and hamsters (Kuhn and Engleman, 1978). The composition of this carbohydrate-rich coat differs between cilia and adjacent nonciliary cell surfaces; periodic acid–thiocarbohydrazide–silver proteinate (PA-TCH-SP) stains human and rat microvilli but not cilia,

Table I. Staining Properties of Luminal Glycocalyx[a]

Cell type		PA-TCH-SP[b] (long chain[c])	Dialyzed iron (sialic acid)	High-iron diamine (sulfated)
Cilia	Human	−	+	+
	Rat	−	+	−
Microvilli	Human	+	+	
	Rat	+	+	
Secretory cell		+		
Alveolar pneumocyte	Human		+	

[a]After Spicer et al. (1983).
[b]Periodic acid–thiocarbohydrazide–silver proteinate.
[c]Specificity of stain.

indicating that short-chain sugars are common only on cilia (Spicer et al., 1983). Species-specific differences in the composition of this luminal surface coat can also be demonstrated (Table I). For example, sulfated glycoconjugates are present on the surface of human but not rat cilia. Table II summarizes differences between species in the composition of the periciliary fluid and mucous blanket (serous and mucous secretions) both of which contribute carbohydrate to the ciliary glycocalyx.

Glycosylation appears to occur on both lipids and proteins on the human ciliary surface as evidenced by the binding of antibodies recognizing various blood group antigens to intact ciliated cells or to glycoprotein and glycolipid ciliary extracts (Tuomanen et al., 1988). For example, antibodies recognizing blood group A, Lewis a, and Lewis b determinants bind to cilia and ciliated cell bodies. Additional antibodies to some Lewis a determinants bind only to cilia. These studies indicate that a variety of fucosylated galactose–N-acetylglucosamine moieties are present on the ciliary surface and that there

Table II. Comparison of Carbohydrate Composition of Secretions Bathing Cilia in Several Species[a]

Secretion	Rat	Mouse	Human
Serous			
Glucose			+
β-Galactose	+	+	+
N-Acetylgalactosamine		+	
N-Acetylglucosamine		+	+
Sialic acid	+	−	+ + + rare neuraminidase but + + + lectin
Sulfate		−	+
Fucose (blood group)	−	−	−
Mucous			
Sialic acid		+ + +	+
Sulfate	+ +	+	+
Fucose (blood group)	−	−	+
Galactose–N-acetylgalactosamine			+

[a]Compiled from Spicer et al. (1983) and Mazzuca et al. (1982).

exists a refined, topographically limited distribution of some of these glycoconjugates along the ciliary membrane. Using lectin binding as a probe for specific carbohydrates, it appears that glycoconjugates prominent on the surface of the cilium may be grouped into regions. Con A, *Ricinus communis* I, and WGA bind to the mucous blanket and ciliary tip, indicating the presence of a combination of mannose, galactose, *N*-acetylglucosamine, and sialic acid (Mazzuca *et al.*, 1982). Con A, mannosylated ferritin, and WGA bind to the ciliary necklace of amphibia but only WGA binds to the shaft (Chailley *et al.*, 1981). Neither Con A nor WGA binds to the ciliary shaft of frogs (Chen and Lancet, 1984). Thus, it appears that species-specific differences in glycosylation are beginning to become evident. This is particularly important in the context of studies of receptors for ciliary pathogens since species specificity of susceptibility to disease may well be dictated by such differences in surface glycosylation.

3. Interactions between Pathogens and Respiratory Cilia

3.1. General Mechanisms of Adherence

The ciliary escalator serves as a highly efficient nonspecific clearing mechanism for the tracheobronchial tree. In order to establish pulmonary infection, pathogens must either sequester in the alveoli as in classical infectious pneumonias, or neutralize cilium-driven bronchial clearance. In general, very few pathogens have developed mechanisms to evade clearance when in contact with cilia. Such mechanisms include inducing ciliostasis, killing the ciliated cell with or without denuding the ciliated epithelium, or adhering to cilia. Given the differentiated ciliary surface, it is not surprising that tropism for cilia infers a high degree of species and site specificity for those pathogens which can infect or intoxicate the ciliated epithelium.

All cilium-directed pathogens thus far described use interactions between surface proteins of the pathogen (herein called adhesins) and surface carbohydrates of the target host cell (herein called receptors) to achieve stable adherence to the ciliated epithelium. Such protein–carbohydrate interactions are also typical for many mucosal infections outside the respiratory tract, i.e., gastrointestinal tract, genitourinary tract, or oropharynx. Frequently, multiple ligands are presented on the pathogen's surface which serve to increase the strength and specificity of adherence when engaged in concert. The specificity of the adherence interaction plays a major role in determining the epidemiology of a disease and the symptoms of host injury characterizing a given infection. For instance, pertussis (whooping cough), caused by *Bordetella pertussis,* is a disease solely of humans; other *Bordetella* species infect animals reflecting differences in the chemistry of adhesin–receptor pairing of pathogen and host (Tuomanen *et al.*, 1983). Within the human host, *B. pertussis* is found adherent only to tracheobronchial ciliated cells and not alveolar cells or nonciliated respiratory epithelial cells, again reflecting the topography of receptors for *B. pertussis* adhesins. It has recently been found that many of the bacteria commonly causing infections of alveoli, i.e., pneumonia (as opposed to bronchi where the ciliated epithelium is located), bind specifically to the *N*-acetylgalactosamine–β1–4 galactose moiety found in the lung glycolipids gangliotetraosylceramide (asialo GM1) and gangliotriaosylceramide (asialo GM2) (Krivan *et al.*, 1988). These bacteria include

Staphylococcus aureus, Pseudomonas aeruginosa, Hemophilus influenzae, Streptococcus pneumoniae, Klebsiella pneumoniae, and mucoid *Escherichia coli.* In contrast, the cilium-specific pathogens *Mycoplasma pneumoniae* and *Bordetella pertussis* bind to different carbohydrate moieties (Loomes *et al.,* 1984; Tuomanen *et al.,* 1988); only pseudomonas appears to bind to both ciliated and nonciliated cells of the lung but by different mechanisms. These results suggest that the differences in topography of infections within the lung correlate with differences in host cell receptors recognized during adherence of the pathogens.

Damage to or killing of ciliated cells by toxic substances from pathogens has not been well studied. It appears that a multiplicity of toxins exist ranging from a ciliated cell-specific tracheal cytotoxin of *Bordetella pertussis* to oxidative cell damage invoked by *Mycoplasma pneumoniae.* While adherence interactions obey strict species-specific rules, it appears that toxicity to ciliated cells occurs across species boundaries.

3.2. Adherence of Specific Pathogens to Respiratory Cilia

3.2.1. *Bordetella pertussis.* Pertussis is an acute respiratory disease of children and adults with several distinctive features (Pittman, 1984). *B. pertussis,* a gram-negative coccobacillus, establishes infection at a single site, the ciliated respiratory epithelial cell. The bacteria remain at this mucosal location throughout the multiweek course of disease, which is characterized by paroxysmal cough and copious mucous production. The lack of invasion following colonization of the respiratory tract distinguishes this bacterium from most pathogens for which mucosal attachment is followed by systemic spread of infection. In order to maintain infection of the surface of the respiratory epithelium for such extended periods, *B. pertussis* has developed particularly effective means of overcoming the nonspecific clearance mechanisms of the ciliary escalator and local cellular defenses. The complex mechanism of adherence of *B. pertussis* to human cilia serves as the prototype for diseases of the respiratory system in which attachment plays an essential role in pathogenesis.

The molecular biology of the process of attachment of *B. pertussis* to human cilia is sufficiently specific and complex so as to provide explanations for many of the unusual features of the disease: specificity for the human host, limitation of infection to a single site, uniform susceptibility of the population to the infection (especially in the newborn), and propensity of infected individuals to develop secondary infections. The adherence process is also remarkable from a microbiological standpoint. It appears that *B. pertussis* secretes its adhesins into the environment from where they can be recaptured. One of the adhesins is also a toxin, and fimbriae, the classic structures mediating adherence for most organisms, seem to play only a minor role, if any, in attaching the bacteria to cilia (Tuomanen and Weiss, 1985; Tuomanen, 1986a).

B. pertussis exhibits an unusually exclusive species specificity for its host. Pertussis is a disease only of man; similar prolonged respiratory diseases in animals (such as kennel cough in dogs, snuffles in rabbits, atrophic rhinitis in pigs, and coryza in birds) are caused by other *Bordetella* species. The ability of *Bordetella* species to adhere to various host ciliated epithelial cells *in vitro* parallels the natural pattern of infection: *B. pertussis* adheres best to human cilia (Tuomanen *et al.,* 1983). Thus, the epidemiological pattern of disease due to *Bordetella* probably arises from differences in the chemistry of the adherence bridge creating highly specific pairing between species of bacteria and species of infected host.

In addition to species specificity, the adherence process also appears to be the basis of the highly restricted affinity of bordetellae for cilia (Fig. 4). Descriptions of the pathology of natural *B. pertussis* infection remark on the consistent presence of the organism enmeshed in respiratory cilia (Rich, 1932; Marks *et al.*, 1980). All bordetellae demonstrate this tropism within their target host. Although *B. pertussis* can attach *in vitro* to ciliated cells of murine oviduct (Opremcak and Rheins, 1983) and microvilli of ventricular ependymal cells (Hopewell *et al.*, 1972), natural infection is restricted to ciliated respiratory epithelial cells. Electron micrographic studies indicate that the bacteria are most frequently seen associated with the base of the ciliary shaft (Fig. 5) (Tuomanen and Hendley, 1983). From studies in hamster models, ciliated cell pathology, including ciliostasis, does not ensue if bacteria are not attached to the cell, signifying that adherence plays a role not only in maintaining the bacterium in the respiratory tract but also in directing local cellular toxicity (Muse *et al.*, 1977, 1978; Bemis and Wilson, 1985). The mechanism of this damage is not known but presumably involves at least one *B. pertussis* toxin, the tracheal cytotoxin (see below; Goldman *et al.*, 1982; Cookson *et al.*, 1988). In hamster tracheal organ culture, infected ciliated epithelial cells develop a marked disarray of the ciliary necklace (Fig. 6) (Muse, 1980). Eventually the ciliary necklace disappears and the ciliated cells are killed and extruded from the epithelium.

The model for *B. pertussis* adherence is most consistent with two types of protein adhesins secreted from virulent bacteria, each acting as a bivalent bridge between the

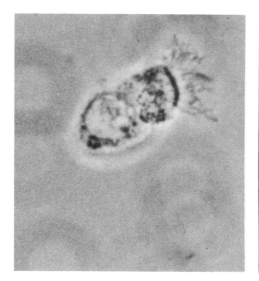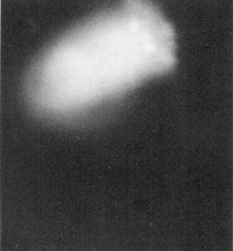

Figure 4. Phase-contrast micrography (left) of human ciliated respiratory epithelial cells (×800). The same cell viewed under a fluorescence microscope (right) demonstrates brightly stained *B. pertussis* associated exclusively with cilia. Reprinted with permission from Tuomanen and Hendley (1983).

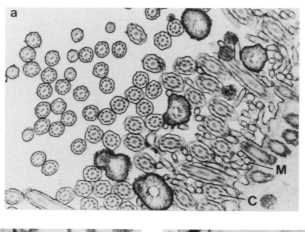

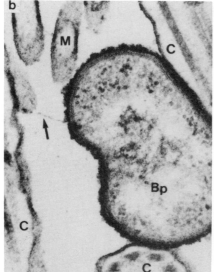
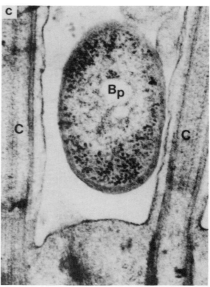

Figure 5. Transmission electron micrographs of *B. pertussis* adherent to cilia of human respiratory epithelial cells. (a) Bacteria located in the proximal portion of the ciliary tuft and associated with cilia (C) and microvilli (M) (ruthenium red stain; ×25,555). (b) *B. pertussis* (Bp) associated with cilia and microvilli. There is direct apposition of the glycocalyx on bacterial and ciliary membranes and a connection between the two by a filament (arrow) (ruthenium red stain; ×81,000). (c) *B. pertussis* apposed to a cilium near the body of the cell. In sections such as this that were not stained with ruthenium red, the space separating the membranes is unstained (×58,750). Reprinted with permission from Tuomanen and Hendley (1983).

bacterium and one or more carbohydrate-containing receptors on cilia. Two proteins, filamentous hemagglutinin (FHA) and pertussis toxin (PT), have been proposed to be adhesins for human ciliated cells (Tuomanen and Weiss, 1985). Adherence is the only known function of FHA. PT, on the other hand, catalyzes the ADP-ribosylation of a membrane-bound protein that is involved in the inhibitory regulation of adenylate cyclase

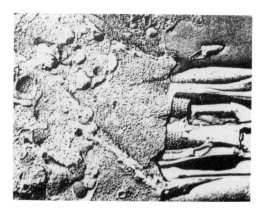

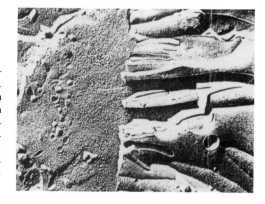

Figure 6. Freeze-fracture replicas of hamster tracheal organ culture. (Top) 60-hr control culture. Note ordered particle array of the ciliary necklace in the P-fracture face. (Middle) 48-hr infection with virulent *B. pertussis*. Note the disruption and absence of the ciliary necklace in both E- and P-fracture face. (Bottom) 48-hr infection with avirulent *B. pertussis*. Note the ciliary necklace and its resemblance to the control culture (×32,000). Compiled from Muse (1980).

within the host cell (Katada and Ui, 1982). Thus, this toxin has a strikingly unique feature—it also acts as a bacterial adhesin (Tuomanen and Weiss, 1985). Current evidence suggests that toxicity of PT may be mediated by the binding of the toxin to sialylated glycoproteins on host cells (Brennan *et al.*, 1988). In contrast, the adherence function of PT seems to involve interactions of the toxin with nonsialylated lactosamine-containing glycoconjugates (Tuomanen *et al.*, 1988). Such multiplicity of ligand binding sites is compatible with the complex multimeric structure of the toxin (Katada and Ui, 1982).

The biochemical details of the ciliary receptor for *B. pertussis* are beginning to emerge. Since the organisms appear to localize preferentially at the ciliary base (Tuomanen and Hendley, 1983), receptor structures may be clustered or localized at this site. Galactose, *N*-acetylglucosamine, lactose, and a large variety of lactosamine-containing complex carbohydrates inhibit adherence of *B. pertussis* to human cilia *in vitro* when preincubated with the bacteria, suggesting that they mimic the natural receptor structure (Tuomanen, 1986a; Tuomanen *et al.*, 1988). Radiolabeled bacteria bind to purified lactosylceramide in thin-layer chromatography binding studies (Tuomanen *et al.*, 1988). Monoclonal antibodies to a variety of blood group determinants containing galactose–*N*-acetylglucosamine residues block adherence of *B. pertussis* to human cilia (Tuomanen, 1986a; Tuomanen *et al.*, 1988). These same antiadherence antibodies recognize a family of galactose–glucose-containing glycolipids in extracts of ciliary membranes (Tuomanen *et al.*, 1988) and their antiadherence activity arises presumably because they bind on or near the galactose–glucose-containing ciliary receptor for *B. pertussis*.

Although the definitive structure and number of receptors for *B. pertussis* on human cilia still require determination, there are several exciting predictions which arise from the existing data. Breast milk, which contains a large variety of lactose-containing oligosaccharides, is a good competitive inhibitor of adherence and may alter susceptibility of infants to colonization of the upper airways and nasopharynx. Blood group antigens are differentiation antigens. The developmental timing of the appearance of galactose–glucose-containing glycoconjugates on ciliary surfaces is not known but may explain the severity of whooping cough in newborns and young infants. Galactose–glucose is a very common terminal or internal sequence of many eukaroytic cell-surface carbohydrate antigens, a fact consistent with the nearly 100% attack rate of whooping cough. In addition to blocking the interaction of *B. pertussis* and human cilia, *N*-acetylglucosamine is also an effective competitive inhibitor of adherence of *B. bronchiseptica* to hamster lung fibroblasts (Plotkin and Bemis, 1984), suggesting that some common structural features may exist between animal and human receptors. However, the preferred receptors for *B. bronchiseptica* and *B. avium* have been shown to be sialic acid-containing gangliosides, clearly different from that of *B. pertussis* (Ishikawa and Isayama, 1987; Arp *et al.*, 1988). This may explain the strict host species specificity of the various *Bordetella* species.

3.2.2. *Mycoplasma pneumoniae*.

M. pneumoniae causes a mild to moderate atypical bronchopneumonia lasting 2–4 weeks (Denny *et al.*, 1971; Murray *et al.*, 1975). Characteristically, the disease is accompanied by (and diagnosed by) the development of autoantibodies to erythrocytes carrying the I blood group determinant. Some evidence suggests that mycoplasmal proteins mimic the host intermediate filament and mitotic spindle leading to autoantibodies during recognition and recovery from infection (Wise and Watson, 1985; Lind *et al.*, 1988).

Like *Bordetella*, *M. pneumoniae* is strictly a human pathogen with affinity for cilia. The organism specifically orients itself towards the host cell by a terminal organelle (Fig. 7). A specific protein is clustered at the tip of the organelle and serves as an adhesin (Hu *et*

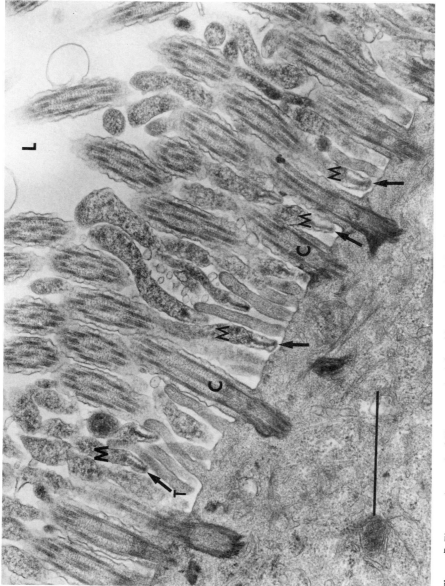

Figure 7. Electron micrograph of *M. pneumoniae*-infected hamster tracheal ring after 72 hr in organ culture. M, mycoplasma; C, cilia; L, lumen; T, tip organelles. Bar = 1 μm. From Collier (1979).

al., 1982). Inspection of freeze-etch replicas of hamster tracheal organ culture infected with *M. pneumoniae* reveals disruption of membrane-associated particles on the ciliary necklace and the ciliary shaft (Fig. 8) (Carson *et al.*, 1979). Within 24 hr of infection, these particles on the ciliary shaft appear clumped and unevenly distributed. Similarly, the rows of evenly spaced particles forming the ciliary necklace are either in complete disarray or completely absent. This is reminiscent of the cytopathology associated with pertussis and in both diseases these changes coincide with ciliostasis and a degenerating epithelium. It remains to be determined if this pathology requires direct contact of the *M. pneumoniae* with the cilium (as it does for *B. pertussis*) or if it is mediated indirectly. Both oxidative damage and parasitism of cilia for nutrients have been proposed as mechanisms of *Mycoplasma*-induced injury to ciliated cells (Almagor *et al.*, 1985; Chen and Krause, 1988).

M. pneumoniae have been shown to adhere to both microvilli and cilia in human nasal polyp tissue culture (Almagor *et al.*, 1985). These organisms also adhere to the nonciliated WiDr colon carcinoma cell line and to hamster tracheal rings (Chandler *et al.*, 1982). The chemistry of the receptor for *Mycoplasma* species on respiratory cells has been inferred from an elegant study by Loomes *et al.* (1984) on the affinity of the organism for various carbohydrates inserted into erythrocyte membranes. Virulent *M. pneumoniae* specifically recognize sialic acid residues α-2–3 linked to galactose such as is found naturally in the Ii antigen of erythrocytes. *M. gallisepticum* has a broader receptor specificity and does not differentiate 2–3 galactose from 2–6 *N*-acetylgalactosamine. Mucin inhibits binding of *M. gallisepticum* but not *M. pneumoniae*. The sialic acid receptor configuration specific for *M. pneumoniae* could potentially be present at both the ciliary crown and the ciliary necklace where surface chemistry studies have shown sialic acid residues to be clustered (see Fig. 3). Thus, the pathological disruption of the ciliary necklace seen during *Mycoplasma* infection could, in principle, relate to adherence of the pathogen to the sialic acid-rich glycocalyx of this region.

3.2.3. *Pseudomonas aeruginosa*. Pseudomonas species are opportunistic pulmonary pathogens. The most striking association of pseudomonas with cilia is the nearly lifelong colonization of the tracheobronchial tree seen in patients with cystic fibrosis. While pseudomonas pulmonary infection of the normal host is a relatively rare, often nosocomial, severe, acute disease, patients with cystic fibrosis experience a chronic smoldering course extending over decades with acute exacerbations of respiratory failure (Thomassen *et al.*, 1987).

Pseudomonas aeruginosa has several mechanisms of adherence to many eukaryotic cell types. Either pili or the surface exopolysaccharide capsule (alginate) can mediate attachment to different host target sites, including buccal epithelial cells, cilia (Fig. 9), and the mucociliary blanket (Table III). Once adherent, *Pseudomonas* produce multiple extracellular products which damage cilia (see below). The biochemical specificity of such toxic interactions has yet to be investigated.

Adherence of *Pseudomonas* to nonciliated oropharyngeal cells is a prelude to pneumonia in severely ill patients (Johanson *et al.*, 1980). This process involves a salivary protease which cleaves fibronectin from the buccal epithelial cell surface allowing pili on the surface of nonmucoid strains of the bacteria to attach to exposed host cell carbohydrates (Marcus and Baker, 1985; Woods *et al.*, 1980). The C-terminal region of the pilin

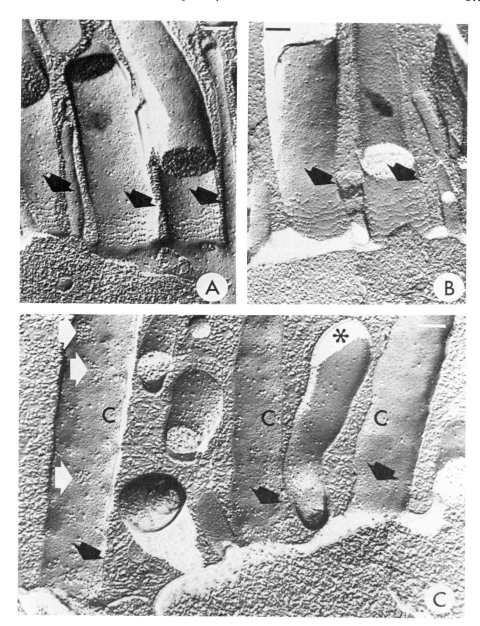

Figure 8. (A) Electron micrograph demonstrating the structure of ciliary necklaces on the P-fracture face of hamster tracheal epithelium fixed immediately upon resection (arrows). The necklace is comprised of four to six, generally five, evenly spaced strands of membrane-associated particles at the base of each cilium; note also the uniform distribution of particles on the shafts of the cilia. (B) Electron micrograph illustrating the retention of the ciliary necklaces of epithelial cell cilia that may be observed after 24 hr of incubation in tracheal organ culture (arrows). (C) Freeze-etch preparation of tracheal epithelial cells incubated in organ culture for 24 hr with virulent *Mycoplasma pneumoniae*. A mycoplasma cell (*) is seen between two cilia (C). Black arrows at the bases of the cilia indicate areas of disrupted and absent ciliary necklaces in both E- and P-fracture faces. White arrows indicate areas of clumping of membrane-associated particles on the ciliary shaft. Bars = 100 nm.

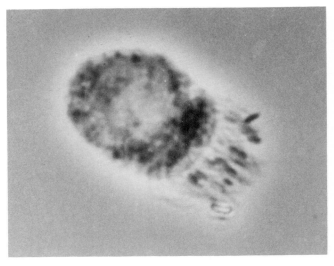

Figure 9. Phase-contrast micrographs of *Pseudomonas aeruginosa* adherent to ciliary tufts of human tracheal epithelial cells. Reprinted with permission from Franklin *et al.* (1987).

contains the binding domain for host cells (Doig *et al.*, 1988). *Pseudomonas* adherence to cilia of the respiratory tract is a more complex process than that for buccal epithelial cells. In contrast to the association of nonmucoid strains with buccal cells, mucoid strains are 10 to 100 times more efficient than nonmucoid strains at adhering to cilia (Marcus and Baker, 1985). Adherence to ciliated cells can far exceed that to buccal cells in the same patient (Niederman *et al.*, 1983) and is significantly affected by alterations in pH (Palmer *et al.*,

Table III. Adhesin–Receptor Interactions of *Pseudomonas* to Various Host Target Sites

Pseudomonas		Host target site		
Strain	Adhesin	Human buccal cell[a]	Hamster cilia[b]	Human mucin[b]
Mucoid	Pili	−	+	−
	Alginate	−	++[c]	++[c] sialic acid (N-acetylglucosamine)
Nonmucoid	Pili	++[c]	+ sialic acid	+ N-acetylglucosamine sialic acid

[a]Marcus and Baker (1985) and Franklin *et al.* (1987).
[b]Vishwanath and Ramphal (1985).
[c]Predominant interaction.

1986). Human cilium-specific binding of *Pseudomonas* is mediated both by pili (numerous proteinaceous filaments on the bacterial surface) and by alginate (the exopolysaccharide of mucoid strains) (Franklin *et al.*, 1987). In a murine *in vitro* assay (Vishwanath and Ramphal, 1985), pili of nonmucoid strains recognized sialic acid-containing glycoconjugates on tracheal cells. This binding was inhibitable by heat-inactivated influenza A virus. Thus, pilus-mediated adherence of *Pseudomonas* appears to involve a ciliary receptor very similar to or the same as that recognized by *Mycoplasma* and influenza A virus. A second, different receptor for mucoid strains has been proposed to involve mucous blanket glycoprotein receptors containing N-acetylglucosamine and sialic acid (Vishwanath and Ramphal, 1985). Two receptor specificities, one for cilia and one for mucous glycoproteins, is consistent with the observation that sublethal concentrations of aminoglycoside antibiotics, such as tobramycin or gentamicin, decrease adherence of mucoid strains to tracheal cells but not to mucin (Geers and Baker, 1987). Treatment with these drugs has been suggested to induce the release of a bacterial alginate depolymerase leading to a loss of exopolysaccharide-mediated ciliary adherence. Clinically, this therapy is associated with improvement in acute exacerbations of respiratory failure in patients with cystic fibrosis, suggesting that modes of therapy directed at altering adherence interactions may alter the course of *Pseudomonas* superinfection in this disease.

3.2.4. Cilium-Associated Respiratory Bacillus.

Laboratory rats often suffer from a chronic respiratory disease (CRD) which is a major cause of morbidity and mortality in conventionally maintained colonies (Ganaway *et al.*, 1985). While it is not clear if there is only one infectious agent which can cause CRD, a filamentous gram-negative bacterium, termed the cilium-associated respiratory (CAR) bacillus, is consistently found in infected animals (van Zwieten *et al.*, 1980). Seroconversion to CAR bacillus antigens occurs during disease (Matsushita *et al.*, 1987), and the disease can be produced by inoculation of the CAR bacillus into experimental animals (Ganaway *et al.*, 1985; Griffith *et al.*, 1988). Infection is associated with heavy colonization of cilia by filamentous bacteria. The bacillus remains poorly characterized as it can only be propagated in embryonated chicken eggs. However, it apparently belongs to the large group of microorganisms termed "gliding bacteria." The association of gliding bacteria with higher organisms is rare and CAR bacillus infection may represent the first proven instance of a gliding bacterium causing disease in a warm-blooded vertebrate (Galaway *et al.*, 1985).

3.2.5. Other Bacteria: Piracy of Adhesins. The only bacteria which have been clearly shown to interact directly with cilia during disease have been discussed in the preceding sections: *Bordetella, Mycoplasma, Pseudomonas,* and CAR bacillus. However, other bacteria have been shown to attach to cilia in a setting which may explain superinfections arising during the course of cilium-specific diseases. *In vitro, Streptococcus pneumoniae, Staphylococcus aureus,* and *Haemophilus influenzae* can capture and become coated with the cilium-specific adhesins secreted by *B. pertussis* (Tuomanen, 1986b). The interaction between heterologous bacteria and the *B. pertussis* adhesins is probably nonspecific (i.e., involves electrostatic or hydrophobic interactions that differ in degree from one capsular type or cell wall to another). Once coated, the pathogens acquire the ability to adhere to a new site in the respiratory tree—the cilia. The *B. pertussis* adhesin–ciliary interaction is known to be specific and affords these bacteria with an anchor to the respiratory epithelium which they normally do not possess. It is reasonable to suggest that as *B. pertussis* grows on the ciliated epithelium during clinical whooping cough, the adhesins it secretes bind to neighboring ciliated cells. These ligands may then serve to either enhance adherence of *B. pertussis* daughter cells or, through the piracy of these adhesins, allow adherence of secondary pathogens such as pneumococci, staphylococci, and *H. influenzae,* prolonging their resident time in the lungs. Thus, the adherence system of the primary pathogen could potentially contribute to secondary infections during whooping cough. Such a mechanism may also contribute to the clinical association of pertussis and adenovirus infection. It is possible that piracy of adhesins may be a common general mechanism of superinfection complicating mucosal diseases (Tuomanen, 1986b).

3.2.6. Viruses. Rhinovirus and influenza A have been found to specifically interact with nasal cilia. Biopsies taken from patients with rhinovirus colds demonstrate viral antigen within ciliated epithelial cells at the time of maximum symptomatology (Turner *et al.*, 1982). However, no morphological changes could be detected in the infected epithelial cells (Winther *et al.*, 1984a,b). The biochemical basis of the rhinovirus–nasal epithelial cell interaction remains to be determined.

Infection with influenza A virus has been studied extensively on a biochemical level. The viral surface contains glycoprotein spikes which act as adhesins recognizing sialic acid residues on target cells. Infectivity is reduced by removing the spikes with trypsin or by addition of competing sialic acid-containing carbohydrates such as fetuin. Rogers *et al.*, (1983) have shown that alteration of even one amino acid in the hemagglutinin can change the adherence specificity to mammalian cells. Carroll *et al.*, (1981) have shown that while the predominant receptor determinant is sialic acid, different hemagglutinins recognize different sialylated carbohydrate backbones. For instance, some influenza A strains can differentiate penultimate 2,3-galactose, 2,6-galactose, or 2,6-*N*-acetylgalactosamine residues on the receptor, adhering most avidly to the former and not at all to the latter. This biochemical specificity suggests that *Mycoplasma* and influenza A share a receptor on respiratory cilia.

4. Toxicity of Microbial Products for Cilia

Several bacteria produce substances toxic to cilia. *Neisseria* colonizing neighboring noncilliated cells (Stephens *et al.*, 1986) and possibly *Haemophilus* (Denny, 1974) elaborate lipopolysaccharide which induces ciliostasis and eventual killing of ciliated cells. The production of hydrogen peroxide and superoxide by *Mycoplasma pneumoniae* results in oxidative damage to cilia (Almagor *et al.*, 1985). Mycoplasma produce a ciliostatic factor (Chandler *et al.*, 1987) and can actually parasitize and thereby destroy ciliated cells (Chen and Krause, 1988).

Poorly characterized heat-sensitive secretory products of *Pseudomonas* and *Haemophilus* induce dyskinesis and ciliostasis (Wilson *et al.*, 1985). *Pseudomonas* elastase and alkaline protease degrade axonemal proteins (Hingley *et al.*, 1986a) and mucin (Poncz *et al.*, 1988). *Pseudomonas* also releases three heat-stable ciliostatic factors: a rhamnolipid, a phenazine derivative, and a hydroxyquinoline (Hingley *et al.*, 1986b,c). The rhamnolipid has been shown to dissolve the ciliary membrane of rabbit tracheal rings (Hingley *et al.*, 1986b,c); at higher concentrations, it removes ciliary dynein (Hastie *et al.*, 1986). The 2-alkyl-4-hydroxyquinoline causes ciliostasis without structural damage to the cilia. The phenazine inhibits the activity of isolated ciliary axonemes. This elaborate array of ciliary toxins presumably interferes with host defenses on the ciliary epithelium and thereby promotes colonization.

In contrast to other pathogens, *Bordetella pertussis* must be adherent to cilia in order to produce ciliary damage. Goldman *et al.* (1982) have identified a tracheal cytotoxin of *B. pertussis* using a hamster tracheal ring *in vitro* assay to define bioactivity. It appears that the toxin is identical to the 1,6-anhydro bisdisaccharide tetrapeptide (Cookson *et al.*, 1988) which forms the ends of the interlocking chains of the bacterial cell wall (Tuomanen *et al.*, 1988). The toxin is released during growth of the organism (Rosenthal *et al.*, 1987). Interestingly, it appears that the same molecule is released by *Neisseria gonorrhoeae* and may contribute to the pathology of infection of the ciliated epithelium of the Fallopian tube (Sinha and Rosenthal, 1980). Thus, a common structural element of all bacterial cell walls appears to be a toxin specifically for ciliated cells. The mechanism of toxicity is unknown. However, the importance of the ciliary toxicity of such a ubiquitous bacterial component is underscored by the fact that the most widely used antibiotics, the β-lactam family of penicillins, cephalosporins, and penems, cause release of cell wall during bacterial killing. In fact, the 1,6-anhydro monomer is a major product released during antibiotic-induced killing of *Escherichia coli* (Kitano *et al.*, 1986) but only a minor product produced during killing of *B. pertussis* (Tuomanen *et al.*, 1988).

5. Summary: Implications of Ciliary Surface Composition to the Therapy of Infections of Cilia

In order to establish infection of the ciliated epithelium, pathogens have evolved remarkably effective mechanisms to remain anchored to the cilia despite ciliary motility and clearance of the mucociliary blanket. Each pathogen has a set of highly specific protein-containing ligands which recognize a narrow range of ciliary carbohydrates. These carbohydrates can be found on both glycolipids and glycoproteins and they may be integral ciliary membrane components, components of the ciliary glycocalyx, or even components of the ciliary mucous blanket. The most broadly used receptor is sialic acid; it is the major determinant of receptor specificity for *Mycoplasma, Pseudomonas,* and influenza A virus. Human ciliary infections with *B. pertussis* involve recognition of galactose–glucose-containing receptors while *Bordetella* infections in animals again involve sialic acid. Taken together, histochemical and histopathological studies indicate that receptors for all of these pathogens are probably clustered either at the ciliary tip or at the base of the ciliary shaft, specifically the sialic acid-rich regions termed the ciliary crown and necklace. Disruption of the architecture of the ciliary necklace and resultant ciliostasis occur following infection with *Mycoplasma* and *Bordetella.*

Development of an accurate map of the ciliary surface in biochemical terms will provide information which is essential to the development of new modes of therapy for infectious diseases of the ciliated epithelium. Currently, none of these diseases can be prevented by vaccination except for pertussis. In the case of pertussis, the vaccine has a high degree of toxicity and is therefore the focus of intensive effort to improve its acceptability without sacrificing efficacy. A desirable activity of any new vaccine against these infections is the ability to prevent adherence; incorporation of such activity by design in a new vaccine requires knowledge of the biochemistry of the attachment process. For instance, incorporation of the epitopes of adhesins which are critical for attachment is actively being considered in efforts to design vaccines for pertussis (Tuomanen *et al.*, 1988), mycoplasma (Hu *et al.*, 1982) and pseudomonas (Doig *et al.*, 1988). The feasibility of these antiadherence-directed strategies is suggested by the fact that antibodies which block adherence appear during convalescence from natural *Bordetella* (Tuomanen *et al.*, 1984) and mycoplasma (Hu *et al.*, 1982) infection.

In addition to vaccine development, knowledge of the biochemistry of the ciliary surface and its involvement in adherence of pathogens is important to the development of two new modes of treatment of these diseases: receptor analogues and antireceptor antibodies. Administration of analogues, for instance by aerosol, might effectively prevent adherence of bacteria and thereby arrest infection. If the receptor analogue were of particularly high affinity, it might actually eliminate the infection by effectively eluting adherent bacteria from the ciliary surface as is true for galactose and *B. pertussis* (Tuomanen *et al.*, 1988). By masking receptor sites, administration of antireceptor antibodies may prevent the establishment of infection in healthy subjects who come in contact with the disease. Such concepts would be of particular use in the respiratory tract since antibiotics penetrate poorly to the mucosal surface. For instance, antibiotic therapy of whooping cough is not effective once paroxysmal cough develops and chronic colonization of patients with cystic fibrosis by *Pseudomonas aeruginosa* is not eradicated by antibiotics. With current techniques, the biochemical mapping of the differentiated, mammalian ciliary cell membrane can be realized and with it will come important new insights into how to design new therapies for surface-oriented ciliary diseases.

References

Adoutte, A., Ramanathan, R., Lewis, R. M., Dute, R. D., Ling, K.-Y., Kung, C., and Nelson, D. L., 1980, Biochemical studies of the excitable membrane of *Paramecium tetraurelia*. III. Proteins of cilia and ciliary membranes, *J. Cell Biol.* **84:**717–738.

Almagor, M., Kahane, I., Wiesel, J. M., and Yatziv, S., 1985, Human ciliated epithelial cells from nasal polyps as an experimental model for *Mycoplasma pneumoniae* infection, *Infect. Immun.* **48:**552–555.

Anderson, R. G. W., and Hein, C. E., 1977, Distribution of anionic sites on the oviduct ciliary membrane, *J. Cell Biol.* **72:**482–492.

Arp, L. H., Hellwig, D. H., and Huffman, E. L., 1988, Sialogangliosides as the putative tracheal mucosal receptors for *Bordetella avium* adhesins, Abstract No. B-164, 88th Annual Meeting of the American Society for Microbiology, Miami Beach.

Bemis, D. A., and Wilson, S. C., 1985, Influence of potential virulence determinants on *Bordetella bronchiseptica*-induced ciliostasis, *Infect. Immun.* **50:**35–42.

Brennan, M. J., David, J. L., Kenimer, J. G., and Manclark, C. R., 1988, Lectin like binding of pertussis toxin to a 165 kilodalton Chinese hamster ovary cell glycoprotein, *J. Biol. Chem.* **263:**4895–4899.

Carroll, S. M., Higa, H. H., and Paulson, J. C., 1981, Different cell-surface receptor determinants of antigenically similar influenza virus hemagglutinins, *J. Biol. Chem.* **256:**8357–8363.

Carson, J. L., Collier, A. M., and Clyde, W. A., Jr., 1979, Ciliary membrane alterations occurring in experimental *Mycoplasma pneumoniae* infection, *Science* **206:**349–351.

Chailley, B., and Boisvieux-Ulrich, E., 1985, Detection of plasma membrane cholesterol by filipin during microvillogenesis and ciliogenesis in quail oviduct, *J. Histochem. Cytochem.* **33:**1–10.

Chailley, B., N'Diaye, A., Boisvieux-Ulrich, E., Sandoz, D. and Delaunay, M.-C., 1981, Comparative study of the distribution of fuzzy coat, lectin receptors, and intramembrane particles of the ciliary membrane, *Eur. J. Clin. Microbiol.* **25:**300–307.

Chandler, D. K., Collier, A. M., and Barile, M. F., 1982, Attachment of *Mycoplasma pneumoniae* to hamster tracheal organ cultures, tracheal outgrowth monolayers, human erythrocytes, and WiDr human tissue culture cells, *Infect. Immun.* **35:**937–942.

Chandler, D. K. F., Grabowski, M. W., Rabson, A. S., and Barile, M. F., 1987, Further studies on the *Mycoplasma pneumoniae* extract: Ciliostatic and cell recruitment activities, *Isr. J. Med. Sci.* **23:**580–584.

Chen, Y.-Y., and Krause, D. C., 1988, Parasitism of hamster trachea epithelial cells by *Mycoplasma pneumoniae*, *Infect. Immun.* **56:**570–576.

Chen, Z., and Lancet, D., 1984, Membrane proteins unique to vertebrate olfactory cilia: Candidates for sensory receptor molecules, *Proc. Natl. Acad. Sci. USA* **81:**1859–1863.

Collier, A. M., 1979, Adherence of mycoplasm to cilia, in: *The Mycoplasmas*, Volume 2 (J. T. Tully and R. F. Whitcomb, eds.), Academic Press, New York, pp. 49–63.

Cookson, B. T., Tyler, A. N., and Goldman, W. E., 1988, Definitive structure of the peptidoglycan-derived tracheal cytotoxin of *Bordetella pertussis*, Abstract No. B-70, 88th Annual Meeting of the American Society for Microbiology, Miami Beach.

Cordier, A. C., 1975, Ultrastructure of the cilia of thymic cysts in "nude" mice, *Anat. Rec.* **181:**127–150.

Denny, F. W., 1974, Effect of a toxin produced by *Haemophilus influenzae* on ciliated respiratory epithelium, *J. Infect. Dis.* **129:**93–100.

Denny, F. W., Clyde, W. A., Jr., and Glezen, W. P., 1971, *Mycoplasma pneumoniae* disease: Clinical spectrum, pathophysiology, epidemiology, and control, *J. Infect. Dis.* **123:**74–92.

Dentler, W. L., 1987, Cilia and flagella, *Int. Rev. Cytol.* Suppl. **17:**391–438.

Dentler, W. L., and LeCluyse, E. L., 1982a, Microtubule capping structures at the tips of tracheal cilia: Evidence for their firm attachment during ciliary bend formation and the restriction of microtubule sliding, *Cell. Motil.* **2:**549–572.

Dentler, W. L., and LeCluyse, E. L., 1982b, The effects of structures attached to the tips of tracheal ciliary microtubules on the nucleation of microtubule assembly *in vitro*, *Cell Motil.* Suppl. **1:**13–18.

Dirksen, E. R., and Satir, P., 1972, Ciliary activity in the mouse oviduct as studied by transmission and scanning electron microscopy, *Tissue Cell* **4:**389–404.

Doig, P., Todd, T., Sastry, P. A., Lee, K. K., Hodges, R. S., Paranchych, W., and Irvin, R. T., 1988, Role of pili in adhesion of *Pseudomonas aeruginosa* to human respiratory epithelial cells, *Infect. Immun.* **56:**1641–1646.

Foliquet, B., and Puchelle, E., 1986, Apical structure of human respiratory cilia, *Bull. Eur. Physiopathol. Respir.* **22:**43–47.

Franklin, A. L., Todd, T., Gurman, G., Black, D., Mankinen-Irvin, P. M., and Irvin, R. T., 1987, Adherence of *Pseudomonas aeruginosa* to cilia of human tracheal epithelial cells, *Infect. Immun.* **55:**1523–1525.

Ganaway, J. R., Spencer, T. H., Moore, T. D., and Allen, A. M., 1985, Isolation, propagation, and characterization of a newly recognized pathogen, cilia-associated respiratory bacillus of rats, an etiological agent of chronic respiratory disease, *Infect. Immun.* **47**:472–479.

Geers, T. A., and Baker, N. R., 1987, The effect of sublethal concentration of aminoglycoside on adherence of *Pseudomonas aeruginosa* to hamster tracheal epithelium, *J. Antimicrob. Chemother.* **19**:561–568.

Gibbons, I. R., 1981, Cilia and flagella of eukaryotes, *J. Cell Biol.* **91**:107s–130s.

Gilula, N. B., and Satir, P., 1972, The ciliary necklace. A Ciliary membrane specialization, *J. Cell Biol.* **53**:494–508.

Goldman, W. E., Klapper, D. G., and Baseman, J. E., 1982, Detection, isolation, and analysis of a released *Bordetella pertussis* product toxic to cultured tracheal cells, *Infect. Immun.* **36**:782–794.

Griffith, J. W., White, W. J., Danneman, P. J., and Lang, C. M., 1988, Cilia-associated respiratory (CAR) bacillus infection of obese mice, *Vet. Pathol.* **25**:72–76.

Hastie, A. T., Hingley, S. T., Higgins, M. L., Kueppers, F., and Shyrock, T., 1986, Rhamnolipid from *Pseudomonas aeruginosa* inactivates mammalian tracheal ciliary axonemes, *Cell Motil. Cytoskel.* **6**:502–509.

Hingley, S. T., Hastie, A. T., Kueppers, F., Higgins, M. L., and Weinbaum, G., 1986a, Bacterial ciliostatic factors; effect on respiratory cilia, *Eur. J. Respir. Dis.* **69**(Suppl. 146):291–293.

Hingley, S. T., Hastie, A. T., Kueppers, F., and Higgins, M. L., 1986b, Disruption of respiratory cilia by proteases including those of *Pseudomonas aeruginosa*, *Infect. Immun.* **54**:379–385.

Hingley, S. T., Hastie, A. T., Kueppers, F., Higgins, M. L., Weinbaum, G., and Shyrock, T., 1986c, Effect of ciliostatic factors from *Pseudomonas aeruginosa* on rabbit respiratory cilia, *Infect. Immun.* **51**:254–262.

Hirano, J., and Watanabe, Y., 1985, Studies on calmodulin-binding proteins (CaMBPs) in the cilia of *Tetrahymena*, *Exp. Cell Res.* **157**:441–450.

Hopewell, J. W., Holt, L. B., and Desombre, T. R., 1972, An electron-microscope study of intracerebral infection of mice with low-virulence *Bordetella pertussis*, *J. Med. Microbiol.* **5**:154–157.

Hu, P. C., Cole, R. M., Huang, Y. S., Graham, J. A., Gardner, D. E., Collier, A. M., and Clyde, W. A., 1982, *Mycoplasma pneumoniae* infection: Role of a surface protein in the attachment organelle, *Science* **216**:313–315.

Ishikawa, H., and Isayama, Y., 1987, Evidence for sialyl glycoconjugates as receptors for *Bordetella bronchiseptica* on swine nasal mucosa, *Infect. Immun.* **55**:1607–1609.

Jeffrey, P. K., and Reid, L., 1975, New observations of rat airway epithelium: A quantitative and electron microscopic study, *J. Anat.* **120**:295–320.

Johanson, W. G., Jr., Higuchi, J. H., Chaudhuri, T. R., and Woods, D. E., 1980, Bacterial adherence to epithelial cells in bacillary colonization of the respiratory tract, *Am. Rev. Respir. Dis.* **121**:55–63.

Katada, T., and Ui, M., 1982, Direct modification of the membrane adenylate cyclase system by islet-activating protein due to ADP-ribosylation of a membrane protein, *Proc. Natl. Acad. Sci. USA* **79**:3129–3133.

Kitano, K., Tuomanen, E., and Tomasz, A., 1986, Transglycosylase and endopeptidase participate in the degradation of murein during antolysis of *Escherichia coli*, *J. Bacteriol.* **167**:759–765.

Krivan, H. C., Roberts, D. D., and Ginsburg, V., 1988, Many pulmonary pathogens bind specifically to the glycolipids gangliotetraosylceramide (asialo GM1) and gangliotriaosylceramide (asialo GM2), Abstract No. B-163, 88th Annual Meeting of the American Society for Microbiology, Miami Beach.

Kuhn, C., III, and Engleman, W., 1978, The structure of the tips of mammalian respiratory cilia, *Cell Tissue Res.* **186**:491–498.

Lind, K., Hoier-Madsen, M., and Wiik, A., 1988, Autoantibodies to the mitotic spindle apparatus in *Mycoplasma pneumoniae* disease, *Infect. Immun.* **56**:714–715.

Loomes, L. M., Uemura, K.-I., Childs, R. A., Paulson, J. C., Rogers, G. N., Scudder, P. R., Michalski, J.-C., Hounsell, E. F., Taylor-Robinson, D., and Feize, T., 1984, Erythrocyte receptors for *Mycoplasma pneumoniae* are sialylated oligosaccharides of Ii antigen type, *Nature* **307**:560–563.

Low, P.-M. P., Luk, C. K., Dulfano, M. J., and Finch, P. J. P., 1984, Ciliary beat frequency of human respiratory tract by different sampling techniques, *Am. Rev. Respir. Dis.* **130**:497–498.

Marcus, H., and Baker, N. R., 1985, Quantitation of adherence of mucoid and nonmucoid *Pseudomonas aeruginosa* to hamster tracheal epithelium, *Infect. Immun.* **47**:723–729.

Marks, M. I., Stacy, T., and Krous, H. F., 1980, Progressive cough associated with lymphocytic leukemoid reaction in an infant, *J. Pediatr.* **97**:156–160.

Matsushita, S., Kashima, M., and Joshima, H., 1987, Serodiagnosis of cilia-associated respiratory bacillus infection by the indirect immunofluorescence assay technique, *Lab. Anim.* **21**:356–359.

Mazzuca, M., Lhermitte, M., Lafitte, J.-J., and Roussel, P., 1982, Use of lectins for detection of glycoconjugates in the glandular cells of the human bronchial mucosa, *J. Histochem. Cytochem.* **30:**956–966.

Menco, M., 1980, Qualitative and quantitative freeze-fracture studies on olfactory and nasal respiratory epithelial surfaces of frog, ox, rat, and dog, *Cell Tissue Res.* **211:**5–29.

Moller, P. C., Chang, J. P., and Partridge, L. R., 1983, Immunocytochemical localization of tubulin with colloidal gold in cilia of rabbit tracheal epithelial cultures, *Tissue Cell* **15:**39–45.

Monk, B. C., Adair, W. S., Cohen, R. A., and Goodenough, U. W., 1983, Topography of *Chlamydomonas:* Fine structure and polypeptide components of the gametic flagellar membrane surface and the cell wall, *Planta* **158:**517–533.

Montesano, R., 1979, Inhomogeneous distribution of filipin–sterol complexes in the ciliary membrane of rat tracheal epithelium (1), *Am. J. Anat.* **156:**139–145.

Murray, H. W., Masur, H., Senterfit, L. B., and Roberts, R. B., 1975, The protean manifestations of *Mycoplasma pneumoniae* infection in adults, *Am. J. Med.* **58:**229–242.

Muse, K. E., 1980, Host cell membrane perturbations during experimental *Bordetella pertussis* infection, *Electron Microsc.* **2:**432–433.

Muse, K. E., Collier, A. M., and Baseman, J. B., 1977, Scanning electron microscopic study of hamster tracheal organ cultures infected with *Bordetella pertussis, J. Infect. Dis.* **136:**768–777.

Muse, K. E., Findley, D., Allen, L., and Collier, A. M., 1978, *In vitro* model of *Bordetella pertussis* infection: Pathogenic and microbicidal interactions, in: *Third International Symposium on Pertussis* (C. R. Manclark and J. C. Hill, eds.), U.S. Department of Health, Education and Welfare, Washington, D.C., pp. 41–50.

Niederman, M. S., Rafferty, T. D., Sasaki, C. T., Merrill, W. W., Matthay, R. A., and Reynolds, H. Y., 1983, Comparison of bacterial adherence to ciliated and squamous epithelial cells obtained from the human respiratory tract, *Am. Rev. Respir. Dis.* **127:**85–90.

Opremcak, L. B., and Rheins, M. S., 1983, Scanning electron microscopy of mouse ciliated oviduct and tracheal epithelium infected *in vitro* with *Bordetella pertussis, Can. J. Microbiol.* **29:**415–420.

Palmer, L. B., Merrill, W. W., Niederman, M. S., Ferranti, R. D., and Reynolds, H. Y., 1986, Bacterial adherence to respiratory tract cells. Relationships between *in vivo* and *in vitro* pH and bacterial attachment, *Am. Rev. Respir. Dis.* **133:**784–788.

Pittman, M., 1984, The concept of pertussis as a toxin-mediated disease, *Pediatr. Infect. Dis.* **3:**467–486.

Plattner, H., 1975, Ciliary granule plaques: Membrane-intercalated particle aggregates associated with Ca²⁺-binding sites in *Paramecium, J. Cell Sci.* **18:**257–269.

Plopper, C. G., Mariassy, A. T., and Lollini, L. O., 1983, Structure as revealed by airway dissection, *Am. Rev. Respir. Dis.* **128:**S4–S7.

Plotkin, B. J., and Bemis, D. A., 1984, Adherence of *Bordetella bronchiseptica* to hamster lung fibroblasts, *Infect. Immun.* **46:**697–702.

Poncz, L., Jentoft, N., Ho, M.-C. D., and Dearborn, D. G., 1988, Kinetics of proteolysis of hog gastric mucin by human neutrophil elastase and by *Pseudomonas aeruginosa* elastase, *Infect. Immun.* **56:**703–704.

Rich, A. R., 1932, On the etiology and pathogenesis of whooping-cough, *Johns Hopkins Hosp. Bull.* **51:**346–363.

Rogers, G. N., Paulson, J. C., Daniels, R. S., Skehel, J. J., Wilson, I. A., and Wiley, D. C., 1983, Single amino acid substitutions in influenza haemagglutinin change receptor binding specificity, *Nature* **304:**76–78.

Rosenthal, R. S., Nogami, W., Cookson, B. T., Goldman, W. E., and Folkening, W. J., 1987, Major fragment of soluble peptidoglycan released from growing *Bordetella pertussis* is tracheal cytotoxin, *Infect. Immun.* **55:**2117–2120.

Sandoz, D., Boisvieux-Ulrich, E., and Chailley, B., 1979, Relationships between intramembrane particles and glycoconjugates in the ciliary membrane of the quail oviduct, *Biol. Cell.* **36:**267–280.

Sinha, R. K., and Rosenthal, R. S., 1980, Release of soluble peptidoglycan from growing gonococci: Demonstration of anhydro-muramyl-containing fragments, *Infect. Immun.* **29:**914–925.

Sleigh, M. A., Blake, J. R., and Liron, N., 1988, The propulsion of mucus by cilia, *Am. Rev. Respir. Dis.* **137:**726–741.

Spicer, S. S., Schulte, B. A., and Thomopoulos, G. N., 1983, Histochemical properties of the respiratory tract epithelium in different species, *Am. Rev. Respir. Dis.* **128:**S20–S26.

Stephens, D. S., Whitney, A. M., Melly, M. A., Hoffman, L. H., Farley, M. M., and Frasch, C. E., 1986,

Analysis of damage to human ciliated nasopharyngeal epithelium by *Neisseria meningitidis, Infect. Immun.* **51:**579–585.

Stephens, R. E., 1977, Major membrane protein differences in cilia and flagella: Evidence for a membrane-associated tubulin, *Biochemistry* **16:**2047–2058.

Stephens, R. E., 1983, Reconstitution of ciliary membranes containing tubulin, *J. Cell Biol.* **96:**68–75.

Stephens, R. E., 1985, Evidence for a tubulin-containing lipid–protein structural complex in ciliary membranes, *J. Cell Biol.* **100:**1082–1090.

Thomassen, M. J., Demko, C. A., and Doershuk, C. F., 1987, Cystic fibrosis: A review of pulmonary infections and interventions, *Pediatr. Pulmonol.* **3:**334–351.

Tuomanen, E., 1986a, Adherence of *Bordetella pertussis* to human cilia: Implications for disease prevention and therapy, in: *Microbiology, 1986* (D. Schlessinger, ed.), American Society for Microbiology, Washington, D.C., pp. 59–64.

Tuomanen, E., 1986b, Piracy of adhesins: Attachment of superinfecting pathogens to respiratory cilia by secreted adhesins of *Bordetella pertussis, Infect. Immun.* **54:**905–908.

Tuomanen, E. I., and Hendley, J. O., 1983, Adherence of *Bordetella pertussis* to human respiratory epithelial cells, *J. Infect. Dis.* **148:**125–130.

Tuomanen, E., and Weiss, A., 1985, Characterization of two adhesins of *Bordetella pertussis* for human ciliated respiratory-epithelial cells, *J. Infect. Dis.* **152:**118–125.

Tuomanen, E. I., Nedelman, J., Hendley, J. O., and Hewlett, E. L., 1983, Species specificity of *Bordetella* adherence to human and animal ciliated respiratory epithelial cells, *Infect. Immun.* **42:**692–695.

Tuomanen, E. I., Zapiain, L. A., Galvan, P., and Hewlett, E. L., 1984, Characterization of antibody inhibiting adherence of *Bordetella pertussis* to human respiratory epithelial cells, *J. Clin. Microbiol.* **20:**167–170.

Tuomanen, E., Towbin, H., Rosenfelder, G., Braun, D., Larson, G., Hansson, G., and Hill, R., 1988, Receptor analogs and monoclonal antibodies that inhibit adherence of *Bordetella pertussis* to human ciliated respiratory epithelial cells, *J. Exp. Med.* **168:**267–278.

Turner, R. B., Hendley, J. O., and Gwaltney, J. M., Jr., 1982, Shedding of infected ciliated epithelial cells in rhinovirus colds, *J. Infect. Dis.* **145:**849–853.

van Zwieten, M. J., Solleveld, H. A., Lindsey, J. R., de Groot, F. G., Zurcher, C., and Hollander, C. F., 1980, Respiratory disease in rats associated with a filamentous bacterium: A preliminary report, *Lab. Anim. Sci.* **30:**215–221.

Vishwanath, S., and Ramphal, R., 1985, Tracheobronchial mucin receptor for *Pseudomonas aeruginosa:* Predominance of amino sugars in binding sites, *Infect. Immun.* **48:**331–335.

Wilson, R., Roberts, D., and Cole, P., 1985, Effect of bacterial products on human ciliary function *in vitro, Thorax* **40:**125–131.

Winther, B., Brofeldt, S., Christensen, B., and Mygind, N., 1984a, Light and scanning electron microscopy of nasal biopsy material from patients with naturally acquired common colds, *Acta Oto-Laryngol.* **97:**309–318.

Winther, B., Farr, B., Turner, R. B., Hendley, J. O., Gwaltney, J. M., and Mygind, N., 1984b, Histopathologic examination and enumeration of polymorphonuclear leukocytes in the nasal mucosa during experimental rhinovirus colds, *Acta Oto-Laryngol.* Suppl. **413:**19–24.

Wise, K. S., and Watson, R. K., 1985, Antigenic mimicry of mammalian intermediate filaments by mycoplasmas, *Infect. Immun.* **48:**587–591.

Woods, D. E., Bass, J. A., Johanson, W. G., Jr., and Straus, D. C., 1980, Role of adherence in the pathogenesis of *Pseudomonas aeruginosa* lung infection in cystic fibrosis patients, *Infect. Immun.* **30:**694–699.

The Photoreceptor Connecting Cilium
A Model for the Transition Zone

Joseph C. Besharse and Cynthia J. Horst

1. Introduction

Vertebrate photoreceptors provide an excellent system for investigating the organization and maintenance of the membrane domains and cell polarity necessary for sensory function. In particular, the connecting cilium of photoreceptors appears to play an important role in photoreceptor organization. This specialized region corresponds structurally to the transition zone of motile cilia. As a consequence, studies of the structure and function of the photoreceptor connecting cilium are likely to provide insight into the function of the transition zone of motile and sensory cilia found more generally in eukaryotic cells. The purpose of this chapter is to review our current understanding of both the structure and function of the photoreceptor connecting cilium with emphasis on features that it may share with other cilia. The principal focus of recent research is on the role of the connecting cilium in the generation and maintenance of the polarized structure of photoreceptors. Details of the many recent advances in photoreceptor development, molecular biology, transduction, and membrane turnover are beyond the scope of this chapter, but are discussed in a series of recent reviews (Papermaster and Schneider, 1982; Bok, 1985; Besharse, 1982, 1986; Besharse *et al.*, 1988; Liebman, 1987; Applebury and Hargrave, 1986; Adler, 1986).

2. Structure of Photoreceptor Cilia

Rod and cone photoreceptors are highly polarized sensory neurons with four functional domains: a photosensitive organelle called an outer segment, an inner segment, a

Joseph C. Besharse and Cynthia J. Horst • Department of Anatomy and Cell Biology, Emory University School of Medicine, Atlanta, Georgia 30322. *Present address* JCB: Department of Anatomy and Cell Biology, The University of Kansas Medical Center, Kansas City, Kansas 66103.

cell body, and a synaptic terminal (Fig. 1). The outer segment consists of membranous disks that are either contained within a surrounding plasma membrane (rods) or are confluent with the plasma membrane (cones). Outer segment disks contain the visual pigment apoprotein, opsin, as an integral membrane component, additional less abundant membrane proteins, and a variety of membrane-associated proteins that play a role in phototransduction. The connecting cilium serves to join the outer segment with the inner segment, which contains most of the endoplasmic reticulum, Golgi complex, and mitochondria. The synaptic terminal is connected to the cell body, which contains the nucleus, by a short process. The synaptic terminal contains synaptic ribbons and vesicles that are involved in neurotransmission.

The structure of the photoreceptor cilium and its relationship to motile cilia can be appreciated by examining immature cells just prior to the time of outer segment formation (Fig. 2). The cilium develops from a basal body–centriole complex at the apical end of the cell (Tokuyasu and Yamada, 1959; De Robertis, 1956a,b, 1960; Eakin and Westfall, 1961; Greiner et al., 1981), and a ciliary rootlet extends into the cell from the basal body, in some species to the level of the nucleus (Spira and Millman, 1982). A similar pattern emerges in cell culture in which nonpolarized retinoblasts gradually establish a polarized

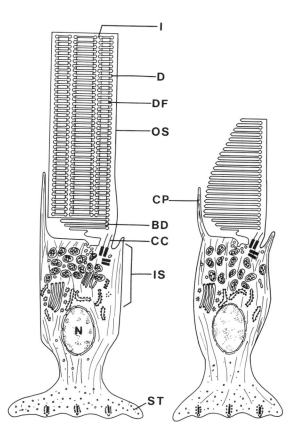

Figure 1. Diagram illustrating the structure of rod (left) and cone (right) photoreceptors. Both cell types are divided into an outer segment (OS), an inner segment (IS), a cell body that contains the nucleus (N), and a synaptic terminal (ST). I, incisures; D, disks; DF, disk filaments connecting adjacent disks; BD, basal open disks; CC, connecting cilium; CP, calycal process or microvillus. Modified from Bok (1985).

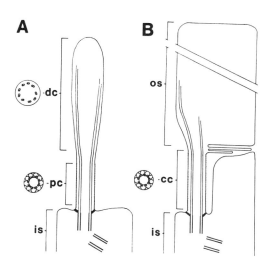

Figure 2. Diagram comparing the organization of the photoreceptor cilium prior to outer segment formation (A) with that of the mature cell (B). The proximal cilium (pc) of the immature cell corresponds to the connecting cilium (cc) of the mature cell. The outer segment (os) of the mature cell develops from the distal cilium (dc) of the immature cell. is, inner segment. Note that the doublet microtubule membrane cross-linkers are present in the proximal cilium of the immature cell and in the connecting cilium of the mature cell but are lacking in the outer segment or distal cilium. From Horst *et al.* (1987).

structure with a distinct ciliary pole (Araki, 1984; Adler, 1986). The immature photoreceptor cilium consists of structurally distinct proximal and distal regions (Fig. 3). In cross section, the proximal region has Y-shaped cross-linkers that reach from the junction of A and B subfibers of each doublet microtubule to the adjacent plasma membrane (Röhlich, 1975; Matsusaka, 1976; Peters *et al.*, 1983; Besharse *et al.*, 1985). In longitudinal section, "beadlike" structures project through the plasma membrane of the proximal region (Röhlich, 1975; Besharse *et al.*, 1985). These beadlike structures may be related to the particle rows or ciliary necklaces (Gilula and Satir, 1972) seen in freeze-fracture replicas (see Chailley *et al.*, this volume); both structures have a similar longitudinal periodicity (Matsusaka, 1974; Röhlich, 1975; Besharse *et al.*, 1985). The distal ciliary region lacks the beadlike structures as well as the microtubule–membrane cross-linkers (Fig. 3). The ciliary axoneme, which lacks a central pair of microtubules (but see Matsusaka, 1976) and dynein arms throughout its length, extends through both proximal and distal cilium. In the proximal cilium, doublets are aligned with the microtubule–membrane cross-linkers, but distally they flair out and are irregularly arranged (see Fig. 3).

The proximal and distal cilium are also distinguishable immunocytochemically (Fig. 4). Opsin immunoreactivity is found in the distal membrane but is sparse or lacking in the proximal membrane (Nir *et al.*, 1984; Besharse *et al.*, 1985; Hicks and Barnstable, 1986). Although opsin immunoreactivity may appear in great abundance in the plasma membrane of the cell body at early developmental stages (Nir *et al.*, 1984; Hicks and Barnstable, 1986; Adler, 1986), its polarized distribution in the distal cilium of both rats and mice occurs prior to the formation of outer segment disks (Besharse *et al.*, 1985; Usukura and Bok, 1987; Jansen *et al.*, 1987). Since this pattern of immunoreactivity is characteristic of mature cells as well (Papermaster and Schneider, 1982; Nir and Papermaster, 1983; Nir *et al.*, 1984; Besharse *et al.*, 1985; Usukura and Bok, 1987), the finding has important implications regarding the mechanism of opsin delivery to either the developing or mature outer segment. That mechanism appears to be characteristic of the cilium itself, and distinct from the process of disk formation (see Sections 4.1 and 4.2).

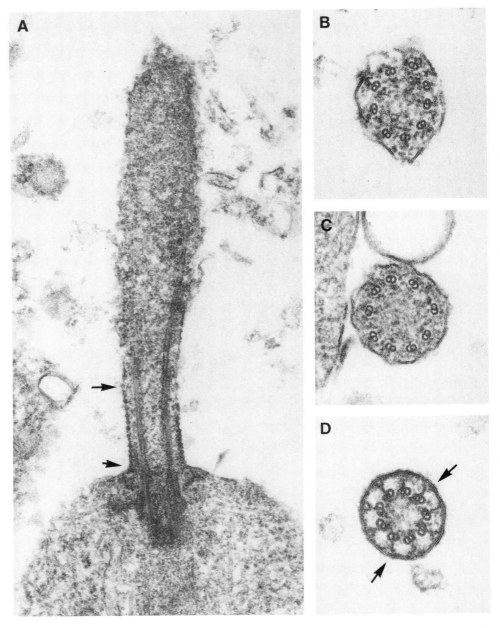

Figure 3. Electron micrographs illustrating the distinct proximal and distal ciliary membrane in an undifferentiated photoreceptor from a 7-day-old rat. Tissue was saponin treated and stained with tannic acid, uranyl acetate, and lead citrate. (A) Longitudinal section through the cilium; proximal region has beadlike structures. (B, C) cross sections through the distal cilium. Microtubule–membrane cross-linkers are lacking. (D) Cross section through the proximal cilium. Arrows indicate microtubule–membrane cross-linkers. Magnification: A, × 72,250; B, × 117,000; C, D, × 119,000; reproduced at 72%. From Besharse *et al.* (1985).

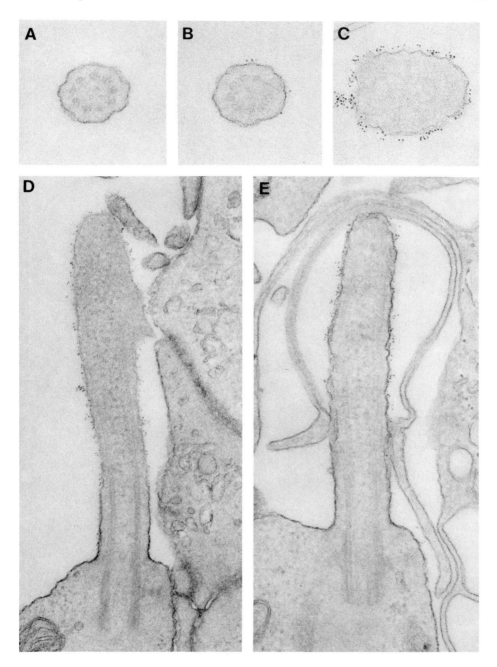

Figure 4. Electron micrographs showing the distribution of immunoreactive opsin in the cilia of developing photoreceptors. Sections stained with bismuth subnitrate. Ferritin grains define the distribution of opsin. (A–E) Three-day-old rat without outer segments. (A) Cross section through proximal cilium. (B, C) Cross sections through more distal parts of cilium. (D, E) Longitudinal sections. Opsin immunoreactivity is found in the distal cilium but is not seen on the proximal cilium. A–C, × 104,500; D, × 68,750; E, × 63,750; reproduced at 75%. From Besharse *et al.* (1985).

The proximal cilium of the immature cell corresponds structurally to the region connecting the inner and outer segments of the mature cell (Fig. 2). In the mature cell this region also has microtubule–membrane cross-linkers, ciliary necklaces, and beadlike membrane structures (Röhlich, 1975; Matsusaka, 1976; Besharse *et al.*, 1985). It is called the connecting cilium (De Robertis, 1956a) and, as first pointed out by Röhlich (1975), corresponds morphologically to the transition zone of motile cilia. As in the immature cell, opsin immunoreactivity is high in the outer segment, but is lacking or at very low density in the plasma membrane of the connecting cilium (Papermaster and Schneider, 1982; Peters *et al.*, 1983; Nir and Papermaster, 1983; Papermaster *et al.*, 1985; Hicks and Barnstable, 1986).

The photosensitive disk membranes of photoreceptors are formed from the membrane of the distal cilium (Tokuyasu and Yamada, 1959; De Robertis, 1956a, 1960; Eakin and Westfall, 1961). This region greatly expands during development to form the outer segment. Mature rod outer segments are typically 25 to 60 μm long, and may contain more than 2000 disks. Doublet microtubules of the axoneme extend into a region between the plasma membrane and disks on one side of the outer segment and eventually take the form of singlet microtubules (Brown *et al.*, 1963; Yacob *et al.*, 1977). In mammalian photoreceptors, axonemal microtubules reportedly extend distally in association with disk incisures (Steinberg and Wood, 1975; Wen *et al.*, 1982).

Recent immunocytochemical studies (Roof and Applebury, 1984; Kaplan *et al.*, 1987; Sale *et al.*, 1988) have confirmed the original observation of Brown *et al.* (1963) that axonemal microtubules extend at least half the length of amphibian rod outer segments and, in some cases, may extend to nearly the full length. As in motile cilia, the axonemal microtubules contain acetylated α-tubulin (Sale *et al.*, 1988) and are not disrupted by antimicrotubule drugs (Kaplan *et al.*, 1987; Sale *et al.*, 1988). The correlation between microtubule stability and acetylated α-tubulin is emphasized by the observation that microtubules adjacent to the ciliary basal body complex in the inner segment do not contain acetylated α-tubulin and are highly labile in the presence of antimicrotubule drugs (Sale *et al.*, 1988). This suggests that the effects of antimicrotubule drugs on outer segment function (Besharse and Dunis, 1982; Bert, 1987) are unlikely to be explained on the basis of axoneme disruption.

The structural and functional relationship of the axoneme to outer segment disk membranes remains poorly defined and is complicated by the existence of unusual microtubulelike structures in the outer segment. In both developing and mature photoreceptors we have observed tubular structures about 30% larger than typical microtubules (25 nm) that are structurally distinct from typical microtubules in the same cell (Fig. 5). In the distal cilium of developing cells it is clear that these structures occur in addition to doublet microtubules and, except for their high degree of curvature, they resemble the adjacent plasma membrane. Similar structures were reported previously in developing guinea pig photoreceptors, but were clearly distinguished by the author from typical microtubules in the same cell (Spira, 1975, cf. Fig. 12 and 13). The existence of these structures suggests the need for caution in interpreting microtubule organization in electron microscopic images of the distal outer segment. Further structural studies and rigorous criteria, including immunocytochemistry, for identification will be required to fully understand the role of these tubular structures in the photoreceptor.

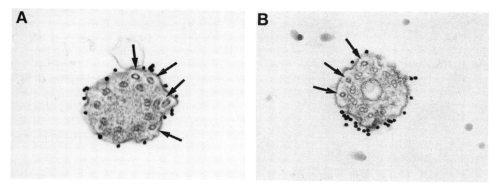

Figure 5. Electron micrographs of cross sections through the distal cilium of 7-day-old rat photoreceptors showing the distribution of doublet microtubules of the axoneme and the additional tubular structures. Micrographs are from a lectin staining experiment. Black dots are colloidal gold particles that show the distribution of Con A. Arrows indicate tubular structures that superficially resemble microtubules. They are distinct from axonemal doublets. A, × 95,500; B, × 98,900; reproduced at 75%. Unpublished micrographs from a previously published study on lectin binding to developing rat photoreceptors (Besharse *et al.*, 1985).

3. Microtubule–Membrane Cross-Linkers of the Connecting Cilium

Since Röhlich's (1975) structural studies on rat photoreceptors, it has been thought that the connecting cilium is homologous to the transition zone of motile cilia. Its structure, including microtubule–membrane cross-linkers, ciliary necklaces, and a richly endowed glycocalyx, is common to all motile and sensory cilia examined to date (Sandoz *et al.*, 1979; Gilula and Satir, 1972; Flower, 1971; Menco, 1980; Ringo, 1967; Wunderlich and Speth, 1972; Dentler, this volume). This region invariably lies between a basal body complex in the cell body and a distal ciliary region or, in the case of photoreceptors, the outer segment. The function of the motile cilium transition zone also appears to be similar to the photoreceptor connecting cilium in that it lies at the junction of distinct membrane domains (Dunlap, 1977; Machemer and Ogura, 1979; Ogura and Takahashi, 1976; Schultz and Klumpp, 1984; Hennessey *et al.*, 1983; Bardele, 1983; Musgrave *et al.*, 1986). Thus, the detailed analysis of the connecting cilium in photoreceptors is likely to provide a basis for understanding the transition zone of cilia in general. Recent observations have revealed a cytoskeletal–membrane assemblage that links glycoconjugates at the cell surface to the doublet microtubules of the axoneme (Horst *et al.*, 1987). Although cross-linkers have been described on the distal region of motile cilia (Dentler, 1980, 1981; Reinhart and Bloodgood, 1988), the microtubule–membrane assemblage of the connecting cilium (and transition zone) appears to be distinct. Since this assemblage confers unique structural stability on the connecting cilium, it is likely to be important as a determinant of photoreceptor membrane domains.

3.1. The Ciliary Surface and Its Transmembrane Assemblage

Photoreceptor surfaces are abundantly endowed with glycoconjugates as revealed by lectin binding studies (Molday, 1976; Hall and Nir, 1976; Nir and Hall, 1979; Bridges and

Fong, 1980; Blanks and Johnson, 1984; Wood *et al.*, 1984; Hicks and Molday, 1985; Wood and Napier-Marshall, 1985a,b; Besharse *et al.*, 1985; Hicks and Barnstable, 1986; Johnson *et al.*, 1986; Koide *et al.*, 1986; Molday and Molday, 1987; Cohen and Nir, 1987; Nir and Cohen, 1987). The most abundant glycoconjugate is opsin, the visual pigment apoprotein, which has two simple oligosaccharides that contain mannose and *N*-acetylglucosamine (Fukuda *et al.*, 1979; Liang *et al.*, 1979). Consequently, opsin binds both wheat germ agglutinin (WGA) and Con A, lectins specific for oligosaccharides containing *N*-acetylglucosamine and mannose, respectively (Steinemann and Stryer, 1973; Molday and Molday, 1979). Opsin's great abundance in the outer segment, including its plasma membrane (Jan and Revel, 1974; Basinger *et al.*, 1976; Papermaster and Schneider, 1982; Defoe and Besharse, 1985), suggests that it is the most abundant WGA and Con A binding protein in photoreceptor membranes. However, immunocytochemical studies have revealed only a sparse distribution of opsin in the connecting cilium (Papermaster and Schneider, 1982; Nir and Papermaster, 1983), even in developing photoreceptors (Nir *et al.*, 1984; Besharse *et al.*, 1985; Hicks and Barnstable, 1986). Despite this, the connecting cilium is richly endowed with both Con A and WGA binding sites (Besharse *et al.*, 1985; Hicks and Molday, 1985; Hicks and Barnsable, 1986; Koide *et al.*, 1986). Therefore, not all WGA and Con A binding can be attributed to opsin oligosaccharides.

The existence of glycoconjugates other than opsin is also indicted by the fact that photoreceptors bind lectins with different oligosaccharide specificities (Bridges and Fong, 1980; Nir and Hall, 1979). For example, in addition to its binding to *N*-acetylglucosamine, WGA also binds to oligosaccharides with terminal *N*-acetylneuraminic acid (lacking in opsin). Such binding sites have been demonstrated in both the outer segment and connecting cilium (Koide *et al.*, 1986; Molday and Molday, 1987; Cohen and Nir, 1987). Treatment with neuraminidase, an enzyme specific for terminal *N*-acetylneuraminic acid, removes some WGA binding sites from the connecting cilium. This indicates the presence of oligosaccharides bearing terminal *N*-acetylneuraminic acid residues. Both neuraminidase-sensitive and -insensitive high-molecular-weight glycoconjugates have been identified on the connecting cilium (see Horst *et al.*, 1987).

Some oligosaccharide components of the connecting cilium surface are part of a transmembrane assemblage linked to the axoneme by the microtubule–membrane cross-linkers (Horst *et al.*, 1987). Early studies designed to isolate an axoneme fraction from bovine rod outer segments revealed that many cilia are uprooted from the cell body and separate with outer segments on sucrose density gradients (Fleischman and Denisevich, 1979). After treatment of the outer segments with Triton X-100, a cytoskeletal preparation enriched in axonemes can be separated from membrane components by a second sucrose density gradient. Electron microscopic observations of that preparation suggested that the microtubule–membrane cross-linkers remained associated with the doublet microtubules during detergent treatment and isolation (Fleischman *et al.*, 1980). Recently, we found that, following Triton X-100 extraction sufficient to remove virtually all of the phospholipid bilayer, the cross-linkers remain associated with the doublets in developing rat retina (Horst *et al.*, 1987; see Fig. 6). In addition, the detergent treatment removes virtually all cell surface WGA binding sites except those in the region of the microtubule–membrane cross-linkers. Here, after detergent extraction, abundant WGA binding sites

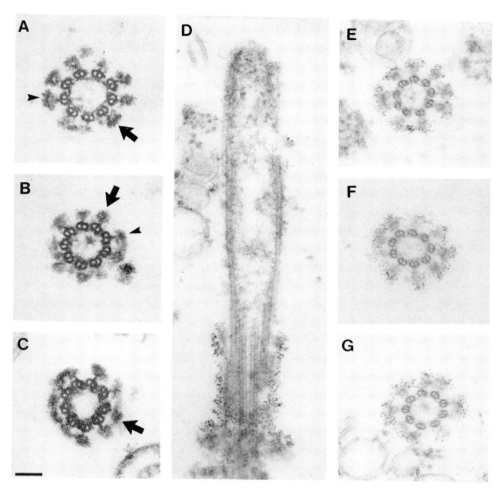

Figure 6. Electron micrographs of Triton X-100-extracted neonatal rat retinas showing the stable microtubule–membrane cross-linkers. (A–C) Cross sections through the proximal cilium of developing photoreceptors which have been detergent-extracted and stained with tannic acid, uranyl acetate, and lead citrate. Amorphous tufts (arrows) remain attached to microtubule doublets by short stalks. Occasionally, remnants of phospholipid bilayer can be seen in the tufts (arrowheads). (D) Longitudinal section through a developing photoreceptor which has been extracted with detergent prior to labeling with ferritin-conjugated WGA. Ferritin particles decorate the proximal cilium in the region of the cross-linkers. Virtually all WGA binding sites have been removed from the distal cilium. (E–G) Cross sections of material treated the same as that in D. Ferritin particles are located at the distal ends of the cross-linkers. Bar = 1 μm. Modified from Horst *et al.* (1987).

remain associated with the distal ends of the cross-linkers. These observations indicate that some cell-surface glycoconjugates are connected to the cross-linkers by a trans-membrane component and are thereby connected to the axonemal microtubules creating an extraordinarily stable microtubule–membrane assemblage.

Similar lectin binding characteristics and cross-linker stability have been described in the transition zone of motile cilia. Like the connecting cilium, quail oviduct cilia have Con A, WGA, and colloidal iron hydroxide binding sites on the transition zone plasma membrane (see Chailley *et al.*, this volume). While neuraminidase treatment does not remove WGA binding sites, it does remove colloidal iron hydroxide binding from the plasma membrane overlying the transition zone (Sandoz *et al.*, 1979). This indicates the presence of both *N*-acetylneuraminic acid- and *N*-acetylglucosamine-containing glycoconjugates in the transition zone. In addition, WGA binding sites were found to be resistant to chloroform–methanol extraction after fixation, similar to the detergent-resistant WGA binding sites of the photoreceptor connecting cilium. Stable microtubule–membrane cross-linkers have also been proposed in the transition zone of motile cilia. Isolated sheets of cilia from newt lung (Hard and Rieder, 1983) or rabbit oviduct (Anderson, 1974) which have been extracted with Triton X-100 retain cross-linker structures in the region of the transition zone. These parallels between the transition zone of motile cilia and the photoreceptor connecting cilium further support the proposal that they are homologous structures.

3.2. Identification of Surface Components of the Assemblage

We have taken advantage of the stability of the cross-linker assemblage to identify the glycoconjugate components of the cross-linkers (Horst *et al.*, 1987). The approach used a subcellular fraction enriched in axonemes as described by Fleishman and Denisevich (1979). The sucrose density gradient centrifugation after Triton X-100 extraction separates the dominant WGA binding membrane glycoconjugate, opsin, from those glycoconjugates that remain associated with the cross-linkers. Protein components of the axoneme fraction were resolved by SDS-PAGE. Transblots of the preparation were probed with WGA revealing the presence of three large glycoconjugates with apparent molecular masses of 425, 600, and 700 kDa. The unusually large size of these glycoconjugates is based on extrapolation from sea urchin sperm dynein heavy chain (450 kDa; Bell, 1983), used as a molecular mass marker. The estimates are considered tentative since currently we do not know the relative abundance of protein as opposed to saccharide in these glycoconjugates. As such, their migration positions may be aberrant.

Further characterization of the glycoconjugates using glycosidases (Horst *et al.*, 1987) indicates that at least one of the glycoconjugates contains terminal *N*-acetylneuraminic acid as suggested in histochemical studies using neuraminidase (Koide *et al.*, 1986; Cohen and Nir, 1987; Nir and Cohen, 1987). WGA binding to the 425-kDa glycoconjugate is eliminated by neuraminidase treatment (Fig. 7) and such enzyme treatment reveals peanut agglutinin (PNA) binding sites (unpublished results). PNA binds specifically to the dimer galactose–*N*-acetylgalactosamine, a disaccharide common to *O*-linked, mucin-type oligosaccharides. In order to further evaluate the possible presence of such a structure, we have recently used the enzyme endo-α-*N*-acetylgalactosaminidase (*O*-glycanase), which is specific for the unsubstituted disaccharide galactose–*N*-acetylgalactosamine (Umemoto *et al.*, 1977). In these experiments, transblots were first treated with neuraminidase to reveal PNA binding sites. Subsequent treatment with *O*-glycanase reduced the PNA binding (unpublished results). These studies suggest that the 425-kDa glycoconjugate is dominated by *O*-linked, mucin-type oligosaccharides. In contrast, the

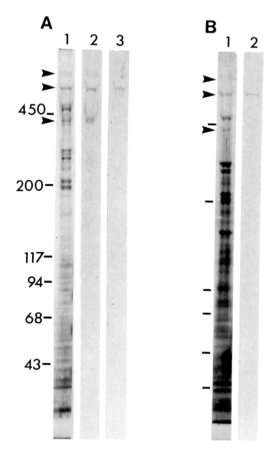

Figure 7. Transblots illustrating the axoneme-associated, high-molecular-weight, WGA- and Con A-binding glycoconjugates. (A) Lane 1: Silver-stained gel of the axoneme-enriched fraction separated on a 3–6% polyacrylamide gradient. Lane 2: WGA-stained transblot of the axoneme fraction. The three major high-molecular-weight WGA-binding glycoconjugates are indicated by the arrowheads. The migration positions are approximately 425, 600, and 700 kDa. Lane 3: Neuraminidase treatment prior to WGA staining reveals that WGA binding to both the 425- and 700-kDa glycoconjugates is neuraminidase sensitive whereas staining of the 600-kDa glycoconjugate remains. Molecular mass markers are given, in kDa, on the left. (B) Lane 1: Silver-stained gel as in part A, lane 1. Lane 2: Con A-stained transblot of the axoneme fraction. Only the 600-kDa glycoconjugate is stained by Con A. Molecular mass markers are as for A. From Horst *et al.* (1987).

600-kDa glycoconjugate exhibits little neuraminidase sensitivity, binds succinyl WGA (a derivative of WGA that binds *N*-acetylglucosamine but not *N*-acetylneuraminic acid; Monsigny *et al.*, 1980), and binds Con A. Thus, it would appear that there are at least two distinct oligosaccharide structures on isolated axonemes that bind WGA.

Although lectins have been useful for preliminary characterization of the connecting cilium plasma membrane, they are relatively nonselective in that they recognize oligosaccharide components displayed on many glycoconjugates. In order to develop more selective probes, we have begun to raise monoclonal antibodies using the crude bovine axoneme preparation (see Horst *et al.*, 1987) as antigen. Initial screening has revealed an antibody that selectively stains only the connecting cilium of intact cells and the region of cross-linkers in the axoneme preparation (Horst, *et al.*, 1989). This antibody binds to the 425-kDa glycoconjugate on transblots and thus confirms the unique ciliary location of this component. The antibody also binds to the transition zone of cilia from bovine oviduct, indicating that the epitope recognized by the antibody is common to both photoreceptor and motile cilia.

3.3. The Periciliary Ridge Complex

The ciliary region of amphibian photoreceptors differs in several details from that of mammalian photoreceptors (Sections 3.1 and 3.2). These structural differences may be related to the fact that amphibian photoreceptors are much larger and synthesize new disk membrane area at a rate at least 10-fold higher than in mammals (see Besharse, 1986). Although the amphibian connecting cilium has structural features like those discussed above, it is much shorter, not exceeding 0.5 μm in length, and is surrounded at its base by an elaborate pattern of grooves and ridges (Peters *et al.*, 1983). The latter, revealed in high-resolution electron micrographs, has been called the periciliary ridge complex (Peters *et al.*, 1983). Its most characteristic feature is a series of grooves (usually nine) arranged irregularly around the ciliary base and confluent with a trough immediately surrounding the cilium. The function of this structure remains unknown. However, it has been suggested that it is involved in the delivery of membrane components to the outer segment (Andrews, 1982; Peters *et al.*, 1983; Defoe and Besharse, 1985; Papermaster *et al.*, 1985). The cytosolic region adjacent to the complex is a site of accumulation of opsin-containing vesicles (Defoe and Besharse, 1985; Papermaster *et al.*, 1985) that are thought to be destined for incorporation into the outer segment. Andrews (1982) has noted the similarity of this structure to the active zones of the presynaptic component of neuromuscular junctions and has suggested that vesicular fusion occurs at a high rate in this region.

4. Functions of the Photoreceptor Connecting Cilium

The connecting cilium of photoreceptors lies at a strategic functional position. In addition to a role in differentiation of the outer segment, the connecting cilium lies in the pathway of information flow between the visual transduction events in the outer segment and the release of synaptic transmitter. As the sole persistent connection between the inner segment and the outer segment, the cilium must play an important role in the delivery and turnover of both enzymes and substrates for visual transduction as well as the delivery and turnover of photosensitive membrane. A further function of the connecting cilium is the maintenance of discrete membrane domains. Photoreceptor function depends on the segregation of visual pigment and a light-regulated cation channel to the outer segment (see Liebman, 1987) and a sodium pump to the cell body (Hagins, 1972). The connecting cilium is the zone of demarcation separating those discrete membrane domains. Recent evidence suggests that the ciliary membrane acts as a barrier which segregates opsin to the outer segment (Spencer *et al.*, 1988).

4.1. Delivery of Membrane Components

An understanding of the mechanism by which membrane lipids and proteins are delivered from the inner segment to the distal cilium is crucial to understanding either outer segment development or renewal. In each case the problem is operationally the same. The developing cell uses distal ciliary membrane to form the outer segment; the mature cell continues to produce new membrane disks at the distal cilium. Quantitative analysis of both processes in mammalian photoreceptors (Young, 1967; LaVail, 1973) indicates that throughout life, disk assembly requires about 0.1 μm^2 of membrane containing some 2000 opsin molecules per minute (see Besharse, 1986, for the calculation). The magnitude of the problem is compounded in the much larger photoreceptors of the amphibian *Xenopus laevis* (Besharse *et al.*, 1977). Here, disk assembly requires about 3.2 μm^2 of membrane containing about 60,000 opsin molecules/min. A large number of studies have demonstrated that both protein and phospholipid components of disk membranes are synthesized in the cell body (reviewed in Young, 1976; Besharse, 1986). Thus, an efficient mechanism for delivery of membrane components to the distal cilium must exist. The view (Papermaster *et al.*, 1975, 1985, Besharse and Pfenninger, 1980; Defoe and Besharse, 1985) that opsin follows a pathway involving rough endoplasmic reticulum, Golgi complex, and post-Golgi vesicles to the periciliary region seems well established (Fig. 8). However, the nature of the final step of delivery from the inner segment to the distal cilium remains an enigma. At present, it is not even known whether phospholipids and membrane proteins are delivered together or through separate pathways. Most studies have emphasized the pathway for opsin delivery.

In the earliest studies of disk assembly in photoreceptors (Young and Droz, 1968; Young, 1968), it was suggested that opsin is transferred through the connecting cilium in the cytosolic compartment. This was based on electron microscope autoradiographic images showing silver grains around the connecting cilium during a period when recently synthesized, radioactive opsin was being incorporated into disks. Since opsin was the dominant radioactive protein (see Hall *et al.*, 1969), the silver grains were interpreted as reflecting the distribution of opsin. Subsequently, it was learned that opsin is incorporated into membrane during its synthesis in endoplasmic reticulum (Papermaster *et al.*, 1975; Goldman and Blobel, 1981), and it has generally been assumed that opsin remains membrane associated during its sojourn through the cell (see Besharse, 1986, for review). Most authors have assumed, therefore, that opsin reaches the forming disks via the plasma membrane of the connecting cilium (Matsusaka, 1974; Röhlich, 1975; Besharse and Pfenninger, 1980). Because of the small size of the connecting cilium and low resolution of the autoradiographic technique, the original images of Young (1968), while clearly demonstrating the accumulation of radioactivity in the periciliary cytoplasm and adjacent disks, do not clearly delineate the pathway for transfer between the two compartments.

The simplest version of the ciliary membrane model for opsin transport is movement of both opsin and membrane lipid via the connecting cilium plasma membrane (Matsusaka, 1974; Röhlich, 1975; Besharse and Pfenninger, 1980; Peters *et al.*, 1983; Papermaster *et al.*, 1985; Besharse *et al.*, 1985). Although this model has intuitive appeal and is commonly cited, there is currently no direct evidence supporting it. In fact, given the vast quantities of membrane that must pass into forming disks, particularly in amphibian photoreceptors, the question has been raised (Besharse, 1986) as to whether a highly ordered ciliary membrane (see Section 3.2) of restricted dimensions (total surface area of 0.33 μm^2 in frog) could support bulk transfer in sufficient quantities for disk assembly. In *X. laevis* this would require that the entire membrane of the connecting cilium be exchanged about ten times per minute. In the rat, which has a longer connecting cilium and generates disc area at a lower rate (see above), a similar mechanism would require complete ciliary membrane turnover once in about 10 min. Such numerical estimates emphasize our limited understanding of molecular mechanisms of membrane transport and suggest the need for alternative models as a guide for experimental analysis.

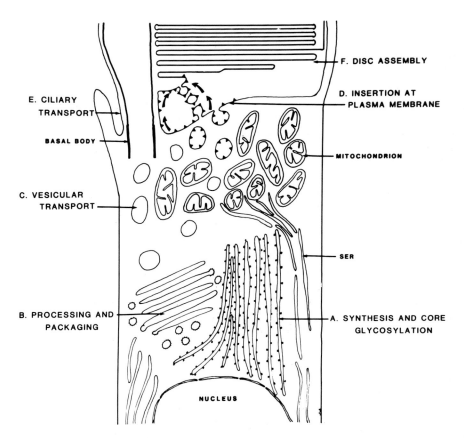

Figure 8. Diagram illustrating a model for the transport and delivery of opsin to frog rod outer segments. Detailed discussion of each aspect of the model is included in a recent review (Besharse, 1986). (A) Opsin is synthesized and glycosylated in rough endoplasmic reticulum. (B) Opsin is processed and packaged in the Golgi complex. (C) A post-Golgi population of vesicles transports opsin to the cell surface (D) where it is thought to be incorporated into the plasma membrane. (E) Transport from the inner segment to the outer segment is poorly understood. Opsin may pass to forming disks via the ciliary plasma membrane, but an alternative pathway involving an extracellular vesicle cannot be eliminated (see arrows). Triangles on the membranes emphasize the topological relationships of vesicular and disk membranes. (F) Disk assembly appears to involve evagination of the plasma membrane of the distal connecting cilium. Modified from Besharse (1986).

Are outer segment membrane proteins and lipids cotransported into forming disks or are they delivered through distinct pathways? Although critical to an understanding of photoreceptor membrane assembly, this question cannot be answered presently. Precedent for the idea of separate pathways comes from studies on motile cilia in which the kinetics of protein and lipid transport differ (see Williams, this volume). Several observations suggest different patterns of transport in photoreceptors as well. Both protein synthesis inhibitors and the Golgi transport inhibitor monensin block opsin transport into outer segments, but neither treatment blocks phospholipid transport (Basinger and Hoffman, 1976; Matheke and Holtzman, 1984; Matheke *et al.*, 1984). Thus, protein and phospholipid transport can be dissociated. Another difference is that outer segment phospholipids appear to turn over at a greater rate than the outer segment disks in which they reside (Anderson et al., 1980a–c). It is well established that the disks are shed from the distal end of rod photoreceptors in groups for degradation in the adjacent pigment epithelium (Young and Bok, 1969), and that outer segment length is maintained through the continuous formation of new disks in the region of the connecting cilium (Young, 1967; see Besharse, 1986, for review). It is also well established that the rate of opsin turnover is predicted precisely by disk turnover (Hall et al., 1969). Once incorporated into a disk, opsin molecules remain in that disk until they are degraded during the process of disk shedding. However, turnover rates for phospholipids indicate that during the lifetime of a disk, a superimposed process of phospholipid turnover must be occurring. This implies that net transfer of phospholipid into the outer segment occurs normally at a higher rate than that predicted from observations on disk assembly. The nature of the molecular mechanisms involved in lipid transport remains undefined (see Besharse, 1986; Fliesler and Anderson, 1983, for review).

Another potential difficulty with the ciliary membrane model for opsin transport is that it does not explain the sparse distribution of opsin immunoreactivity in the connecting cilium which would be expected to contain membrane components in transit. This could be accounted for if opsin were localized in restricted tracts that rarely appear in the plane of section (Papermaster and Schneider, 1982; Nir and Papermaster, 1983; Besharse et al., 1985). Alternatively, antigenic sites on opsin could be masked by the highly developed cytoskeleton–membrane assemblage of the connecting cilium (Besharse et al., 1985; Bok, 1985), or could occur in a nonimmunoreactive conformation (Hicks and Barnstable, 1986, 1987). The latter two suggestions are of some interest because freeze-fracture analysis of the PF leaflets of the connecting cilium has revealed large intramembranous particles distinct from ciliary necklace particles but similar to those associated with opsin in PF leaflets of the outer segment (Matsusaka, 1974; Röhlich, 1975; Besharse and Pfenninger, 1980; Besharse et al., 1985). Such particles have been interpreted as reflecting the ciliary distribution of opsin, and leave open the possibility that significant ciliary content of opsin may yet be detected.

Although much emphasis has been placed on a ciliary membrane pathway for opsin delivery, three alternative mechanisms have been suggested that cannot be rigorously excluded. The first is that opsin is transferred through the ciliary cytosol (Young, 1967). This mechanism is inconsistent with the integral membrane nature of opsin, the lack of conventional membrane structures in the ciliary cytosol, and the lack of opsin immunoreactivity (see Besharse, 1986). The second is transfer via a cytosolic bridge between inner and outer segment separate from the connecting cilium (Richardson, 1969). This mechanism was suggested on the basis of electron microscopic observation of connections between inner and outer segments. Such connections have been reported by others anecdotally. They have been seen in degenerating photoreceptors (Nir et al., 1987) and in cells obtained by mechanical or enzymatic dissociation (Townes-Anderson et al., 1985; Spencer et al., 1988; Hicks et al., 1988). They are not thought to be a common feature of photoreceptor structure and may represent an artifact of fixation (see Besharse, 1986). Furthermore, this mechanism would provide no obvious means to prevent back diffusion of opsin into inner segment membranes which is known to occur when inner and outer segments are artificially fused (Spencer et al., 1988; see Section 4.3). The third is that opsin and phospholipids are transferred in an extracellular vesicle that buds from the inner segment and fuses in the disk-forming region through a process similar to viral budding (Fig. 8; Besharse, 1986).

The model involving transport via an extracellular vesicle deserves some comment. It was originally suggested as a conceptual alternative because of ambiguities in the ciliary membrane model (Besharse, 1986). Because of the close juxtaposition of inner segment plasma membrane and the disk-forming region of the distal connecting cilium, extra-cellular transfer would be difficult to distinguish from the ciliary membrane model (see Fig. 8). Such a mechanism has at least three advantages. First, transport of membrane protein and lipid would not be quantitatively limited by structural constraints in the connecting cilium. Second, vectorial transport would be ensured; budding occurs at the inner segment and insertion occurs in the forming disk. Third, the connecting cilium would serve primarily as a barrier separating membrane domains (see Section 4.3) rather than as a transport corridor. A principal objection to the model is that vesicles normally are not seen in the extracellular space, at least in large numbers. However, the quantitative requirements for membrane transfer call for relatively few transfer vesicles at any given time. Based on estimates of the area of membrane delivered to support disk assembly (Besharse, 1986), such a mechanism in rat would require only one vesicle with a radius of 0.1 μm per minute. If the radius of the vesicle were increased to 0.25 μm, only one vesicle delivered in 10 min would be required. The actual detection of a vesicle in the extracellular space in electron microscopic images would depend on the transit time of the vesicle and the plane of section. It is clear that if the transit time were fast, such vesicles would be rarely seen.

In summary, it should be emphasized that we do not understand how large amounts of membrane are transferred from the inner segment to forming disks. None of the proposed models are directly supported by experimental evidence. Although the ciliary membrane model is intuitively appealing, future research directed at understanding bulk lipid transport and experimental designs that include clear alternative possibilities will be required to solve this problem.

4.2. The Connecting Cilium in Disk Morphogenesis

For many years, disks were thought to be formed from outer segment plasma membrane through a process of invagination (Sjöstrand, 1959; Eakin and Westfall, 1961; Nilsson, 1964). However, disks increase in size as they are displaced distally. This has led to a model (Steinberg et al., 1980) in which disks form through evagination of the plasma membrane of the connecting cilium (Fig. 9). Since disks are formed throughout life, it is common to see a series of disks at the base of the outer segment of gradually increasing size in a proximal-to-distal direction. It is helpful to recall that, in rod cells, disks are closed and surrounded by a plasma membrane while most disks remain open to the extracellular space in cones (Fig. 1). Closed disks incorporate the extracellular space between two adjacent evaginations into the intradiskal space of the mature disk. The mechanism of closure must involve separation of the disk from the plasma membrane. Structural observations suggest that this is a regular process that begins on the ciliary side of the outer segment and extends gradually to the opposite side of the cell (Fig. 9). Furthermore, the fact that in mature rod cells, substantial numbers of opened disks may

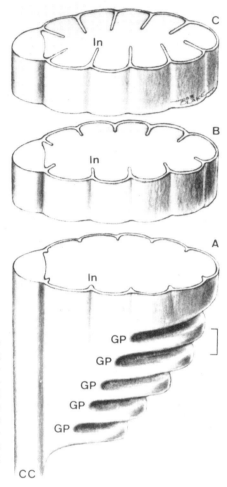

Figure 9. Diagram illustrating the evagination model for disk membrane assembly. Membrane of the cilium forms a series of shelves that increase in diameter in a proximal-to-distal direction. The evaginations are thought to be converted into disks at growth points (GP) which gradually extend from the ciliary side around the periphery of the disk. The extracellular space between adjacent evaginations becomes the intradiskal space of the disk. Incisures which are extensions of the disk edge toward the disk center increase in depth as the disk matures (B and C). CC, connecting cilium; In, incisures. From Corless and Fetter (1987).

transiently accumulate (Besharse *et al.*, 1977; Hollyfield *et al.*, 1982) suggests that the processes of evagination and disk closure are separate and discrete.

Recently, a refinement of the model of Steinberg *et al.* (1980) has been proposed in which mature disks act as templates in the maturation and alignment of new disks (Corless and Fetter, 1987; Corless *et al.*, 1987a,b). This model addresses two issues: the highly regular alignment of disks and the unique structural features of the highly curved edge region. The disk edge appears to lack opsin (Corless *et al.*, 1987a; Molday *et al.*, 1987) and to contain at least two membrane proteins referred to as large intrinsic membrane protein (Papermaster *et al.*, 1978) and peripherin (Molday *et al.*, 1987). Fibrillar connections between adjacent disks near their edge also have been shown to be a regular feature of outer segment structure (Yamada, 1982; Roof and Heuser, 1982), and may serve to align adjacent disks. The model suggests that the significance of the disk edge with its unique chemical and structural features is at least in part morphogenetic. Although the model provides a clear conceptual account of the processes that must operate to form disks, it provides no insight into the molecular processes that serve to segregate disk membrane components into lamellar and edge domains or the motive forces involved in disk evagination and closure.

At the earliest stage of disk formation in the developing photoreceptor, the distal membrane of the cilium forms irregularly arranged lamellae that often are oriented parallel to the ciliary axis (De Robertis, 1956a,b; Tokuyasu and Yamada, 1959; Weidman and Kuwabara, 1968, 1969; Spira, 1975; Besharse et al., 1985). The irregularity of the forming disks is surprising because disk formation is a highly ordered process in the mature cell (Steinberg et al., 1980). Possible explanations are that newly formed disks are labile and difficult to fix or that a preformed stack of disks is necessary as a template to guide orderly disk assembly like that seen in the adult cell (Corless and Fetter, 1987). Another feature of initial disk formation is the appearance of tubular structures around the forming disks that resemble microtubules (Fig. 5). These structures are larger than typical microtubules, stain more densely, and are clearly distinguishable from the doublets of the distal cilium. These structures stain like membrane and are reminiscent of the edge region of mature disks. Such observations raise the question as to whether the disk edge and disk lamella form as independent structures.

The motive forces involved in disk evagination and the molecular mechanism of edge formation remain completely undefined. One possibility is that evagination occurs as a consequence of the mechanism that inserts membrane components into the growing disk. It seems likely that the process involves cytoskeletal components. Actin has been localized to the region of disk formation including the cytosol of membrane evaginations of both developing and mature cells (Chaitin et al., 1984; Chaitin and Bok, 1986). Actin is also found in motile cilia in a region distal to the transition zone (Sandoz et al., 1982), emphasizing the similarity of this region to the analogous region of motile cilia. The presence of actin in the evagination, and its absence from the region between mature disks, is consistent with a morphogenetic role. Furthermore, cytochalasin D, a drug that prevents actin polymerization and disrupts actin filaments, causes the formation of irregular and enlarged evaginations of membrane at the distal connecting cilium (Williams et al., 1988). These results suggest a role for actin in shaping the disk. However, the fact that an evagination forms, even though aberrant, in the presence of cytochalasin D suggests that actin does not provide the force necessary for evagination. The force may be a direct consequence of new membrane expansion.

Disk morphogenesis is also disrupted after treatment with tunicamycin, an inhibitor of dolicholphosphate-dependent glycosylation of N-linked glycoproteins (Fliesler et al., 1985). Tunicamycin blocks opsin glycosylation but permits transport of unglycosylated opsin through the endoplasmic reticulum and Golgi complex to the region of disk assembly (Plantner et al., 1980; Fliesler and Basinger, 1985; Fliesler et al., 1985). This shows that opsin's oligosaccharides are not necessary for its intracellular transport to the periciliary region. However, tunicamycin severely disrupts disk assembly (Fliesler et al., 1985). Instead of flattened disks, large numbers of vesicles and tubules accumulate in the space between inner and outer segments. The extracellular vesicles have the membrane structure characteristic of newly formed disks (Defoe et al., 1986). It has been suggested that they form through vesiculation of the evaginating disk at the distal connecting cilium. However, an origin from inner segment plasma membrane, perhaps reflecting an extracellular intermediate in disk assembly (see Fig. 8), cannot be ruled out. Whatever their origin, the formation of extracellular vesicles suggests that oligosaccharide structure plays a crucial role in either the final stage of opsin transport into the forming disk or the formation of the disk itself. Whether opsin's oligosaccharides or those of some other less abundant glycoprotein are involved remains to be determined.

4.3. The Connecting Cilium as a Barrier between Membrane Domains

Substantial evidence favors the idea that the connecting cilium serves as a barrier between the unique membrane domains of the inner and outer segment. First, opsin immunoreactivity occurs at low density in the inner segment membrane and at high density in the outer segment but is sparse in the connecting cilium (Papermaster and Schneider, 1982; Peters et al., 1983; Nir and Papermaster, 1983; Papermaster et al., 1985; Defoe and Besharse, 1985; Besharse et al., 1985; Usukura and Bok, 1987; Hicks and Barnstable, 1986). In contrast, a Na^+/K^+-ATPase, essential for maintenance of the photoreceptor dark current (Hagins, 1972), is localized to the inner segment and is not observed in the outer segment (Stirling and Lee, 1980; Spencer et al., 1988; Schneider et al., 1987). Recently, the study of dissociated outer segments has provided direct evidence that the connecting cilium indeed serves as a barrier (Spencer et al., 1988). Amphibian outer segment preparations generally contain some outer segments that retain a portion of the inner segment (Biernbaum and Bownds, 1985). Such cells are viable in vitro and have provided an important model for studies on visual transduction. However, in some cells, inner and outer segment membranes spontaneously fuse in a region outside the ciliary domain (Spencer et al., 1988). Studies using antiopsin antibodies show that, after fusion, opsin redistributes into the inner segment (Fig. 10). This does not occur in those cells that maintain the connecting cilium as the only link between inner and outer segment. The redistribution of opsin indicates that it is free to diffuse throughout the membrane but is normally prevented from doing so. The connecting cilium with its stable cytoskeletal–membrane assemblage (see Section 3.2) may play a crucial role in establishing the barrier.

While the evidence for opsin redistribution is unequivocal, it is of some interest that the Na^+/K^+-ATPase, detected immunocytochemically in the inner segment, does not appear to redistribute in cells with fused inner and outer segments (Spencer et al., 1988; see Fig. 10). This implies that factors other than the connecting cilium play a role in maintaining photoreceptor membrane domains. It has been suggested that the Na^+/K^+-ATPase is anchored by a membrane cytoskeleton in the inner segment, which seems likely given the recent evidence for such anchorage in other cell types (Jesaitis and Yguerabide, 1986; Nelson and Veshnock, 1986). In fact, recent observations on cultured chicken photoreceptors show colocalization of a "spectrinlike" protein and Na^+/K^+-ATPase (Madreperla, et al., 1989). The anchorage of the Na^+/K^+-ATPase in the inner segment may be of great significance in preventing its codelivery to the outer segment with opsin and other outer segment membrane components (Madreperla et al., 1989). Restriction of its mobility by a membrane cytoskeleton in the inner segment may prevent it from entering the inner segment plasma membrane through which opsin transport occurs.

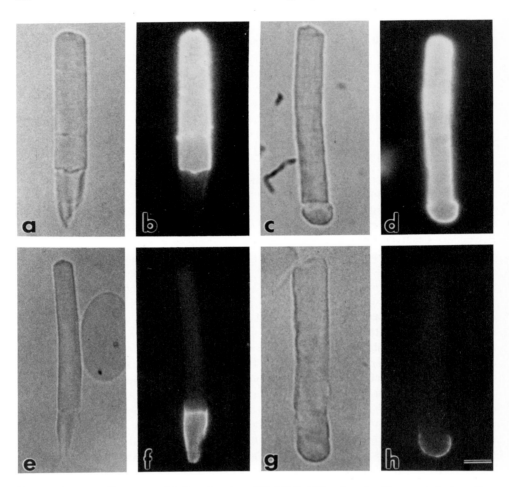

Figure 10. Immunofluorescence labeling of opsin and Na$^+$/K$^+$-ATPase in rods with separate (a, b, e, f) or fused inner segments (c, d, g, h). (a) Phase-contrast image of outer segment isolated with the inner segment. The tapered inner segment is connected to the outer segment only by the connecting cilium (not visible). (b) Opsin immunoreactivity is restricted to the outer segment. (c) Phase-contrast image of outer segment isolated with a fused inner segment. The rounded inner segment is attached to the outer segment by membrane fusion outside the ciliary region. (d) Opsin immunoreactivity is distributed throughout the cell. (e, f) Pair similar to a and b labeled with antibody to Na$^+$/K$^+$-ATPase. Staining is restricted to inner segment. Bar = 5 μm. From Spencer *et al.* (1988).

Recent studies of opsin distribution in the RCS dystrophic rat (Nir *et al.*, 1987) and the rds mouse (Nir and Papermaster, 1986; Usukura and Bok, 1987; Jansen *et al.*, 1987), two animal models for hereditary photoreceptor degeneration, indicate that opsin delivery to the outer segment is aberrant in degenerating photoreceptors. Such studies may ultimately prove useful if specific defects in the disk-forming mechanism are identified. In the RCS rat, photoreceptor degeneration is secondary to a defect in the pigment epithelium that limits photoreceptor membrane phagocytosis (Mullen and LaVail, 1976).

Thus, the observations on opsin distribution in this model do not reflect the primary cause of degeneration. In fact, photoreceptors initially develop normal outer segments with the expected, polarized distribution of opsin. However, as outer segment degeneration progresses, the polarized distribution is lost and opsin accumulates in the inner segment (Nir *et al.*, 1987). Although no structural defects in the connecting cilium were detected, these observations suggest either that opsin accumulation results from breakdown of the barrier function of the connecting cilium or from continued opsin synthesis and delivery to the inner segment after delivery to the outer segment is aborted. It was also reported (Nir *et al.*, 1987) that some degenerating rods had fused inner and outer segment membranes outside the ciliary region similar to the inner segment–outer segment fragments of amphibian rods (Fig. 10), but retained a polarized distribution of opsin. This observation was interpreted to mean that an extraciliary barrier prevented mixing of membrane components in those cells. However, it is also possible that membrane fusion in the degenerating cells was a recent event, perhaps induced by fixation, and that insufficient time had elapsed for opsin redistribution. In support of this, it has recently been reported that opsin redistributes after inner–outer segment fusion in enzymatically dissociated rat photoreceptors (Hicks *et al.*, 1989).

The rds mouse differs from the RCS rat in that photoreceptors fail to develop normal disk structure. Although an aborted attempt to form an outer segment is suggested by the appearance of irregularly arranged opsin-containing membranes in the distal cilium, the degeneration is characterized by the accumulation of large amounts of opsin in the inner segment plasma membrane and in vesicular structures in the extracellular space (Nir and Papermaster, 1986; Usukura and Bok, 1987; Jansen *et al.*, 1987). The ultrastructure suggests a possible defect in either the membrane delivery mechanism or the mechanism involved in disk morphogenesis. The accumulation of extracellular, opsin-containing membranes is of some interest. It has been suggested that such membranes pinch off from the distal ciliary membrane in a process reminiscent of normal disk shedding and are phagocytosed by adjacent pigment epithelium (Jansen *et al.*, 1987; Nir and Papermaster, 1986). It is also possible, however, that the extracellular membranes reflect an extracellular intermediate (see Section 4.1) in the delivery pathway that is inefficiently incorporated into the developing outer segment. In any case, it is evident that an understanding of the molecular basis for the defect in this mutant may provide insight into the normal mechanism of disk formation.

4.4. Delivery of Cytosolic Components to the Outer Segment

The photoreceptor outer segment transduces photon absorption into a membrane voltage change that controls synaptic transmitter release. The connecting cilium is in the pathway of the passive propagation of the outer segment hyperpolarizing light response to the synaptic terminal. In the opposite direction, it is the probable conduit for delivery of both soluble proteins and metabolites essential for outer segment function. The general dependence of outer segment function on nucleoside triphosphates and enzymes synthesized in the inner segment (see Liebman, 1987, for review) suggests that flow of components through the ciliary cytosol may be immense. Although passive diffusion is often assumed to mediate such transport, several observations raise questions about its role and emphasize the need for experimental work in this area. First, the restricted size of the ciliary cytosolic compartment relative to that of the inner and outer segments (a factor of about 75 in frog) suggests that it would be a bottleneck that would limit diffusion. Evidence for this comes from experiments in which either a fluorescent dye (carboxyfluorescein) or cGMP analogues were injected into the inner segment (Zimmerman *et al.*, 1985; Hestrin and Korenbrot, 1987). Delay on the order of minutes occurred before either the dye or the effect of cGMP could be detected in the outer segment. Second, many outer segment enzymatic components may be present at higher concentration in the outer segment than in the inner segment (see below). Thus, passive diffusion would account for their transport only if they are bound in the outer segment or if a barrier to back diffusion exists.

In the early studies of outer segment membrane renewal, a significant level of diffusely distributed, radioactive protein was detected in the outer segment. Since newly synthesized opsin was restricted to disks, the diffusely distributed radioactivity was interpreted as reflecting the distribution of newly synthesized soluble protein (Young, 1967). Autoradiographic analysis of the diffusely distributed protein showed that it was reduced half maximally in about 2 weeks (Bok and Young, 1972). If soluble proteins were degraded only during the disk shedding–phagocytosis process by random inclusion in shed photoreceptor tips, one would predict that about 35% would remain after a complete turnover period of outer segment disks (~ 40 days; see Anderson *et al.*, 1980a, for the calculation). The fact that only half remained after 2 weeks implies that some of the soluble protein either is degraded locally or is returned to the inner segment for degradation. It also implies that newly synthesized soluble proteins must be delivered to the outer segment in order to compensate for degradation.

Many of the soluble proteins have now been identified and antibodies to some are available, making it possible to determine turnover rates for individual proteins. Such information is not yet available, but will be essential for the full interpretation of recent immunocytochemical studies which show large and rapid shifts in the immunoreactivity of outer segments (Broekhuyse *et al.*, 1985; Brann and Cohen, 1987; Mangini and Pepperberg, 1987; Philp *et al.*, 1987). Four proteins that are components of the cGMP regulatory system of the outer segment (see review by Liebman, 1987) have now been shown to exhibit light-modulated immunoreactivity that has been interpreted as involving rapid movement between inner and outer segments. Antibodies to the α subunit of transducin, a GTP binding protein involved in light activation of cGMP phosphodiesterase, show reduced outer segment labeling and increased inner segment labeling in light (Brann and Cohen, 1987; Philp *et al.*, 1987). An additional 33-kDa protein involved in transduction and the β subunit of transducin also exhibit a similar pattern of labeling in light and darkness (Whelan *et al.*, 1988). A reversed pattern of immunoreactivity is seen in the case of the outer segment 48-kDa protein (called S-antigen or arrestin) that binds phophorylated rhodopsin (Broekhuyse *et al.*, 1985; Philp *et al.*, 1987). The 48-kDa protein exhibits high outer segment and low inner segment labeling in light and a reversal in darkness. In contrast, the outer segment phosphodiesterase that hydrolyzes cGMP shows no changes in immunoreactivity in light or darkness (Philp *et al.*, 1987). Since transblot analysis also reveals changes in the actual amount of the α subunit of transducin and the 48-kDa protein in outer segments prepared in light and darkness, it has been suggested that the changes in immunoreactivity involve transport between inner and outer segment, presumably through the connecting cilium (see Philp *et al.*, 1987). Although it remains possible that light-modulated binding, protein conformational changes, or masking of antigenic sites could account for the light-modulated staining of photoreceptors, current data suggest that protein transport between inner and outer segment occurs in both directions. Further studies on the cellular mechanisms involved in transport and the relationship of regulated transport to protein turnover should prove interesting.

5. Summary

A striking feature of the photoreceptor connecting cilium is that it lies at the boundary of distinct cellular domains. As such, it is often ascribed functions in the generation and maintenance of photoreceptor polarity. In this review, we have described the experimental basis for current concepts of ciliary function in photoreceptor development, membrane turnover, and physiology. Based largely on structural homology, similar functions have been ascribed to the transition zone of other sensory and motile cilia. Although the transition zone appears to be common to all cilia, the distal part of the cilium has evolved to perform different functions including phototransduction, olfaction, movement, and reproduction. Because of its role in delineating distinct membrane domains, the transition zone is probably important in generating a diverse array of distal ciliary membrane specializations. However, our knowledge of the factors controlling photoreceptor and motile ciliary function is strikingly incomplete. By recognizing the similarities between motile and sensory cilia, we may be better able to exploit their individual advantages to gain a better understanding of ciliary physiology.

ACKNOWLEDGMENTS. We thank Win Sale, Pal Röhlich, Dean Bok, Itzak Nir, Maribeth Spencer, Ann Bunt-Milam, Ruben Adler, Lincoln Johnson, Michael Kaplan, and Joe Corless for helpful discussion and Dean Bok and Maribeth Spencer for providing Figs. 1 and 10. We also greatfully acknowledge Donna Forestner for her many contributions to tne research on ciliary membranes and the production of this review. Research from the authors' laboratory was supported by NIH Grants EY 03222 and EY 02414.

References

Adler, R., 1986, The differentiation of retinal photoreceptors and neurons in vitro, *Prog. Retin. Res.* **6**:1–27.

Anderson, R. E., Maude, M. B., Kelleher, P. A., and Basinger, S. F., 1980a, Metabolism of phosphatidylcholine in the frog retina, *Biochim. Biophys. Acta* **620**:212–226.

Anderson, R. E., Kelleher, P. A., and Maude, M. B., 1980b, Metabolism of phosphatidylethanolamine in the frog retina, *Biochim. Biophys. Acta* **620**:227–235.

Anderson, R. E., Maude, M. B., and Kelleher, P. A., 1980c, Metabolism of phosphatidylethanolamine in the frog retina, *Biochim. Biophys. Acta* **620**:236–246.

Anderson, R. G. W., 1974, Isolation of ciliated or unciliated basal bodies from rabbit oviduct, *J. Cell Biol.* **60**:393–404.

Andrews, L. D., 1982, Freeze-fracture studies of vertebrate photoreceptor membranes, in: *The Structure of the Eye* (J. G. Hollyfield, ed.), Elsevier, Amsterdam, pp. 11–23.

Applebury, M. L., and Hargrave, P. A., 1986, Molecular biology of the visual pigments, *Vision Res.* **26**:1881–1895.

Araki, M., 1984, Immunocytochemical study on photoreceptor cell differentiation in the cultured retina of the chick, *Dev. Biol.* **103**:313–318.

Bardele, C. F., 1983, Mapping of highly ordered membrane domains in the plasma membrane of the ciliate *Cyclidium glaucoma*, *J. Cell Sci.* **61**:1–30.

Basinger, S., and Hoffman, R., 1976, Phosphatidylcholine metabolism in the frog rod photoreceptor, *Exp. Eye Res.* **23**:117–126.

Basinger, S., Bok, D., and Hall, M., 1976, Rhodopsin in the rod outer segment plasma membrane, *J. Cell Biol.* **69**:29–42.

Bell, C. W., 1983, The molecular weight of dynein heavy chains, *J. Submicrosc. Cytol.* **15**:201–202.

Bert, R. J., 1987, Colchicine's effects on the light-sensitive conductance in toad rods are due to inhibition of sodium–calcium exchange, *Invest. Ophthalmol. Vis. Sci. Abstr. Suppl.* **28**:353.

Besharse, J. C., 1982, The daily light–dark cycle and rhythmic metabolism in the photoreceptor-pigment epithelial complex, *Prog. Retin. Res.* **1**:81–124.

Besharse, J. C., 1986, Photosensitive membrane turnover: Differentiated membrane domains and cell–cell interaction, in: *The Retina: A Model for Cell Biological Studies,* Part I (R. Adler and D. Farber, eds.), Academic Press, New York, pp. 297–352.

Besharse, J. C., and Dunis, D. A., 1982, Rod photoreceptor disc shedding in vitro: Inhibition by cytochalasins and activation by colchicine, in: *The Structure of the Eye* (J. G. Hollyfield, ed.), Elsevier/North-Holland, Amsterdam, pp. 85–96.

Besharse, J. C., and Pfenninger, K. H., 1980, Membrane assembly in retinal photoreceptors. I. Freeze-fracture analysis of cytoplasmic vesicles in relationship to disc assembly, *J. Cell Biol.* **87**:451–463.

Besharse, J. C., Hollyfield, J. G., and Rayborn, M. E., 1977, Turnover of rod photoreceptor outer segments. II. Membrane addition and loss in relationships to light, *J. Cell Biol.* **75**:507–527.

Besharse, J. C., Forestner, D. M., and Defoe, D. M., 1985, Membrane assembly in retinal photoreceptors. III. Distinct membrane domains of the connecting cilium of developing rods, *J. Neurosci.* **5**:1035–1048.

Besharse, J. C., Iuvone, P. M., and Pierce, M. E., 1988, Regulation of rhythmic photoreceptor metabolism: A role for post-receptoral neurons, *Prog. Retin. Res.* **7**:21–61.

Biernbaum, M. S., and Bownds, M. D., 1985, Frog rod outer segments with attached inner segments ellipsoids as an in vitro model for photoreceptors on the retina, *J. Gen. Physiol.* **85**:83–105.

Blanks, J. C., and Johnson, L. V., 1984, Specific binding of peanut lectin to a class of retinal photoreceptor cells. A species comparison, *Invest. Ophthalmol. Vis. Sci.* **25**:546–555.

Bok, D., 1985, Retinal photoreceptor-pigment epithelium interactions, *Invest. Ophthalmol. Vis. Sci.* **26**:1659–1694.

Bok, D., and Young, R. W., 1972, The renewal of diffusely distributed protein in the outer segments of rods and cones, *Vision Res.* **12**:161–168.

Brann, M. R., and Cohen, L. V., 1987, Diurnal expression of transducin mRNA and translocation of transducin in rods of rat retina, *Science* **235**:585–587.

Bridges, C. D. B., and Fong, S. L., 1980, Lectins as probes of glycoprotein and glycolipid oligosaccharides in rods and cones, in: *Neurochemistry of the Retina* (N. G. Bazan and R. N. Lolley, eds.), Pergamon Press, Elmsford, N.Y., pp. 255–267.

Broekhuyse, R. M., Tolhuizen, E. F. J., Janssen, A. P. M., and Winkens, H. J., 1985, Light induced shift and binding of S-antigen in retinal rods, *Curr. Eye Res.* **4**:613–618.

Brown, P. K., Gibbons, I. R., and Wald, G., 1963, The visual cells and visual pigment of the mudpuppy, *Necturus, J. Cell Biol.* **19**:79–106.

Chaitin, M. H., and Bok, D., 1986, Immunoferritin localization of actin in retinal photoreceptors, *Invest. Ophthalmol. Vis. Sci.* **27**:1764–1767.

Chaitin, M. H., Schneider, B. G., Hall, M. O., and Papermaster, D. S., 1984, Actin in the photoreceptor connecting cilium: Immunocytochemical localization to the site of outer segment disk formation, *J. Cell Biol.* **99**:239–247.

Cohen, D., and Nir, I., 1987, Cytochemical characterization of sialoglycoconjugates on rat photoreceptor cell surface, *Invest. Ophthalmol. Vis. Sci.* **28**:640–645.

Corless, J. M., and Fetter, R. D., 1987, Structural features of the terminal loop region of frog retinal rod outer segment disk membranes: III. Implications of the terminal loop complex for disk morphogenesis, membrane fusion, and cell surface interactions, *J. Comp. Neurol.* **257**:24–38.

Corless, J. M., Fetter, R. D., and Costello, M. J., 1987a, Structural features of the terminal loop region of frog retinal rod outer segment disk membranes: I. Organization of lipid components, *J. Comp. Neurol.* **257**:1–8.

Corless, J. M., Fetter, R. D., Zampighi, O. B., Costello, M. J., and Wall-Buford, D. L., 1987b, Structural features of the terminal loop region of frog retinal rod outer segment disk membranes: II. Organization of the terminal loop complex, *J. Comp. Neurol.* **257**:9–23.

Defoe, D. M., and Besharse, J. C., 1985, Membrane assembly in retinal photoreceptors. II. Immunocytochemical analysis of freeze-fractured rod photoreceptor membranes using anti-opsin antibodies, *J. Neurosci.* **5**:1023–1034.

Defoe, D. M., Besharse, J. C., and Fliesler, S. J., 1986, Tunicamycin-induced dysgenesis of retinal rod outer segment membranes. II. Quantitative freeze-fracture analysis, *Invest. Ophthalmol. Vis. Sci.* **27**:1595–1601.

Dentler, W. L., 1980, Microtubule–membrane interactions in cilia: I. Isolation and characterization of ciliary membranes from Tetrahymena pyriformis, *J. Cell Biol.* **84**:364–380.

Dentler, W. L., 1981, Microtubule–membrane interactions in ctenophore swimming plate cilia, *Tissue Cell* **13**:197–208.

De Robertis, E., 1956a, Electron microscope observations on the submicroscopic organization of the retinal rods, *J. Biophys. Biochem. Cytol.* **2**:319–329.

De Robertis, E., 1956b, Morphogenesis of retinal rods: An electron microscope study, *J. Biophys. Biochem. Cytol. Suppl.* **2**:209–216.

De Robertis, E., 1960, Some observations on the ultrastructure and morphogenesis of photoreceptors, *J. Gen. Physiol.* **43**:1–13.

Dunlap, K., 1977, Localization of calcium channels in *Paramecium caudatum*, *J. Physiol. (London)* **271**:119–133.

Eakin, R. M., and Westfall, J. A., 1961, The development of photoreceptors in the stirnorgan of the treefrog, Hyla regilla, *Embryologia* **6**:84–98.

Fleischman, D., and Denisevich, M., 1979, Guanylate cyclase of isolated bovine retinal rod axonemes, *Biochemistry* **18**:5060–5066.

Fleischman, D., Denisevich, M., Raveed, D., and Pannbacker, R. G., 1980, Association of guanylate cyclase with the axoneme of retinal rods, *Biochim. Biophys. Acta* **630**:176–186.

Fliesler, S. J., and Anderson, R. E., 1983, Chemistry and metabolism of lipids in the vertebrate retina, *Prog. Lipid Res.* **22**:79–131.

Fliesler, S. J., and Basinger, S. F., 1985, Tunicamycin blocks the incorporation of opsin into rod outer segment membranes, *Proc. Natl. Acad. Sci. USA* **82**:1116–1120.

Fliesler, S. J., Rayborn, M. E., and Hollyfield, J. G., 1985, Membrane morphogenesis in retinal rod outer segments. Inhibition by tunicamycin, *J. Cell Biol.* **100**:574–587.

Flower, N. E., 1971, Particles within membranes: A freeze-etch view, *J. Cell Sci.* **9**:435–441.

Fukuda, M. N., Papermaster, D. S., and Hargrave, P. A., 1979, Rhodopsin carbohydrate: Structure of small oligosaccharides attached at two sites near the amino terminus, *J. Biol. Chem.* **254**:8201–8207.

Gilula, N. B., and Satir, P., 1972, The ciliary necklace: A ciliary membrane specialization, *J. Cell Biol.* **53**:494–509.

Goldman, B. M., and Blobel, G., 1981, In vitro biosynthesis, core glycosylation, and membrane integration of opsin, *J. Cell Biol.* **90**:236–242.

Greiner, J. V., Weidman, T. A., Bodley, H. D., and Greiner, C. A. M., 1981, Ciliogenesis in photoreceptor cells of the retina, *Exp. Eye Res.* **33**:433–446.

Hagins, W. A., 1972, The visual process: Excitatory mechanisms in the primary receptor cells, *Annu. Rev. Biophys. Bioeng.* **1**:131–158.

Hall, M. O., and Nir, I., 1976, The binding of concanavalin A to the rod outer segments and pigment epithelium of normal and RCS rats, *Exp. Eye Res.* **22**:469–476.

Hall, M. O., Bok, D., and Bacharach, A. D. E., 1969, Biosynthesis and assembly of the rod outer segment membrane system. Formation and fate of visual pigment in the frog retina, *J. Mol. Biol.* **45**:397–406.

Hard, R., and Rieder, C. L., 1983, Muciliary transport in newt lungs: The ultrastructure of the ciliary apparatus in isolated epithelial sheets and in functional Triton-extracted models, *Tissue Cell* **15**:227–243.

Hennessey, T. N., Andrews, D., and Nelson, D. L., 1983, Biochemical studies of the excitable membrane of *Paramecium tetraurelia*. VII. Sterols and other neutral lipids of cells and cilia, *J. Lipid Res.* **24**:575–587.

Hestrin, S., and Korenbrot, J. I., 1987, Effects of cyclic GMP on the kinetics of the photocurrent in rods and detached rod outer segments, *J. Gen. Physiol.* **90**:527–551.

Hicks, D., and Barnstable, C. J., 1986, Lectin and antibody labelling of developing rat photoreceptor cells: An electron microscope immunocytochemical study, *J. Neurocytol.* **15**:219–230.

Hicks, D., and Barnstable, C. J., 1987, Different rhodopsin monoclonal antibodies reveal different binding patterns on developing and adult rat retina, *J. Histochem. Cytochem.* **35**:1317–1328.

Hicks, D., and Molday, R. S., 1985, Localization of lectin receptors on bovine photoreceptor cells using dextran–gold markers, *Invest. Ophthalmol. Vis. Sci.* **26**:1002–1013.

Hicks, D., Sparrow, J., and Barnstable, C. J., 1989, Immunoelectron microscopical examination of the surface distribution of opsin in rat rod photoreceptor cells, *Exp. Eye Res.* in press.

Hollyfield, J. G., Rayborn, M. E., Verner, G. E., Maude, M. B., and Anderson, R. E., 1982, Membrane addition to rod photoreceptor outer segments: Light stimulates membrane assembly in the absence of increased membrane biosynthesis, *Invest. Ophthalmol. Vis. Sci.* **32**:417–427.

Horst, C. J., Forestner, D. M., and Besharse, J. C., 1987, Cytoskeletal–membrane interactions: A stable interaction between cell surface glycoconjugates and doublet microtubules of the photoreceptor connecting cilium, *J. Cell Biol.* **105**:2973–2987.

Horst, C. J., Johnson, L. V., and Besharse, J. C., 1989, A 425 kd glycoconjugate restricted to the photoreceptor connecting cilium is also found at the motile cilium transition zone, *Invest. Ophthalmol. Vis. Sci.* (ARVO Abstract Suppl) **30**: 157.

Jan, L. Y., and Revel, J. P., 1974, Ultrastructural localization of rhodopsin in the vertebrate retina, *J. Cell Biol.* **62**:257–273.

Jansen, H. G., Sanyal, S., DeGrip, W. J., and Schalken, J. J., 1987, Development and degeneration of retina in rds mutant mice: Ultraimmunohistochemical localization of opsin, *Exp. Eye Res.* **44**:347–361.

Jesaitis, A. J., and Yguerabide, J., 1986, The lateral mobility of the (Na,K)-dependent ATPase in Madin–Darby canine kidney epithelial cells, *J. Cell Biol.* **102**:1256–1263.

Johnson, L. V., Hageman, G. S., and Blanks, J. C., 1986, Interphotoreceptor matrix domains ensheath vertebrate cone photoreceptor cells, *Invest. Ophthalmol. Vis. Sci.* **27**:129–135.

Kaplan, M. W., Iwata, R. T., and Sears, R. C., 1987, Lengths of immunolabeled ciliary microtubules in frog photoreceptor outer segments, *Exp. Eye Res.* **44**:623–632.

Koide, H., Suganuma, T., Fusayoshi, M., and Ohba, N., 1986, Ultrastructural localization of lectin receptors in the monkey retinal photoreceptors and pigment epithelium: Application of lectin–gold complexes on thin sections, *Exp. Eye Res.* **43**:343–354.

LaVail, M. M., 1973, Kinetics of rod outer segment renewal in the developing mouse retina, *J. Cell Biol.* **58**: 650–661.

Liang, D.-J., Yamashita, K., Muellenberg, C. G., Shichi, H., and Kobata, A., 1979, Structure of the carbohydrate moieties of bovine rhodopsin, *J. Biol. Chem.* **254**:6414–6418.

Liebman, P. A., 1987, The molecular mechanism of visual excitation and its relation to the structure and composition of the rod outer segment, *Annu. Rev. Physiol.* **49**:765–791.

Machemer, H., and Ogura, A., 1979, Ionic conductances of membranes in ciliated and deciliated Paramecium, *J. Physiol. (London)* **296**:49–60.

Madreperla, S. A., Edidin, M., and Adler, R., 1989, Na$^+$, K$^+$-ATPase polarity in retinal photoreceptors: A role for cytoskeletal attachments, *J. Cell Biol.* in press.

Mangini, N. J., and Pepperberg, D. R., 1987, Localization of retinal "48K" (s-antigen) by electron microscopy, *Jpn. J. Ophthalmol.* **31**:207–217.

Matheke, M. L., and Holtzman, E., 1984, The effects of monensin and puromycin on transport of membrane components in the frog retinal photoreceptor. II. Electron microscope autoradiography of proteins and glycerolipids, *J. Neurosci.* **4**:1093–1103.

Matheke, M. L., Fliesler, S. J., Basinger, S. F., and Holtzman, E., 1984, The effects of monensin on transport of membrane components in the frog retinal photoreceptor. I. Light microscope autoradiography and biochemical analysis, *J. Neurosci.* **4**:1086–1092.

Matsusaka, T., 1974, Membrane particles of the connecting cilium, *J. Ultrastruct. Res.* **48**:305–312.

Matsusaka, T., 1976, Cytoplasmic fibrils of the connecting cilium, *J. Ultrastruct. Res.* **54**:318–324.

Menco, B. M., 1980, Qualitative and quantitative freeze-fracture studies on olfactory and respiratory epithelial surfaces of frog, ox, rat and dog, *Cell Tissue Res.* **212**:1–16.

Molday, L. L., and Molday, R. S., 1987, Glycoproteins specific for the retinal rod outer segment plasma membrane, *Biochim. Biophys. Acta* **897**:335–340.

Molday, R. S., 1976, A scanning electron microscope study of concanavalin A receptors on retinal rod cells labeled with latex microspheres, *J. Supramol. Struct.* **4**:549–557.

Molday, R. S., and Molday, L. L., 1979, Identification and characterization of multiple forms of rhodopsin and minor proteins in frog and bovine rod outer segment disc membranes, *J. Biol. Chem.* **254**:4653–4660.

Molday, R. S., Hicks, D., and Molday, L., 1987, Peripherin. A rim-specific membrane protein of rod outer segment discs, *Invest. Ophthalmol. Vis. Sci.* **28**:50–61.

Monsigny, M., Roche, A.-C., Sene, C., Maget-Dana, R., and Delmotte, R., 1980, Sugar–lectin interactions: How does wheat-germ agglutinin bind sialoglycoconjugates? *Eur. J. Biochem.* **104**:147–153.

Mullen, R. J., and LaVail, M. M., 1976, Inherited retinal dystrophy: Primary defect in pigment epithelium determined with experimental rat chimeras, *Science* **192**:799–801.

Musgrave, A., de Wildt, P., van Etton, I., Scholma, C., Kooyman, R., Homan, W., and van den Ende, H., 1986, Evidence for a functional membrane barrier in the transition zone between the flagellum and cell body of Chlamydomonas eugametos gametes, *Planta* **167**:544–553.

Nelson, W. J., and Veshnock, P. J., 1986, Dynamics of membrane-skeleton (fodrin) organization during development and polarity in Madin–Darby canine kidney epithelial cells, *J. Cell Biol.* **103**:1751–1765.

Nilsson, S. E. G., 1964, Receptor cell outer segment development and ultrastructure of the disk membranes in the retina of the tadpole (*Rana pipens*), *J. Ultrastruct. Res.* **11**:581–620.

Nir, I., and Cohen, D., 1987, Surface glycoconjugates on rat photoreceptor cilium: Effect of neuraminidase, *Invest. Ophthalmol. Vis. Sci.* **28**:1070–1077.

Nir, I., and Hall, M. O., 1979, Ultrastructural localization of lectin binding sites on the surface of retinal photoreceptors and pigment epithelium, *Exp. Eye Res.* **29**:181–194.

Nir, I., and Papermaster, D. S., 1983, Differential distribution of opsin in the plasma membrane of frog photoreceptors: An immunocytochemical study, *Invest. Ophthalmol. Vis. Sci.* **24**:868–878.

Nir, I., and Papermaster, D. S., 1986, Immunocytochemical localization of opsin in the inner segment and ciliary plasma membrane of photoreceptors in retinas of rds mutant mice, *Invest. Ophthalmol. Vis. Sci.* **27**:836–840.

Nir, I., Cohen, D., and Papermaster, D. S., 1984, Immunocytochemical localization of opsin in the cell membrane of developing rat retinal photoreceptors, *J. Cell Biol.* **98**:1788–1795.

Nir, I., Sagie, G., and Papermaster, D. S., 1987, Opsin accumulation in photoreceptor inner segment plasma membranes of dystrophic RCS rats, *Invest. Ophthalmol. Vis. Sci.* **28**:62–69.

Ogura, A., and Takahashi, K., 1976, Artificial deciliation causes loss of calcium-dependent responses in *Paramecium*, *Nature* **264**:170–172.

Papermaster, D. S., and Schneider, B. G., 1982, Biosynthesis and morphogenesis of outer segment membranes in vertebrate photoreceptor cells, in: *Cell Biology of the Eye* (D. S. McDevitt, ed.), Academic Press, New York, pp. 475–531.

Papermaster, D. S., Converse, C. A., and Siu, J., 1975, Membrane biosynthesis in the retina: Opsin transport in the photoreceptor cell, *Biochemistry* **14**:1343–1352.

Papermaster, D. S., Schneider, B. G., Zorn, M. A., and Kraehenbuhl, J. P., 1978, Immunocytochemical localization of a large intrinsic membrane protein to the incisures and margins of frog rod outer segment disks, *J. Cell Biol.* **78**:415–425.

Papermaster, D. S., Schneider, B. G., and Besharse, J. C., 1985, Vesicular transport of newly synthesized opsin from the Golgi apparatus toward the rod outer segment: Ultrastructural immunocytochemical and autoradiographic evidence in *Xenopus* retinas, *Invest. Ophthalmol. Vis. Sci.* **26**:1386–1404.

Peters, K. R., Palade, G. E., Schneider, B. G., and Papermaster, D. S., 1983, Fine structure of a periciliary ridge complex of frog retinal rod cells revealed by ultrahigh resolution scanning electron microscopy, *J. Cell Biol.* **96**:265–276.

Philp, N. J., Chang, W., and Long, K., 1987, Light-stimulated protein movement in rod photoreceptor cells of the rat retina, *FEBS Lett.* **225**:127–132.

Plantner, J. J., Poncz, L., and Kean, E. L., 1980, Effect of tunicamycin on glycosylation of rhodopsin, *Arch. Biochem. Biophys.* **201**:527–532.

Reinhardt, F. D., and Bloodgood, R. A., 1988, Membrane–cytoskeleton interactions in the flagellum: A 240 kDa surface-exposed glycoprotein is tightly associated with the axoneme in *Chlamydomonas moewusii*, *J. Cell Sci.* **89:** 521–530.

Richardson, T. M., 1969, Cytoplasmic and ciliary connections between the inner and outer segments of mammalian visual receptors, *Vision Res.* **9:**727–731.

Ringo, D. L., 1967, Flagellar motion and fine structure of the flagellar apparatus in Chlamydomonas, *J. Cell Biol.* **33:**543–571.

Röhlich, P., 1975, The sensory cilium of retinal rods is analogous to the transitional zone of motile cilia, *Cell Tissue Res.* **161:**421–430.

Roof, D., and Applebury, M., 1984, Localization of calmodulin and characterization of calmodulin binding proteins in the vertebrate rod outer segment, *Biophys. J.* **45:**1a.

Roof, D. J., and Heuser, J. E., 1982, Surfaces of rod photoreceptor disk membranes: Integral membrane components, *J. Cell Biol.* **95:**487–500.

Sale, W. S., Besharse, J. C., and Piperno, G., 1988, Distribution of acetylated α-tubulin in retina and in in vitro-assembled microtubules, *Cell Motil. Cytoskel.* **9:**243–253.

Sandoz, D., Boisvieux-Ulrich, E., and Chailley, B., 1979, Relationships between intramembrane particles and glycoconjugates in the ciliary membrane of the quail oviduct, *Biol. Cell.* **36:**267–280.

Sandoz, D., Gounon, P., Karsenti, E., and Sauron, M.-E., 1982, Immunocytochemical localization of tubulin, actin, and myosin in axonemes of ciliated cells from quail oviduct, *Proc. Natl. Acad. Sci. USA* **79:**3198–3202.

Schneider, B. G., Papermaster, D. S., and Sweadner, K. J., 1987, Localization of the alpha + subunit of (Na,K)ATPase in the rod inner segment and axonal plasma membrane, *Invest. Ophthalmol. Vis. Sci. Abstr. Suppl.* **28:**340.

Schultz, J. E., and Klumpp, S., 1984, Calcium/calmodulin-regulated guanylate cyclases in the ciliary membranes from *Paramecium* and *Tetrahymena*, *Adv. Cyclic Nucleotide Protein Phosphorylation Res.* **17:**275–283.

Sjöstrand, F. S., 1959, Fine structure of cytoplasm: The organization of membranous layers, *Rev. Mod. Phys.* **31:**301–318.

Spencer, M., Detwiler, P. B., and Bunt-Milam, A. H., 1988, Distribution of membrane proteins in mechanically dissociated retinal rods, *Invest. Ophthalmol. Vis. Sci.* **29:**1012–1020.

Spira, A. W., 1975, In utero development and maturation of the retina of a nonprimate mammal: A light and electron microscopic study of the guinea pig, *Anat. Embryol.* **146:**279–300.

Spira, A. W., and Millman, G. E., 1982, Filament arrays in the photoreceptor cell of the human, monkey, and guinea pig retina, in: *The Structure of the Eye* (J. G. Hollyfield, ed.), Elsevier Biomedical, New York, pp. 1–10.

Steinberg, R. H., and Wood, I., 1975, Clefts and microtubules of photoreceptor outer segments in the retina of the domestic cat, *J. Ultrastruct. Res.* **51:**397–403.

Steinberg, R. H., Fisher, S. K., and Anderson, D. H., 1980, Disc morphogenesis in vertebrate photoreceptors, *J. Comp. Neurol.* **190:**501–518.

Steinemann, A., and Stryer, L., 1973, Accessibility of the carbohydrate moiety of rhodopsin, *Biochemistry* **12:**1499–1502.

Stirling, C. E., and Lee, A., 1980, [^{3}H]-ouabain autoradiography of frog retina, *J. Cell Biol.* **85:**313–324.

Tokuyasu, K., and Yamada, E., 1959, The fine structure of the retina studied with the electron microscope. IV. Morphogenesis of outer segments of retinal rods, *J. Biophys. Biochem. Cytol.* **6:**225–230.

Townes-Anderson, E., MacLeish, P. R., and Raviola, E., 1985, Rod cells dissociated from mature salamander retina: Ultrastructure and uptake of horseradish peroxidase, *J. Cell Biol.* **100:**175–188.

Umemoto, J., Bhavanandan, V.P., and Davidson, E. A., 1977, Purification and properties of an endo-α-N-acetyl-*D*-galactosaminidase from Diplococcus pneumoniae, *J. Biol. Chem.* **252:**8609–8614.

Usukura, J., and Bok, D., 1987, Changes in the localization and content of opsin during retinal development in the rds mutant mouse: Immunocytochemistry and immunoassay, *Exp. Eye Res.* **45:**501–515.

Weidman, T. A., and Kuwabara, T., 1968, Postnatal development of the rat retina: An electron microscopic study, *Arch. Ophthalmol.* **79:**470–484.

Weidman, T. A., and Kuwabara, T., 1969, Development of the rat retina, *Invest. Ophthalmol.* **8:**60–69.

Wen, G. Y., Soifer, D., and Wisniewski, H. M., 1982, The doublet microtubules of rods of the rabbit retina, *Anat. Embryol.* **165:**315–328.

Whelan, J. P., Lee, R. H., Lolley, R. N., and McGinnis, J. F., 1988, Light mediated relocalization of photoreceptor proteins coincides with protein movement, *Invest. Ophthalmol. Vis. Sci. Abstr. Suppl.* **29:** 106.

Williams, D. S., Linberg, K. A., Vaughan, D. K., Fariss, R. N., and Fisher, S. K., 1988, Disruption of microfilament organization and deregulation of disk membrane morphogenesis by cytochalasin D in rod and cone photoreceptors, *J. Comp. Neurol.* **272:**161–172.

Wood, J. G., and Napier-Marshall, L., 1985a, Differential effects of protease digestion on photoreceptor lectin binding sites, *J. Histochem. Cytochem.* **33:**642–646.

Wood, J. G., and Napier-Marshall, L., 1985b, Cytochemical analysis of oligosaccharide processing in frog photoreceptors, *Histochem. J.* **17:**585–594.

Wood, J. G., Besharse, J. C., and Napier-Marshall, L., 1984, Partial characterization of lectin binding sites of retinal photoreceptor outer segments and interphotoreceptor matrix, *J. Comp. Neurol.* **228:**299–307.

Wunderlich, F., and Speth, V., 1972, Membranes in *Tetrahymena.* I. The cortical pattern, *J. Ultrastruct. Res.* **41:**258–269.

Yacob, A., Wise, C., and Kunz, Y. W., 1977, The accessory outer segment of rods and cones in the retina of the guppy, *Poecilia reticulata* P. (Teleostei). An electron microscopical study, *Cell Tissue Res.* **177:**181–193.

Yamada, E., 1982, Morphology of vertebrate photoreceptors, *Methods Enzymol.* **81:**3–17.

Young, R. W., 1967, The renewal of photoreceptor cell outer segments, *J. Cell Biol.* **33:**61–72.

Young, R. W., 1968, Passage of newly formed protein through the connecting cilium of retinal rods in the frog, *J. Ultrastruct. Res.* **23:**462–473.

Young, R. W., 1976, Visual cells and the concept of renewal, *Invest. Ophthalmol.* **15:**700–725.

Young, R. W., and Bok, D., 1969, Participation of the retinal pigment epithelium in the rod outer segment renewal process, *J. Cell Biol.* **42:**392–403.

Young, R. W., and Droz, B., 1968, The renewal of protein in retinal rods and cones, *J. Cell Biol.* **39:**169–184.

Zimmerman, A. L., Yamanaka, G., Eckstein, F., Baylor, D. A., and Stryer, L., 1985, Interaction of hydrolysis-resistant analogs of cyclic GMP with the phosphodiesterase and light-sensitive channel of retinal rod outer segments, *Proc. Natl. Acad. Sci. USA* **82:**8813–8817.

Index